Modern Discrete Probability

Providing a graduate-level introduction to discrete probability and its applications, this book develops a toolkit of essential techniques for analyzing stochastic processes on graphs, other random discrete structures, and algorithms.

The topics covered in this book include the first and second moment methods, concentration inequalities, coupling and stochastic domination, martingales and potential theory, spectral methods, and branching processes. Each chapter expands on a fundamental technique, outlining common uses and showing them in action on simple examples and more substantial classical results. The focus is predominantly on non-asymptotic methods and results.

All chapters provide a detailed background review section, plus exercises and signposts to the wider literature. Readers are assumed to have undergraduate-level linear algebra and basic real analysis, while prior exposure to graduate-level probability is recommended.

This much-needed broad overview of discrete probability could serve as a textbook or as a reference for researchers in mathematics, statistics, data science, computer science, and engineering.

SÉBASTIEN ROCH is Professor of Mathematics at the University of Wisconsin–Madison. He has received an NSF CAREER award, an Alfred P. Sloan Fellowship, and a Simons Fellowship in Mathematics, and is a Fellow of the Institute for Mathematical Statistics.

Modern Discrete Probability

An Essential Toolkit

Sébastien Roch

University of Wisconsin–Madison

CAMBRIDGE
UNIVERSITY PRESS

Shaftesbury Road, Cambridge CB2 8EA, United Kingdom

One Liberty Plaza, 20th Floor, New York, NY 10006, USA

477 Williamstown Road, Port Melbourne, VIC 3207, Australia

314–321, 3rd Floor, Plot 3, Splendor Forum, Jasola District Centre,
New Delhi – 110025, India

103 Penang Road, #05–06/07, Visioncrest Commercial, Singapore 238467

Cambridge University Press is part of Cambridge University Press & Assessment,
a department of the University of Cambridge.

We share the University's mission to contribute to society through the pursuit of
education, learning and research at the highest international levels of excellence.

www.cambridge.org
Information on this title: www.cambridge.org/9781009305112

DOI: 10.1017/9781009305129

First published 2024

A catalogue record for this publication is available from the British Library

Library of Congress Cataloging-in-Publication Data
Names: Roch, Sébastien, author.
Title: Modern discrete probability : an essential toolkit / Sébastien
Roch, University of Wisconsin, Madison.
Description: Cambridge, United Kingdom ; New York, NY : Cambridge
University Press, 2024. | Series: Cambridge series in statistical and
probabilistic mathematics ; 55 | Includes bibliographical references and index.
Identifiers: LCCN 2023044218 | ISBN 9781009305112 (hardback) | ISBN
9781009305129 (ebook)
Subjects: LCSH: Stochastic processes. | Graph theory.
Classification: LCC QA274.2 .R63 2024 | DDC 519.2/3–dc23/eng/20231023
LC record available at https://lccn.loc.gov/2023044218

ISBN 978-1-009-30511-2 Hardback

To Betsy

Contents

Preface

This book arose from a set of lecture notes prepared for a one-semester topics course I taught at the University of Wisconsin–Madison in 2014, 2017, 2020, and 2023, which attracted a wide spectrum of students in mathematics, computer sciences, engineering, and statistics.

What Is It About?

The purpose of the book is to provide a graduate-level introduction to discrete probability. Topics covered are drawn primarily from stochastic processes on graphs: percolation, random graphs, Markov random fields, random walks on graphs, and so on. No attempt is made at covering these broad areas in depth. Rather, the emphasis is on illustrating important techniques used to analyze such processes. Along the way, many standard results regarding discrete probability models are worked out.

The "modern" in the title refers to the (nonexclusive) focus on non-asymptotic methods and results, reflecting the impact of the theoretical computer science literature on the trajectory of this field. In particular, several applications in randomized algorithms, probabilistic analysis of algorithms, and theoretical machine learning are used throughout to motivate the techniques described (although, again, these areas are not covered exhaustively).

Of course, the selection of topics is somewhat arbitrary and driven in part by personal interests. But the choice was guided by a desire to introduce techniques that are widely used across discrete probability and its applications. The material discussed here is developed in much greater depth in the following (incomplete list of) excellent textbooks and expository monographs, many of which influenced various sections of this book:

- Agarwal, Jiang, Kakade, Sun. *Reinforcement Learning: Theory and Algorithms.* [AJKS22]
- Aldous, Fill. *Reversible Markov Chains and Random Walks on Graphs.* [AF]
- Alon, Spencer. *The Probabilistic Method.* [AS11]
- B. Bollobás. *Random graphs.* [Bol01]
- Boucheron, Lugosi, Massart. *Concentration Inequalities: A Nonasymptotic Theory of Independence.* [BLM13]
- Chung, Lu. *Complex Graphs and Networks.* [CL06]
- Durrett. *Random Graph Dynamics.* [Dur06]
- Frieze and Karoński. *Introduction to Random Graphs.* [FK16]
- Grimmett. *Percolation.* [Gri10b]
- Janson, Luczak, Rucinski. *Random Graphs.* [JLR11]
- Lattimore, Szepesvári. *Bandit Algorithms.* [LS20]

- Levin, Peres, Wilmer. *Markov Chains and Mixing Times.* [LPW06]
- Lyons, Peres. *Probability on Trees and Networks.* [LP16]
- Mitzenmacher, Upfal. *Probability and Computing: Randomized Algorithms and Probabilistic Analysis.* [MU05]
- Motwani, Raghavan. *Randomized Algorithms.* [MR95]
- Rassoul-Agha, Seppäläinen. *A Course on Large Deviations with an Introduction to Gibbs Measures.* [RAS15]
- S. Shalev-Shwartz and S. Ben-David. *Understanding Machine Learning: From Theory to Algorithms.* [SSBD14]
- van Handel. *Probability in High Dimension.* [vH16]
- van der Hofstad. *Random Graphs and Complex Networks. Vol. 1.* [vdH17]
- Vershynin. *High-Dimensional Probability: An Introduction with Applications in Data Science.* [Ver18]

In fact, the book is meant as a first foray into the basic results and/or toolkits detailed in these more specialized references. My hope is that, by the end, the reader will have picked up sufficient fundamental background to learn advanced material on their own with some ease. I should add that I used many additional helpful sources; they are acknowledged in the "Bibliographic Remarks" at the end of each chapter. It is impossible to cover everything. Some notable omissions include, for example, graph limits [Lov12], influence [KS05], and group-theoretic methods [Dia88], among others. Much of the material covered here (and more) can also be found in [HMRAR98], [Gri10a], and [Bre17] with a different emphasis and scope.

Prerequisites

It is assumed throughout that the reader is fluent in undergraduate linear algebra, for example, at the level of [Axl15], and basic real analysis, for example, at the level of [Mor05]. In addition, it is recommended that the reader has taken at least one semester of graduate probability at the level of [Dur10]. I am also particularly fond of [Wil91], which heavily influenced Appendix B, where measure-theoretic background is reviewed. Some familiarity with countable Markov chain theory is necessary, as covered for instance in [Dur10, chapter 6]. An advanced undergraduate or Master's level treatment such as [Dur12], [Nor98], [GS20], [Law06], or [Bre20] will suffice, however.

Organization of the Book

The book is organized around five major "tools." The reader will have likely encountered those tools in prior probability courses. The goal here is to develop them further, specifically with their application to discrete random structures in mind, and to illustrate them in this setting on a variety of major classical results and applications.

In the interest of keeping the book relatively self-contained and serving the widest spectrum of readers, each chapter begins with a "background" section that reviews the basic material on which the rest of the chapter builds. The remaining sections then proceed to expand on two or three important specializations of the tools. While the chapters are meant to be somewhat modular, results from previous chapters do occasionally make an appearance.

The techniques are illustrated throughout with simple examples first, and then with more substantial ones in separate sections marked with the symbol ▷. I have attempted to provide applications from many areas of discrete probability and theoretical computer science, although some techniques are better suited for certain types of models or questions. The examples and applications are important: many of the tools are quite straightforward (or even elementary), and it is only when seen in action that their full power can be appreciated. Moreover, the ▷ sections serve as an excuse to introduce the reader to classical results and important applications – beyond their reliance on specific tools.

Chapter 1 introduces some of the main probability on graph models we come back to repeatedly throughout the book. It begins with a brief review of graph theory and Markov chain theory.

Chapter 2 starts out with the probabilistic method, including the first moment principle and second moment method, and then it moves on to concentration inequalities for sums of independent random variables, mostly sub-Gaussian and sub-exponential variables. It also discusses techniques to analyze the suprema of random processes.

Chapter 3 turns to martingales. The first main topic there is the Azuma–Hoeffding inequality and the method of bounded differences with applications to random graphs and stochastic bandit problems. The second main topic is electrical network theory for random walks on graphs.

Chapter 4 introduces coupling. It covers stochastic domination and correlation inequalities as well as couplings of Markov chains with applications to mixing. It also discusses the Chen–Stein method for Poisson approximation.

Chapter 5 is concerned with spectral methods. A major topic there is the use of the spectral theorem and geometric bounds on the spectral gap to control the mixing time of a reversible Markov chain. The chapter also introduces spectral methods for community recovery in network analysis.

Chapter 6 ends the book with applications of branching processes. Among other applications, an introduction to the reconstruction problem on trees is provided. The final section gives a detailed analysis of the phase transition of the Erdős–Rényi graph, where techniques from all chapters of the book are brought to bear.

Acknowledgments

The lecture notes on which this book is based were influenced by graduate courses of David Aldous, Steve Evans, Elchanan Mossel, Yuval Peres, Alistair Sinclair, and Luca Trevisan at UC Berkeley, where I learned much of this material. In particular, scribe notes for some of these courses helped shape early iterations of this book.

I have also learned a lot over the years from my collaborators and mentors as well as my former and current Ph.D. students and postdocs. I am particularly grateful to Elchanan Mossel and Allan Sly for encouragements to finish this project and to the UW–Madison students who have taken the various iterations of the course that inspired the book for their invaluable feedback.

Warm thanks to everyone in the departments of mathematics at UCLA and UW–Madison who have provided the stimulating environments that made this project possible. Beyond my current departments, I am particularly indebted to my colleagues in the NSF-funded Institute for Foundations of Data Science (IFDS) who have significantly expanded my knowledge of applications of this material in machine learning and statistics.

This book would have not have been possible without the hard work and advice of various people at Cambridge University Press and Integra, including Diana Gillooly, Natalie Tomlinson, Anna Scriven, Rebecca Grainger, Clare Dennison, Bhavani Vijayamani, and Sajukrishnan Balakrishnan, as well as several anonymous reviewers.

Finally, I thank my parents, my wife, and my son for their love, patience, and support.

Notation

Throughout the book, we will use the following notation.

- The real numbers are denoted by $\mathbb{R}$, the non-negative reals are denoted by $\mathbb{R}_+$, the integers are denoted by $\mathbb{Z}$, the non-negative integers are denoted by $\mathbb{Z}_+$, the natural numbers (i.e., positive integers) are denoted by $\mathbb{N}$, and the rational numbers are denoted by $\mathbb{Q}$. We will also use the notation $\overline{\mathbb{Z}}_+ := \{0, 1, \ldots, +\infty\}$.
- For two reals $a, b \in \mathbb{R}$,

$$a \wedge b := \min\{a, b\}, \quad a \vee b := \max\{a, b\},$$

and

$$a^+ = 0 \vee a, \quad a^- = 0 \vee (-a).$$

- For a real a, $\lfloor a \rfloor$ is the largest integer that is smaller than or equal to a and $\lceil a \rceil$ is the smallest integer that is larger than or equal to a.
- For $x \in \mathbb{R}$, the natural (i.e., base e) logarithm of x is denoted by $\log x$. We also let $\exp(x) = e^x$. NATURAL LOGARITHM
- For a positive integer $n \in \mathbb{N}$, we let

$$[n] := \{1, \ldots, n\}.$$

- The cardinality of a set A is denoted by $|A|$. The powerset of A is denoted by 2^A.
- For two sets A, B, their Cartesian product is denoted by $A \times B$.
- We will use the following notation for standard vectors: $\mathbf{0}$ is the all-zero vector, $\mathbf{1}$ is the all-one vector, and $\mathbf{e}_i$ is the standard basis vector with a 1 in coordinate i and 0 elsewhere. In each case, the dimension is implicit, as well as whether it is a row or column vector.
- For a vector $\mathbf{u} = (u_1, \ldots, u_n) \in \mathbb{R}^n$ and real $p > 0$, its *p-norm* (or ℓ^p-*norm*) is p-NORM

$$\|\mathbf{u}\|_p := \left(\sum_{i=1}^{n} |u_i|^p \right)^{1/p}.$$

When $p = +\infty$, we have

$$\|\mathbf{u}\|_\infty := \max_i |u_i|.$$

We also use the notation $\|\mathbf{u}\|_0$ to denote the number of non-zero coordinates of $\mathbf{u}$ (although it is not a norm; see Exercise 1.1). For two vectors $\mathbf{u} = (u_1, \ldots, u_n), \mathbf{v} = (v_1, \ldots, v_n) \in \mathbb{R}^n$, their *inner product* is INNER PRODUCT

$$\langle \mathbf{u}, \mathbf{v} \rangle := \sum_{i=1}^{n} u_i v_i.$$

The same notations apply to row vectors.

- For a matrix A, we denote the entries of A either by $A(i,j)$ or by $A_{i,j}$ (unless otherwise specified). The ith row of A is denoted by $A(i,\cdot)$ or $A_{i,\cdot}$. The jth column of A is denoted by $A(\cdot,j)$ or $A_{\cdot,j}$. The transpose of A is A^T.

- For a vector $\mathbf{z} = (z_1, \ldots, z_d)$, we let $\mathrm{diag}(\mathbf{z})$ be the diagonal matrix with diagonal entries $z_1, \ldots, z_d$.

BINOMIAL
COEFFICIENTS

- The *binomial coefficients* are defined as

$$\binom{n}{k} = \frac{n!}{k!(n-k)!},$$

where $k, n \in \mathbb{N}$ with $k \le n$ and $n! = 1 \times 2 \times \cdots \times n$ is the factorial of n. Some standard approximations for $\binom{n}{k}$ and $n!$ are listed in Appendix A. See also Exercises 1.2, 1.3, and 1.4.

- We use the abbreviation "a.s." for "almost surely," that is, with probability 1. We use "w.p." for "with probability."

- Convergence in probability is denoted as $\to_p$. Convergence in distribution is denoted as $\xrightarrow{d}$.

- For a random variable X and a probability distribution μ, we write $X \sim \mu$ to indicate that X has distribution μ. We write $X \overset{d}{=} Y$ if the random variables X and Y have the same distribution.

- For an event A, the random variable $\mathbf{1}_A$ is the indicator of A, that is, it is 1 if A occurs and 0 otherwise. We also use $\mathbf{1}\{A\}$.

TOTAL
VARIATION
DISTANCE

- For probability measures μ, ν on a countable set S, their *total variation distance* is

$$\|\mu - \nu\|_{\mathrm{TV}} := \sup_{A \subseteq S} |\mu(A) - \nu(A)|.$$

- For non-negative functions $f(n)$, $g(n)$ of $n \in \mathbb{Z}_+$, we write $f(n) = O(g(n))$ if there exists a positive constant $C > 0$ such that $f(n) \le Cg(n)$ for all n large enough. Similarly, $f(n) = \Omega(g(n))$ means that $f(n) \ge cg(n)$ for some constant $c > 0$ for all n large enough. The notation $f(n) = \Theta(g(n))$ indicates that both $f(n) = O(g(n))$ and $f(n) = \Omega(g(n))$ hold. We also write $f(n) = o(g(n))$ or $g(n) = \omega(f(n))$ or $f(n) \ll g(n)$ or $g(n) \gg f(n)$ if $f(n)/g(n) \to 0$ as $n \to +\infty$. If $f(n)/g(n) \to 1$ we write $f(n) \sim g(n)$. The same notations are used for functions of a real variable x as $x \to +\infty$.

1

Introduction

In this chapter, we describe a few discrete probability models to which we will come back repeatedly throughout the book. While there exists a vast array of well-studied random combinatorial structures (permutations, partitions, urn models, Boolean functions, polytopes, etc.), our focus is primarily on a limited number of graph-based processes, namely, percolation, random graphs, the Ising model, and random walks on networks. We will not attempt to derive the theory of these models exhaustively here. Instead, we will employ them to illustrate some essential techniques from discrete probability. Note that the toolkit developed in this book is meant to apply to other probabilistic models of interest as well, and in fact many more will be encountered along the way. After a brief review of graph basics and Markov chains theory in Section 1.1, we formally introduce our main models in Section 1.2. We also formulate various questions about these models that will be answered (at least partially) later on. We assume that the reader is familiar with the measure-theoretic foundations of probability. A refresher of all required concepts and results is provided in Appendix B.

1.1 Background

We start with a brief review of graph terminology and standard countable-space Markov chains results.

1.1.1 Review of Graph Theory

Basic definitions An *undirected graph* (or *graph* for short) is a pair $G = (V, E)$, where V is GRAPH the set of *vertices* (or *nodes* or *sites*) and

$$E \subseteq \{\{u, v\} : u, v \in V\}$$

is the set of *edges* (or *bonds*). See Figure 1.1 for an example. We occasionally write $V(G)$ and $E(G)$ for the vertices and edges of the graph G. The set of vertices V is either finite or countably infinite. Edges of the form $\{u\}$ are called *self-loops*. In general, we do not allow E to be a multiset unless otherwise stated. But, when E is a multiset, G is called a *multigraph*. MULTIGRAPH

A vertex $v \in V$ is *incident* with an edge $e \in E$ (or vice versa) if $v \in e$. The incident vertices of an edge are called *endvertices*. Two vertices $u, v \in V$ are *adjacent* (or *neighbors*), denoted by $u \sim v$, if $\{u, v\} \in E$. The set of adjacent vertices of v, denoted by $N(v)$, is called the *neighborhood* of v and its size, that is, $\delta(v) := |N(v)|$, is the *degree* of v. A vertex v with $\delta(v) = 0$ is called *isolated*. A graph is called *d-regular* if all its degrees are d. A countable graph is *locally finite* if all its vertices have a finite degree.

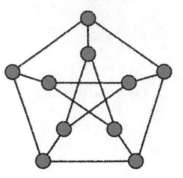

Figure 1.1 Petersen graph.

Example 1.1.1 (Petersen graph). All vertices in the Petersen graph in Figure 1.1 have degree 3, that is, it is a 3-regular graph. In particular, it has no isolated vertex.   ◀

A convenient (and mathematically useful) way to specify a graph is the following matrix representation. Assume the graph $G = (V, E)$ has $n = |V|$ vertices. Assume that the vertices are numbered $1, \ldots, n$. The *adjacency matrix* A of G is the $n \times n$ symmetric matrix defined as

ADJACENCY
MATRIX

$$A_{xy} = \begin{cases} 1 & \text{if } \{x, y\} \in E, \\ 0 & \text{otherwise.} \end{cases}$$

Example 1.1.2 (Triangle). The adjacency matrix of a *triangle*, that is, a 3-vertex graph with all possible non-loop edges, is

$$A = \begin{bmatrix} 0 & 1 & 1 \\ 1 & 0 & 1 \\ 1 & 1 & 0 \end{bmatrix}.$$

                    ◀

There exist other matrix representations. Here is one. Let $m = |E|$ and assume that the edges are labeled arbitrarily as $e_1, \ldots, e_m$. The *incidence matrix* of an undirected graph $G = (V, E)$ is the $n \times m$ matrix B such that $B_{ij} = 1$ if vertex i and edge e_j are incident and 0 otherwise.

INCIDENCE
MATRIX

Subgraphs, paths, and cycles A *subgraph* of $G = (V, E)$ is a graph $G' = (V', E')$ with $V' \subseteq V$ and $E' \subseteq E$. Implied in this definition is the fact that the edges in E' are incident only to V'. The subgraph G' is said to be *induced* if

$$E' = \{\{x, y\} : x, y \in V', \{x, y\} \in E\},$$

that is, it contains exactly those edges of G that are between vertices in V'. In that case the notation $G' := G[V']$ is used. A subgraph is said to be *spanning* if $V' = V$. A subgraph containing all possible non-loop edges between its vertices is called a *clique* (or *complete subgraph*). A clique with k nodes is referred to as a k-clique.

CLIQUE

Example 1.1.3 (Petersen graph (continued)). The Petersen graph contains no triangle, that is, 3-clique, induced or not.   ◀

A *walk* in G is a sequence of (not necessarily distinct) vertices $x_0 \sim x_1 \sim \cdots \sim x_k$. Note the requirement that consecutive vertices of a walk are adjacent. The number $k \geq 0$ is the *length* of the walk. If the *endvertices* x_0, x_k coincide, that is, $x_0 = x_k$, we refer to the walk as *closed*. If the vertices of a walk are all distinct, we call it a *path* (or *self-avoiding walk*). If the vertices of a closed walk are all distinct except for the endvertices and its length is at least 3, we call it a *cycle*. A path or cycle can be seen as a (not necessarily induced) subgraph of G. The length of the shortest path connecting two distinct vertices u, v is the *graph distance* GRAPH between u and v, denoted by $d_G(u, v)$. It can be checked that the graph distance is a metric DISTANCE (and that, in particular, it satisfies the triangle inequality; see Exercise 1.6). The minimum length of a cycle in a graph is its *girth*.

We write $u \leftrightarrow v$ if there is a path between u and v. It can be checked that the binary relation $\leftrightarrow$ is an equivalence relation (i.e., it is reflexive, symmetric, and transitive; see Exercise 1.6). Its equivalence classes are called *connected components*. A graph is *connected* if any two vertices are linked by a path, that is, if $u \leftrightarrow v$ for all $u, v \in V$. Or put differently, if there is only one connected component.

Example 1.1.4 (Petersen graph (continued)). The Petersen graph is connected. ◀

Trees A *forest* is a graph with no cycle, or an *acyclic* graph. A *tree* is a connected forest. TREE Vertices of degree 1 are called *leaves*. A spanning tree of G is a subgraph which is a tree and is also spanning. A tree is said to be *rooted* if it has a single distinguished vertex called the *root*.

Trees will play a key role and we collect several important facts about them (mostly without proofs). The following characterizations of trees will be useful. The proof is left as an exercise (see Exercise 1.8). We write $G + e$ (respectively $G - e$) to indicate the graph G with edge e added (respectively removed).

Theorem 1.1.5 (Trees: characterizations). *The following are equivalent.*

(i) *The graph T is a tree.*
(ii) *For any two vertices in T, there is a unique path between them.*
(iii) *The graph T is connected, but $T - e$ is not for any edge e in T.*
(iv) *The graph T is acyclic, but $T + \{x, y\}$ is not for any pair of non-adjacent vertices x, y.*

Here are two important implications.

Corollary 1.1.6 *If G is connected, then it has at least one spanning tree.*

Proof Indeed, from Theorem 1.1.5, a graph is a tree if and only if it is minimally connected, in the sense that removing any of its edges disconnects it. So a spanning tree can be obtained by removing edges of G that do not disconnect it until it is not possible anymore. ∎

The following characterization is proved in Exercise 1.7.

Corollary 1.1.7 *A connected graph with n vertices is a tree if and only if it has $n - 1$ edges.*

And here is a related fact.

Corollary 1.1.8 *Let G be a graph with n vertices. If an acyclic subgraph H has n vertices and $n - 1$ edges, then it is a spanning tree of G.*

Proof If H is not connected, then it has at least two connected components. Each of them is acyclic and therefore a tree. By applying Corollary 1.1.7 to the connected components and summing up, we see that the total number of edges in H is $\leq n - 2$, a contradiction. So H is connected and therefore a spanning tree.   ∎

Finally, a classical formula:

Theorem 1.1.9 (Cayley's formula). *There are k^{k-2} trees on a set of k labeled vertices.*

We give a proof of Cayley's formula based on branching processes in Exercise 6.19.

Some standard graphs Here are a few more examples of finite graphs.

- *Complete graph K_n*: This graph is made of n vertices with all possible non-loop edges.
- *Cycle graph C_n (or n-cycle)*: The vertex set is $\{0, 1, \ldots, n-1\}$ and two vertices $i \neq j$ are adjacent if and only if $|i - j| = 1$ or $n - 1$.
- *Torus $\mathbb{L}_n^d$*: The vertex set is $\{0, 1, \ldots, n-1\}^d$ and two vertices $x \neq y$ are adjacent if and only if there is a coordinate i such that $|x_i - y_i| = 1$ or $n - 1$ and all other coordinates $j \neq i$ satisfy $x_j = y_j$.
- *Hypercube $\mathbb{Z}_2^n$ (or n-dimensional hypercube)*: The vertex set is $\{0, 1\}^n$ and two vertices $x \neq y$ are adjacent if and only if $\|x - y\|_1 = 1$.
- *Rooted b-ary tree $\widehat{\mathbb{T}}_b^\ell$*: This graph is a tree with ℓ levels. The unique vertex on level 0 is called the root. For $j = 1, \ldots, \ell - 1$, level j has b^j vertices, each of which has exactly one neighbor on level $j - 1$ (its parent) and b neighbors on level $j + 1$ (its children). The b^ℓ vertices on level ℓ are leaves.

Here are a few examples of *infinite graphs*, that is, a graph with a countably infinite number of vertices and edges.

- *Infinite d-regular tree $\mathbb{T}_d$*: This is an infinite tree where each vertex has exactly d neighbors. The rooted version, that is, $\widehat{\mathbb{T}}_b^\ell$ with $\ell = +\infty$ levels, is denoted by $\widehat{\mathbb{T}}_b$.
- *Lattice $\mathbb{L}^d$*: The vertex set is $\mathbb{Z}^d$ and two vertices $x \neq y$ are adjacent if and only if $\|x - y\|_1 = 1$.

A *bipartite graph* $G = (L \cup R, E)$ is a graph whose vertex set is composed of the union of two disjoint sets L, R and whose edge set E is a subset of $\{\{\ell, r\} : \ell \in L, r \in R\}$. That is, there is no edge between vertices in L, and likewise for R.

Example 1.1.10 (Some bipartite graphs). The cycle graph C_{2n} is a bipartite graph. So is the *complete bipartite graph $K_{n,m}$* with vertex set $\{\ell_1, \ldots, \ell_n\} \cup \{r_1, \ldots, r_m\}$ and edge set $\{\{\ell_i, r_j\} : i \in [n], j \in [m]\}$.   ◀

In a bipartite graph $G = (L \cup R, E)$, a *perfect matching* is a collection of edges $M \subseteq E$ such that each vertex is incident to exactly one edge in M.

An *automorphism* of a graph $G = (V, E)$ is a bijection ϕ of V to itself that preserves the edges, that is, such that $\{x, y\} \in E$ if and only if $\{\phi(x), \phi(y)\} \in E$. A graph $G = (V, E)$ is *vertex-transitive* if for any $u, v \in V$ there is an automorphism mapping u to v.

Example 1.1.11 (Petersen graph (continued)). For any $\ell \in \mathbb{Z}$, a $(2\pi\ell/5)$-rotation of the planar representation of the Petersen graph in Figure 1.1 corresponds to an automorphism. ◀

Example 1.1.12 (Trees). The graph $\mathbb{T}_d$ is vertex-transitive. The graph $\widehat{\mathbb{T}}_b^\ell$ on the other hand has many automorphisms, but is not vertex-transitive. ◀

Flows Let $G = (V, E)$ be a connected graph with two distinguished disjoint vertex sets, a *source-set* (or *source* for short) $A \subseteq V$ and a *sink-set* (or *sink* for short) Z. Let $\kappa : E \to \mathbb{R}_+$ be a *capacity* function.

Definition 1.1.13 (Flow). *A flow from source A to sink Z is a function $f : V \times V \to \mathbb{R}$ such that:* FLOW

F1 (Antisymmetry) $f(x, y) = -f(y, x), \forall x, y \in V$.
F2 (Capacity constraint) $|f(x, y)| \leq \kappa(e), \forall e = \{x, y\} \in E$, *and* $f(x, y) = 0$ *otherwise.*
F3 (Flow-conservation constraint)

$$\sum_{y : y \sim x} f(x, y) = 0, \qquad \forall x \in V \setminus (A \cup Z).$$

For $U, W \subseteq V$, let $f(U, W) := \sum_{u \in U, w \in W} f(u, w)$. The strength *of f is $\|f\| := f(A, A^c)$.*

One useful consequence of antisymmetry is that, for any $U \subseteq V$, we have $f(U, U) = 0$ since each distinct pair $x \neq y \in U$ appears exactly twice in the sum, once in each ordering. Also if W_1 and W_2 are disjoint, then $f(U, W_1 \cup W_2) = f(U, W_1) + f(U, W_2)$. In particular, combining both observations, $f(U, W) = f(U, W \setminus U) = -f(W \setminus U, U)$.

For $F \subseteq E$, let $\kappa(F) := \sum_{e \in F} \kappa(e)$. We call F a *cutset separating A and Z* (or *cutset* for short) if all paths connecting A and Z include an edge in F. For such an F, let A_F be the set of vertices not separated from A by F, that is, vertices from which there is a path to A not crossing an edge in F. Clearly, $A \subseteq A_F$ but $A_F \cap Z = \emptyset$.

Lemma 1.1.14 (Max flow ≤ min cut). *For any flow f and cutset F,*

$$\|f\| = f(A_F, A_F^c) \leq \sum_{\{x,y\} \in F} |f(x, y)| \leq \kappa(F). \tag{1.1.1}$$

Proof Since F is a cutset, $(A_F \setminus A) \cap (A \cup Z) = \emptyset$. So, by (F3),

$$\begin{aligned}
f(A, A^c) &= f(A, A^c) + \sum_{u \in A_F \setminus A} f(u, V) \\
&= f(A, A_F \setminus A) + f(A, A_F^c) \\
&\quad + f(A_F \setminus A, A_F) + f(A_F \setminus A, A_F^c) \\
&= f(A, A_F \setminus A) + f(A, A_F^c) \\
&\quad + f(A_F \setminus A, A) + f(A_F \setminus A, A_F^c) \\
&= f(A_F, A_F^c) \\
&\leq \sum_{\{x,y\} \in F} |f(x, y)|,
\end{aligned}$$

where we used (F1) twice. The last line is justified by the fact that the edges between a vertex in A_F and a vertex in A_F^c have to be in F by definition of A_F. That proves the equality and the first inequality in the claim. Condition (F2) implies the second inequality.   ∎

Remarkably, this bound is tight, in the following sense.

Theorem 1.1.15 (Max-flow min-cut theorem). *Let G be a finite connected graph with source A and sink Z, and let κ be a capacity function. Then the following holds*

$$\sup\{\|f\| : \text{flow } f\} = \min\{\kappa(F) : \text{cutset } F\}.$$

Proof Note that, by compactness, the supremum on the left-hand side is achieved. Let f be an optimal flow. The idea of the proof is to construct a "matching" cutset.

An *augmentable path* is a path $x_0 \sim \cdots \sim x_k$ with $x_0 \in A$, $x_i \notin A \cup Z$ for all $i \neq 0$ or k, and $f(x_{i-1}, x_i) < \kappa(\{x_{i-1}, x_i\})$ for all $i \neq 0$. By default, each vertex in A is an augmentable path. Moreover, by the optimality of f there cannot be an augmentable path with $x_k \in Z$. Indeed, otherwise, we could "push more flow through that path" and increase the strength of f – a contradiction.

Let $B \subseteq V$ be the set of all final vertices in some augmentable path and let F be the edge set between B and $B^c := V \setminus B$. Note that, again by contradiction, all vertices in B can be reached from A without crossing F and that $f(x, y) = \kappa(e)$ for all $e = \{x, y\} \in F$ with $x \in B$ and $y \in B^c$. Furthermore, F is a cutset separating A from Z: trivially $A \subseteq B$; $Z \subseteq B^c$ as argued above, and any path from A to Z must exit B and enter B^c through an edge in F. Thus, $A_F = B$ and we have equality in (1.1.1). That concludes the proof.   ∎

COLORING **Colorings, independent sets, and matchings** A *coloring* of a graph $G = (V, E)$ is an assignment of colors to each vertex in G. In a coloring, two vertices may share the same color. A coloring is *proper* if for every edge e in G the endvertices of e have distinct colors. The smallest number of colors in a proper coloring of a graph G is called the *chromatic number* $\chi(G)$ of G.

INDEPENDENT SET An *independent vertex set* (or *independent set* for short) of $G = (V, E)$ is a subset of vertices $W \subseteq V$ such that all pairs of vertices in W are non-adjacent. Likewise, two edges are independent if they are not incident to the same vertex. A *matching* is a set of pairwise MATCHING independent edges. A matching F is *perfect* if every vertex in G is incident to an edge of F.

Edge-weighted graphs We refer to an edge-weighted graph $G = (V, E, w)$ as a *network*. Here $w \colon E \to \mathbb{R}_+$ is a function that assigns positive real weights to the edges. Definitions can be generalized naturally. In particular, one defines the degree of a vertex i as

$$\delta(i) = \sum_{j : e = \{i, j\} \in E} w_e.$$

The adjacency matrix A of G is the $n \times n$ symmetric matrix defined as

$$A_{ij} = \begin{cases} w_e & \text{if } e = \{i, j\} \in E, \\ 0 & \text{otherwise,} \end{cases}$$

where we denote the vertices $\{1, \ldots, n\}$.

Directed graphs A *directed graph* (or *digraph* for short) is a pair $G = (V, E)$ where V is a DIGRAPH
set of *vertices* (or *nodes* or *sites*) and $E \subseteq V^2$ is a set of *directed edges* (or *arcs*). A directed
edge from x to y is typically denoted by (x, y), or occasionally by $\langle x, y \rangle$. A *directed path* is
a sequence of vertices $x_0, \ldots, x_k$, all distinct, with $(x_{i-1}, x_i) \in E$ for all $i = 1, \ldots, k$. We
write $u \to v$ if there is such a directed path with $x_0 = u$ and $x_k = v$. We say that $u, v \in V$
communicate, denoted by $u \leftrightarrow v$, if $u \to v$ and $v \to u$. In particular, we always have $u \leftrightarrow u$
for every state u. The binary relation $\leftrightarrow$ relation is an equivalence relation (see Exercise 1.6).
The equivalence classes of $\leftrightarrow$ are called the *strongly connected components* of G.

The following definition will prove useful.

Definition 1.1.16 (Oriented incidence matrix). *Let $G = (V, E)$ be an undirected graph.
Assume that the vertices of $G = (V, E)$ are numbered $1, \ldots, |V|$ and that the edges are
labeled arbitrarily as $e_1, \ldots, e_{|E|}$. An* orientation *of G is the choice of a direction $\vec{e}_i$ for each
edge e_i, turning it into a digraph $\vec{G}$. An* oriented incidence matrix *of G is the incidence* ORIENTED
matrix of an orientation, that is, the matrix B such that $B_{ij} = -1$ if edge $\vec{e}_j$ leaves vertex i, INCIDENCE
$B_{ij} = 1$ if edge $\vec{e}_j$ enters vertex i, and 0 otherwise. MATRIX

1.1.2 Review of Markov Chain Theory

Informally, a *Markov chain* (or *Markov process*) is a time-indexed stochastic process sat- MARKOV
isfying the property: conditioned on the present, the future is independent of the past. We CHAIN
restrict ourselves to the discrete-time, time-homogeneous, countable-space case, where such
a process is characterized by its initial distribution and a transition matrix.

Construction of a Markov chain For our purposes, it will suffice to "define" a Markov
chain through a particular construction. Let V be a finite or countable space. Recall that a
stochastic matrix on V is a non-negative matrix $P = (P(i,j))_{i,j \in V}$ satisfying STOCHASTIC
MATRIX

$$\sum_{j \in V} P(i,j) = 1, \qquad \forall i \in V.$$

We think of $P(i, \cdot)$ as a probability distribution on V. In particular, for a set of states $A \subseteq V$,
we let

$$P(i, A) = \sum_{j \in A} P(i,j).$$

Let μ be a probability measure on V and let P be a stochastic matrix on V. One way to
construct a Markov chain $(X_t)_{t \geq 0}$ on V with *transition matrix* P and *initial distribution* μ is
the following:

- Pick $X_0 \sim \mu$ and let $(Y(i, n))_{i \in V, n \geq 1}$ be a mutually independent array of random variables
  with $Y(i, n) \sim P(i, \cdot)$.
- Set inductively $X_n := Y(X_{n-1}, n)$, $n \geq 1$.

So in particular:

$$\mathbb{P}[X_0 = x_0, \ldots, X_t = x_t] = \mu(x_0) P(x_0, x_1) \cdots P(x_{t-1}, x_t).$$

We use the notation $\mathbb{P}_x, \mathbb{E}_x$ for the probability distribution and expectation under the chain started at x. Similarly for $\mathbb{P}_\mu, \mathbb{E}_\mu$, where μ is a probability distribution.

Example 1.1.17 (Simple random walk on a graph). Let $G = (V, E)$ be a finite or infinite, locally finite graph. *Simple random walk* on G is the Markov chain on V, started at an arbitrary vertex, which at each time picks a uniformly chosen neighbor of the current state. (Exercise 1.9 asks for the transition matrix.) ◀

Markov property Let $(X_t)_{t \geq 0}$ be a Markov chain (or *chain* for short) with transition matrix P and initial distribution μ. Define the filtration $(\mathcal{F}_t)_{t \geq 0}$ with $\mathcal{F}_t = \sigma(X_0, \ldots, X_t)$ (see Appendix B). As mentioned above, the defining property of Markov chains, known as the Markov property, is that given the present, the future is independent of the past. In its simplest form, that can be interpreted as $\mathbb{P}[X_{t+1} = y \mid \mathcal{F}_t] = \mathbb{P}_{X_t}[X_{t+1} = y] = P(X_t, y)$. More generally:

Theorem 1.1.18 (Markov property). *Let $f: V^\infty \to \mathbb{R}$ be bounded, measurable and let $F(x) := \mathbb{E}_x[f((X_t)_{t \geq 0})]$, then*

$$\mathbb{E}[f((X_{s+t})_{t \geq 0}) \mid \mathcal{F}_s] = F(X_s) \qquad a.s.$$

Remark 1.1.19 *We will come back to the "strong" Markov property in Chapter 3.*

We define $P^t(x, y) := \mathbb{P}_x[X_t = y]$. An important consequence of the Markov property (Theorem 1.1.18) is the following.

Theorem 1.1.20 (Chapman–Kolmogorov).

$$P^t(x, z) = \sum_{y \in V} P^s(x, y) P^{t-s}(y, z), \qquad s \in \{0, 1, \ldots, t\}.$$

Proof This follows from the Markov property. Indeed, note that $\mathbb{P}_x[X_t = z \mid \mathcal{F}_s] = F(X_s)$ with $F(y) := \mathbb{P}_y[X_{t-s} = z]$ and take $\mathbb{E}_x$ on each side. ∎

Example 1.1.21 (Random walk on $\mathbb{Z}$). Let (X_t) be simple random walk on $\mathbb{Z}$ interpreted as a graph (i.e., $\mathbb{L}$) where $i \sim j$ if $|i - j| = 1$.[1] Then $P(0, x) = 1/2$ if $|x| = 1$. And $P^2(0, x) = 1/4$ if $|x| = 2$ and $P^2(0, 0) = 1/2$. ◀

If we write μ_s for the law of X_s as a row vector, then

$$\mu_s = \mu_0 P^s,$$

where P^s is the matrix product of P by itself s times. As is conventional in Markov chain theory, we think of probability distributions over the state space as row vectors. We will typically denote them by Greek letters (e.g., μ, π).

Stationarity The *transition graph* of a chain is the directed graph on V whose edges are the transitions with *strictly positive probability*. A chain is *irreducible* if V is the unique (strongly) connected component of its transition graph, that is, if all pairs of states have a directed path between them in the transition graph.

---

[1] On $\mathbb{Z}$, simple random walk often refers to any nearest-neighbor random walk, whereas the example here is called simple symmetric random walk. We will not adopt this terminology here.

Example 1.1.22 (Simple random walk on a graph (continued)). Simple random walk on G is irreducible if and only if G is connected.   ◀

A *stationary measure* π is a measure on V such that

$$\sum_{x \in V} \pi(x)P(x,y) = \pi(y), \qquad \forall y \in V,$$

or in matrix form $\pi = \pi P$. We say that π is a *stationary distribution* if in addition π is a probability measure.

STATIONARY
DISTRIBUTION

Example 1.1.23 (Random walk on $\mathbb{Z}^d$). The all-one measure $\pi \equiv 1$ is stationary for simple random walk on $\mathbb{L}^d$.   ◀

Finite, irreducible chains always have a unique stationary distribution.

Theorem 1.1.24 (Existence and uniqueness: finite case). *If P is irreducible and has a finite state space, then:*

   (i) (Existence) *it has a stationary distribution which, furthermore, is strictly positive;*
   (ii) (Uniqueness) *the stationary distribution is unique.*

This result follows from Perron–Frobenius theory (a version of which is stated as Theorem 6.1.17). We give a self-contained proof.

Proof of Theorem 1.1.24 (i)   We begin by proving existence. Denote by n the number of states. Because P is stochastic, we have by definition that $P\mathbf{1} = \mathbf{1}$, where $\mathbf{1}$ is the all-one vector. Put differently,

$$(P - I)\mathbf{1} = \mathbf{0}.$$

In particular, the columns of $P - I$ are linearly dependent, that is, the rank of $P - I$ is $< n$. That, in turn, implies that the rows of $P - I$ are linearly dependent since row rank and column rank are equal. Hence, there exists a non-zero row vector $\mathbf{z} \in \mathbb{R}^n$ such that $\mathbf{z}(P - I) = \mathbf{0}$, or after rearranging,

$$\mathbf{z}P = \mathbf{z}. \tag{1.1.2}$$

The rest of the proof is broken up into a series of lemmas. To take advantage of irreducibility, we first construct a positive stochastic matrix with $\mathbf{z}$ as a left eigenvector with eigenvalue 1. We then show that all entries of $\mathbf{z}$ have the same sign. Finally, we normalize $\mathbf{z}$.

Lemma 1.1.25 (Existence: Step 1). *There exists a non-negative integer h such that*

$$R = \frac{1}{h+1}[I + P + P^2 + \cdots + P^h]$$

is a stochastic matrix with strictly positive entries which satisfies

$$\mathbf{z}R = \mathbf{z}. \tag{1.1.3}$$

Lemma 1.1.26 (Existence: Step 2). *The entries of $\mathbf{z}$ are either all non-negative or all non-positive.*

Lemma 1.1.27 (Existence: Step 3). *Let*

$$\pi = \frac{\mathbf{z}}{\mathbf{z}\mathbf{1}}.$$

Then π is a strictly positive stationary distribution.

We denote the entries of R and P^s by $R_{x,y}$ and $P^s_{x,y}$, $x, y = 1, \ldots, n$, respectively.

Proof of Lemma 1.1.25   By irreducibility (see Exercise 1.10), for any $x, y \in [n]$ there is h_{xy} such that $P^{h_{xy}}_{x,y} > 0$. Now define

$$h = \max_{x,y \in [n]} h_{xy}.$$

The matrix P^s, as a product of stochastic matrices, is a stochastic matrix for all s (see Exercise 1.11). In particular, it has non-negative entries. Hence, for each x, y,

$$R_{x,y} = \frac{1}{h+1}[I_{x,y} + P_{x,y} + P^2_{x,y} + \cdots + P^h_{x,y}]$$

$$\geq \frac{1}{h+1}P^{h_{xy}}_{x,y} > 0.$$

Moreover, the matrix R, as a convex combination of stochastic matrices, is a stochastic matrix (see Exercise 1.11).

Since $\mathbf{z}P = \mathbf{z}$, it follows by induction that $\mathbf{z}P^s = \mathbf{z}$ for all s. Therefore,

$$\mathbf{z}R = \frac{1}{h+1}[\mathbf{z}I + \mathbf{z}P + \mathbf{z}P^2 + \cdots + \mathbf{z}P^h]$$

$$= \frac{1}{h+1}[\mathbf{z} + \mathbf{z} + \mathbf{z} + \cdots + \mathbf{z}]$$

$$= \mathbf{z}.$$

That concludes the proof.   ∎

Proof of Lemma 1.1.26   We argue by contradiction. Suppose that two entries of $\mathbf{z} = (z_1, \ldots, z_n)$ have different signs, say $z_i > 0$, while $z_j < 0$. By the previous lemma, $R_{x,y} > 0$ for all x, y. Therefore,

$$|z_y| = \left| \sum_x z_x R_{x,y} \right|$$

$$= \left| \sum_{x:z_x \geq 0} z_x R_{x,y} + \sum_{x:z_x < 0} z_x R_{x,y} \right|.$$

The first term on the last line is strictly positive (since it is at least $z_i R_{i,y} > 0$), while the second term is strictly negative (since it is at most $z_j R_{j,y} < 0$). Hence, because of cancellations (see Exercise 1.13), the expression above is strictly smaller than the sum of the absolute values, that is,

$$|z_y| < \sum_x |z_x| R_{x,y}.$$

Since R is stochastic by the previous lemma, we deduce after summing over y that

$$\sum_y |z_y| < \sum_y \sum_x |z_x| R_{x,y} = \sum_x |z_x| \sum_y R_{x,y} = \sum_x |z_x|,$$

a contradiction, thereby proving the claim. ∎

Proof of Lemma 1.1.27 Now define π entrywise by

$$\pi_x = \frac{z_x}{\sum_i z_i} = \frac{|z_x|}{\sum_i |z_i|} \geq 0,$$

where the second equality comes from the previous lemma. We also used the fact that $\mathbf{z} \neq \mathbf{0}$.

For all y, by definition of $\mathbf{z}$,

$$\sum_x \pi_x P_{x,y} = \sum_x \frac{z_x}{\sum_i z_i} P_{x,y} = \frac{1}{\sum_i z_i} \sum_x z_x P_{x,y} = \frac{z_y}{\sum_i z_i} = \pi_y.$$

The same holds with $P_{x,y}$ replaced by $R_{x,y}$ from (1.1.3). Since $R_{x,y} > 0$ and $\mathbf{z} \neq \mathbf{0}$, it follows that $\pi_y > 0$ for all y. That proves the claim. ∎

That concludes the proof of the existence claim. ∎

It remains to prove uniqueness. See Exercise 1.14 for an alternative proof based on the maximum principle (to which we come back in Theorem 3.3.9 and Exercise 3.12).

Proof of Theorem 1.1.24 (ii) Suppose there are two distinct stationary distributions π_1 and π_2 (which must be strictly positive). Since they are distinct, they are not a multiple of each other and therefore are linearly independent. Apply the Gram–Schmidt procedure:

$$\mathbf{q}_1 = \frac{\pi_1}{\|\pi_1\|_2} \quad \text{and} \quad \mathbf{q}_2 = \frac{\pi_2 - \langle \pi_2, \mathbf{q}_1 \rangle \mathbf{q}_1}{\|\pi_2 - \langle \pi_2, \mathbf{q}_1 \rangle \mathbf{q}_1\|_2}.$$

Then

$$\mathbf{q}_1 P = \frac{\pi_1}{\|\pi_1\|_2} P = \frac{\pi_1 P}{\|\pi_1\|_2} = \frac{\pi_1}{\|\pi_1\|_2} = \mathbf{q}_1,$$

and all entries of $\mathbf{q}_1$ are strictly positive.

Similarly,

$$\begin{aligned} \mathbf{q}_2 P &= \frac{\pi_2 - \langle \pi_2, \mathbf{q}_1 \rangle \mathbf{q}_1}{\|\pi_2 - \langle \pi_2, \mathbf{q}_1 \rangle \mathbf{q}_1\|_2} P \\ &= \frac{\pi_2 P - \langle \pi_2, \mathbf{q}_1 \rangle \mathbf{q}_1 P}{\|\pi_2 - \langle \pi_2, \mathbf{q}_1 \rangle \mathbf{q}_1\|_2} \\ &= \frac{\pi_2 - \langle \pi_2, \mathbf{q}_1 \rangle \mathbf{q}_1}{\|\pi_2 - \langle \pi_2, \mathbf{q}_1 \rangle \mathbf{q}_1\|_2} \\ &= \mathbf{q}_2. \end{aligned}$$

Since $\mathbf{z} := \mathbf{q}_2$ satisfies (1.1.2), by Lemmas 1.1.25–1.1.27 there is a multiple of $\mathbf{q}_2$, say $\mathbf{q}_2' = \alpha \mathbf{q}_2$ with $\alpha \neq 0$, such that $\mathbf{q}_2' P = \mathbf{q}_2'$ and all entries of $\mathbf{q}_2'$ are strictly positive. By the Gram–Schmidt procedure,

$$\langle \mathbf{q}_1, \mathbf{q}_2' \rangle = \langle \mathbf{q}_1, \alpha \mathbf{q}_2 \rangle = \alpha \langle \mathbf{q}_1, \mathbf{q}_2 \rangle = 0.$$

But this is a contradiction since both vectors are strictly positive. That concludes the proof of the uniqueness claim.   ∎

Reversibility A transition matrix P is *reversible* with respect to (w.r.t.) a measure η if

$$\eta(x)P(x,y) = \eta(y)P(y,x)$$

for all $x, y \in V$. These equations are known as *detailed balance*. Here is the key observation: by summing over y and using the fact that P is stochastic, *such a measure is necessarily stationary*. (Exercise 1.12 explains the name.)

Example 1.1.28 (Random walk on $\mathbb{Z}^d$ (continued)). The measure $\eta \equiv 1$ is reversible for simple random walk on $\mathbb{L}^d$.   ◄

Example 1.1.29 (Simple random walk on a graph (continued)). Let (X_t) be simple random walk on a connected graph $G = (V, E)$. Then (X_t) is reversible with respect to $\eta(v) := \delta(v)$, where recall that $\delta(v)$ is the degree of v. Indeed, for all $\{u, v\} \in E$,

$$\delta(u)P(u, v) = \delta(u)\frac{1}{\delta(u)} = 1 = \delta(v)\frac{1}{\delta(v)} = \delta(v)P(v, u).$$

(See Exercise 1.9 for the transition matrix of simple random walk on a graph.)   ◄

Example 1.1.30 (Metropolis chain). The Metropolis algorithm modifies an irreducible, symmetric (i.e., whose transition matrix is a symmetric matrix) chain Q to produce a new chain P with the same transition graph and a prescribed positive stationary distribution π. The idea is simple. For each pair $x \neq y$, either we multiply $Q(x, y)$ by $\pi(y)/\pi(x)$ and leave $Q(y, x)$ intact or vice versa. Detailed balance immediately follows. To ensure that the new transition matrix remains stochastic, for each pair we make the choice that lowers the transition probabilities; then we add the lost probability to the diagonal (i.e., to the probability of staying put).

Formally, the definition of the new chain is

$$P(x, y) := \begin{cases} Q(x, y)\left[\frac{\pi(y)}{\pi(x)} \wedge 1\right] & \text{if } x \neq y, \\ 1 - \sum_{z \neq x} Q(x, z)\left[\frac{\pi(z)}{\pi(x)} \wedge 1\right] & \text{otherwise.} \end{cases}$$

Note that, by definition of P and the fact that Q is stochastic, we have $P(x, y) \leq Q(x, y)$ for all $x \neq y$, so

$$\sum_{y \neq x} P(x, y) \leq 1,$$

and hence P is well defined as a transition matrix. We claim further that P is reversible with respect to π. Suppose $x \neq y$, and assume without loss of generality that $\pi(x) \geq \pi(y)$. Then, by definition of P, we have

$$\pi(x)P(x,y) = \pi(x)Q(x,y)\frac{\pi(y)}{\pi(x)}$$
$$= Q(x,y)\pi(y)$$
$$= Q(y,x)\pi(y)$$
$$= P(y,x)\pi(y),$$

where we used the symmetry of Q. ◄

Convergence and mixing time A key property of Markov chains is that, under suitable assumptions, they converge to a stationary regime. We need one more definition before stating the theorem. A chain is said to be *aperiodic* if, for all $x \in V$, the greatest common divisor of APERIODIC $\{t \colon P^t(x,x) > 0\}$ is 1.

Example 1.1.31 (Lazy random walk on a graph). The *lazy simple random walk* on G is the LAZY Markov chain such that, at each time, it stays put with probability $1/2$ or chooses a uniformly random neighbor of the current state otherwise. Such a chain is aperiodic. ◄

Lemma 1.1.32 (Consequence of aperiodicity). *If P is aperiodic, irreducible, and has a finite state space, then there is a positive integer t_0 such that for all $t \geq t_0$ the matrix P^t has strictly positive entries.*

We can now state the convergence theorem. For probability measures μ, ν on V, their *total* TOTAL *variation distance* is VARIATION DISTANCE

$$\|\mu - \nu\|_{\mathrm{TV}} := \sup_{A \subseteq V} |\mu(A) - \nu(A)|. \tag{1.1.4}$$

Theorem 1.1.33 (Convergence theorem). *Suppose P is irreducible, aperiodic, and has stationary distribution π. Then, for all x,*

$$\|P^t(x, \cdot) - \pi(\cdot)\|_{\mathrm{TV}} \to 0,$$

as $t \to +\infty$.

We give a proof in the finite case in Example 4.3.3. In particular, the convergence theorem implies that for all x, y,

$$P^t(x,y) \to \pi(y).$$

Without aperiodicity, we have the weaker claim

$$\frac{1}{t} \sum_{s=1}^{t} P^s(x,y) \to \pi(y), \tag{1.1.5}$$

as $t \to +\infty$.

We will be interested in quantifying the speed of convergence in Theorem 1.1.33. For this purpose, we define

$$d(t) := \sup_{x \in V} \|P^t(x, \cdot) - \pi(\cdot)\|_{\mathrm{TV}}. \tag{1.1.6}$$

Lemma 1.1.34 (Monotonicity of $d(t)$). *The function $d(t)$ is non-increasing in t.*

Proof Note that, by definition of P^{t+1},

$$d(t+1) = \sup_{x \in V} \sup_{A \subseteq V} |P^{t+1}(x, A) - \pi(A)|$$

$$= \sup_{x \in V} \sup_{A \subseteq V} \left| \sum_z P(x, z)(P^t(z, A) - \pi(A)) \right|$$

$$\leq \sup_{x \in V} \sum_z P(x, z) \sup_{A \subseteq V} |P^t(z, A) - \pi(A)|$$

$$\leq \sup_{z \in V} \sup_{A \subseteq V} |P^t(z, A) - \pi(A)|$$

$$= d(t),$$

where on the second and fourth line we used that P is a stochastic matrix.   ∎

The following concept will play a key role.

Definition 1.1.35 (Mixing time). *For a fixed $\varepsilon > 0$, the* mixing time *is defined as*

$$t_{\text{mix}}(\varepsilon) := \inf\{t \geq 0 \colon d(t) \leq \varepsilon\}.$$

1.2 Some Discrete Probability Models

With the necessary background covered, we are now in a position to define formally a few important discrete probability models that will be ubiquitous in this book. These are all graph-based processes. Many more interesting random discrete structures and other related probabilistic models will be encountered throughout (and defined where needed).

Percolation Percolation processes are meant to model the movement of a fluid through a porous medium. There are several types of percolation models. We focus here on bond percolation. In words, edges of a graph are "open" at random, indicating that fluid is passing through. We are interested in the "open clusters," that is, the regions reached by the fluid.

Definition 1.2.1 (Bond percolation). *Let $G = (V, E)$ be a finite or infinite graph. The* bond percolation *process on G with density $p \in [0, 1]$, whose measure is denoted by $\mathbb{P}_p$, is defined as follows: each edge of G is independently set to* open *with probability p, otherwise it is set to* closed*. Write $x \Leftrightarrow y$ if $x, y \in V$ are connected by a path all of whose edges are open. The* open cluster *of x is*

$$\mathcal{C}_x := \{y \in V \colon x \Leftrightarrow y\}.$$

We will mostly consider bond percolation on the infinite graphs $\mathbb{L}^d$ or $\mathbb{T}_d$. The main question we will ask is: *For which values of p is there an infinite open cluster?*

Random graphs Random graphs provide a natural framework to study complex networks. Different behaviors are observed depending on the modeling choices made. Perhaps the simplest and most studied is the Erdős–Rényi random graph model. We consider the version due to Gilbert. Here the edges are present independently with a fixed probability. Despite its simplicity, this model exhibits a rich set of phenomena that make it a prime example for the use of a variety of probabilistic techniques.

Definition 1.2.2 (Erdős–Rényi graph model). *Let $n \in \mathbb{N}$ and $p \in [0,1]$. Set $V := [n]$. Under the* Erdős–Rényi *graph model on n vertices with density p, a random graph $G = (V, E)$ is generated as follows: for each pair $x \neq y$ in V, the edge $\{x, y\}$ is in E with probability p independently of all other edges. We write $G \sim \mathbb{G}_{n,p}$ and we denote the corresponding measure by $\mathbb{P}_{n,p}$.* ERDŐS–RÉNYI GRAPH MODEL

Typical questions regarding the Erdős–Rényi graph model (and random graphs more generally) include: *How are degrees distributed? Is G connected? What is the (asymptotic) probability of observing a particular subgraph, for example, a triangle? What is the typical chromatic number?*

As one alternative to the Erdős–Rényi model, we will also encounter preferential attachment graphs. These are meant to model the growth of a network where new edges are more likely to be incident with vertices of high degree, a reasonable assumption in some applied settings. Such a process produces graphs with properties that differ from those of the Erdős–Rényi model; in particular, they tend to have a "fatter" degree distribution tail. In the definition of preferential attachment graphs, we restrict ourselves to the tree case (see Exercise 1.15).

Definition 1.2.3 (Preferential attachment graph). *The* preferential attachment graph process *produces a sequence of graphs $(G_t)_{t \geq 1}$ as follows: we start at time 1 with two vertices, denoted v_0 and v_1, connected by an edge. At time t, we add vertex v_t with a single edge connecting it to an old vertex, which is picked proportionally to its degree. We write $(G_t)_{t \geq 1} \sim \mathrm{PA}_1$.* PREFERENTIAL ATTACHMENT GRAPHS

Markov random fields Another common class of graph-based processes involves the assignment of random "states" to the vertices of a *fixed graph*. The state distribution is typically specified through "interactions between neighboring vertices." Such models are widely studied in statistical physics and also have important applications in statistics. We focus on models with a Markovian (i.e., conditional independence) structure that makes them particularly amenable to rigorous analysis and computational methods. We start with Gibbs random fields, a broad class of such models.

Definition 1.2.4 (Gibbs random field). *Let S be a finite set and let $G = (V, E)$ be a finite graph. Denote by $\mathcal{K}$ the set of all cliques of G. A positive probability measure μ on $\mathcal{X} := S^V$ is called a* Gibbs random field *if there exist* clique potentials *$\phi_K : S^K \to \mathbb{R}$, $K \in \mathcal{K}$, such that* GIBBS RANDOM FIELD

$$\mu(x) = \frac{1}{\mathcal{Z}} \exp\left(\sum_{K \in \mathcal{K}} \phi_K(x_K) \right),$$

where x_K is x restricted to the vertices of K and $\mathcal{Z}$ is a normalizing constant.

The following example introduces the primary Gibbs random field we will encounter.

Example 1.2.5 (Ising model). *For $\beta > 0$, the (ferromagnetic)* Ising model *with inverse temperature β is the Gibbs random field with $S := \{-1, +1\}$, $\phi_{\{i,j\}}(\sigma_{\{i,j\}}) = \beta \sigma_i \sigma_j$ and $\phi_K \equiv 0$ if $|K| \neq 2$. The function $\mathcal{H}(\sigma) := -\sum_{\{i,j\} \in E} \sigma_i \sigma_j$ is known as the* Hamiltonian. *The normalizing constant $\mathcal{Z} := \mathcal{Z}(\beta)$ is called the* partition function. *The states $(\sigma_i)_{i \in V}$ are referred to as* spins. ISING MODEL ◀

Typical questions regarding Ising models include: *How fast is correlation decaying down the graph? How well can one guess the state at an unobserved vertex?* We will also consider certain Markov chains related to Ising models (see Definition 1.2.8).

Random walks on graphs and reversible Markov chains The last class of processes we focus on are random walks on graphs and their generalizations. Recall the following definition.

SIMPLE
RANDOM
WALK ON A
GRAPH
Definition 1.2.6 (Simple random walk on a graph). *Let $G = (V, E)$ be a finite or countable, locally finite graph. Simple random walk on G is the Markov chain on V, started at an arbitrary vertex, which at each time picks a uniformly chosen neighbor of the current state.*

We generalize the definition by adding weights to the edges. In this context, we denote edge weights by $c(e)$ for "conductance" (see Section 3.3).

RANDOM
WALK ON A
NETWORK
Definition 1.2.7 (Random walk on a network). *Let $G = (V, E)$ be a finite or countably infinite graph. Let $c \colon E \to \mathbb{R}_+$ be a positive edge weight function on G. Recall that we call $\mathcal{N} = (G, c)$ a network. We assume that for all $u \in V$,*

$$c(u) := \sum_{e = \{u, v\} \in E} c(e) < +\infty.$$

Random walk on network $\mathcal{N}$ is the Markov chain on V, started at an arbitrary vertex, which at each time picks a neighbor of the current state proportionally to the weight of the corresponding edge. That is, the transition matrix is given by

$$P(u, v) = \begin{cases} \frac{c(\{u,v\})}{c(u)} & \text{if } \{u, v\} \in E, \\ 0 & \text{otherwise.} \end{cases}$$

By definition of P, it is immediate that this Markov chain is reversible with respect to the measure $\eta(u) := c(u)$. In fact, conversely, any countable reversible Markov chain can be seen as a random walk on a network by setting $c(e) := \pi(x)P(x, y) = \pi(y)P(y, x)$ for all x, y such that $P(x, y) > 0$.

Typical questions include: *How long does it take to visit all vertices at least once or a particular subset of vertices for the first time? How fast does the walk approach stationarity? How often does the walk return to its starting point?*

We will also encounter a particular class of Markov chains related to Ising models, the Glauber dynamics.

GLAUBER
DYNAMICS
Definition 1.2.8 (Glauber dynamics of the Ising model). *Let μ_β be the Ising model with inverse temperature $\beta > 0$ on a graph $G = (V, E)$. The* (single-site) *Glauber dynamics is the Markov chain on $\mathcal{X} := \{-1, +1\}^V$, which at each time*

- *selects a site $i \in V$ uniformly at random, and*
- *updates the spin at i according to μ_β conditioned on agreeing with the current state at all sites in $V \setminus \{i\}$.*

Specifically, for $\gamma \in \{-1, +1\}$, $i \in V$, and $\sigma \in \mathcal{X}$, let $\sigma^{i,\gamma}$ be the configuration σ with the spin at i being set to γ. Let $n = |V|$ and $S_i(\sigma) := \sum_{j \sim i} \sigma_j$. The non-zero entries of the transition matrix are

$$Q_\beta(\sigma, \sigma^{i,\gamma}) := \frac{1}{n} \cdot \frac{e^{\gamma \beta S_i(\sigma)}}{e^{-\beta S_i(\sigma)} + e^{\beta S_i(\sigma)}}.$$

This chain is irreducible since we can flip each site one by one to go from any state to any other state. It is straightforward to check that $Q_\beta(\sigma, \sigma^{i,\gamma})$ is a stochastic matrix. The next theorem shows that μ_β is its stationary distribution.

Theorem 1.2.9 *The Glauber dynamics is reversible with respect to μ_β.*

Proof For all $\sigma \in \mathcal{X}$ and $i \in V$, let

$$S_{\neq i}(\sigma) := \mathcal{H}(\sigma^{i,+}) + S_i(\sigma) = \mathcal{H}(\sigma^{i,-}) - S_i(\sigma).$$

We have

$$\begin{aligned}
\mu_\beta(\sigma^{i,-}) Q_\beta(\sigma^{i,-}, \sigma^{i,+}) &= \frac{e^{-\beta S_{\neq i}(\sigma)} e^{-\beta S_i(\sigma)}}{\mathcal{Z}(\beta)} \cdot \frac{e^{\beta S_i(\sigma)}}{n[e^{-\beta S_i(\sigma)} + e^{\beta S_i(\sigma)}]} \\
&= \frac{e^{-\beta S_{\neq i}(\sigma)}}{n \mathcal{Z}(\beta)[e^{-\beta S_i(\sigma)} + e^{\beta S_i(\sigma)}]} \\
&= \frac{e^{-\beta S_{\neq i}(\sigma)} e^{\beta S_i(\sigma)}}{\mathcal{Z}(\beta)} \cdot \frac{e^{-\beta S_i(\sigma)}}{n[e^{-\beta S_i(\sigma)} + e^{\beta S_i(\sigma)}]} \\
&= \mu_\beta(\sigma^{i,+}) Q_\beta(\sigma^{i,+}, \sigma^{i,-}).
\end{aligned}$$

That concludes the proof. ∎

Exercises

Exercise 1.1 (0-norm). Show that $\|\mathbf{u}\|_0$ does not define a norm.

Exercise 1.2 (A factorial bound: one way). Let ℓ be a positive integer.

(i) Use the bound $1 + x \leq e^x$ to show that

$$\frac{k+1}{k} \leq e^{1/k}$$

and

$$\frac{k}{k+1} \leq e^{1/(k+1)}$$

for all positive integers k.

(ii) Use part (i) and the quantity

$$\prod_{k=1}^{\ell-1} \frac{(k+1)^k}{k^k}$$

to show that

$$\ell! \geq \frac{\ell^\ell}{e^{\ell-1}}.$$

(iii) Use part (i) and the quantity

$$\prod_{k=1}^{\ell-1} \frac{k^{k+1}}{(k+1)^{k+1}}$$

to show that

$$\ell! \leq \frac{\ell^{\ell+1}}{e^{\ell-1}}.$$

Exercise 1.3 (A factorial bound: another way). Let ℓ be a positive integer. Show that

$$\frac{\ell^\ell}{e^{\ell-1}} \leq \ell! \leq \frac{\ell^{\ell+1}}{e^{\ell-1}}$$

by considering the logarithm of $\ell!$, interpreting the resulting quantity as a Riemann sum, and bounding above and below by an integral.

Exercise 1.4 (A binomial bound). Show that for integers $0 < d \leq n$,

$$\sum_{k=0}^{d} \binom{n}{k} \leq \left(\frac{en}{d}\right)^d.$$

(Hint: Multiply the left-hand side of the inequality by $(d/n)^d \leq (d/n)^k$ and use the binomial theorem.)

Exercise 1.5 (Powers of the adjacency matrix). Let A^n be the nth matrix power of the adjacency matrix A of a graph $G = (V, E)$. Prove that the (i,j)th entry a_{ij}^n is the number of walks of length exactly n between vertices i and j in G. (Hint: Use induction on n.)

Exercise 1.6 (Paths). Let u, v be vertices of a graph $G = (V, E)$.

  (i) Show that the graph distance $d_G(u, v)$ is a metric.
 (ii) Show that the binary relation $u \leftrightarrow v$ is an equivalence relation.
(iii) Prove (ii) in the directed case.

Exercise 1.7 (Trees: number of edges). Prove that a connected graph with n vertices is a tree if and only if it has $n - 1$ edges. (Hint: Proceed by induction. Then use Corollary 1.1.6.)

Exercise 1.8 (Trees: characterizations). Prove Theorem 1.1.5.

Exercise 1.9 (Simple random walk on a graph). Let $(X_t)_{t \geq 0}$ be simple random walk on a finite graph $G = (V, E)$. Suppose the vertex set is $V = [n]$. Write down an expression for the transition matrix of (X_t).

Exercise 1.10 (Communication lemma). Let (X_t) be a finite Markov chain. Show that if $x \to y$, then there is an integer $r \geq 1$ such that

$$\mathbb{P}[X_r = y \,|\, X_0 = x] = (P^r)_{x,y} > 0.$$

Exercise 1.11 (Stochastic matrices from stochastic matrices). Let

$$P^{(1)}, P^{(2)}, \ldots, P^{(r)} \in \mathbb{R}^{n \times n},$$

be stochastic matrices.

(i) Show that $P^{(1)}P^{(2)}$ is a stochastic matrix. That is, a product of stochastic matrices is a stochastic matrix.
(ii) Show that for any $\alpha_1, \ldots, \alpha_r \in [0,1]$ with $\sum_{i=1}^r \alpha_i = 1$,

$$\sum_{i=1}^r \alpha_i P^{(i)}$$

is stochastic. That is, a convex combination of stochastic matrices is a stochastic matrix.

Exercise 1.12 (Reversing time). Let (X_t) be a finite Markov chain with transition matrix P. Assume P is reversible with respect to a probability distribution π. Assume that the initial distribution is π. Show that for any sequence of states $z_0, \ldots, z_s$, the reversed sequence has the same probability, that is,

$$\mathbb{P}[X_s = z_0, \ldots, X_0 = z_s] = \mathbb{P}[X_s = z_s, \ldots, X_0 = z_0].$$

Exercise 1.13 (A strict inequality). Let $a, b \in \mathbb{R}$ with $a < 0$ and $b > 0$. Show that

$$|a + b| < |a| + |b|.$$

(Hint: Consider the cases $a + b \geq 0$ and $a + b < 0$ separately.)

Exercise 1.14 (Uniqueness: maximum principle). Let $P = (P_{i,j})_{i,j=1}^n \in \mathbb{R}^n$ be a transition matrix.

(i) Let $\alpha_1, \ldots, \alpha_m > 0$ such that $\sum_{i=1}^m \alpha_i = 1$. Let $\mathbf{x} = (x_1, \ldots, x_m) \in \mathbb{R}^n$. Show that

$$\sum_{i=1}^m \alpha_i x_i \leq \max_i x_i,$$

and that equality holds if and only if $x_1 = x_2 = \cdots = x_m$.
(ii) Let $\mathbf{0} \neq \mathbf{y} \in \mathbb{R}^n$ be a right eigenvector of P with eigenvalue 1, that is, $P\mathbf{y} = \mathbf{y}$. Assume that $\mathbf{y}$ is not a constant vector, that is, there is $i \neq j$ such that $y_i \neq y_j$. Let k be such that $y_k = \max_{i \in [n]} y_i$. Show that for any ℓ such that $P_{k,\ell} > 0$, we necessarily have $y_\ell = y_k$. (Hint: Use that $y_k = \sum_{i=1}^n P_{k,i} y_i$ and apply (i).)
(iii) Assume that P is irreducible. Let $\mathbf{0} \neq \mathbf{y} \in \mathbb{R}^n$ again be a right eigenvector of P with eigenvalue 1. Use (ii) to show that $\mathbf{y}$ is necessarily a constant vector.
(iv) Use (iii) to conclude that the dimension of the null space of $P^T - I$ is exactly 1. (Hint: Use the Rank–Nullity Theorem.)

Exercise 1.15 (Preferential attachment trees). Let $(G_t)_{t\geq1} \sim \text{PA}_1$ as in Definition 1.2.3. Show that G_t is a tree with $t + 1$ vertices for all $t \geq 1$.

Exercise 1.16 (Warm-up: a little calculus). Prove the following inequalities which we will encounter throughout. (Hint: Basic calculus should do.)

(i) Show that $1 - x \leq e^{-x}$ for all $x \in \mathbb{R}$.
(ii) Show that $1 - x \geq e^{-x-x^2}$ for $x \in [0, 1/2]$.
(iii) Show that $\log(1 + x) \leq x - x^2/4$ for $x \in [0, 1]$.
(iv) Show that $\log x \leq x - 1$ for all $x \in \mathbb{R}_+$.

(v) Show that $\cos x \leq e^{-x^2/2}$ for $x \in [0, \pi/2)$. (Hint: Consider the function $h(x) = \log(e^{x^2/2} \cos x)$ and recall that the derivative of $\tan x$ is $1 + \tan^2 x$.)

Bibliographic Remarks

Section 1.1 For an introduction to graphs, see, for example, [Die10] or [Bol98]. Four different proofs of Cayley's formula are detailed in the delightful [AZ18]. Markov chain theory is covered in detail in [Dur10, chapter 6]. For a more gentle introduction, see, for example, [Dur12, chapter 1], [Nor98, chapter 1], or [Res92, chapter 2].

Section 1.2 For book-length treatments of percolation theory, see [BR06a, Gri10b]. The version of the Erdős–Rényi random graph model we consider here is due to Gilbert [Gil59]. For much deeper accounts of the theory of random graphs and related processes, see, for example, [Bol01, Dur06, JLR11, vdH17, FK16]. Two standard references on finite Markov chains and mixing times are [AF, LPW06].

2

Moments and Tails

In this chapter, we look at the moments of a random variable. Specifically, we demonstrate that moments capture useful information about the tail of a random variable while often being simpler to compute or at least bound. Several well-known inequalities quantify this intuition. Although they are straightforward to derive, such inequalities are surprisingly powerful. Through a range of applications, we illustrate the utility of controlling the tail of a random variable, typically by allowing one to dismiss certain "bad events" as rare. We begin in Section 2.1 by recalling the classical Markov and Chebyshev inequalities. Then we discuss three of the most fundamental tools in discrete probability and probabilistic combinatorics. In Sections 2.2 and 2.3, we derive the complementary *first* and *second moment methods*, and give several standard applications, especially to threshold phenomena in random graphs and percolation. In Section 2.4, we develop the *Chernoff–Cramér method*, which relies on the moment-generating function and is the building block for a large class of tail bounds. Two key applications in data science are briefly introduced: sparse recovery and empirical risk minimization.

2.1 Background

We start with a few basic definitions and standard inequalities. See Appendix B for a refresher on random variables and their expectation.

2.1.1 Definitions

Moments As a quick reminder, let X be a random variable with $\mathbb{E}|X|^k < +\infty$ for some non-negative integer k. In that case we write $X \in L^k$. Recall that the quantities $\mathbb{E}[X^k]$ and $\mathbb{E}[(X - \mathbb{E}X)^k]$, which are well defined when $X \in L^k$, are called, respectively, the *kth moment* and *kth central moment* of X. The first moment and the second central moment are known MOMENTS as the *mean* and *variance*, the square root of which is the *standard deviation*. A random variable is said to be *centered* if its mean is 0. Recall that for a non-negative random variable X, the kth moment can be expressed as

$$\mathbb{E}[X^k] = \int_0^{+\infty} k x^{k-1} \mathbb{P}[X > x] \, dx. \qquad (2.1.1)$$

The *moment-generating function* (or *exponential moment*) of X is the function MOMENT-GENERATING FUNCTION

$$M_X(s) := \mathbb{E}\left[e^{sX}\right],$$

21

defined for all $s \in \mathbb{R}$ where it is finite, which includes at least $s = 0$. If $M_X(s)$ is defined on $(-s_0, s_0)$ for some $s_0 > 0$, then X has finite moments of all orders, for any $k \in \mathbb{Z}$,

$$\frac{\mathrm{d}^k}{\mathrm{d}s} M_X(s) = \mathbb{E}\left[X^k e^{sX}\right], \tag{2.1.2}$$

and the following expansion holds

$$M_X(s) = \sum_{k \geq 0} \frac{s^k}{k!} \mathbb{E}[X^k], \qquad |s| < s_0.$$

The moment-generating function plays nicely with sums of independent random variables. Specifically, if X_1 and X_2 are independent random variables with M_{X_1} and M_{X_2} defined over a joint interval $(-s_0, s_0)$, then for s in that interval,

$$\begin{aligned}
M_{X_1+X_2}(s) &= \mathbb{E}\left[e^{s(X_1+X_2)}\right] \\
&= \mathbb{E}\left[e^{sX_1} e^{sX_2}\right] \\
&= \mathbb{E}\left[e^{sX_1}\right] \mathbb{E}\left[e^{sX_2}\right] \\
&= M_{X_1}(s) M_{X_2}(s), \tag{2.1.3}
\end{aligned}$$

where we used independence in the third equality.

One more piece of notation: if A is an event and $X \in L^1$, then we use the shorthand

$$\mathbb{E}[X; A] = \mathbb{E}[X \mathbf{1}_A].$$

Tails We refer to a probability of the form $\mathbb{P}[X \geq x]$ as an *upper tail* (or *right tail*) probability. Typically, x is (much) greater than the mean or median of X. Similarly, we refer to $\mathbb{P}[X \leq x]$ as a *lower tail* (or *left tail*) probability. Our general goal in this chapter is to bound tail probabilities using moments and moment-generating functions.

Tail bounds arise naturally in many contexts, as events of interest can often be framed in terms of a random variable being unusually large or small. Such probabilities are often hard to compute directly however. As we will see in this chapter, moments offer an effective means to control tail probabilities for two main reasons: (i) moments contain information about the tails of a random variable, as (2.1.1) makes explicit for instance; and (ii) they are typically easier to compute – or, at least, to approximate.

As we will see, tail bounds are also useful to study the maximum of a collection of random variables.

2.1.2 Basic Inequalities

Markov's inequality Our first bound on the tail of a random variable is *Markov's inequality*. In words, for a non-negative random variable, the heavier the tail, the larger the expectation. This simple inequality is in fact a key ingredient in more sophisticated tail bounds, as we will see.

Theorem 2.1.1 (Markov's inequality). *Let X be a non-negative random variable. Then, for all $b > 0$,*

$$\mathbb{P}[X \geq b] \leq \frac{\mathbb{E}X}{b}. \tag{2.1.4}$$

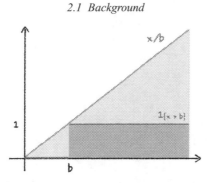

Figure 2.1 Proof of Markov's inequality: taking expectations of the two functions depicted in the figure yields the inequality.

Proof

$$\mathbb{E}\,X \geq \mathbb{E}[X; X \geq b] \geq \mathbb{E}[b; X \geq b] = b\,\mathbb{P}[X \geq b]. \qquad \blacksquare$$

See Figure 2.1 for a proof by picture. Note that this inequality is non-trivial only when $b > \mathbb{E}\,X$.

Chebyshev's inequality An application of Markov's inequality (Theorem 2.1.1) to $|X - \mathbb{E}\,X|^2$ gives a classical tail bound featuring the second moment of a random variable.

Theorem 2.1.2 (Chebyshev's inequality). *Let X be a random variable with $\mathbb{E}\,X^2 < +\infty$. Then, for all $\beta > 0$,* CHEBYSHEV'S INEQUALITY

$$\mathbb{P}[|X - \mathbb{E}\,X| > \beta] \leq \frac{\text{Var}[X]}{\beta^2}. \qquad (2.1.5)$$

Proof This follows immediately by applying (2.1.4) to $|X - \mathbb{E}\,X|^2$ with $b = \beta^2$. $\qquad \blacksquare$

Of course, this bound is non-trivial only when β is larger than the standard deviation. Results of this type that quantify the probability of deviating from the mean are referred to as *concentration inequalities*. Chebyshev's inequality is perhaps the simplest instance – we CONCENTRATION will derive many more. To bound the variance, the following standard formula is sometimes INEQUALITIES useful:

$$\text{Var}\left[\sum_{i=1}^{n} X_i\right] = \sum_{i=1}^{n} \text{Var}[X_i] + 2\sum_{i<j} \text{Cov}[X_i, X_j], \qquad (2.1.6)$$

where we recall that the *covariance* of X_i and X_j is COVARIANCE

$$\text{Cov}[X_i, X_j] := \mathbb{E}[X_i X_j] - \mathbb{E}[X_i]\,\mathbb{E}[X_j].$$

When X_i and X_j are independent, then $\text{Cov}[X_i, X_j] = 0$.

Example 2.1.3 Let X be a Gaussian random variable with mean μ and variance σ^2, that is, GAUSSIAN whose density is

$$f_X(x) = \frac{1}{\sqrt{2\pi\sigma^2}} \exp\left(-\frac{(x-\mu)^2}{2\sigma^2}\right), \qquad x \in \mathbb{R}.$$

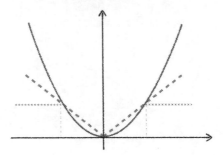

Figure 2.2 Comparison of the Markov and Chebyshev inequalities: the squared deviation from the mean (solid) gives a better approximation of the indicator function (dotted) close to the mean than the absolute deviation (dashed).

We write $X \sim N(\mu, \sigma^2)$. A direct computation shows that $\mathbb{E}|X - \mu| = \sigma\sqrt{\frac{2}{\pi}}$. Hence Markov's inequality gives

$$\mathbb{P}[|X - \mu| \geq b] \leq \frac{\mathbb{E}|X - \mu|}{b} = \sqrt{\frac{2}{\pi}} \cdot \frac{\sigma}{b},$$

while Chebyshev's inequality (Theorem 2.1.2) gives

$$\mathbb{P}[|X - \mu| \geq b] \leq \left(\frac{\sigma}{b}\right)^2.$$

Hence, for b large enough, Chebyshev's inequality produces a stronger bound. See Figure 2.2 for some insight.   ◄

UNIFORM **Example 2.1.4** (Coupon collector's problem). Let $(X_t)_{t \in \mathbb{N}}$ be i.i.d. uniform random variables over $[n]$, that is, that are equally likely to take any value in $[n]$. Let $T_{n,i}$ be the first time that i elements of $[n]$ have been picked, that is,

$$T_{n,i} = \inf\{t \geq 1 : |\{X_1, \ldots, X_t\}| = i\},$$

COUPON
COLLECTOR
with $T_{n,0} := 0$. We prove that the time it takes to pick all elements at least once – or "collect each coupon" – has the following tail. For any $\varepsilon > 0$, we have as $n \to +\infty$:

Claim 2.1.5

$$\mathbb{P}\left[\left|T_{n,n} - n\sum_{j=1}^{n} j^{-1}\right| \geq \varepsilon n \log n\right] \to 0.$$

To prove this claim we note that the time elapsed between $T_{n,i-1}$ and $T_{n,i}$, which we denote by $\tau_{n,i} := T_{n,i} - T_{n,i-1}$, is geometric with success probability $1 - \frac{i-1}{n}$. And all $\tau_{n,i}$s are independ-
GEOMETRIC
ent. Recall that a geometric random variable Z with success probability p has probability mass function $\mathbb{P}[Z = z] = (1-p)^{z-1}p$ for $z \in \mathbb{N}$ and has mean $1/p$ and variance $(1-p)/p^2$. So, the expectation and variance of $T_{n,n} = \sum_{i=1}^{n} \tau_{n,i}$ are

$$\mathbb{E}[T_{n,n}] = \sum_{i=1}^{n}\left(1 - \frac{i-1}{n}\right)^{-1} = n\sum_{j=1}^{n} j^{-1} = \Theta(n \log n) \qquad (2.1.7)$$

and

$$\text{Var}[T_{n,n}] \leq \sum_{i=1}^{n} \left(1 - \frac{i-1}{n}\right)^{-2} = n^2 \sum_{j=1}^{n} j^{-2} \leq n^2 \sum_{j=1}^{+\infty} j^{-2} = \Theta(n^2). \tag{2.1.8}$$

So by Chebyshev's inequality

$$\mathbb{P}\left[\left|T_{n,n} - n\sum_{j=1}^{n} j^{-1}\right| \geq \varepsilon\, n\log n\right] \leq \frac{\text{Var}[T_{n,n}]}{(\varepsilon\, n\log n)^2}$$

$$\leq \frac{n^2 \sum_{j=1}^{+\infty} j^{-2}}{(\varepsilon\, n\log n)^2}$$

$$\to 0,$$

by (2.1.7) and (2.1.8). ◄

A classical implication of Chebyshev's inequality is (a version of) the law of large numbers. Recall that a sequence of random variables $(X_n)_{n\geq 1}$ converges in probability to a random variable X, denoted by $X_n \to_p X$, if for all $\varepsilon > 0$,

$$\lim_{n\to+\infty} \mathbb{P}[|X_n - X| \geq \varepsilon] \to 0.$$

Theorem 2.1.6 (L^2 *weak law of large numbers*). *Let* $X_1, X_2, \ldots$ *be* uncorrelated *random variables, that is,* $\mathbb{E}[X_i X_j] = \mathbb{E}[X_i]\,\mathbb{E}[X_j]$ *for* $i \neq j$, *with* $\mathbb{E}[X_i] = \mu < +\infty$ *and* $\sup_i \text{Var}[X_i] < +\infty$. *Then*

$$\frac{1}{n}\sum_{k\leq n} X_k \to_p \mu.$$

See Exercise 2.5 for a proof. When the X_ks are i.i.d. and integrable (but not necessarily square integrable), convergence is almost sure. That result, the strong law of large numbers, also follows from Chebyshev's inequality (and other ideas), but we will not prove it here.

2.2 First Moment Method

In this section, we develop some techniques based on the first moment. Recall that the expectation of a random variable has an elementary, yet handy, property: linearity. That is, if random variables $X_1, \ldots, X_k$ defined on a joint probability space have finite first moments, then

$$\mathbb{E}[X_1 + \cdots + X_k] = \mathbb{E}[X_1] + \cdots + \mathbb{E}[X_k] \tag{2.2.1}$$

without any further assumption. In particular, linearity holds whether or not the X_is are independent.

2.2.1 The Probabilistic Method

A key technique of probabilistic combinatorics is the so-called *probabilistic method*. The idea is that one can establish the existence of an object satisfying a certain property – without having to construct one explicitly. Instead, one argues that a randomly chosen object exhibits the given property with positive probability. The following "obvious" observation, sometimes referred to as the *first moment principle*, plays a key role in this context.

FIRST
MOMENT
PRINCIPLE
Theorem 2.2.1 (First moment principle). *Let X be a random variable with finite expectation. Then, for any $\mu \in \mathbb{R}$,*

$$\mathbb{E}X \leq \mu \implies \mathbb{P}[X \leq \mu] > 0.$$

Proof We argue by contradiction. Assume $\mathbb{E}X \leq \mu$ and $\mathbb{P}[X \leq \mu] = 0$. We can write $\{X \leq \mu\} = \bigcap_{n \geq 1}\{X < \mu + 1/n\}$. That implies by monotonicity (see Lemma B.2.6) that, for any $\varepsilon \in (0, 1)$, it holds that $\mathbb{P}[X < \mu + 1/n] < \varepsilon$ for n large enough. Hence, because we assume that $\mathbb{P}[X \leq \mu] = 0$,

$$
\begin{aligned}
\mu &\geq \mathbb{E}X \\
&= \mathbb{E}[X; X < \mu + 1/n] + \mathbb{E}[X; X \geq \mu + 1/n] \\
&\geq \mu\,\mathbb{P}[X < \mu + 1/n] + (\mu + 1/n)(1 - \mathbb{P}[X < \mu + 1/n]) \\
&= \mu + n^{-1}(1 - \mathbb{P}[X < \mu + 1/n]) \\
&> \mu + n^{-1}(1 - \varepsilon) \\
&> \mu,
\end{aligned}
$$

a contradiction.   ∎

The power of this principle is easier to appreciate through an example.

Example 2.2.2 (Balancing vectors). Let $v_1, \ldots, v_n$ be arbitrary unit vectors in $\mathbb{R}^n$. How small can we make the 2-norm of the linear combination

$$x_1 v_1 + \cdots + x_n v_n$$

by appropriately choosing $x_1, \ldots, x_n \in \{-1, +1\}$? We claim that it can be as small as $\sqrt{n}$, for *any* collection of v_is. At first sight, this may appear to be a complicated geometry problem. But the proof is trivial once one thinks of choosing the x_is *at random*. Let $X_1, \ldots, X_n$ be independent random variables uniformly distributed in $\{-1, +1\}$. Then, since $\mathbb{E}[X_iX_j] = \mathbb{E}[X_i]\,\mathbb{E}[X_j] = 0$ for all $i \neq j$ but $\mathbb{E}[X_i^2] = 1$ for all i,

$$
\begin{aligned}
\mathbb{E}\|X_1 v_1 + \cdots + X_n v_n\|_2^2 &= \mathbb{E}\left[\sum_{i,j} X_iX_j\langle v_i, v_j\rangle\right] \\
&= \sum_{i,j} \mathbb{E}[X_iX_j\langle v_i, v_j\rangle] \\
&= \sum_{i,j} \langle v_i, v_j\rangle\,\mathbb{E}[X_iX_j] \\
&= \sum_i \|v_i\|_2^2 \\
&= n,
\end{aligned}
$$

where we used the linearity of expectation on the second line. Hence, random variable $Z = \|X_1 v_1 + \cdots + X_n v_n\|^2$ has expectation $\mathbb{E}Z = n$ and must take a value $\leq n$ with positive probability by the first moment principle (Theorem 2.2.1). In other words, there must be a choice of X_is such that $Z \leq n$. That proves the claim. ◄

Here is a slightly more subtle example of the probabilistic method, where one has to *modify* the original random choice.

Example 2.2.3 (Independent sets). For $d \in \mathbb{N}$, let $G = (V, E)$ be a d-regular graph with n vertices. Such a graph necessarily has $m = nd/2$ edges. Our goal is to derive a lower bound on the size, $\alpha(G)$, of the largest independent set in G. Recall that an independent set is a set of vertices in a graph, no two of which are adjacent. Again, at first sight, this may seem like a rather complicated graph-theoretic problem. But an appropriate random choice gives a non-trivial bound. Specifically:

Claim 2.2.4

$$\alpha(G) \geq \frac{n}{2d}.$$

Proof The proof proceeds in two steps:

1. We first prove the existence of a subset S of vertices with relatively few edges.
2. We remove vertices from S to obtain an independent set.

Step 1. Let $0 < p < 1$ to be chosen below. To form the set S, pick each vertex in V independently with probability p. Letting X be the number of vertices in S, we have by the linearity of expectation that

$$\mathbb{E}X = \mathbb{E}\left[\sum_{v \in V} \mathbf{1}_{v \in S}\right] = np,$$

where we used $\mathbb{E}[\mathbf{1}_{v \in S}] = p$. Letting Y be the number of edges between vertices in S, we have by the linearity of expectation

$$\mathbb{E}Y = \mathbb{E}\left[\sum_{\{i,j\} \in E} \mathbf{1}_{i \in S} \mathbf{1}_{j \in S}\right] = \frac{nd}{2}p^2,$$

where we also used that $\mathbb{E}[\mathbf{1}_{i \in S} \mathbf{1}_{j \in S}] = p^2$ by independence. Hence, subtracting,

$$\mathbb{E}[X - Y] = np - \frac{nd}{2}p^2,$$

which, as a function of p, is maximized at $p = 1/d$, where it takes the value $n/(2d)$. As a result, by the first moment principle applied to $X - Y$, there must exist a set S of vertices in G such that

$$|S| - |\{\{i, j\} \in E : i, j \in S\}| \geq \frac{n}{2d}. \tag{2.2.2}$$

Step 2. For each edge e connecting two vertices in S, remove one of the endvertices of e. By construction, the remaining set of vertices (i) forms an independent set, and (ii) has a size larger than or equal to the left-hand side of (2.2.2). That inequality implies the claim. ∎

Note that a graph G made of $n/(d + 1)$ cliques of size $d + 1$ (with no edge between the cliques) has $\alpha(G) = n/(d + 1)$, showing that our bound is tight up to a constant. This is known as a Turán graph.   ◄

Remark 2.2.5 *The previous result can be strengthened to*

$$\alpha(G) \geq \sum_{v \in V} \frac{1}{\delta(v) + 1}$$

for a general graph $G = (V, E)$, where $\delta(v)$ is the degree of v. This bound is achieved for Turán graphs. See, for example, [AS11, The probabilistic lens: Turán's theorem].

INDICATOR
TRICK

The previous example also illustrates the important *indicator trick*, that is, writing a random variable as a sum of indicators, which is naturally used in combination with the linearity of expectation.

2.2.2 Boole's Inequality

One implication of the first moment principle (Theorem 2.2.1) is that if a *non-negative, integer-valued* random variable X has expectation strictly smaller than 1, then its value is 0 with positive probability. The following application of Markov's inequality (Theorem 2.1.1) adds a quantitative twist: if that same X has a "small" expectation, then its value is 0 with "large" probability.

Theorem 2.2.6 (First moment method). *If X is a non-negative, integer-valued random variable, then*

$$\mathbb{P}[X > 0] \leq \mathbb{E}X. \tag{2.2.3}$$

Proof   Take $b = 1$ in Markov's inequality.   ∎

This simple fact is typically used in the following manner: one wants to show that a certain "bad event" *does not occur* with probability approaching 1; the random variable X then counts the number of such "bad events." In that case, X is a sum of indicators and Theorem 2.2.6 reduces simply to the standard *union bound*, also known as *Boole's inequality*. We record one useful version of this setting in the next corollary.

UNION
BOUND

Corollary 2.2.7 *Let $B_n = A_{n,1} \cup \cdots \cup A_{n,m_n}$, where $A_{n,1}, \ldots, A_{n,m_n}$ is a collection of events for each n. Then, letting*

$$\mu_n := \sum_{i=1}^{m_n} \mathbb{P}[A_{n,i}],$$

we have

$$\mathbb{P}[B_n] \leq \mu_n.$$

In particular, if $\mu_n \to 0$ as $n \to +\infty$, then $\mathbb{P}[B_n] \to 0$.

Proof   Take $X := X_n = \sum_{i=1}^{m_n} \mathbf{1}_{A_{n,i}}$ in Theorem 2.2.6.   ∎

A useful generalization of the union bound is given in Exercise 2.2.

FIRST
MOMENT
METHOD

We will refer to applications of Theorem 2.2.6 as the *first moment method*. We give a few examples.

Example 2.2.8 (Random k-SAT threshold). For $r \in \mathbb{R}_+$, let $\Phi_{n,r} \colon \{0,1\}^n \to \{0,1\}$ be a random k-CNF formula on n Boolean variables $z_1, \ldots, z_n$ with $\lceil rn \rceil$ clauses. That is, $\Phi_{n,r}$ is an AND of $\lceil rn \rceil$ ORs, each obtained by picking independently k literals uniformly at random (with replacement). Recall that a literal is a variable z_i or its negation $\bar{z}_i$. The formula $\Phi_{n,r}$ is said to be satisfiable if there exists an assignment $z = (z_1, \ldots, z_n)$ such that $\Phi_{n,r}(z) = 1$. Clearly, the higher the value of r, the less likely it is for $\Phi_{n,r}$ to be satisfiable. In fact, it is natural to conjecture that a sharp transition takes place, that is, that there exists an $r_k^* \in \mathbb{R}_+$ (depending on k but not on n) such that

$$\lim_{n \to \infty} \mathbb{P}[\Phi_{n,r} \text{ is satisfiable}] = \begin{cases} 0 \text{ if } r > r_k^*, \\ 1 \text{ if } r < r_k^*. \end{cases} \tag{2.2.4}$$

Studying such *threshold phenomena* is a major theme of modern discrete probability. Using the first moment method (Theorem 2.2.6), we give an upper bound on the threshold. Formally:

THRESHOLD
PHENOMENON

Claim 2.2.9

$$r > 2^k \log 2 \implies \limsup_{n \to \infty} \mathbb{P}[\Phi_{n,r} \text{ is satisfiable}] = 0.$$

Proof How to start the proof should be obvious: let X_n be the number of satisfying assignments of $\Phi_{n,r}$. Applying the first moment method, since

$$\mathbb{P}[\Phi_{n,r} \text{ is satisfiable}] = \mathbb{P}[X_n > 0],$$

it suffices to show that $\mathbb{E} X_n \to 0$. To compute $\mathbb{E} X_n$, we use the indicator trick

$$X_n = \sum_{z \in \{0,1\}^n} \mathbf{1}_{\{z \text{ satisfies } \Phi_{n,r}\}}.$$

There are 2^n possible assignments. Each fixed assignment satisfies the random choice of clauses $\Phi_{n,r}$ with probability $(1 - 2^{-k})^{\lceil rn \rceil}$. Indeed, note that the rn clauses are picked independently and each clause literal picked is satisfied with probability $1/2$. Therefore, by the assumption on r, for $\varepsilon > 0$ small enough and n large enough,

$$\begin{aligned} \mathbb{E} X_n &= 2^n (1 - 2^{-k})^{\lceil rn \rceil} \\ &\leq 2^n (1 - 2^{-k})^{(2^k \log 2)(1+\varepsilon)n} \\ &\leq 2^n e^{-(\log 2)(1+\varepsilon)n} \\ &= 2^{-\varepsilon n} \\ &\to 0, \end{aligned}$$

where we used $(1 - 1/\ell)^\ell \leq e^{-1}$ for all $\ell \in \mathbb{N}$ (see Exercise 1.16). Theorem 2.2.6 implies the claim. $\blacksquare$

Remark 2.2.10 *Bounds in the other direction are also known. For instance, for $k \geq 3$, it has been shown that if $r < 2^k \log 2 - k$, then*

$$\liminf_{n \to \infty} \mathbb{P}[\Phi_{n,r} \text{ is satisfiable}] = 1.$$

See [ANP05]. For the $k = 2$ case, it is known that (2.2.4) in fact holds with $r_2^* = 1$ [CR92].
A breakthrough of [DSS22] also establishes (2.2.4) for large k; the threshold r_k^* is charac-
terized as the root of a certain equation coming from statistical physics. ◄

2.2.3 ▷ *Random Permutations: Longest Increasing Subsequence*

In this section, we bound the expected length of a longest increasing subsequence in a
RANDOM random permutation. Let $\sigma_n = (\sigma_n(1), \ldots, \sigma_n(n))$ be a uniformly random permutation of
PERMUTATION $[n] := \{1, \ldots, n\}$ (i.e., a bijection of $[n]$ to itself chosen uniformly at random among all such
mappings) and let L_n be the length of a longest increasing subsequence of σ_n (i.e., a sequence
of indices $i_1 < \cdots < i_k$ such that $\sigma_n(i_1) < \cdots < \sigma_n(i_k)$).

Claim 2.2.11

$$\mathbb{E}L_n = \Theta(\sqrt{n}).$$

Proof We first prove that

$$\limsup_{n\to\infty} \frac{\mathbb{E}L_n}{\sqrt{n}} \le e, \tag{2.2.5}$$

which implies half of the claim. Bounding the expectation of L_n is not straightforward as it
is the expectation of a *maximum*. A natural way to proceed is to find a value ℓ for which
$\mathbb{P}[L_n \ge \ell]$ is "small." More formally, we bound the expectation as

$$\mathbb{E}L_n \le \ell\,\mathbb{P}[L_n < \ell] + n\,\mathbb{P}[L_n \ge \ell] \le \ell + n\,\mathbb{P}[L_n \ge \ell] \tag{2.2.6}$$

for an ℓ chosen below. To bound the probability on the right-hand side, we appeal to the first
moment method (Theorem 2.2.6) by letting X_n be the number of increasing subsequences of
length ℓ. We also use the indicator trick, that is, we think of X_n as a sum of indicators over
subsequences (not necessarily increasing) of length ℓ.

There are $\binom{n}{\ell}$ such subsequences, each of which is increasing with probability $1/\ell!$. Note
that these subsequences are not independent. Nevertheless, by the linearity of expectation
and the first moment method,

$$\mathbb{P}[L_n \ge \ell] = \mathbb{P}[X_n > 0] \le \mathbb{E}X_n = \frac{1}{\ell!}\binom{n}{\ell} \le \frac{n^\ell}{(\ell!)^2} \le \frac{n^\ell}{e^2[\ell/e]^{2\ell}} \le \left(\frac{e\sqrt{n}}{\ell}\right)^{2\ell},$$

where we used a standard bound on factorials recalled in Appendix A. Note that, in order
for this bound to go to 0, we need $\ell > e\sqrt{n}$. Then (2.2.5) follows by taking $\ell = (1+\delta)e\sqrt{n}$
in (2.2.6), for an arbitrarily small $\delta > 0$.

For the other half of the claim, we show that

$$\frac{\mathbb{E}L_n}{\sqrt{n}} \ge 1.$$

This part does not rely on the first moment method (and may be skipped). We seek a lower
bound on the expected length of a longest increasing subsequence. The proof uses the following two ideas. First observe that there is a natural symmetry between the lengths of the
longest *increasing* and *decreasing* subsequences – they are identically distributed. Moreover,
if a permutation has a "short" longest increasing subsequence, then intuitively it must have

a "long" decreasing subsequence, and vice versa. Combining these two observations gives a lower bound on the expectation of L_n. Formally, let D_n be the length of a longest decreasing subsequence. By symmetry and the arithmetic mean-geometric mean inequality, note that

$$\mathbb{E}L_n = \mathbb{E}\left[\frac{L_n + D_n}{2}\right] \geq \mathbb{E}\sqrt{L_n D_n}.$$

We show that $L_n D_n \geq n$, which proves the claim. Let $L_n^{(k)}$ be the length of a longest increasing subsequence ending at position k, and similarly for $D_n^{(k)}$. It suffices to show that the pairs $(L_n^{(k)}, D_n^{(k)})$, $1 \leq k \leq n$, are *distinct*. Indeed, noting that $L_n^{(k)} \leq L_n$ and $D_n^{(k)} \leq D_n$, the number of pairs in $[L_n] \times [D_n]$ is at most $L_n D_n$, which must then be at least n.

Let $1 \leq j < k \leq n$. If $\sigma_n(k) > \sigma_n(j)$, then we see that $L_n^{(k)} > L_n^{(j)}$ by appending $\sigma_n(k)$ to the subsequence ending at position j achieving $L_n^{(j)}$. If the opposite holds, then we have instead $D_n^{(k)} > D_n^{(j)}$. Either way, $(L_n^{(j)}, D_n^{(j)})$ and $(L_n^{(k)}, D_n^{(k)})$ must be distinct. This clever combinatorial argument is known as the *Erdős–Szekeres Theorem*. That concludes the proof of the second claim. $\blacksquare$

Remark 2.2.12 *It has been shown that in fact*

$$\mathbb{E}L_n = 2\sqrt{n} + cn^{1/6} + o(n^{1/6}),$$

as $n \to +\infty$, where $c = -1.77\ldots$ [BDJ99].

2.2.4 ▷ *Percolation: Existence of a Non-Trivial Threshold on $\mathbb{Z}^2$*

In this section, we use the first moment method (Theorem 2.2.6) to prove the existence of a non-trivial threshold in bond percolation on the two-dimensional lattice. We begin with some background.

Threshold in bond percolation Consider bond percolation (Definition 1.2.1) on the two-dimensional lattice $\mathbb{L}^2$ (see Section 1.1.1) with density p. Let $\mathbb{P}_p$ denote the corresponding measure. Recall that paths are "self-avoiding" by definition (see Section 1.1.1). We say that a path is open if all edges in the induced subgraph are open. Writing $x \Leftrightarrow y$ if $x, y \in \mathbb{L}^2$ are OPEN PATH connected by an open path; recall that the open cluster of x is

$$\mathcal{C}_x := \{y \in \mathbb{Z}^2 : x \Leftrightarrow y\}.$$

The *percolation function* is defined as PERCOLATION FUNCTION

$$\theta(p) := \mathbb{P}_p[|\mathcal{C}_0| = +\infty],$$

that is, $\theta(p)$ is the probability that the origin is connected by open paths to infinitely many vertices. It is intuitively clear that the function $\theta(p)$ is non-decreasing. Indeed, consider the following alternative representation of the percolation process: to each edge e, assign a uniform $[0, 1]$ random variable U_e and declare the edge open if $U_e \leq p$. Using the same U_es for densities $p_1 < p_2$, it follows immediately from the monotonicity of the construction that $\theta(p_1) \leq \theta(p_2)$. (We will have much more to say about this type of "coupling" argument in Chapter 4.) Moreover, note that $\theta(0) = 0$ and $\theta(1) = 1$. The *critical value* is defined as CRITICAL VALUE

$$p_c(\mathbb{L}^2) := \sup\{p \geq 0 : \theta(p) = 0\},$$

the point at which the probability that the origin is contained in an infinite open cluster becomes positive. Note that by a union bound over all vertices, when $\theta(p) = 0$, we have that $\mathbb{P}_p[\exists x, |\mathcal{C}_x| = +\infty] = 0$. Conversely, because $\{\exists x, |\mathcal{C}_x| = +\infty\}$ is a tail event (see Definition B.3.9) for any enumeration of the edges, by Kolmogorov's 0-1 law (Theorem B.3.11) it holds that $\mathbb{P}_p[\exists x, |\mathcal{C}_x| = +\infty] = 1$ when $\theta(p) > 0$.

Using the first moment method we show that the critical value is non-trivial, that is, it is *strictly* between 0 and 1. This is a different example of a threshold phenomenon.

Claim 2.2.13

$$p_c(\mathbb{L}^2) \in (0, 1).$$

Proof We first show that for any $p < 1/3$, $\theta(p) = 0$. In order to apply the first moment method, roughly speaking, we need to reduce the problem to counting the number of instances of an appropriately chosen substructure. The key observation is the following:

An infinite $\mathcal{C}_0$ contains an open path starting at 0 of *infinite* length and, as a result, of *all* lengths.

Hence, we let X_n be the number of open paths of length n starting at 0. Then, by monotonicity,

$$\mathbb{P}_p[|\mathcal{C}_0| = +\infty] \le \mathbb{P}_p[\cap_n\{X_n > 0\}] = \lim_n \mathbb{P}_p[X_n > 0] \le \limsup_n \mathbb{E}_p[X_n], \qquad (2.2.7)$$

where the last inequality follows from Theorem 2.2.6. We bound the number of paths by noting that they *cannot backtrack*. That gives four choices at the first step, and at most three choices at each subsequent step. Hence, we get the following bound

$$\mathbb{E}_p X_n \le 4(3^{n-1})p^n.$$

The right-hand side goes to 0 for all $p < 1/3$. When combined with (2.2.7), that proves half of the claim:

$$p_c(\mathbb{L}^2) > 0.$$

For the other direction, we show that $\theta(p) > 0$ for p close enough to 1. This time, we count "dual cycles." This type of proof is known as a contour argument, or Peierls' argument, DUAL LATTICE and is based on the following construction. Consider the *dual lattice* $\widetilde{\mathbb{L}}^2$ whose vertices are $\mathbb{Z}^2 + (1/2, 1/2)$ and whose edges connect vertices u, v with $\|u - v\|_1 = 1$. See Figure 2.3.

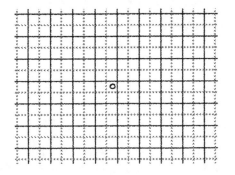

Figure 2.3 Primal (solid) and dual (dotted) lattices.

Note that each edge in the *primal lattice* $\mathbb{L}^2$ has a unique corresponding edge in the dual lattice which crosses it perpendicularly. We make the same assignment, open or closed, for corresponding primal and dual edges. The following graph-theoretic lemma, whose proof is sketched below, forms the basis of contour arguments. Recall that cycles are "self-avoiding" by definition (see Section 1.1.1). We say that a cycle is closed if all edges in the induced subgraph are closed, that is, are not open.

Lemma 2.2.14 (Contour lemma). *If $|\mathcal{C}_0| < +\infty$, then there is a closed cycle around the origin in the dual lattice $\widetilde{\mathbb{L}}^2$.*

CONTOUR LEMMA

To prove that $\theta(p) > 0$ for p close enough to 1, the idea is to apply the first moment method to Z_n equal to the number of closed dual cycles of length n surrounding the origin. We bound from above the number of dual cycles of length n around the origin by the number of choices for the starting edge across the upper y-axis and for each $n - 1$ subsequent non-backtracking choices. Namely,

$$
\begin{aligned}
\mathbb{P}[|\mathcal{C}_0| < +\infty] &\leq \mathbb{P}[\exists n \geq 4, Z_n > 0] \\
&\leq \sum_{n \geq 4} \mathbb{P}[Z_n > 0] \\
&\leq \sum_{n \geq 4} \mathbb{E} Z_n \\
&\leq \sum_{n \geq 4} \frac{n}{2} 3^{n-1} (1 - p)^n \\
&= \frac{3^3 (1 - p)^4}{2} \sum_{m \geq 1} (m + 3)(3(1 - p))^{m-1} \\
&= \frac{3^3 (1 - p)^4}{2} \left(\frac{1}{(1 - 3(1 - p))^2} + 3 \frac{1}{1 - 3(1 - p)} \right)
\end{aligned}
$$

when $p > 2/3$, where the first term in parentheses on the last line comes from differentiating with respect to q the geometric series $\sum_{m \geq 0} q^m$ and setting $q := 1 - p$. This expression can be taken smaller than 1 if we let p approach 1. We have shown that $\theta(p) > 0$ for p close enough to 1, and that concludes the proof. (Exercise 2.3 sketches a proof that $\theta(p) > 0$ for all $p > 2/3$.) ∎

It is straightforward to extend the claim to $\mathbb{L}^d$. (Exercise 2.4 asks for the details.)

Proof of the contour lemma We conclude this section by sketching the proof of the contour lemma, which relies on topological arguments.

Proof of Lemma 2.2.14 Assume $|\mathcal{C}_0| < +\infty$. Imagine identifying each vertex in $\mathbb{L}^2$ with a square of side 1 centered around it so that the sides line up with dual edges. Paint green the squares of vertices in $\mathcal{C}_0$. Paint red the squares of vertices in $\mathcal{C}_0^c$ *which share a side with a green square*. Leave the other squares white. Let u_0 be a highest vertex in $\mathcal{C}_0$ along the y-axis and let v_0 and v_1 be the dual vertices corresponding to the upper left and right corners, respectively, of the square of u_0. Because u_0 is highest, it must be that the square above it is red. Walk along the dual edge $\{v_0, v_1\}$, separating the squares of u_0 and $u_0 + (0, 1)$ from

v_0 to v_1. Notice that this edge satisfies what we call the *red-green property*: as you traverse it from v_0 to v_1, a red square sits on your left and a green square is on your right. Proceed further by iteratively walking along an incident dual edge with the following rule. Choose an edge satisfying the red-green property, with the edges to your left, straight ahead, and to your right in decreasing order of priority. Stop when a previously visited dual vertex is reached. The claim is that this procedure constructs the desired cycle. Let $v_0, v_1, v_2, \ldots$ be the dual vertices visited. By construction $\{v_{i-1}, v_i\}$ is a dual edge for all i.

- *A dual cycle is produced.* We first argue that this procedure cannot get stuck. Let $\{v_{i-1}, v_i\}$ be the edge just crossed and assume that it has the red-green property. If there is a green square to the left ahead, then the edge to the left, which has highest priority, has the red-green property. If the left square ahead is not green, but the right one is, then the left square must in fact be red by construction (i.e., it cannot be white). In that case, the edge straight ahead has the red-green property. Finally, if neither square ahead is green, then the right square must in fact be red because the square behind to the right is green by assumption. That implies that the edge to the right has the red-green property. Hence, we have shown that the procedure does not get stuck. Moreover, because by assumption the number of green squares is finite, this procedure must eventually terminate when a previously visited dual vertex is reached, forming a cycle (of length at least 4).

- *The origin lies within the cycle.* The inside of a cycle in the plane is well defined by the Jordan curve theorem. So the dual cycle produced above has its adjacent green squares either on the inside (negative orientation) or on the outside (positive orientation). In the former case the origin must lie inside the cycle as otherwise the vertices corresponding to the green squares on the inside would not be in $\mathcal{C}_0$, as they could not be connected to the origin with open paths.

  So it remains to consider the latter case, where through a similar reasoning the origin must lie outside the cycle. Let v_j be the repeated dual vertex. Assume first that $v_j \neq v_0$ and let v_{j-1} and v_{j+1} be the dual vertices preceding and following v_j during the first visit to v_j. Let v_k be the dual vertex preceding v_j on the second visit. After traversing the edge from v_{j-1} to v_j, v_k cannot be to the left or to the right because in those cases the red-green properties of the two corresponding edges (i.e., $\{v_{j-1}, v_j\}$ and $\{v_k, v_j\}$) are not compatible. So v_k is straight ahead and, by the priority rules, v_{j+1} must be to the left upon entering v_j the first time. But in that case, for the origin to lie outside the cycle as we are assuming and for the cycle to avoid the path $v_0, \ldots, v_{j-1}$, we must traverse the cycle with a negative orientation, that is, the green squares adjacent to the cycle must be on the inside, a contradiction.

  So, finally, assume v_0 is the repeated vertex. If the cycle is traversed with a positive orientation and the origin is on the outside, it must be that the cycle crosses the y-axis at least once *above* $u_0 + (0, 1)$, again a contradiction.

  Hence, we have shown that the origin is inside the cycle.

That concludes the proof.   ∎

Remark 2.2.15 *It turns out that $p_c(\mathbb{L}^2) = 1/2$. We will prove $p_c(\mathbb{L}^2) \geq 1/2$, known as Harris' Theorem, in Section 4.2.5. The other direction is due to Kesten [Kes80].*

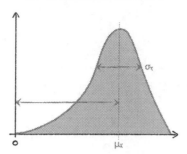
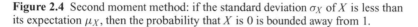

Figure 2.4 Second moment method: if the standard deviation σ_X of X is less than its expectation μ_X, then the probability that X is 0 is bounded away from 1.

2.3 Second Moment Method

The first moment method (Theorem 2.2.6) gives an *upper bound* on the probability that a non-negative, integer-valued random variable is positive – which is non-trivial provided its expectation is small enough. In this section, we seek a *lower bound* on that probability. We first note that a large expectation does not suffice in general. Say, X_n is n^2 with probability $1/n$, and 0 otherwise. Then, $\mathbb{E} X_n = n \to +\infty$, yet $\mathbb{P}[X_n > 0] \to 0$. That is, although the expectation diverges, the probability that X_n is positive can be arbitrarily small.

So we turn to the second moment. Intuitively, the basis for the so-called second moment method is that if the expectation of X_n is large *and* its variance is relatively small, then we can bound the probability that X_n is close to 0. As we will see in applications, the first and second moment methods often work hand in hand.

2.3.1 Paley–Zygmund Inequality

As an immediate corollary of Chebyshev's inequality (Theorem 2.1.2), we get a first version of the *second moment method*: if the standard deviation of X is less than its expectation, then the probability that X is 0 is bounded away from 1 (see Figure 2.4). Formally, let X be a non-negative random variable (not identically zero). Then

$$\mathbb{P}[X > 0] \geq 1 - \frac{\text{Var}[X]}{(\mathbb{E} X)^2}. \tag{2.3.1}$$

Indeed, by (2.1.5),

$$\mathbb{P}[X = 0] \leq \mathbb{P}[|X - \mathbb{E} X| \geq \mathbb{E} X] \leq \frac{\text{Var}[X]}{(\mathbb{E} X)^2}.$$

The following tail bound, a simple application of Cauchy–Schwarz (Theorem B.4.8), leads to an improved version of this inequality.

Theorem 2.3.1 (Paley–Zygmund inequality). *Let X be a non-negative random variable. For all $0 < \theta < 1$,*

$$\mathbb{P}[X \geq \theta \, \mathbb{E} X] \geq (1 - \theta)^2 \frac{(\mathbb{E} X)^2}{\mathbb{E}[X^2]}. \tag{2.3.2}$$

PALEY–
ZYGMUND
INEQUALITY

Proof   We have

$$\mathbb{E}X = \mathbb{E}[X\mathbf{1}_{\{X<\theta\mathbb{E}X\}}] + \mathbb{E}[X\mathbf{1}_{\{X\geq\theta\mathbb{E}X\}}]$$
$$\leq \theta\mathbb{E}X + \sqrt{\mathbb{E}[X^2]\mathbb{P}[X\geq\theta\mathbb{E}X]},$$

where we used Cauchy–Schwarz. Rearranging gives the result.   ∎

As an immediate application:

Theorem 2.3.2 (Second moment method). *Let X be a non-negative random variable (not identically zero). Then*

$$\mathbb{P}[X > 0] \geq \frac{(\mathbb{E}X)^2}{\mathbb{E}[X^2]}. \tag{2.3.3}$$

Proof   Take $\theta \downarrow 0$ in (2.3.2).   ∎

Since

$$\frac{(\mathbb{E}X)^2}{\mathbb{E}[X^2]} = 1 - \frac{\mathrm{Var}[X]}{(\mathbb{E}X)^2 + \mathrm{Var}[X]},$$

we see that (2.3.3) is stronger than (2.3.1).

We typically apply the second moment method to a sequence of random variables (X_n). The previous theorem gives a uniform lower bound on the probability that $\{X_n > 0\}$ when $\mathbb{E}[X_n^2] \leq C\mathbb{E}[X_n]^2$ for some $C > 0$. Just like the first moment method, the second moment method is often applied to a sum of indicators (but see Section 2.3.3 for a weighted case). We record in the next corollary a convenient version of the method.

Corollary 2.3.3   *Let $B_n = A_{n,1} \cup \cdots \cup A_{n,m_n}$, where $A_{n,1}, \ldots, A_{n,m_n}$ is a collection of events for each n. Write $i \overset{n}{\sim} j$ if $i \neq j$ and $A_{n,i}$ and $A_{n,j}$ are not independent. Then, letting*

$$\mu_n := \sum_{i=1}^{m_n} \mathbb{P}[A_{n,i}], \qquad \gamma_n := \sum_{i\overset{n}{\sim}j} \mathbb{P}[A_{n,i} \cap A_{n,j}],$$

where the second sum is over ordered pairs, we have $\lim_n \mathbb{P}[B_n] > 0$ whenever $\mu_n \to +\infty$ and $\gamma_n \leq C\mu_n^2$ for some $C > 0$. If moreover $\gamma_n = o(\mu_n^2)$, then $\lim_n \mathbb{P}[B_n] = 1$.

Proof   We apply the second moment method to $X_n := \sum_{i=1}^{m_n} \mathbf{1}_{A_{n,i}}$ so that $B_n = \{X_n > 0\}$. Note that

$$\mathrm{Var}[X_n] = \sum_i \mathrm{Var}[\mathbf{1}_{A_{n,i}}] + \sum_{i\neq j} \mathrm{Cov}[\mathbf{1}_{A_{n,i}}, \mathbf{1}_{A_{n,j}}],$$

where

$$\mathrm{Var}[\mathbf{1}_{A_{n,i}}] = \mathbb{E}[(\mathbf{1}_{A_{n,i}})^2] - (\mathbb{E}[\mathbf{1}_{A_{n,i}}])^2 \leq \mathbb{P}[A_{n,i}],$$

and, if $A_{n,i}$ and $A_{n,j}$ are independent,

$$\mathrm{Cov}[\mathbf{1}_{A_{n,i}}, \mathbf{1}_{A_{n,j}}] = 0,$$

whereas, if $i \overset{n}{\sim} j$,

$$\mathrm{Cov}[\mathbf{1}_{A_{n,i}}, \mathbf{1}_{A_{n,j}}] = \mathbb{E}[\mathbf{1}_{A_{n,i}}\mathbf{1}_{A_{n,j}}] - \mathbb{E}[\mathbf{1}_{A_{n,i}}]\mathbb{E}[\mathbf{1}_{A_{n,j}}] \leq \mathbb{P}[A_{n,i} \cap A_{n,j}].$$

Hence,

$$\frac{\text{Var}[X_n]}{(\mathbb{E} X_n)^2} \leq \frac{\mu_n + \gamma_n}{\mu_n^2} = \frac{1}{\mu_n} + \frac{\gamma_n}{\mu_n^2}.$$

Noting

$$\frac{(\mathbb{E} X_n)^2}{\mathbb{E}[X_n^2]} = \frac{(\mathbb{E} X_n)^2}{(\mathbb{E} X_n)^2 + \text{Var}[X_n]} = \frac{1}{1 + \text{Var}[X_n]/(\mathbb{E} X_n)^2}$$

and applying Theorem 2.3.2 gives the result. ∎

2.3.2 ▷ *Random Graphs: Subgraph Containment and Connectivity in the Erdős–Rényi Model*

Threshold phenomena are also common in random graphs. We consider here the Erdős–Rényi random graph model (Definition 1.2.2). In this context, a *threshold function for a graph property* P is a function $r(n)$ such that

THRESHOLD FUNCTION

$$\lim_n \mathbb{P}_{n,p_n}[G_n \text{ has property } P] = \begin{cases} 0 & \text{if } p_n \ll r(n), \\ 1 & \text{if } p_n \gg r(n), \end{cases}$$

where $G_n \sim \mathbb{G}_{n,p_n}$ is a random graph with n vertices and density p_n. In this section, we illustrate this type of phenomenon on two properties: the containment of small subgraphs and connectivity.

Subgraph containment

We first consider the clique number, then we turn to more general subgraphs.

Cliques Let $\omega(G)$ be the *clique number* of a graph G, that is, the size of its largest clique.

CLIQUE NUMBER

Claim 2.3.4 *The property* $\omega(G_n) \geq 4$ *has threshold function* $n^{-2/3}$.

Proof Let X_n be the number of 4-cliques in the random graph $G_n \sim \mathbb{G}_{n,p_n}$. Then, noting that there are $\binom{4}{2} = 6$ edges in a 4-clique,

$$\mathbb{E}_{n,p_n}[X_n] = \binom{n}{4} p_n^6 = \Theta(n^4 p_n^6),$$

which goes to 0 when $p_n \ll n^{-2/3}$. Hence, the first moment method (Theorem 2.2.6) gives one direction: $\mathbb{P}_{n,p_n}[\omega(G_n) \geq 4] \to 0$ in that case.

For the other direction, we apply the second moment method for sums of indicators, that is, Corollary 2.3.3. We use the notation from that corollary. For an enumeration $S_1, \ldots, S_{m_n}$ of the 4-tuples of vertices in G_n, let $A_{n,1}, \ldots, A_{n,m_n}$ be the events that the corresponding 4-clique is present. By the calculation above we have $\mu_n = \Theta(n^4 p_n^6)$, which goes to $+\infty$ when $p_n \gg n^{-2/3}$. Also $\mu_n^2 = \Theta(n^8 p_n^{12})$, so it suffices to show that $\gamma_n = o(n^8 p_n^{12})$. Note that two 4-cliques with disjoint edge sets (but possibly sharing one vertex) are independent (i.e., their presence or absence is independent). Suppose S_i and S_j share three vertices. Then, $i \overset{n}{\sim} j$ and

$$\mathbb{P}_{n,p_n}[A_{n,i} \mid A_{n,j}] = p_n^3,$$

as the event $A_{n,j}$ implies that all edges between three of the vertices in S_i are already present, and there are three edges between the remaining vertex and the rest of S_i. Similarly, if $|S_i \cap S_j| = 2$, we have again $i \overset{n}{\sim} j$ and this time $\mathbb{P}_{n,p_n}[A_{n,i} | A_{n,j}] = p_n^5$. Putting these together, we get by the definition of the conditional probability (see Appendix B) and the fact that $\mathbb{P}_{n,p_n}[A_{n,j}] = p_n^6$

$$
\begin{aligned}
\gamma_n &= \sum_{i \overset{n}{\sim} j} \mathbb{P}[A_{n,i} \cap A_{n,j}] \\
&= \sum_{i \overset{n}{\sim} j} \mathbb{P}_{n,p_n}[A_{n,j}] \, \mathbb{P}_{n,p_n}[A_{n,i} | A_{n,j}] \\
&= \sum_{j} \mathbb{P}_{n,p_n}[A_{n,j}] \sum_{i:i \overset{n}{\sim} j} \mathbb{P}_{n,p_n}[A_{n,i} | A_{n,j}] \\
&= \binom{n}{4} p_n^6 \left[\binom{4}{3}(n-4)p_n^3 + \binom{4}{2}\binom{n-4}{2}p_n^5 \right] \\
&= O(n^5 p_n^9) + O(n^6 p_n^{11}) \\
&= O\left(\frac{n^8 p_n^{12}}{n^3 p_n^3}\right) + O\left(\frac{n^8 p_n^{12}}{n^2 p_n}\right) \\
&= o(n^8 p_n^{12}) \\
&= o(\mu_n^2),
\end{aligned}
$$

where we used that $p_n \gg n^{-2/3}$ (so that for example $n^3 p_n^3 \gg 1$). Corollary 2.3.3 gives the result: $\mathbb{P}_{n,p_n}[\cup_i A_{n,i}] \to 1$ when $p_n \gg n^{-2/3}$.   ∎

Roughly speaking, the first and second moments suffice to pinpoint the threshold in this case because the indicators in X_n are "mostly" pairwise independent and, as a result, the sum is "concentrated around its mean."

General subgraphs The methods of Claim 2.3.4 can be applied to more general subgraphs. However, the situation is somewhat more complicated than it is for cliques. For a graph H_0, let v_{H_0} and e_{H_0} be the number of vertices and edges of H_0, respectively. Let X_n be the number of (not necessarily induced) copies of H_0 in $G_n \sim \mathbb{G}_{n,p_n}$. By the first moment method,

$$
\mathbb{P}[X_n > 0] \leq \mathbb{E}[X_n] = \Theta(n^{v_{H_0}} p_n^{e_{H_0}}) \to 0,
$$

when $p_n \ll n^{-v_{H_0}/e_{H_0}}$. The constant factor, which does not play a role in the asymptotics, accounts in particular for the number of automorphisms of H_0. Indeed, note that a fixed set of v_{H_0} vertices can contain several distinct copies of H_0, depending on its structure.

From the proof of Claim 2.3.4, one might guess that the threshold function is $n^{-v_{H_0}/e_{H_0}}$. That is not the case in general. To see what can go wrong, consider the graph H_0 in Figure 2.5 EDGE DENSITY whose *edge density* is $\frac{e_{H_0}}{v_{H_0}} = \frac{6}{5}$. When $p_n \gg n^{-5/6}$, the expected number of copies of H_0 in G_n tends to $+\infty$. But observe that the subgraph H of H_0 has the *higher* density $5/4$ and, hence, when $n^{-5/6} \ll p_n \ll n^{-4/5}$ the expected number of copies of H tends to 0. By the first moment method, the probability that a copy of H_0 – and therefore H – is present in

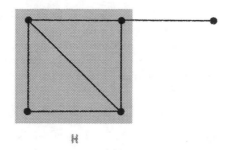

H

Figure 2.5 Graph H_0 and subgraph H.

that regime is asymptotically negligible despite its diverging expectation. This leads to the following definition

$$r_{H_0} := \max \left\{ \frac{e_H}{v_H} : \text{subgraphs } H \subseteq H_0, \ e_H > 0 \right\}.$$

Assume H_0 has at least one edge.

Claim 2.3.5 *"Having a copy of H_0" has threshold $n^{-1/r_{H_0}}$.*

Proof We proceed as in Claim 2.3.4. Let H_0^* be a subgraph of H_0 achieving r_{H_0}. When $p_n \ll n^{-1/r_{H_0}}$, the probability that a copy of H_0^* is in G_n tends to 0 by the argument above. Therefore, the same conclusion holds for H_0 itself.

Assume $p_n \gg n^{-1/r_{H_0}}$. Let $S_1, \ldots, S_{m_n}$ be an enumeration of the *copies (as subgraphs) of H_0 in a complete graph* on the vertices of G_n. Let $A_{n,i}$ be the event that $S_i \subseteq G_n$. Using again the notation of Corollary 2.3.3,

$$\mu_n = \Theta(n^{v_{H_0}} p_n^{e_{H_0}}) = \Omega(\Phi_{H_0}(n)),$$

where

$$\Phi_{H_0}(n) := \min \left\{ n^{v_H} p_n^{e_H} : \text{subgraphs } H \subseteq H_0, e_H > 0 \right\}.$$

Note that $\Phi_{H_0}(n) \to +\infty$ when $p_n \gg n^{-1/r_{H_0}}$ by definition of r_{H_0}. The events $A_{n,i}$ and $A_{n,j}$ are independent if S_i and S_j share no edge. Otherwise we write $i \overset{n}{\sim} j$. Note that there are $\Theta(n^{v_H} n^{2(v_{H_0} - v_H)})$ pairs S_i, S_j whose intersection is isomorphic to H. The probability that both S_i and S_j of such a pair are present in G_n is $\Theta(p_n^{e_H} p_n^{2(e_{H_0} - e_H)})$. Hence,

$$\gamma_n = \sum_{i \overset{n}{\sim} j} \mathbb{P}[A_{n,i} \cap A_{n,j}]$$

$$= \sum_{H \subseteq H_0, e_H > 0} \Theta \left(n^{2v_{H_0} - v_H} p_n^{2e_{H_0} - e_H} \right)$$

$$\leq \frac{\Theta(\mu_n^2)}{\Theta(\Phi_{H_0}(n))}$$

$$= o(\mu_n^2),$$

where we used that $\Phi_{H_0}(n) \to +\infty$. The result follows from Corollary 2.3.3. ∎

Going back to the example of Figure 2.5, the proof above confirms that when $n^{-5/6} \ll p_n \ll n^{-4/5}$ the second moment method fails for H_0 since $\Phi_{H_0}(n) \to 0$. In that regime, although there is in expectation a large number of copies of H_0, those copies are *highly correlated* as they are produced from a small (vanishing in expectation) number of copies of H – producing a large variance that helps to explain the failure of the second moment method.

Connectivity threshold

Next we use the second moment method to show that the threshold function for connectivity in the Erdős–Rényi random graph model is $\frac{\log n}{n}$. In fact, we prove this result by deriving the threshold function for the presence of isolated vertices. The connection between the two is obvious in one direction. Isolated vertices imply a disconnected graph. What is less obvious is that it also works the other way in the following sense: the two thresholds actually *coincide*.

Isolated vertices We begin with isolated vertices.

Claim 2.3.6 *"Not having an isolated vertex" has threshold function $\frac{\log n}{n}$.*

Proof Let X_n be the number of isolated vertices in the random graph $G_n \sim \mathbb{G}_{n,p_n}$. Using $1 - x \le e^{-x}$ for all $x \in \mathbb{R}$ (see Exercise 1.16),

$$\mathbb{E}_{n,p_n}[X_n] = n(1 - p_n)^{n-1} \le e^{\log n - (n-1)p_n} \to 0, \tag{2.3.4}$$

when $p_n \gg \frac{\log n}{n}$. So the first moment method gives one direction: $\mathbb{P}_{n,p_n}[X_n > 0] \to 0$.

For the other direction, we use the second moment method. Let $A_{n,j}$ be the event that vertex j is isolated. By the computation above, using $1 - x \ge e^{-x-x^2}$ for $x \in [0, 1/2]$ (see Exercise 1.16),

$$\mu_n = \sum_i \mathbb{P}_{n,p_n}[A_{n,i}] = n(1 - p_n)^{n-1} \ge e^{\log n - np_n - np_n^2}, \tag{2.3.5}$$

which goes to $+\infty$ when $p_n \ll \frac{\log n}{n}$. Note that $A_{n,i}$ and $A_{n,j}$ are not independent for all $i \neq j$ (because the absence of an edge between i and j is part of both events) and

$$\mathbb{P}_{n,p_n}[A_{n,i} \cap A_{n,j}] = (1 - p_n)^{2(n-2)+1},$$

so that

$$\gamma_n = \sum_{i \neq j} \mathbb{P}_{n,p_n}[A_{n,i} \cap A_{n,j}] = n(n-1)(1 - p_n)^{2n-3}.$$

Because γ_n is *not* $o(\mu_n^2)$, we cannot apply Corollary 2.3.3. Instead we use Theorem 2.3.2 directly. We have

$$\begin{aligned}
\frac{\mathbb{E}_{n,p_n}[X_n^2]}{\mathbb{E}_{n,p_n}[X_n]^2} &= \frac{\mu_n + \gamma_n}{\mu_n^2} \\
&\le \frac{n(1 - p_n)^{n-1} + n^2(1 - p_n)^{2n-3}}{n^2(1 - p_n)^{2n-2}} \\
&\le \frac{1}{n(1 - p_n)^{n-1}} + \frac{1}{1 - p_n}, \tag{2.3.6}
\end{aligned}$$

which is $1 + o(1)$ when $p_n \ll \frac{\log n}{n}$ by (2.3.5). The second moment method implies that $\mathbb{P}_{n,p_n}[X_n > 0] \to 1$ in that case. ∎

Connectivity We use Claim 2.3.6 to study the threshold for connectivity.

Claim 2.3.7 *Connectivity has threshold function $\frac{\log n}{n}$.*

Proof We start with the easy direction. If $p_n \ll \frac{\log n}{n}$, Claim 2.3.6 implies that the graph has at least one isolated vertex – and therefore is necessarily disconnected – with probability going to 1 as $n \to +\infty$.

Assume now that $p_n \gg \frac{\log n}{n}$. Let $\mathcal{D}_n$ be the event that G_n is disconnected. To bound $\mathbb{P}_{n,p_n}[\mathcal{D}_n]$, we let Y_k be the number of subsets of k vertices that are disconnected from all other vertices in the graph for $k \in \{1, \ldots, n/2\}$. Then, by the first moment method,

$$\mathbb{P}_{n,p_n}[\mathcal{D}_n] \le \mathbb{P}_{n,p_n}\left[\sum_{k=1}^{n/2} Y_k > 0\right] \le \sum_{k=1}^{n/2} \mathbb{E}_{n,p_n}[Y_k].$$

The expectation of Y_k is straightforward to bound. Using $k \le n/2$ and $\binom{n}{k} \le n^k$,

$$\mathbb{E}_{n,p_n}[Y_k] = \binom{n}{k}(1 - p_n)^{k(n-k)} \le \left(n(1 - p_n)^{n/2}\right)^k.$$

The expression in parentheses is $o(1)$ when $p_n \gg \frac{\log n}{n}$ by a calculation similar to (2.3.4). Summing over k,

$$\mathbb{P}_{n,p_n}[\mathcal{D}_n] \le \sum_{k=1}^{+\infty} \left(n(1 - p_n)^{n/2}\right)^k = O(n(1 - p_n)^{n/2}) = o(1),$$

where we used that the geometric series (started at $k = 1$) is dominated asymptotically by its first term. So the probability of being disconnected goes to 0 when $p_n \gg \frac{\log n}{n}$ and we have proved the claim. ∎

A closer look We have shown that connectivity and the absence of isolated vertices have the same threshold function. In fact, in a sense, isolated vertices are the "last obstacle" to connectivity. A slight modification of the proof above leads to the following more precise result. For $k \in \{1, \ldots, n/2\}$, let Z_k be the number of connected components of size k in G_n. In particular, Z_1 is the number of isolated vertices. We consider the "critical window" $p_n = \frac{c_n}{n}$, where $c_n := \log n + s$ for some fixed $s \in \mathbb{R}$. We show that, in that regime, asymptotically the graph is typically composed of a large connected component together with some isolated vertices. Formally, we prove Claim 2.3.8, which says that with probability close to 1, either the graph is connected or there are some isolated vertices together with a (necessarily unique) connected component of size greater than $n/2$.

Claim 2.3.8

$$\mathbb{P}_{n,p_n}[Z_1 > 0] \ge \frac{1}{1 + e^s} + o(1) \quad and \quad \mathbb{P}_{n,p_n}\left[\sum_{k=2}^{n/2} Z_k > 0\right] = o(1).$$

The limit of $\mathbb{P}_{n,p_n}[Z_1 > 0]$ can be computed explicitly using the method of moments. See Exercise 2.19.

Proof of Claim 2.3.8   We first consider isolated vertices. From (2.3.5), (2.3.6), and the second moment method,

$$\mathbb{P}_{n,p_n}[Z_1 > 0] \geq \left(e^{-\log n + np_n + np_n^2} + \frac{1}{1-p_n} \right)^{-1} = \frac{1}{1+e^s} + o(1),$$

as $n \to +\infty$ by our choice of p_n.

To bound the number of components of size $k > 1$, we note first that the random variable Y_k used in the previous claim (which imposes no condition on the edges *between* the vertices in the subsets of size k) is too loose to provide a suitable bound. Instead, to bound the probability that a subset of k vertices forms a connected component, we observe that a connected component is characterized by two properties: it is disconnected from the rest of the graph; and it contains a spanning tree. Formally, for $k = 2, \ldots, n/2$, we let Z_k' be the number of (not necessarily induced) maximal trees of size k or, put differently, the number of spanning trees of connected components of size k. Then, by the first moment method, the probability that a connected component of size > 1 is present in G_n is bounded by

$$\mathbb{P}_{n,p_n}\left[\sum_{k=2}^{n/2} Z_k > 0 \right] \leq \mathbb{P}_{n,p_n}\left[\sum_{k=2}^{n/2} Z_k' > 0 \right] \leq \sum_{k=2}^{n/2} \mathbb{E}_{n,p_n}[Z_k']. \tag{2.3.7}$$

To bound the expectation of Z_k', we use Cayley's formula, which states that there are k^{k-2} trees on a set of k labeled vertices. Recall further that a tree on k vertices has $k-1$ edges (see Exercise 1.7). Hence,

$$\mathbb{E}_{n,p_n}[Z_k'] = \underbrace{\binom{n}{k} k^{k-2}}_{(a)} \underbrace{p_n^{k-1}}_{(b)} \underbrace{(1-p_n)^{k(n-k)}}_{(c)},$$

where (a) is the number of trees of size k (as subgraphs) in a complete graph of size n, (b) is the probability that such a tree is present in the graph, and (c) is the probability that this tree is disconnected from every other vertex in the graph. Using that $k! \geq (k/e)^k$ (see Appendix A) and $1 - x \leq e^{-x}$ for all $x \in \mathbb{R}$ (see Exercise 1.16),

$$\mathbb{E}_{n,p_n}[Z_k'] \leq \frac{n^k}{k!} k^{k-2} p_n^{k-1} (1-p_n)^{k(n-k)}$$

$$\leq \frac{n^k e^k}{k^k} k^k n p_n^k e^{-p_n k(n-k)}$$

$$\leq n \left(e c_n e^{-\left(1-\frac{k}{n}\right)c_n} \right)^k$$

$$= n \left(e(\log n + s) e^{-\left(1-\frac{k}{n}\right)(\log n + s)} \right)^k.$$

For $k \leq n/2$, the expression in parentheses is $o(1)$. In fact, for $2 \leq k \leq n/2$, $\mathbb{E}_{n,p_n}[Z_k'] = o(1)$. Furthermore, summing over $k > 2$,

$$\sum_{k=3}^{n/2} \mathbb{E}_{n,p_n}[Z_k'] \leq \sum_{k=3}^{+\infty} n \left(e(\log n + s) e^{-\frac{1}{2}(\log n + s)} \right)^k = O(n^{-1/2} \log^3 n) = o(1).$$

Plugging this back into (2.3.7) gives the second claim in the statement.   ∎

2.3.3 ▷ *Percolation: Critical Value on Trees and Branching Number*

Consider bond percolation (see Definition 1.2.1) on the infinite d-regular tree $\mathbb{T}_d$. Root the tree arbitrarily at a vertex 0 and let $\mathcal{C}_0$ be the open cluster of the root. In this section, we illustrate the use of the first and second moment methods on the identification of the critical value

$$p_c(\mathbb{T}_d) = \sup\{p \in [0,1]: \theta(p) = 0\},$$

where recall that the percolation function is $\theta(p) = \mathbb{P}_p[|\mathcal{C}_0| = +\infty]$. We then consider general trees, introduce the branching number, and present a weighted version of the second moment method.

Regular tree Our main result for $\mathbb{T}_d$ is the following.

Claim 2.3.9

$$p_c(\mathbb{T}_d) = \frac{1}{d-1}.$$

Proof Let ∂_n be the nth level of $\mathbb{T}_d$, that is, the set of vertices at graph distance n from 0. Let X_n be the number of vertices in $\partial_n \cap \mathcal{C}_0$. In order for the open cluster of the root to be infinite, there must be at least one vertex on the nth level connected to the root by an open path. By the first moment method (Theorem 2.2.6),

$$\theta(p) = \mathbb{P}_p[|\mathcal{C}_0| = +\infty] \le \mathbb{P}_p[X_n > 0] \le \mathbb{E}_p X_n = d(d-1)^{n-1}p^n \to 0, \qquad (2.3.8)$$

as $n \to +\infty$, for any $p < \frac{1}{d-1}$. Here we used that there is a unique path between 0 and any vertex in the tree to deduce that $\mathbb{P}_p[x \in \mathcal{C}_0] = p^n$ for $x \in \partial_n$. Equation (2.3.8) implies half of the claim: $p_c(\mathbb{T}_d) \ge \frac{1}{d-1}$.

The second moment method gives a lower bound on $\mathbb{P}_p[X_n > 0]$. To simplify the notation, it is convenient to introduce the "branching ratio" $b := d - 1$. We say that x is a *descendant* of z if the path between 0 and x goes through z. Each $z \ne 0$ has $d - 1$ *descendant subtrees*, that is, subtrees of $\mathbb{T}_d$ rooted at z made of all descendants of z. Let $x \wedge y$ be the *most recent common ancestor* of x and y, that is, the furthest vertex from 0 that lies on both the path from 0 to x and the path from 0 to y; see Figure 2.6. Letting

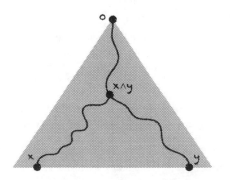

Figure 2.6 Most recent common ancestor of x and y.

$$\mu_n := \mathbb{E}_p[X_n] = \mathbb{E}_p\left[\sum_{x \in \partial_n} \mathbf{1}_{\{x \in \mathcal{C}_0\}}\right] = (b+1)b^{n-1}p^n,$$

we have

$$\mathbb{E}_p[X_n^2] = \mathbb{E}_p\left[\left(\sum_{x \in \partial_n} \mathbf{1}_{\{x \in \mathcal{C}_0\}}\right)^2\right]$$

$$= \sum_{x,y \in \partial_n} \mathbb{P}_p[x, y \in \mathcal{C}_0]$$

$$= \sum_{x \in \partial_n} \mathbb{P}_p[x \in \mathcal{C}_0] + \sum_{m=0}^{n-1} \sum_{x,y \in \partial_n} \mathbf{1}_{\{x \wedge y \in \partial_m\}} p^m p^{2(n-m)}$$

$$= \mu_n + (b+1)b^{n-1} \sum_{m=0}^{n-1} (b-1)b^{(n-m)-1}p^{2n-m}$$

$$\leq \mu_n + (b+1)(b-1)b^{2n-2}p^{2n} \sum_{m=0}^{+\infty} (bp)^{-m}$$

$$= \mu_n + \mu_n^2 \cdot \frac{b-1}{b+1} \cdot \frac{1}{1-(bp)^{-1}},$$

where, on the fourth line, we used that all vertices on the nth level are equivalent and that, for a fixed x, the set $\{y: x \wedge y \in \partial_m\}$ is composed of those vertices in ∂_n that are descendants of $x \wedge y$ but not in the descendant subtree of $x \wedge y$ containing x. When $p > \frac{1}{d-1} = \frac{1}{b}$, dividing by $(\mathbb{E}_p X_n)^2 = \mu_n^2 \to +\infty$, we get

$$\frac{\mathbb{E}_p[X_n^2]}{(\mathbb{E}_p X_n)^2} \leq \frac{1}{\mu_n} + \frac{b-1}{b+1} \cdot \frac{1}{1-(bp)^{-1}} \tag{2.3.9}$$

$$\leq 1 + \frac{b-1}{b+1} \cdot \frac{1}{1-(bp)^{-1}}$$

$$=: C_{b,p}.$$

By the second moment method (Theorem 2.3.2) and monotonicity,

$$\theta(p) = \mathbb{P}_p[|\mathcal{C}_0| = +\infty] = \mathbb{P}_p[\forall n, X_n > 0] = \lim_n \mathbb{P}_p[X_n > 0] \geq C_{b,p}^{-1} > 0,$$

which concludes the proof. (Note that the version of the second moment method in (2.3.1) does not work here. Subtract 1 in (2.3.9) and take p close to $1/b$.) ∎

The argument in the proof of Claim 2.3.9 relies crucially on the fact that, in a tree, any two vertices are connected by a unique path. For instance, approximating $\mathbb{P}_p[x \in \mathcal{C}_0]$ is much harder on a lattice. Note furthermore that, intuitively, the reason why the first moment captures the critical threshold exactly in this case is that bond percolation on $\mathbb{T}_d$ is a "branching process" (defined formally and studied at length in Chapter 6), where X_n represents the "population size at generation n." The qualitative behavior of a branching process is governed by its expectation: when the mean number of children bp exceeds 1, the process grows exponentially on average and "explodes" with positive probability (see Theorem 6.1.6). We will come back to this point of view in Section 6.2.4 where branching processes are used to give a more refined analysis of bond percolation on $\mathbb{T}_d$.

General trees Let $\mathscr{T}$ be a locally finite tree (i.e., all its degrees are finite) with root 0. For an edge e, let v_e be the endvertex of e furthest from the root. We denote by $|e|$ the graph distance between 0 and v_e. Generalizing a previous definition from Section 1.1.1 to infinite, locally finite graphs, a cutset separating 0 and $+\infty$ is a finite set of edges Π such that all infinite paths (which, recall, are self-avoiding by definition) starting at 0 go through Π. (For our purposes, it will suffice to assume that cutsets are finite by default.) For a cutset Π, we let $\Pi_v := \{v_e \colon e \in \Pi\}$. Repeating the argument in (2.3.8), for any cutset Π, by the first moment method (i.e., union bound),

$$
\begin{aligned}
\theta(p) &= \mathbb{P}_p[|\mathcal{C}_0| = +\infty] \\
&\leq \mathbb{P}_p[\mathcal{C}_0 \cap \Pi_v \neq \emptyset] \\
&\leq \sum_{u \in \Pi_v} \mathbb{P}_p[u \in \mathcal{C}_0] \\
&= \sum_{e \in \Pi} p^{|e|}.
\end{aligned}
\tag{2.3.10}
$$

This bound naturally leads to the next definition.

Definition 2.3.10 (Branching number). *The* branching number *of $\mathscr{T}$ is given by*

BRANCHING
NUMBER

$$
\mathrm{br}(\mathscr{T}) = \sup \left\{ \lambda \geq 1 \colon \inf_{cutset\ \Pi} \sum_{e \in \Pi} \lambda^{-|e|} > 0 \right\}.
\tag{2.3.11}
$$

Using the max-flow min-cut theorem (Theorem 1.1.15), the branching number can also be characterized in terms of a "flow to $+\infty$." We will not do this here. (But see Theorem 3.3.30.)

Equation (2.3.10) implies that $p_c(\mathscr{T}) \geq \frac{1}{\mathrm{br}(\mathscr{T})}$. Remarkably, this bound is tight. The proof is based on a "weighted" second moment method.

Claim 2.3.11 *For any rooted, locally finite tree $\mathscr{T}$,*

$$
p_c(\mathscr{T}) = \frac{1}{\mathrm{br}(\mathscr{T})}.
$$

Proof Suppose $p < \frac{1}{\mathrm{br}(\mathscr{T})}$. Then, $p^{-1} > \mathrm{br}(\mathscr{T})$ and the sum in (2.3.10) can be made arbitrarily small by definition of the branching number, that is, $\theta(p) = 0$. Hence we have shown that $p_c(\mathscr{T}) \geq \frac{1}{\mathrm{br}(\mathscr{T})}$.

To argue in the other direction, let $p > \frac{1}{\mathrm{br}(\mathscr{T})}$, $p^{-1} < \lambda < \mathrm{br}(\mathscr{T})$, and $\varepsilon > 0$ such that

$$
\sum_{e \in \Pi} \lambda^{-|e|} \geq \varepsilon
\tag{2.3.12}
$$

for all cutsets Π. The existence of such an ε is guaranteed by the definition of the branching number. As in the proof of Claim 2.3.9, we use that $\theta(p)$ is the limit as $n \to +\infty$ of the probability that $\mathcal{C}_0$ reaches the nth level (i.e., the vertices at graph distance n from the root 0, which is necessarily a finite set in a locally finite tree). However, this time, we use a *weighted* count on the nth level. Let $\mathcal{T}_n$ be the first n levels of $\mathcal{T}$ and, as before, let ∂_n be the vertices on the nth level. For a probability measure ν_n on ∂_n, we define the weighted count

$$X_n = \sum_{z \in \partial_n} \mathbf{1}_{\{z \in \mathcal{C}_0\}} \frac{\nu_n(z)}{\mathbb{P}_p[z \in \mathcal{C}_0]}.$$

The purpose of the denominator is normalization, that is,

$$\mathbb{E}_p X_n = \sum_{z \in \partial_n} \nu_n(z) = 1.$$

Observe that, while $\nu_n(z)$ may be 0 for some zs (but not all), we still have that $X_n > 0, \forall n$ implies $\{|\mathcal{C}_0| = +\infty\}$, which is what we need to apply the second moment method.

Because of (2.3.12), a natural choice of ν_n follows from the max-flow min-cut theorem (Theorem 1.1.15) applied to $\mathcal{T}_n$ with source 0, sink ∂_n, and capacity constraint $|\phi(x,y)| \leq \kappa(e) := \varepsilon^{-1} \lambda^{-|e|}$ for all edges $e = \{x, y\}$. Indeed, for all cutsets Π in $\mathcal{T}_n$ separating 0 and ∂_n, we have $\sum_{e \in \Pi} \kappa(e) = \sum_{e \in \Pi} \varepsilon^{-1} \lambda^{-|e|} \geq 1$ by (2.3.12). That then guarantees by Theorem 1.1.15 the existence of a unit flow ϕ from 0 to ∂_n satisfying the capacity constraints. Define $\nu_n(z)$ to be the flow entering $z \in \partial_n$ under ϕ. In particular, because ϕ is a unit flow, ν_n defines a probability measure. It remains to bound the second moment of X_n under this choice. We have

$$\mathbb{E}_p X_n^2 = \mathbb{E}_p \left[\left(\sum_{z \in \partial_n} \mathbf{1}_{\{z \in \mathcal{C}_0\}} \frac{\nu_n(z)}{\mathbb{P}_p[z \in \mathcal{C}_0]} \right)^2 \right]$$

$$= \sum_{x,y \in \partial_n} \nu_n(x) \nu_n(y) \frac{\mathbb{P}_p[x, y \in \mathcal{C}_0]}{\mathbb{P}_p[x \in \mathcal{C}_0] \mathbb{P}_p[y \in \mathcal{C}_0]}$$

$$= \sum_{m=0}^{n} \sum_{x,y \in \partial_n} \mathbf{1}_{\{x \wedge y \in \partial_m\}} \nu_n(x) \nu_n(y) \frac{p^m p^{2(n-m)}}{p^{2n}}$$

$$= \sum_{m=0}^{n} p^{-m} \sum_{z \in \partial_m} \left(\sum_{x,y \in \partial_n} \mathbf{1}_{\{x \wedge y = z\}} \nu_n(x) \nu_n(y) \right).$$

In the expression in parentheses, for each x descendant of z, the sum over y is at most $\nu_n(x) \nu_n(z)$ by the definition of a flow; then the sum over those xs gives at most $\nu_n(z)^2$. So

$$\mathbb{E}_p X_n^2 \leq \sum_{m=0}^{n} p^{-m} \sum_{z \in \partial_m} \nu_n(z)^2$$

$$\leq \sum_{m=0}^{n} p^{-m} \sum_{z \in \partial_m} (\varepsilon^{-1} \lambda^{-m}) \nu_n(z)$$

$$\leq \varepsilon^{-1} \sum_{m=0}^{+\infty} (p\lambda)^{-m}$$

$$= \frac{\varepsilon^{-1}}{1 - (p\lambda)^{-1}} =: C_{\varepsilon, \lambda, p} < +\infty,$$

where the second line follows from the capacity constraint, and we used $p\lambda > 1$ on the last line. From the second moment method (recalling that $\mathbb{E}_p X_n = 1$),

$$\theta(p) = \mathbb{P}_p[|\mathcal{C}_0| = +\infty] \geq \mathbb{P}_p[\forall n, \, X_n > 0] = \lim_n \mathbb{P}_p[X_n > 0] \geq C_{\varepsilon,\lambda,p}^{-1} > 0.$$

It follows that

$$\theta(p) \geq C_{\varepsilon,\lambda,p}^{-1} > 0,$$

and $p_c(\mathcal{T}) \leq \frac{1}{\text{br}(\mathcal{T})}$. That concludes the proof. ∎

Note that Claims 2.3.9 and 2.3.11 imply that $\text{br}(\mathbb{T}_d) = d - 1$. The next example is more striking and insightful.

Example 2.3.12 (The 3–1 tree). The 3–1 tree $\widehat{\mathcal{T}_{3-1}}$ is an infinite rooted tree. We give a planar description. The root ρ (level 0) is at the top. It has two children below it (level 1). Then on level n, for $n \geq 1$, the first 2^{n-1} vertices starting from the left have exactly 1 child and the next 2^{n-1} vertices have exactly 3 children. In particular, level n has 2^n vertices, which we denote by $u_{n,1}, \ldots, u_{n,2^n}$. For vertex $u_{n,j}$ we refer to $j/2^n$ as its *relative position* (on level n). So vertices have 1 or 3 children according to whether their relative position is $\leq 1/2$ or $> 1/2$. RELATIVE POSITION

Because the level size is growing at rate 2, it is tempting to conjecture that the branching number is 2 – but that turns out to be way off.

Claim 2.3.13 $\text{br}(\widehat{\mathcal{T}_{3-1}}) = 1$.

What makes this tree entirely different from the infinite 2-ary tree, despite having the same level growth, is that each infinite path from the root in $\widehat{\mathcal{T}_{3-1}}$ eventually "stops branching," with the sole exception of the rightmost path which we refer to as the *main path*. Indeed, let MAIN PATH $\Gamma = v_0 \sim v_1 \sim v_2 \sim \cdots$ with $v_0 = \rho$ be an infinite path distinct from the main path. Let x_i be the relative position of v_i, $i \geq 1$. Let v_k be the first vertex of Γ *not* on the main path. It lies on the kth level.

Lemma 2.3.14 *Let v be a vertex that is not on the main path with relative position x and assume that $0 \leq x \leq \alpha < 1$. Let w be a child of v and denote by y its relative position. Then,*

$$y \leq \begin{cases} \frac{1}{2}x & \text{if } x \leq 1/2, \\ x - \frac{1}{2}(1 - \alpha) & \text{otherwise.} \end{cases}$$

Proof Assume without loss of generality that $v = u_{n,j}$ for some n and $j < 2^n$. If $j \leq 2^{n-1}$, then by construction v has exactly one child with relative position

$$y = \frac{j}{2^{n+1}} = \frac{1}{2}x.$$

That proves the first claim.

If $j > 2^{n-1}$, then all vertices to the right of v have 3 children, all of whom are to the right of the children of v. Hence, the children of v have relative position at most

$$y \leq \frac{2^{n+1} - 3(2^n - j)}{2^{n+1}} = \frac{3j - 2^n}{2^{n+1}} = \frac{3}{2}x - \frac{1}{2}.$$

Subtracting x and using $x \leq \alpha$ gives the second claim. ∎

We now apply Lemma 2.3.14 to v_k as defined above and its descendants on Γ with $\alpha = 1 - 1/2^k$. We get that the relative position decreases from v_k by $1/2^{k+1}$ on each level until it falls below $1/2$ at which point it gets cut in half at each level. Once this last regime is reached, each vertex on Γ from then on has exactly one child – that is, there is no more branching.

We are now ready to prove the claim.

Proof of Claim 2.3.13   Take any $\lambda > 1$. From the definition of the branching number (Definition 2.3.10), it suffices to find a sequence of cutsets $(\Pi_n)_n$ such that

$$\sum_{e \in \Pi_n} \lambda^{-|e|} \to 0,$$

as $n \to +\infty$. What does *not* work is to choose $\Pi_n := \Lambda_n$ to be the edges between level $n - 1$ and level n, since we then have

$$\sum_{e \in \Lambda_n} \lambda^{-|e|} = 2^n \lambda^{-n},$$

which diverges whenever $\lambda < 2$. Instead, we construct a new cutset Φ_n based on Λ_n as follows. We divide up Λ_n into the disjoint union $\Lambda_n^- \cup \Lambda_n^+$, where Λ_n^- are the edges whose endvertex on level n has relative position $\leq 1/2$ and Λ_n^+ are the rest of the edges. Start with $\Phi_n := \emptyset$.

Step 1. For each edge e in Λ_n^-, letting v be the endvertex of e on level n, add to Φ_n the edge $\{v', v''\}$ where v' and v'' are the unique descendants of v on level $m_n - 1$ and m_n, respectively. The value of $m_n \geq n$ is chosen so that

$$2^n \lambda^{-m_n} \leq \frac{1}{2n}. \tag{2.3.13}$$

Any infinite path from the root going through one of the edges in Λ_n^- has to go through the edge that replaced it in Φ_n since there is no branching below that point by Lemma 2.3.14.

Step 2. We also add to Φ_n the edge $\{w', w''\}$ on the main path where $w' = u_{\ell_n-1, 2^{\ell_n-1}}$ is on level $\ell_n - 1$ and $w'' = u_{\ell_n, 2^{\ell_n}}$ is on level ℓ_n. We mean for the value of ℓ_n to be such that any infinite path going through an edge in Λ_n^+ has to go through $\{w', w''\}$ first. That is, we need all vertices of level n with relative position $> 1/2$ to be a descendant of w''. The number of descendants of w'' on level $J > \ell_n$ is $3^{J-\ell_n}$ until the last J such that it is $\leq 2^{J-1}$, which we denote by J^*. A quick calculation gives

$$J^* = \left\lfloor \frac{\ell_n \log 3 - \log 2}{\log 3 - \log 2} \right\rfloor.$$

After level J^*, the leftmost descendant of w'' has relative position $\leq 1/2$ by Lemma 2.3.14. Therefore, we need $n > J^*$. Taking

$$\ell_n = \left\lfloor \frac{\log 3/2}{\log 3} n \right\rfloor, \tag{2.3.14}$$

will do for n large enough, say $n \geq n_0$.

Finishing up. By construction, Φ_n is a cutset for all $n \geq n_0$. Moreover,

$$\sum_{e \in \Phi_n} \lambda^{-|e|} = 2^{n-1} \lambda^{-m_n} + \lambda^{-\ell_n} < \frac{1}{n}$$

for n large enough, where we used (2.3.13) and (2.3.14). Taking $n \to +\infty$ gives the claim.   ∎

As a consequence of Claims 2.3.11 and 2.3.13, $|\mathcal{C}_\rho| < +\infty$ almost surely for all $p < 1$ on $\widehat{\mathscr{T}}_{3-1}$. ◀

2.4 Chernoff–Cramér Method

Chebyshev's inequality (Theorem 2.1.2) gives a bound on the concentration around its mean of a square integrable random variable. It is, in general, best possible. Indeed, take X to be $\mu + b\sigma$ or $\mu - b\sigma$ with probability $(2b^2)^{-1}$ each, and μ otherwise. Then $\mathbb{E}X = \mu$, $\mathrm{Var}X = \sigma^2$, and for $\beta = b\sigma$,

$$\mathbb{P}[|X - \mathbb{E}X| \geq \beta] = \mathbb{P}[|X - \mathbb{E}X| = \beta] = \frac{1}{b^2} = \frac{\mathrm{Var}X}{\beta^2}.$$

However, in many cases, much stronger bounds can be derived. For instance, if $X \sim \mathrm{N}(0, 1)$, by the following lemma

$$\mathbb{P}[|X - \mathbb{E}X| \geq \beta] \sim \sqrt{\frac{2}{\pi}}\beta^{-1}\exp(-\beta^2/2) \ll \frac{1}{\beta^2}, \tag{2.4.1}$$

as $\beta \to +\infty$. Indeed:

Lemma 2.4.1 *For $x > 0$,*

$$(x^{-1} - x^{-3})e^{-x^2/2} \leq \int_x^{+\infty} e^{-y^2/2}\mathrm{d}y \leq x^{-1}e^{-x^2/2}.$$

Proof By the change of variable $y = x + z$ and using $e^{-z^2/2} \leq 1$

$$\int_x^{+\infty} e^{-y^2/2}\mathrm{d}y \leq e^{-x^2/2}\int_0^{+\infty} e^{-xz}\mathrm{d}z = e^{-x^2/2}x^{-1}.$$

For the other direction, by differentiation,

$$\int_x^{+\infty} (1 - 3y^{-4})e^{-y^2/2}\mathrm{d}y = (x^{-1} - x^{-3})e^{-x^2/2}. \qquad \blacksquare$$

In this section, we discuss the Chernoff–Cramér method, which produces *exponential* tail bounds, provided the moment-generating function (see Section 2.1.1) is finite in a neighborhood of 0.

2.4.1 Tail Bounds via the Moment-Generating Function

Under a finite variance, squaring within Markov's inequality (Theorem 2.1.1) produces Chebyshev's inequality (Theorem 2.1.2). This "boosting" can be pushed further when stronger integrability conditions hold.

Chernoff–Cramér We refer to (2.4.2) in the next lemma as the *Chernoff–Cramér bound.*

Lemma 2.4.2 (Chernoff–Cramér bound). *Assume X is a random variable such that $M_X(s) < +\infty$ for $s \in (-s_0, s_0)$ for some $s_0 > 0$. For any $\beta > 0$ and $s > 0$,*

$$\mathbb{P}[X \geq \beta] \leq \exp\left[-\{s\beta - \Psi_X(s)\}\right], \tag{2.4.2}$$

Moments and Tails

where

$$\Psi_X(s) := \log M_X(s)$$

is the cumulant-generating function of X.

Proof Exponentiating within Markov's inequality gives for $s > 0$,

$$\mathbb{P}[X \geq \beta] = \mathbb{P}[e^{sX} \geq e^{s\beta}] \leq \frac{M_X(s)}{e^{s\beta}} = \exp\left[-\{s\beta - \Psi_X(s)\}\right]. \qquad \blacksquare$$

Returning to the Gaussian case, let $X \sim N(0, v)$, where $v > 0$ is the variance and note that

$$\begin{aligned}
M_X(s) &= \int_{-\infty}^{+\infty} e^{sx} \frac{1}{\sqrt{2\pi v}} e^{-\frac{x^2}{2v}} \, dx \\
&= \int_{-\infty}^{+\infty} e^{\frac{s^2 v}{2}} \frac{1}{\sqrt{2\pi v}} e^{-\frac{(x-sv)^2}{2v}} \, dx \\
&= \exp\left(\frac{s^2 v}{2}\right).
\end{aligned}$$

By straightforward calculus, the optimal choice of s in (2.4.2) gives the exponent

$$\sup_{s>0}(s\beta - s^2 v/2) = \frac{\beta^2}{2v}, \qquad (2.4.3)$$

achieved at $s_\beta = \beta/v$. For $\beta > 0$, this leads to the bound

$$\mathbb{P}[X \geq \beta] \leq \exp\left(-\frac{\beta^2}{2v}\right), \qquad (2.4.4)$$

which is much sharper than Chebyshev's inequality for large β – compare to (2.4.1).

As another toy example, we consider simple random walk on $\mathbb{Z}$.

Lemma 2.4.3 (Chernoff bound for simple random walk on $\mathbb{Z}$). *Let $Z_1, \ldots, Z_n$ be independ- ent Rademacher variables, that is, they are $\{-1, 1\}$-valued random variables with $\mathbb{P}[Z_i = 1] = \mathbb{P}[Z_i = -1] = 1/2$. Let $S_n = \sum_{i \leq n} Z_i$. Then, for any $\beta > 0$,*

RADEMACHER VARIABLE

$$\mathbb{P}[S_n \geq \beta] \leq e^{-\beta^2/2n}. \qquad (2.4.5)$$

Proof The moment-generating function of Z_1 can be bounded as follows

$$M_{Z_1}(s) = \frac{e^s + e^{-s}}{2} = \sum_{j \geq 0} \frac{s^{2j}}{(2j)!} \leq \sum_{j \geq 0} \frac{(s^2/2)^j}{j!} = e^{s^2/2}. \qquad (2.4.6)$$

Taking $s = \beta/n$ in the Chernoff–Cramér bound (2.4.2), we get

$$\begin{aligned}
\mathbb{P}[S_n \geq \beta] &\leq \exp\left(-s\beta + n\Psi_{Z_1}(s)\right) \\
&\leq \exp\left(-s\beta + ns^2/2\right) \\
&= e^{-\beta^2/2n},
\end{aligned}$$

which concludes the proof. $\qquad \blacksquare$

Observe the similarity between (2.4.5) and the Gaussian bound (2.4.4) if one takes v to be the variance of S_n, that is,

$$v = \text{Var}[S_n] = n\text{Var}[Z_1] = n\mathbb{E}[Z_1^2] = n,$$

where we used that Z_1 is centered. The central limit theorem says that simple random walk is well approximated by a Gaussian in the "bulk" of the distribution; the bound above extends the approximation in the "large deviation" regime. The bounding technique used in the proof of Lemma 2.4.3 will be substantially extended in Section 2.4.2.

Example 2.4.4 (Set balancing). Let $v_1, \ldots, v_m$ be arbitrary non-zero vectors in $\{0, 1\}^n$. Think of $v_i = (v_{i,1}, \ldots, v_{i,n})$ as representing a subset of $[n] = \{1, \ldots, n\}$: $v_{i,j} = 1$ indicates that j is in subset i. Suppose we want to partition $[n]$ into two groups such that the subsets corresponding to the v_is are as balanced as possible, that is, are as close as possible to having the same number of elements from each group. More formally, we seek a vector $x = (x_1, \ldots, x_n) \in \{-1, +1\}^n$ such that $B^* = \max_{i=1,\ldots,m} |x \cdot v_i|$ is as small as possible.

A simple random choice does well: select each x_i independently, uniformly at random in $\{-1, +1\}$. Fix $\varepsilon > 0$. We claim that

$$\mathbb{P}\left[B^* \geq \sqrt{2n(\log m + \log(2\varepsilon^{-1}))}\right] \leq \varepsilon. \tag{2.4.7}$$

Indeed, by (2.4.5) (considering only the non-zero entries of v_i),

$$\mathbb{P}\left[|x \cdot v_i| \geq \sqrt{2n(\log m + \log(2\varepsilon^{-1}))}\right]$$
$$\leq 2\exp\left(-\frac{2n(\log m + \log(2\varepsilon^{-1}))}{2\|v_i\|_1}\right)$$
$$\leq \frac{\varepsilon}{m},$$

where we used that $\|v_i\|_1 \leq n$. Taking a union bound over the m vectors gives the result. In (2.4.7), the $\sqrt{n}$ term on the right-hand side of the inequality is to be expected since it is the standard deviation of $|x \cdot v_i|$ in the worst case. The power of the exponential tail bound (2.4.5) appears in the logarithmic terms, which would have been much larger if one had used Chebyshev's inequality instead. ◀

The Chernoff–Cramér bound is particularly useful for sums of independent random variables as the moment-generating function then factorizes; see (2.1.3). Let

$$\Psi_X^*(\beta) = \sup_{s \in \mathbb{R}_+}(s\beta - \Psi_X(s))$$

be the *Fenchel–Legendre dual* of the cumulant-generating function of X.

FENCHEL–
LEGENDRE
DUAL

Theorem 2.4.5 (Chernoff–Cramér method). *Let* $S_n = \sum_{i \leq n} X_i$, *where the* X_is *are i.i.d. random variables. Assume* $M_{X_1}(s) < +\infty$ *on* $s \in (-s_0, s_0)$ *for some* $s_0 > 0$. *For any* $\beta > 0$,

$$\mathbb{P}[S_n \geq \beta] \leq \exp\left(-n\Psi_{X_1}^*\left(\frac{\beta}{n}\right)\right). \tag{2.4.8}$$

In particular, in the large deviations regime, that is, when $\beta = bn$ for some $b > 0$, we have

$$-\limsup_n \frac{1}{n} \log \mathbb{P}[S_n \geq bn] \geq \Psi_{X_1}^*(b).$$ (2.4.9)

Proof By independence, we get

$$\Psi_{S_n}^*(\beta) = \sup_{s>0}(s\beta - n\Psi_{X_1}(s)) = \sup_{s>0} n\left(s\left(\frac{\beta}{n}\right) - \Psi_{X_1}(s)\right) = n\Psi_{X_1}^*\left(\frac{\beta}{n}\right),$$

and then we optimize over s in (2.4.2).   ∎

We use the Chernoff–Cramér method to derive a few standard bounds.

POISSON **Poisson variables** We start with the Poisson case. Let $Z \sim \text{Poi}(\lambda)$ be Poisson with mean λ, where we recall that

$$\mathbb{P}[Z = k] = e^{-\lambda}\frac{\lambda^k}{k!}, \quad k \in \mathbb{Z}_+.$$

Letting $X = Z - \lambda$,

$$\begin{aligned}
\Psi_X(s) &= \log\left(\sum_{\ell \geq 0} e^{-\lambda}\frac{\lambda^\ell}{\ell!}e^{s(\ell-\lambda)}\right) \\
&= \log\left(e^{-(1+s)\lambda}\sum_{\ell \geq 0}\frac{(e^s\lambda)^\ell}{\ell!}\right) \\
&= \log\left(e^{-(1+s)\lambda}e^{e^s\lambda}\right) \\
&= \lambda(e^s - s - 1),
\end{aligned}$$

so that straightforward calculus gives for $\beta > 0$,

$$\begin{aligned}
\Psi_X^*(\beta) &= \sup_{s>0}(s\beta - \lambda(e^s - s - 1)) \\
&= \lambda\left[\left(1 + \frac{\beta}{\lambda}\right)\log\left(1 + \frac{\beta}{\lambda}\right) - \frac{\beta}{\lambda}\right] \\
&=: \lambda\, h\left(\frac{\beta}{\lambda}\right),
\end{aligned}$$

achieved at $s_\beta = \log\left(1 + \frac{\beta}{\lambda}\right)$, where h is defined as the expression in square brackets in the above display. Plugging $\Psi_X^*(\beta)$ into Theorem 2.4.5 leads for $\beta > 0$ to the bound

$$\mathbb{P}[Z \geq \lambda + \beta] \leq \exp\left(-\lambda\, h\left(\frac{\beta}{\lambda}\right)\right).$$ (2.4.10)

A similar calculation for $-(Z - \lambda)$ gives for $\beta < 0$,

$$\mathbb{P}[Z \leq \lambda + \beta] \leq \exp\left(-\lambda\, h\left(\frac{\beta}{\lambda}\right)\right).$$ (2.4.11)

If S_n is a sum of n i.i.d. Poi(λ) variables, then by (2.4.9) for $a > \lambda$,

$$-\limsup_n \frac{1}{n} \log \mathbb{P}[S_n \geq an] \geq \lambda h\left(\frac{a - \lambda}{\lambda}\right)$$

$$= a \log\left(\frac{a}{\lambda}\right) - a + \lambda$$

$$=: I_\lambda^{\mathrm{Poi}}(a), \tag{2.4.12}$$

and similarly for $a < \lambda$,

$$-\limsup_n \frac{1}{n} \log \mathbb{P}[S_n \leq an] \geq I_\lambda^{\mathrm{Poi}}(a). \tag{2.4.13}$$

In fact, these bounds follow immediately from (2.4.10) and (2.4.11) by noting that $S_n \sim$ Poi$(n\lambda)$ (see, for example, Exercise 6.7).

Binomial variables and Chernoff bounds Let $Z \sim$ Bin(n, p) be a binomial random variable with parameters n and p. Recall that Z is a sum of i.i.d. indicators $Y_1, \ldots, Y_n$ equal to 1 with probability p. The Y_is are also known as Bernoulli random variables or Bernoulli trials, and their law is denoted by Ber(p). We also refer to p as the success probability. Letting $X_i = Y_i - p$ and $S_n = Z - np$,

$$\Psi_{X_1}(s) = \log\left(pe^s + (1 - p)\right) - ps.$$

For $b \in (0, 1 - p)$, letting $a = b + p$, direct calculation gives

$$\Psi_{X_1}^*(b) = \sup_{s>0}(sb - (\log[pe^s + (1 - p)] - ps))$$

$$= (1 - a) \log \frac{1 - a}{1 - p} + a \log \frac{a}{p} =: D(a\|p), \tag{2.4.14}$$

achieved at $s_b = \log \frac{(1-p)a}{p(1-a)}$. The function $D(a\|p)$ in (2.4.14) is the so-called *Kullback–Leibler divergence* or *relative entropy* between two Bernoulli variables with parameters a and p, respectively. By (2.4.8) for $\beta > 0$,

$$\mathbb{P}[Z \geq np + \beta] \leq \exp\left(-n D\left(p + \beta/n \| p\right)\right).$$

Applying the same argument to $Z' = n - Z$ gives a bound in the other direction.

Remark 2.4.6 *In the large deviations regime, it can be shown that the previous bound is tight in the sense that*

$$-\frac{1}{n} \log \mathbb{P}[Z \geq np + bn] \to D\left(p + b\|p\right) =: I_{n,p}^{\mathrm{Bin}}(b),$$

as $n \to +\infty$. The theory of large deviations provides general results of this type. See, for example, [Dur10, section 2.6]. Upper bounds will be enough for our purposes.

The following related bounds, proved in Exercise 2.7, are often useful.

Theorem 2.4.7 (Chernoff bounds for Poisson trials). *Let $Y_1, \ldots, Y_n$ be independent $\{0, 1\}$-valued random variables with $\mathbb{P}[Y_i = 1] = p_i$ and $\mu = \sum_i p_i$. These are called* Poisson *trials. Let $Z = \sum_i Y_i$. Then:*

(i) **Above the mean**

   (a) For any $\delta > 0$,

$$\mathbb{P}[Z \geq (1+\delta)\mu] \leq \left(\frac{e^{\delta}}{(1+\delta)^{(1+\delta)}}\right)^{\mu}.$$

   (b) For any $0 < \delta \leq 1$,

$$\mathbb{P}[Z \geq (1+\delta)\mu] \leq e^{-\mu\delta^2/3}.$$

(ii) **Below the mean**

   (a) For any $0 < \delta < 1$,

$$\mathbb{P}[Z \leq (1-\delta)\mu] \leq \left(\frac{e^{-\delta}}{(1-\delta)^{(1-\delta)}}\right)^{\mu}.$$

   (b) For any $0 < \delta < 1$,

$$\mathbb{P}[Z \leq (1-\delta)\mu] \leq e^{-\mu\delta^2/2}.$$

2.4.2 Sub-Gaussian and Sub-Exponential Random Variables

The bounds in Section 2.4.1 were obtained by computing the moment-generating function explicitly (possibly with some approximations). This is not always possible. In this section, we give some important examples of tail bounds derived from the Chernoff–Cramér method for broad classes of random variables under natural conditions on their distributions.

Sub-Gaussian random variables

We begin with sub-Gaussian random variables which, as the name suggests, have a tail that is bounded by that of a Gaussian.

General case Here is our key definition.

SUB-
GAUSSIAN
VARIABLE

Definition 2.4.8 (Sub-Gaussian random variables). *We say that a random variable X with mean μ is sub-Gaussian with variance factor v if*

$$\Psi_{X-\mu}(s) \leq \frac{s^2 v}{2} \quad \forall s \in \mathbb{R} \tag{2.4.15}$$

for some $v > 0$. We use the notation $X \in s\mathcal{G}(v)$.

Note that the right-hand side in (2.4.15) is the cumulant-generating function of a $N(0, v)$. By the Chernoff–Cramér method and (2.4.3) it follows immediately that

$$\mathbb{P}[X - \mu \leq -\beta] \vee \mathbb{P}[X - \mu \geq \beta] \leq \exp\left(-\frac{\beta^2}{2v}\right), \tag{2.4.16}$$

where we used that $X \in s\mathcal{G}(v)$ implies $-X \in s\mathcal{G}(v)$. As a quick example, note that this is the approach we took in Lemma 2.4.3, that is, we showed that a uniform random variable in $\{-1, 1\}$ (i.e., a Rademacher variable) is sub-Gaussian with variance factor 1.

When considering (weighted) sums of independent sub-Gaussian random variables, we get the following.

Theorem 2.4.9 (General Hoeffding inequality). *Suppose $X_1, \ldots, X_n$ are independent random variables where, for each i, $X_i \in \mathrm{s}\mathcal{G}(v_i)$ with $0 < v_i < +\infty$. For $w_1, \ldots, w_n \in \mathbb{R}$, let $S_n = \sum_{i \leq n} w_i X_i$. Then*

$$S_n \in \mathrm{s}\mathcal{G}\left(\sum_{i=1}^{n} w_i^2 v_i\right).$$

In particular, for all $\beta > 0$,

$$\mathbb{P}[S_n - \mathbb{E}S_n \geq \beta] \leq \exp\left(-\frac{\beta^2}{2\sum_{i=1}^{n} w_i^2 v_i}\right).$$

Proof Assume the X_is are centered. By independence and (2.1.3),

$$\Psi_{S_n}(s) = \sum_{i \leq n} \Psi_{w_i X_i}(s) = \sum_{i \leq n} \Psi_{X_i}(s w_i) \leq \sum_{i \leq n} \frac{(s w_i)^2 v_i}{2} = \frac{s^2 \sum_{i \leq n} w_i^2 v_i}{2}. \qquad \blacksquare$$

Bounded random variables For bounded random variables, the previous inequality reduces to a standard bound.

Theorem 2.4.10 (Hoeffding's inequality for bounded variables). *Let $X_1, \ldots, X_n$ be independent random variables where, for each i, X_i takes values in $[a_i, b_i]$ with $-\infty < a_i \leq b_i < +\infty$. Let $S_n = \sum_{i \leq n} X_i$. For all $\beta > 0$,*

$$\mathbb{P}[S_n - \mathbb{E}S_n \geq \beta] \leq \exp\left(-\frac{2\beta^2}{\sum_{i \leq n}(b_i - a_i)^2}\right).$$

By Theorem 2.4.9, it suffices to show that $X_i - \mathbb{E}X_i \in \mathrm{s}\mathcal{G}(v_i)$ with $v_i = \frac{1}{4}(b_i - a_i)^2$. We first give a quick proof of a weaker version that uses a trick called *symmetrization*. Suppose the X_is are centered and satisfy $|X_i| \leq c_i$ for some $c_i > 0$. Let X_i' be an independent copy of X_i and let Z_i be an independent uniform random variable in $\{-1, 1\}$. For any s,

SYMMETRIZA-
TION

$$\begin{aligned}
\mathbb{E}\left[e^{s X_i}\right] &= \mathbb{E}\left[e^{s \mathbb{E}[X_i - X_i' \mid X_i]}\right]\\
&\leq \mathbb{E}\left[\mathbb{E}\left[e^{s(X_i - X_i')} \mid X_i\right]\right]\\
&= \mathbb{E}\left[e^{s(X_i - X_i')}\right],
\end{aligned}$$

where the first line comes from the taking out what is known lemma (Lemma B.6.16) and the fact that X_i' is centered and independent of X_i, the second line follows from the conditional Jensen's inequality (Lemma B.6.12), and the third line uses the tower property (Lemma B.6.16). Observe that $X_i - X_i'$ is *symmetric*, that is, identically distributed to $-(X_i - X_i')$. Hence, using that Z_i is independent of both X_i and X_i', we get

$$\begin{aligned}
\mathbb{E}\left[e^{s(X_i - X_i')}\right] &= \mathbb{E}\left[\mathbb{E}\left[e^{s(X_i - X_i')} \mid Z_i\right]\right]\\
&= \mathbb{E}\left[\mathbb{E}\left[e^{s Z_i(X_i - X_i')} \mid Z_i\right]\right]\\
&= \mathbb{E}\left[e^{s Z_i(X_i - X_i')}\right]\\
&= \mathbb{E}\left[\mathbb{E}\left[e^{s Z_i(X_i - X_i')} \mid X_i - X_i'\right]\right].
\end{aligned}$$

From (2.4.6) (together with Lemma B.6.15), the last line is

$$\leq \mathbb{E}\left[e^{(s(X_i - X_i'))^2/2} \right]$$

$$\leq e^{-4c_i^2 s^2/2}$$

since $|X_i|, |X_i'| \leq c_i$. Putting everything together, we arrive at

$$\mathbb{E}\left[e^{sX_i} \right] \leq e^{-4c_i^2 s^2/2}.$$

That is, X_i is sub-Gaussian with variance factor $4c_i^2$. By Theorem 2.4.9, S_n is sub-Gaussian with variance factor $\sum_{i \leq n} 4c_i^2$ and

$$\mathbb{P}[S_n \geq t] \leq \exp\left(-\frac{t^2}{8 \sum_{i \leq n} c_i^2} \right).$$

Proof of Theorem 2.4.10   As pointed out above, it suffices to show that $X_i - \mathbb{E}X_i$ is sub-Gaussian with variance factor $\frac{1}{4}(b_i - a_i)^2$. This is the content of Hoeffding's lemma below (which we will use again in Chapter 3). First an observation:

Lemma 2.4.11 (Variance of bounded random variables).   *For any random variable Z taking values in $[a, b]$ with $-\infty < a \leq b < +\infty$, we have*

$$\mathrm{Var}[Z] \leq \frac{1}{4}(b - a)^2.$$

Proof   Indeed,

$$\left| Z - \frac{a+b}{2} \right| \leq \frac{b-a}{2}$$

and

$$\mathrm{Var}[Z] = \mathrm{Var}\left[Z - \frac{a+b}{2} \right] \leq \mathbb{E}\left[\left(Z - \frac{a+b}{2} \right)^2 \right] \leq \left(\frac{b-a}{2} \right)^2. \qquad \blacksquare$$

Lemma 2.4.12 (Hoeffding's lemma).   *Let X be a random variable taking values in $[a, b]$ for $-\infty < a \leq b < +\infty$. Then, $X \in s\mathcal{G}\left(\frac{1}{4}(b-a)^2 \right)$.*

Proof   Note first that $X - \mathbb{E}X \in [a - \mathbb{E}X, b - \mathbb{E}X]$ and $\frac{1}{4}((b - \mathbb{E}X) - (a - \mathbb{E}X))^2 = \frac{1}{4}(b-a)^2$. So without loss of generality we assume that $\mathbb{E}X = 0$. Because X is bounded, $M_X(s)$ is finite for all $s \in \mathbb{R}$. Hence, by (2.1.2),

$$\Psi_X(0) = \log M_X(0) = 0, \qquad \Psi_X'(0) = \frac{M_X'(0)}{M_X(0)} = \mathbb{E}X = 0,$$

and by a Taylor expansion,

$$\Psi_X(s) = \Psi_X(0) + s\Psi_X'(0) + \frac{s^2}{2}\Psi_X''(s^*) = \frac{s^2}{2}\Psi_X''(s^*)$$

for some $s^* \in [0, s]$. Therefore, it suffices to show that for all s,

$$\Psi_X''(s) \leq \frac{1}{4}(b-a)^2. \tag{2.4.17}$$

Note that

$$\Psi_X''(s) = \frac{M_X''(s)}{M_X(s)} - \left(\frac{M_X'(s)}{M_X(s)}\right)^2$$

$$= \frac{1}{M_X(s)} \mathbb{E}\left[X^2 e^{sX}\right] - \left(\frac{1}{M_X(s)} \mathbb{E}\left[X e^{sX}\right]\right)^2$$

$$= \mathbb{E}\left[X^2 \frac{e^{sX}}{M_X(s)}\right] - \left(\mathbb{E}\left[X \frac{e^{sX}}{M_X(s)}\right]\right)^2.$$

The trick to conclude is to notice that $\frac{e^{sx}}{M_X(s)}$ defines a density on $[a, b]$ with respect to the law of X. The variance under this density – the last line above – is less than $\frac{1}{4}(b - a)^2$ by Lemma 2.4.11. This establishes (2.4.17) and concludes the proof. ∎

Remark 2.4.13 *The change of measure above is known as* tilting *and is a standard trick in large deviation theory. See, for example, [Dur10, section 2.6].*

Since we have shown that $X_i - \mathbb{E} X_i$ is sub-Gaussian with variance factor $\frac{1}{4}(b_i - a_i)^2$, Theorem 2.4.10 follows from Theorem 2.4.9. ∎

Sub-exponential random variables

Unfortunately, not every random variable of interest is sub-Gaussian. A simple example is the square of a Gaussian variable. Indeed, suppose $X \sim N(0, 1)$. Then $W = X^2$ is χ^2-distributed and its moment-generating function can be computed explicitly. Using the change of variable $u = x\sqrt{1 - 2s}$, for $s < 1/2$,

$$M_W(s) = \frac{1}{\sqrt{2\pi}} \int_{-\infty}^{+\infty} e^{sx^2} e^{-x^2/2}\, dx$$

$$= \frac{1}{\sqrt{1 - 2s}} \times \frac{1}{\sqrt{2\pi}} \int_{-\infty}^{+\infty} e^{-u^2/2}\, du$$

$$= \frac{1}{(1 - 2s)^{1/2}}. \tag{2.4.18}$$

When $s \geq 1/2$, however, we clearly have $M_W(s) = +\infty$. In particular, W cannot be sub-Gaussian for any variance factor $\nu > 0$. (Note that centering W produces an additional factor of e^{-s} in the moment-generating function which does not prevent it from diverging.) Further confirming this observation, arguing as in (2.4.1), the upper tail of W decays as

$$\mathbb{P}[W \geq \beta] = \mathbb{P}[X \geq \sqrt{\beta}]$$

$$\sim \sqrt{\frac{1}{2\pi}} [\sqrt{\beta}]^{-1} \exp(-[\sqrt{\beta}]^2/2)$$

$$\sim \sqrt{\frac{1}{2\pi\beta}} \exp(-\beta/2),$$

as $\beta \to +\infty$. That is, it decays exponentially with β, but much slower than the Gaussian tail.

General case We now define a broad class of distributions which have such exponential tail decay.

Definition 2.4.14 (Sub-exponential random variable). *We say that a random variable X with mean μ is* sub-exponential *with parameters* (v, α) *if*

$$\Psi_{X-\mu}(s) \le \frac{s^2 v}{2} \quad \forall |s| \le \frac{1}{\alpha} \tag{2.4.19}$$

for some $v, \alpha > 0$. *We write* $X \in s\mathcal{E}(v, \alpha)$.[1]

Observe that the key difference between (2.4.15) and (2.4.19) is the interval of s over which it holds. As we will see, the parameter α dictates the exponential decay rate of the tail. The specific form of the bound in (2.4.19) is natural once one notices that, as $|s| \to 0$, a centered random variable with variance v should roughly satisfy

$$\log \mathbb{E}[e^{sX}] \approx \log \left\{ 1 + s\mathbb{E}[X] + \frac{s^2}{2}\mathbb{E}[X^2] \right\} \approx \log \left\{ 1 + \frac{s^2 v}{2} \right\} \approx \frac{s^2 v}{2}.$$

Returning to the χ^2 distribution, note that from (2.4.18) we have for $|s| \le 1/4$:

$$\Psi_{W-1}(s) = -s - \frac{1}{2}\log(1 - 2s)$$

$$= -s - \frac{1}{2}\left[-\sum_{i=1}^{+\infty} \frac{(2s)^i}{i} \right]$$

$$= \frac{s^2}{2}\left[4\sum_{i=2}^{+\infty} \frac{(2s)^{i-2}}{i} \right]$$

$$\le \frac{s^2}{2}\left[2\sum_{i=2}^{+\infty} |1/2|^{i-2} \right]$$

$$\le \frac{s^2}{2} \times 4.$$

Hence, $W \in s\mathcal{E}(4, 4)$.

Using the Chernoff–Cramér bound (Lemma 2.4.2), we obtain the following tail bound for sub-exponential variables.

Theorem 2.4.15 (Sub-exponential tail bound). *Suppose the random variable X with mean μ is sub-exponential with parameters* (v, α). *Then, for all* $\beta \in \mathbb{R}_+$,

$$\mathbb{P}[X - \mu \ge \beta] \le \begin{cases} \exp(-\frac{\beta^2}{2v}) & \text{if } 0 \le \beta \le v/\alpha, \\ \exp(-\frac{\beta}{2\alpha}) & \text{if } \beta > v/\alpha. \end{cases} \tag{2.4.20}$$

In words, the tail decays exponentially fast at large deviations but behaves as in the sub-Gaussian case for smaller deviations. We will see that this awkward double-tail allows to extrapolate naturally between different regimes. First we prove the claim.

Proof of Theorem 2.4.15 We start by applying the Chernoff–Cramér bound. For any $\beta > 0$ and $|s| \le 1/\alpha$,

$$\mathbb{P}[X - \mu \ge \beta] \le \exp\left(-s\beta + \Psi_X(s)\right) \le \exp\left(-s\beta + s^2 v/2\right).$$

---

[1] More commonly, "sub-exponential" refers to the case $\alpha = \sqrt{v}$.

At this point, the proof diverges from the sub-Gaussian case because the optimal choice of *s depends on* β because of the additional constraint $|s| \leq 1/\alpha$. When $s^* = \beta/v$ satisfies $s^* \leq 1/\alpha$, the quadratic function of s in the exponent is minimized at s^*, giving the bound

$$\mathbb{P}[X \geq \beta] \leq \exp\left(-\frac{\beta^2}{2v}\right)$$

for $0 \leq \beta \leq v/\alpha$.

On the other hand, when $\beta > v/\alpha$, the exponent is strictly decreasing over the interval $s \leq 1/\alpha$. Hence, the optimal choice is $s^* = 1/\alpha$, which produces the bound

$$\mathbb{P}[X \geq \beta] \leq \exp\left(-\frac{\beta}{\alpha} + \frac{v}{2\alpha^2}\right)$$

$$< \exp\left(-\frac{\beta}{\alpha} + \frac{\beta}{2\alpha}\right)$$

$$= \exp\left(-\frac{\beta}{2\alpha}\right),$$

where we used that $v < \beta\alpha$ on the second line. ∎

For (weighted) sums of independent sub-exponential random variables, we get the following.

Theorem 2.4.16 (General Bernstein inequality). *Suppose* $X_1, \ldots, X_n$ *are independent random variables where, for each* i, $X_i \in s\mathcal{E}(v_i, \alpha_i)$ *with* $0 < v_i, \alpha_i < +\infty$. *For* $w_1, \ldots, w_n \in \mathbb{R}$, *let* $S_n = \sum_{i \leq n} w_i X_i$. *Then,*

$$S_n \in s\mathcal{E}\left(\sum_{i=1}^{n} w_i^2 v_i, \max_i |w_i|\alpha_i\right).$$

In particular, for all $\beta > 0$,

$$\mathbb{P}[S_n - \mathbb{E}S_n \geq \beta] \leq \begin{cases} \exp\left(-\frac{\beta^2}{2\sum_{i=1}^{n} w_i^2 v_i}\right) & \text{if } 0 \leq \beta \leq \frac{\sum_{i=1}^{n} w_i^2 v_i}{\max_i |w_i|\alpha_i}, \\ \exp\left(-\frac{\beta}{2\max_i |w_i|\alpha_i}\right) & \text{if } \beta > \frac{\sum_{i=1}^{n} w_i^2 v_i}{\max_i |w_i|\alpha_i}. \end{cases}$$

Proof Assume the X_is are centered. By independence and (2.1.3),

$$\Psi_{S_n}(s) = \sum_{i \leq n} \Psi_{w_i X_i}(s) = \sum_{i \leq n} \Psi_{X_i}(sw_i) \leq \sum_{i \leq n} \frac{(sw_i)^2 v_i}{2} = \frac{s^2 \sum_{i \leq n} w_i^2 v_i}{2},$$

provided $|sw_i| \leq 1/\alpha_i$ for all i, that is,

$$|s| \leq \frac{1}{\max_i |w_i|\alpha_i}.$$
∎

Bounded random variables: revisited We apply the previous result to bounded random variables.

Theorem 2.4.17 (Bernstein's inequality for bounded variables). *Let* $X_1, \ldots, X_n$ *be independent random variables, where, for each* i, X_i *has mean* μ_i, *variance* v_i, *and satisfies* $|X_i - \mu_i| \leq c$ *for some* $0 < c < +\infty$. *Let* $S_n = \sum_{i \leq n} X_i$. *For all* $\beta > 0$,

$$\mathbb{P}\left[S_n - \mathbb{E}S_n \geq \beta\right] \leq \begin{cases} \exp\left(-\frac{\beta^2}{4\sum_{i=1}^n v_i}\right) & \text{if } 0 \leq \beta \leq \frac{\sum_{i=1}^n v_i}{c}, \\ \exp\left(-\frac{\beta}{4c}\right) & \text{if } \beta > \frac{\sum_{i=1}^n v_i}{c}. \end{cases}$$

Proof We claim that $X_i \in \mathrm{s}\mathcal{E}(2v_i, 2c)$. To establish the claim, we derive a bound on all moments of X_i. Note that for all integers $k \geq 2$,

$$\mathbb{E}|X_i - \mu_i|^k \leq c^{k-2}\mathbb{E}|X_i - \mu_i|^2 = c^{k-2}v_i.$$

Hence, first applying the dominated convergence theorem (Proposition B.4.14) to establish the limit, we have for $|s| \leq \frac{1}{2c}$,

$$\begin{aligned}
\mathbb{E}[e^{s(X_i - \mu_i)}] &= \sum_{k=0}^{+\infty} \frac{s^k}{k!}\mathbb{E}[(X_i - \mu_i)^k] \\
&\leq 1 + s\,\mathbb{E}[(X_i - \mu_i)] + \sum_{k=2}^{+\infty} \frac{s^k}{k!}c^{k-2}v_i \\
&\leq 1 + \frac{s^2 v_i}{2} + \frac{s^2 v_i}{3!}\sum_{k=3}^{+\infty}(cs)^{k-2} \\
&= 1 + \frac{s^2 v_i}{2}\left\{1 + \frac{1}{3}\frac{cs}{1-cs}\right\} \\
&\leq 1 + \frac{s^2 v_i}{2}\left\{1 + \frac{1}{3}\frac{1/2}{1-1/2}\right\} \\
&\leq 1 + \frac{s^2}{2}2v_i \\
&\leq \exp\left(\frac{s^2}{2}2v_i\right).
\end{aligned}$$

Using the general Bernstein inequality (Theorem 2.4.16) gives the result. ∎

It may seem counter-intuitive to derive a tail bound based on the sub-exponential property of bounded random variables when we have already done so using their sub-Gaussian behavior. After all, the latter is on the surface a strengthening of the former. However, note that we have obtained a *better bound* in Theorem 2.4.17 than we did in Theorem 2.4.10 – *when β is not too large.* That improvement stems from the use of the (actual) variance for moderate deviations. This is easier to appreciate through an example.

Example 2.4.18 (Erdős–Rényi: maximum degree). Let $G_n = (V_n, E_n) \sim \mathbb{G}_{n,p_n}$ be a random graph with n vertices and density p_n under the Erdős–Rényi model (Definition 1.2.2). Recall that two vertices $u, v \in V_n$ are adjacent if $\{u, v\} \in E_n$ and that the set of adjacent vertices of v, denoted by $N(v)$, is called the neighborhood of v. The degree of v is the size of its neighborhood, that is, $\delta(v) = |N(v)|$. Here we study the maximum degree of G_n,

$$D_n = \max_{v \in V_n} \delta(v).$$

We focus on the regime $np_n = \omega(\log n)$. Note that for any vertex $v \in V_n$, its degree is $\mathrm{Bin}(n-1, p_n)$ by independence of the edges. In particular, its expected degree is $(n-1)p_n$. To prove a high-probability upper bound on the maximum D_n, we need to control the deviation of the degree of each vertex from its expectation. Observe that the degrees are not independent. Instead, we apply a union bound over all vertices, after using a tail bound.

Claim 2.4.19 *For any $\varepsilon > 0$, as $n \to +\infty$,*

$$\mathbb{P}\left[|D_n - np_n| \geq 2\sqrt{(1+\varepsilon)np_n \log n}\right] \to 0.$$

Proof For a fixed vertex v, think of $\delta(v) = S_{n-1} \sim \mathrm{Bin}(n-1, p_n)$ as a sum of $n-1$ independent $\{0,1\}$-valued random variables, one for each possible edge. That is, $S_{n-1} = \sum_{i=1}^{n-1} X_i$, where X_i is a bounded random variable. The mean of X_i is p_n and its variance is $p_n(1 - p_n)$. So in Bernstein's inequality (Theorem 2.4.17), we can take $\mu_i := p_n$, $\nu_i := p_n(1 - p_n)$, and $c := 1$ for all i. We get

$$\mathbb{P}\left[S_{n-1} \geq (n-1)p_n + \beta\right] \leq \begin{cases} \exp\left(-\frac{\beta^2}{4\nu}\right) & \text{if } 0 \leq \beta \leq \nu, \\ \exp\left(-\frac{\beta}{4}\right) & \text{if } \beta > \nu, \end{cases}$$

where $\nu = (n-1)p_n(1 - p_n) = \omega(\log n)$ by assumption. We choose β to be the smallest value that will produce a tail probability less than $n^{-1-\varepsilon}$ for $\varepsilon > 0$, that is,

$$\beta = \sqrt{4(n-1)p_n(1 - p_n)} \times \sqrt{(1+\varepsilon)\log n} = o(\nu),$$

which falls in the lower regime of the tail bound. In particular, $\beta = o(np_n)$ (i.e., the deviation is much smaller than the expectation). Finally, by a union bound over $v \in V_n$,

$$\mathbb{P}\left[D_n \geq (n-1)p_n + \sqrt{4(1+\varepsilon)p_n(1-p_n)(n-1)\log n}\right] \leq n \times \frac{1}{n^{1+\varepsilon}} \to 0.$$

The same holds in the other direction. That proves the claim. ∎

Had we used Hoeffding's inequality (Theorem 2.4.10) in the proof of Claim 2.4.19 we would have had to take $\beta = \sqrt{(1+\varepsilon)n \log n}$. That would have produced a much weaker bound when $p_n = o(1)$. Indeed, the advantage of Bernstein's inequality is that it makes explicit use of the variance, which when $p_n = o(1)$ is much smaller than the worst case for bounded variables. ◄

2.4.3 ▷ *Probabilistic Analysis of Algorithms: Knapsack Problem*

In a knapsack problem, we have n items. Item i has weight W_i and value V_i. Given a weight bound $\mathcal{W}$, we want to pack as valuable a collection of items in the knapsack under the constraint that the total weight is less than or equal to $\mathcal{W}$. Formally, we seek a solution to the optimization problem

$$Z^* = \max\left\{\sum_{j=1}^{n} x_j V_j : x_1, \ldots, x_n \in [0,1], \sum_{j=1}^{n} x_j W_j \leq \mathcal{W}\right\}. \tag{2.4.21}$$

This is the *fractional knapsack problem*, where we allow a fraction of an item to be added to the knapsack. KNAPSACK PROBLEM

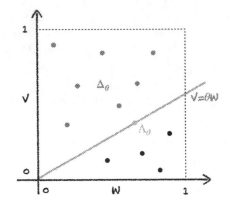

Figure 2.7 Vizualization of the greedy algorithm.

It is used as a computationally tractable relaxation of the 0-1 *knapsack problem*, which also includes the combinatorial constraint $x_j \in \{0, 1\}, \forall j$. Indeed, it turns out that the optimization problem (2.4.21) is solved exactly by a simple greedy solution (see Exercise 2.8 for a formal proof of correctness): let π be a permutation of $\{1, \ldots, n\}$ that puts the items in decreasing order of value per unit weight

$$\frac{V_{\pi(1)}}{W_{\pi(1)}} \geq \frac{V_{\pi(2)}}{W_{\pi(2)}} \geq \cdots \geq \frac{V_{\pi(n)}}{W_{\pi(n)}};$$

add the items in that order until the first time the weight constraints is violated; include whatever fraction of that last item that will fit. This greedy algorithm has a natural geometric interpretation, depicted in Figure 2.7, that will be useful. We associate item j to a point $(W_j, V_j) \in [0, 1]^2$ and keep only those items falling on or above a line with slope θ chosen to satisfy the total weight constraint. Specifically, let

$$\Delta_\theta = \left\{ j \in [n] \colon V_j > \theta W_j \right\},$$

$$\Lambda_\theta = \left\{ j \in [n] \colon V_j = \theta W_j \right\},$$

and

$$\Theta^* = \inf \left\{ \theta \geq 0 \colon W_{\Delta_\theta} < \mathcal{W} \right\},$$

where, for a subset of items $J \subset [n]$, $W_J = \sum_{j \in J} W_j$.

We consider a stochastic version of the fractional knapsack problem where the weights and values are i.i.d. random variables picked uniformly at random in $[0, 1]$. Characterizing Z^* (e.g., its moments or distribution) is not straightforward. Here we show that Z^* is highly concentrated around a natural quantity. Observe that, under our probabilistic model, almost surely $|\Lambda_\theta| \in \{0, 1\}$ for any $\theta \geq 0$. Hence, there are two cases. Either $\Theta^* = 0$, in which case all items fit in the knapsack so that $Z^* = \sum_{j=1}^n W_j$, or $\Theta^* > 0$, in which case $|\Lambda_{\Theta^*}| = 1$ and

$$Z^* = V_{\Delta_{\Theta^*}} + (\mathcal{W} - W_{\Delta_{\Theta^*}}) V_{\Lambda_{\Theta^*}}. \tag{2.4.22}$$

One interesting regime is $\mathcal{W} = \tau n$ for some constant $\tau > 0$. Clearly, $\tau > 1$ is trivial. In fact, because

$$\mathbb{E}\left[\sum_{j=1}^{n} W_j\right] = n\,\mathbb{E}[W_1] = \frac{1}{2}n,$$

we assume that $\tau \leq 1/2$. To further simplify the calculations, we restrict ourselves to the case $\tau \in (1/6, 1/2)$. (See Exercise 2.8 for the remaining case.) In this regime, we show that Z^* grows linearly with n and give a bound on its deviation.

Although Z^* is technically a sum of random variables, the choice of Θ^* correlates them and we cannot apply our concentration bounds directly. Instead, we show that Θ^* itself can be controlled well. It is natural to conjecture that Θ^* is approximately equal to a solution θ_τ of the expected constraint equation $\mathbb{E}[W_{\Delta_{\theta_\tau}}] = \mathcal{W}$, that is,

$$n\bar{w}_{\theta_\tau} = n\tau, \tag{2.4.23}$$

where $\bar{w}_\theta$ is defined through

$$\begin{aligned}
\mathbb{E}[W_{\Delta_\theta}] &= \mathbb{E}\left[\sum_{j \in \Delta_\theta} W_j\right] \\
&= \mathbb{E}\left[\sum_{j=1}^{n} \mathbf{1}\{V_j > \theta W_j\}\, W_j\right] \\
&= n\,\mathbb{E}\left[\mathbf{1}\{V_1 > \theta W_1\}\, W_1\right] \\
&=: n\bar{w}_\theta.
\end{aligned}$$

Similarly, we define

$$\bar{v}_\theta := \mathbb{E}\left[\mathbf{1}\{V_1 > \theta W_1\}\, V_1\right].$$

We see directly from the definitions that both $\bar{w}_\theta$ and $\bar{v}_\theta$ are monotone as functions of θ.

Our main claim is the following.

Claim 2.4.20 *There is a constant $c > 0$ such that for any $\delta > 0$,*

$$\mathbb{P}\left[|Z^* - n\bar{v}_{\theta_\tau}| \geq \sqrt{cn \log \delta^{-1}}\right] \leq \delta$$

for all n large enough.

Proof Because all weights and values are in $[0, 1]$, it follows from (2.4.22) that

$$V_{\Delta_{\Theta^*}} \leq Z^* \leq V_{\Delta_{\Theta^*}} + 1, \tag{2.4.24}$$

and it will suffice to work with $V_{\Delta_{\Theta^*}}$. The idea of the proof is to show that Θ^* is close to θ_τ by establishing that W_{Δ_θ} is highly likely to be less than τn when $\theta > \theta_\tau$, while the opposite holds when $\theta < \theta_\tau$. For this, we view W_{Δ_θ} as a sum of independent bounded random variables and use Hoeffding's inequality (Theorem 2.4.10).

Controlling Θ^*. First, it will be useful to compute $\bar{w}_\theta$ and θ_τ analytically. By definition,

$$
\begin{aligned}
\bar{w}_\theta &= \mathbb{E}\left[\mathbf{1}\{V_1 > \theta W_1\} W_1\right] \\
&= \int_0^1 \int_0^1 \mathbf{1}\{y > \theta x\} x \, dy \, dx \\
&= \int_0^{1 \wedge 1/\theta} (1 - \theta x) x \, dx \\
&= \begin{cases} \frac{1}{2} - \frac{1}{3}\theta & \text{if } \theta \le 1, \\ \frac{1}{6\theta^2} & \text{otherwise.} \end{cases}
\end{aligned} \tag{2.4.25}
$$

Plugging back into (2.4.23), we get the unique solution

$$
\theta_\tau := 3\left(\frac{1}{2} - \tau\right) \in (0, 1)
$$

for the range $\tau \in (1/6, 1/2)$.

Now observe that, for each fixed θ, the quantity

$$
W_{\Delta_\theta} = \sum_{j=1}^n \mathbf{1}\{V_j > \theta W_j\} W_j
$$

is a sum of independent random variables taking values in $[0, 1]$. Hence, for any $\beta > 0$, Hoeffding's inequality gives

$$
\mathbb{P}\left[W_{\Delta_\theta} - n\bar{w}_\theta \ge \beta\right] \le \exp\left(-\frac{2\beta^2}{n}\right).
$$

Using this inequality with $\theta = \theta_\tau - \frac{C}{\sqrt{n}}$ (with n large enough that $\theta < 1$) and $\beta = 3C\sqrt{n}$ gives

$$
\mathbb{P}\left[W_{\Delta_{\theta_\tau - \frac{C}{\sqrt{n}}}} - 3n\left(\frac{1}{2} - \theta_\tau + \frac{C}{\sqrt{n}}\right) \ge 3C\sqrt{n}\right] \le \exp\left(-2(3C)^2\right),
$$

where we used (2.4.25). After rearranging and using that $3n\left(\frac{1}{2} - \theta_\tau\right) = n\tau$ by (2.4.23) and (2.4.25), we get

$$
\mathbb{P}\left[\Theta^* \ge \theta_\tau - \frac{C}{\sqrt{n}}\right] = \mathbb{P}\left[W_{\Delta_{\theta_\tau - \frac{C}{\sqrt{n}}}} \ge n\tau\right] \le \exp\left(-2(3C)^2\right).
$$

Applying the same argument to $-W_{\Delta_\theta}$ with $\theta = \theta_\tau + \frac{C}{\sqrt{n}}$ and combining with the previous inequality gives

$$
\mathbb{P}\left[|\Theta^* - \theta_\tau| > \frac{C}{\sqrt{n}}\right] \le 2\exp\left(-2(3C)^2\right). \tag{2.4.26}
$$

Controlling Z^*. We conclude by applying Hoeffding's inequality to V_{Δ_θ}. Arguing as above with the same θ's and β, we obtain

$$
\mathbb{P}\left[V_{\Delta_{\theta_\tau - \frac{C}{\sqrt{n}}}} - n\bar{v}_{\theta_\tau - \frac{C}{\sqrt{n}}} \ge 3C\sqrt{n}\right] \le \exp\left(-2(3C)^2\right) \tag{2.4.27}
$$

and

$$\mathbb{P}\left[V_{\Delta_{\theta_\tau + \frac{C}{\sqrt{n}}}} - n\bar{v}_{\theta_\tau + \frac{C}{\sqrt{n}}} \le -3C\sqrt{n}\right] \le \exp\left(-2(3C)^2\right). \tag{2.4.28}$$

Again, it will be useful to compute $\bar{v}_\theta$ analytically. By definition,

$$\begin{aligned}
\bar{v}_\theta &= \mathbb{E}\left[\mathbf{1}\{V_1 > \theta W_1\} V_1\right] \\
&= \int_0^1 \int_0^1 \mathbf{1}\{y > \theta x\} y \, dx \, dy \\
&= \int_0^{1 \wedge \theta} \frac{y^2}{\theta} \, dy + \int_{1 \wedge \theta}^1 y \, dy \\
&= \begin{cases} \frac{1}{2} - \frac{1}{6}\theta^2 & \text{if } \theta \le 1, \\ \frac{1}{3\theta} & \text{otherwise.} \end{cases}
\end{aligned}$$

Assuming n is large enough that $\theta_\tau + C/\sqrt{n} < 1$ (recall that $\theta_\tau < 1$), we get

$$\bar{v}_{\theta_\tau} - \bar{v}_{\theta_\tau + \frac{C}{\sqrt{n}}} = \frac{1}{6}\left(2\frac{C}{\sqrt{n}}\theta_\tau + \frac{C^2}{n}\right) \le \frac{C}{\sqrt{n}}.$$

A quick check reveals that, similarly, $\bar{v}_{\theta_\tau - \frac{C}{\sqrt{n}}} - \bar{v}_{\theta_\tau} \le \frac{C}{\sqrt{n}}$. Plugging back into (2.4.27) and (2.4.28) gives

$$\mathbb{P}\left[V_{\Delta_{\theta_\tau - \frac{C}{\sqrt{n}}}} \ge n\bar{v}_{\theta_\tau} + 4C\sqrt{n}\right] \le \exp\left(-2(3C)^2\right) \tag{2.4.29}$$

and

$$\mathbb{P}\left[V_{\Delta_{\theta_\tau + \frac{C}{\sqrt{n}}}} \le n\bar{v}_{\theta_\tau} - 4C\sqrt{n}\right] \le \exp\left(-2(3C)^2\right). \tag{2.4.30}$$

Observe that the following monotonicity property holds almost surely

$$\theta_0 \le \theta_1 \le \theta_2 \implies V_{\Delta_{\theta_0}} \ge V_{\Delta_{\theta_1}} \ge V_{\Delta_{\theta_2}}. \tag{2.4.31}$$

Combining (2.4.24), (2.4.26), (2.4.29), (2.4.30), and (2.4.31), we obtain

$$\mathbb{P}\left[|Z^* - n\bar{v}_{\theta_\tau}| > 5C\sqrt{n}\right] \le 4\exp\left(-2(3C)^2\right)$$

for n large enough. Choosing C appropriately gives the claim. ∎

A similar bound is proved for the 0-1 knapsack problem in Exercise 2.9.

2.4.4 Epsilon-Nets and Chaining

Suppose we are interested in bounding the expectation or tail of the *supremum* of a stochastic process

$$\sup_{t \in \mathcal{T}} X_t,$$

where $\mathcal{T}$ is an arbitrary index set and the X_ts are real-valued random variables. To avoid measurability issues, we assume throughout that $\mathcal{T}$ is countable.[2] Note that t does not in general need to be a "time" index.

So far we have developed tools that can handle cases where $\mathcal{T}$ is finite. When the supremum is over an infinite index set, however, new ideas are required. One way to proceed is to apply a tail inequality to a sufficiently dense finite subset of the index set and then extend the resulting bound by a Lipschitz continuity argument. We present this type of approach in this section, as well as a multi-scale version known as chaining.

First we summarize one important special case that will be useful below: $\mathcal{T}$ is finite and X_t is sub-Gaussian.

Theorem 2.4.21 (Maximal inequalities: sub-Gaussian case). *Let $\{X_t\}_{t \in \mathcal{T}}$ be a stochastic process with finite index set $\mathcal{T}$. Assume that there is $\nu > 0$ such that, for all t, $X_t \in s\mathcal{G}(\nu)$ and $\mathbb{E}[X_t] = 0$. Then,*

$$\mathbb{E}\left[\sup_{t \in \mathcal{T}} X_t\right] \leq \sqrt{2\nu \log |\mathcal{T}|},$$

and, for all $\beta > 0$,

$$\mathbb{P}\left[\sup_{t \in \mathcal{T}} X_t \geq \sqrt{2\nu \log |\mathcal{T}|} + \beta\right] \leq \exp\left(-\frac{\beta^2}{2\nu}\right).$$

Proof For the expectation, we apply a variation on the Chernoff–Cramér method (Section 2.4). Naively, we could bound the supremum $\sup_{t \in \mathcal{T}} X_t$ by the sum $\sum_{t \in \mathcal{T}} |X_t|$, but that would lead to a bound growing linearly with the cardinality $|\mathcal{T}|$. Instead we first take an exponential, which tends to amplify the largest term and produces a much stronger bound. Specifically, by Jensen's inequality (Theorem B.4.15), for any $s > 0$,

$$\mathbb{E}\left[\sup_{t \in \mathcal{T}} X_t\right] = \frac{1}{s}\mathbb{E}\left[\sup_{t \in \mathcal{T}} sX_t\right] \leq \frac{1}{s}\log \mathbb{E}\left[\exp\left(\sup_{t \in \mathcal{T}} sX_t\right)\right].$$

Since $e^{a \vee b} \leq e^a + e^b$ by the non-negativity of the exponential, we can bound

$$\mathbb{E}\left[\sup_{t \in \mathcal{T}} X_t\right] \leq \frac{1}{s}\log\left[\sum_{t \in \mathcal{T}} \mathbb{E}\left[\exp\left(sX_t\right)\right]\right]$$

$$= \frac{1}{s}\log\left[\sum_{t \in \mathcal{T}} M_{X_t}(s)\right]$$

$$\leq \frac{1}{s}\log\left[|\mathcal{T}|\, e^{\frac{s^2 \nu}{2}}\right]$$

$$= \frac{\log|\mathcal{T}|}{s} + \frac{s\nu}{2}.$$

---

[2] Technically, it suffices to assume that there is a countable $\mathcal{T}_0 \subseteq \mathcal{T}$ such that $\sup_{t \in \mathcal{T}} X_t = \sup_{t \in \mathcal{T}_0} X_t$ almost surely.

The optimal choice of s (i.e., leading to the least bound) is when the two terms in the sum above are equal, that is, $s = \sqrt{2v^{-1}\log|\mathscr{T}|}$, which gives finally

$$\mathbb{E}\left[\sup_{t\in\mathscr{T}} X_t\right] \leq \sqrt{2v\log|\mathscr{T}|},$$

as claimed.

For the tail inequality, we use a union bound and (2.4.16):

$$\mathbb{P}\left[\sup_{t\in\mathscr{T}} X_t \geq \sqrt{2v\log|\mathscr{T}|} + \beta\right] \leq \sum_{t\in\mathscr{T}} \mathbb{P}\left[X_t \geq \sqrt{2v\log|\mathscr{T}|} + \beta\right]$$

$$\leq |\mathscr{T}|\exp\left(-\frac{(\sqrt{2v\log|\mathscr{T}|}+\beta)^2}{2v}\right)$$

$$\leq \exp\left(-\frac{\beta^2}{2v}\right),$$

as claimed, where we used that $\beta > 0$ on the last line. ∎

Epsilon-nets and covering numbers

Moving on to infinite index sets, we first define the notion of an ε-net. This notion requires that a pseudometric ρ (i.e., $\rho\colon \mathscr{T}\times\mathscr{T}\to\mathbb{R}_+$ is symmetric and satisfies the triangle inequality) be defined over $\mathscr{T}$.

Definition 2.4.22 (ε-net). *Let $\mathscr{T}$ be a subset of a pseudometric space (M,ρ) and let $\varepsilon > 0$. The collection of points $N \subseteq M$ is called an ε-net of $\mathscr{T}$ if* ε-NET

$$\mathscr{T} \subseteq \bigcup_{t\in N} B_\rho(t,\varepsilon),$$

where $B_\rho(t,\varepsilon) = \{s\in\mathscr{T} : \rho(s,t)\leq\varepsilon\}$, that is, each element of $\mathscr{T}$ is within distance ε of an element in N. The smallest cardinality of an ε-net of $\mathscr{T}$ is called the covering number COVERING
NUMBER

$$\mathcal{N}(\mathscr{T},\rho,\varepsilon) = \inf\{|N| : N \text{ is an } \varepsilon\text{-net of } \mathscr{T}\}.$$

A natural way to construct an ε-net is the following algorithm. Start with $N = \emptyset$ and successively add a point from $\mathscr{T}$ to N at distance at least ε from all other previous points until it is not possible to do so anymore. Provided $\mathscr{T}$ is compact, this procedure will terminate after a finite number of steps. This leads to the following dual perspective.

Definition 2.4.23 (ε-packing). *Let $\mathscr{T}$ be a subset of a pseudometric space (M,ρ) and let $\varepsilon > 0$. The collection of points $N \subseteq \mathscr{T}$ is called an ε-packing of $\mathscr{T}$ if*

$$t \notin B_\rho(t',\varepsilon) \quad \forall t \neq t' \in N,$$

that is, every pair of elements of N is at distance strictly greater than ε. The largest cardinality of an ε-packing of $\mathscr{T}$ is called the packing number PACKING
NUMBER

$$\mathcal{P}(\mathscr{T},\rho,\varepsilon) = \sup\{|N| : N \text{ is an } \varepsilon\text{-packing of } \mathscr{T}\}.$$

Lemma 2.4.24 (Covering and packing numbers). *For any $\mathscr{T} \subseteq M$ and all $\varepsilon > 0$,*

$$\mathcal{N}(\mathscr{T},\rho,\varepsilon) \leq \mathcal{P}(\mathscr{T},\rho,\varepsilon).$$

Proof   Observe that a maximal ε-packing N is an ε-net. Indeed, by maximality, any element of $\mathscr{T} \setminus N$ is at distance at most ε from an element of N.   ■

Example 2.4.25 (Sphere in $\mathbb{R}^k$).   We let $\mathbb{B}^k(x, \varepsilon)$ be the ball of radius ε around $x \in \mathbb{R}^k$ with the Euclidean metric. We let $S := \mathbb{S}^{k-1}$ be the sphere of radius 1 centered around the origin **0**, that is, the surface of $\mathbb{B}^k(\mathbf{0}, 1)$. Let $0 < \varepsilon < 1$.

Claim 2.4.26

$$\mathcal{N}(S, \rho, \varepsilon) \leq \left(\frac{3}{\varepsilon}\right)^k .$$

Proof   Let N be any maximal ε-packing of S. We show that $|N| \leq (3/\varepsilon)^k$, which implies the claim by Lemma 2.4.24. The balls of radius $\varepsilon/2$ around points in N, $\{\mathbb{B}^k(x_i, \varepsilon/2) : x_i \in N\}$, satisfy two properties:

1. They are pairwise disjoint: if $z \in \mathbb{B}^k(x_i, \varepsilon/2) \cap \mathbb{B}^k(x_j, \varepsilon/2)$, then $\|x_i - x_j\|_2 \leq \|x_i - z\|_2 + \|x_j - z\|_2 \leq \varepsilon$, a contradiction.
2. They are included in the ball of radius $3/2$ around the origin: if $z \in \mathbb{B}^k(x_i, \varepsilon/2)$, then $\|z\|_2 \leq \|z - x_i\|_2 + \|x_i\| \leq \varepsilon/2 + 1 \leq 3/2$.

The volume of a ball of radius $\varepsilon/2$ is $\frac{\pi^{k/2}(\varepsilon/2)^k}{\Gamma(k/2+1)}$ and that of a ball of radius $3/2$ is $\frac{\pi^{k/2}(3/2)^k}{\Gamma(k/2+1)}$. Dividing one by the other proves the claim.   ■

This bound will be useful later.   ◀

The basic approach to use an ε-net for controlling the supremum of a stochastic process is the following. We say that a stochastic process $\{X_t\}_{t \in \mathscr{T}}$ is Lipschitz for pseudometric ρ on $\mathscr{T}$ if there is a *random variable* $0 < K < +\infty$ such that

LIPSCHITZ
PROCESS

$$|X_t - X_s| \leq K\rho(s, t), \quad \forall s, t \in \mathscr{T}.$$

If in addition X_t is sub-Gaussian for all t, then we can bound the expectation or tail probability of the supremum of $\{X_t\}_{t \in \mathscr{T}}$ – if we can bound the expectation or tail probability of the (random) Lipschitz constant K itself. To see this, let N be an ε-net of $\mathscr{T}$ and, for each $t \in \mathscr{T}$, let $\pi(t)$ be the closest element of N to t. We will refer to π as the projection map of N. We then have the inequality

$$\sup_{t \in \mathscr{T}} X_t \leq \sup_{t \in \mathscr{T}}(X_t - X_{\pi(t)}) + \sup_{t \in \mathscr{T}} X_{\pi(t)} \leq K\varepsilon + \sup_{s \in N} X_s, \tag{2.4.32}$$

where we can use Theorem 2.4.21 to bound the last term. We give an example of this type of argument next (although we do not apply the above bound directly). Another example (where (2.4.32) is used this time) can be found in Section 2.4.5.

SPECTRAL
NORM

Example 2.4.27 (Spectral norm of a random matrix).   For an $m \times n$ matrix $A \in \mathbb{R}^{m \times n}$, the *spectral norm* (or *induced 2-norm*, or *2-norm* for short) is defined as

$$\|A\|_2 := \sup_{\mathbf{x} \in \mathbb{R}^n \setminus \{0\}} \frac{\|A\mathbf{x}\|_2}{\|\mathbf{x}\|_2} = \sup_{\mathbf{x} \in \mathbb{S}^{n-1}} \|A\mathbf{x}\|_2 = \sup_{\substack{\mathbf{x} \in \mathbb{S}^{n-1} \\ \mathbf{y} \in \mathbb{S}^{m-1}}} \langle A\mathbf{x}, \mathbf{y} \rangle, \tag{2.4.33}$$

where $\mathbb{S}^{n-1}$ is the sphere of Euclidean radius 1 around the origin in $\mathbb{R}^n$. The rightmost expression, which is central to our developments, is justified in Exercise 5.4.

We will be interested in the case where A is a random matrix with independent entries. One key observation is that the quantity $\langle A\mathbf{x}, \mathbf{y} \rangle$ can then be seen as a linear combination of independent random variables

$$\langle A\mathbf{x}, \mathbf{y} \rangle = \sum_{i,j} x_i y_j A_{ij}.$$

Hence we will be able to apply our previous tail bounds. *However,* we also need to deal with the supremum.

Theorem 2.4.28 (Upper tail of the spectral norm). *Let $A \in \mathbb{R}^{m \times n}$ be a random matrix whose entries are centered, independent, and sub-Gaussian with variance factor v. Then there exists a constant $0 < C < +\infty$ such that, for all $t > 0$,*

$$\|A\|_2 \leq C\sqrt{v}(\sqrt{m} + \sqrt{n} + t),$$

with probability at least $1 - e^{-t^2}$.

Without the independence assumption, the norm can be much larger in general (see Exercise 2.15).

Proof Fix $\varepsilon = 1/4$. By Claim 2.4.26, there is an ε-net N (respectively M) of $\mathbb{S}^{n-1}$ (respectively $\mathbb{S}^{m-1}$) with $|N| \leq 12^n$ (respectively $|M| \leq 12^m$). We proceed in two steps:

1. We first apply the general Hoeffding inequality (Theorem 2.4.9) to control the deviations of the supremum in (2.4.33) *restricted to N and M*.
2. We then extend the bound to the full supremum by Lipschitz continuity.

Formally, the result follows from the following two lemmas.

Lemma 2.4.29 (Spectral norm: ε-net). *Let N and M be as above. For C large enough, for all $t > 0$,*

$$\mathbb{P}\left[\max_{\substack{\mathbf{x} \in N \\ \mathbf{y} \in M}} \langle A\mathbf{x}, \mathbf{y} \rangle \geq \frac{1}{2}C\sqrt{v}(\sqrt{m} + \sqrt{n} + t) \right] \leq e^{-t^2}.$$

Lemma 2.4.30 (Spectral norm: Lipschitz constant). *For any ε-nets N and M of $\mathbb{S}^{n-1}$ and $\mathbb{S}^{m-1}$, respectively, the following inequalities hold:*

$$\sup_{\substack{\mathbf{x} \in N \\ \mathbf{y} \in M}} \langle A\mathbf{x}, \mathbf{y} \rangle \leq \|A\|_2 \leq \frac{1}{1 - 2\varepsilon} \sup_{\substack{\mathbf{x} \in N \\ \mathbf{y} \in M}} \langle A\mathbf{x}, \mathbf{y} \rangle.$$

Proof of Lemma 2.4.29 Recall that

$$\langle A\mathbf{x}, \mathbf{y} \rangle = \sum_{i,j} x_i y_j A_{ij}$$

is a linear combination of independent random variables. By the general Hoeffding inequality, $\langle A\mathbf{x}, \mathbf{y} \rangle$ is sub-Gaussian with variance factor

$$\sum_{i,j} (x_i y_j)^2 v = \|\mathbf{x}\|_2^2 \|\mathbf{y}\|_2^2 v = v$$

for all $\mathbf{x} \in N$ and $\mathbf{y} \in M$. In particular, for all $\beta > 0$,

$$\mathbb{P}[\langle A\mathbf{x}, \mathbf{y} \rangle \geq \beta] \leq \exp\left(-\frac{\beta^2}{2\nu}\right).$$

Hence, by a union bound over N and M,

$$\mathbb{P}\left[\max_{\substack{\mathbf{x} \in N \\ \mathbf{y} \in M}} \langle A\mathbf{x}, \mathbf{y} \rangle \geq \frac{1}{2} C \sqrt{\nu} (\sqrt{m} + \sqrt{n} + t)\right]$$

$$\leq \sum_{\substack{\mathbf{x} \in N \\ \mathbf{y} \in M}} \mathbb{P}\left[\langle A\mathbf{x}, \mathbf{y} \rangle \geq \frac{1}{2} C \sqrt{\nu} (\sqrt{m} + \sqrt{n} + t)\right]$$

$$\leq |N||M| \exp\left(-\frac{1}{2\nu}\left\{\frac{1}{2} C \sqrt{\nu} (\sqrt{m} + \sqrt{n} + t)\right\}^2\right)$$

$$\leq 12^{n+m} \exp\left(-\frac{C^2}{8}\left\{m + n + t^2\right\}\right)$$

$$\leq e^{-t^2}$$

for $C^2/8 = \log 12 \geq 1$, where in the third inequality we ignored all cross-products since they are non-negative. ∎

Proof of Lemma 2.4.30 The first inequality is immediate by definition of the spectral norm. For the second inequality, we will use the following observation:

$$\langle A\mathbf{x}, \mathbf{y} \rangle - \langle A\mathbf{x}_0, \mathbf{y}_0 \rangle = \langle A\mathbf{x}, \mathbf{y} - \mathbf{y}_0 \rangle + \langle A(\mathbf{x} - \mathbf{x}_0), \mathbf{y}_0 \rangle. \qquad (2.4.34)$$

Fix $\mathbf{x} \in \mathbb{S}^{n-1}$ and $\mathbf{y} \in \mathbb{S}^{m-1}$ such that $\langle A\mathbf{x}, \mathbf{y} \rangle = \|A\|_2$ (which exist by compactness), and let $\mathbf{x}_0 \in N$ and $\mathbf{y}_0 \in M$ such that

$$\|\mathbf{x} - \mathbf{x}_0\|_2 \leq \varepsilon \qquad \text{and} \qquad \|\mathbf{y} - \mathbf{y}_0\|_2 \leq \varepsilon.$$

Then (2.4.34), Cauchy–Schwarz and the definition of the spectral norm imply

$$\|A\|_2 - \langle A\mathbf{x}_0, \mathbf{y}_0 \rangle \leq \|A\|_2 \|\mathbf{x}\|_2 \|\mathbf{y} - \mathbf{y}_0\|_2 + \|A\|_2 \|\mathbf{x} - \mathbf{x}_0\|_2 \|\mathbf{y}_0\|_2 \leq 2\varepsilon \|A\|_2.$$

Rearranging gives the claim. ∎

Putting the two lemmas together concludes the proof of Theorem 2.4.28. ∎

We will give an application of this bound in Section 5.1.4. ◄

Chaining method

We go back to the inequality

$$\sup_{t \in \mathcal{T}} X_t \leq \sup_{t \in \mathcal{T}} (X_t - X_{\pi(t)}) + \sup_{t \in \mathcal{T}} X_{\pi(t)}. \qquad (2.4.35)$$

Previously we controlled the first term on the right-hand side with a random Lipschitz constant and the second term with a maximal inequality for finite sets. Now we consider cases where we may not have a good almost sure bound on the Lipschitz constant, but where we can control increments uniformly in the following probabilistic sense. We say that a

stochastic process $\{X_t\}_{t\in\mathscr{T}}$ has sub-Gaussian increments on $(\mathscr{T},\rho)$ if there exists a *deter-*
ministic constant $0 < \mathcal{K} < +\infty$ such that

$$X_t - X_s \in s\mathcal{G}(\mathcal{K}^2\rho(s,t)^2) \quad \forall s,t \in \mathscr{T}.$$

Even with this assumption, in (2.4.35) the first term on the right-hand side remains a supre-
mum over an infinite set. To control it, the *chaining method* repeats the argument above at
progressively smaller scales, leading to the following inequality. The diameter of $\mathscr{T}$, denoted
by diam($\mathscr{T}$), is defined as

$$\mathrm{diam}(\mathscr{T}) = \sup\{\rho(s,t)\colon s,t,\in\mathscr{T}\}.$$

Theorem 2.4.31 (Discrete Dudley inequality). *Let $\{X_t\}_{t\in\mathscr{T}}$ be a zero-mean stochastic process with sub-Gaussian increments on $(\mathscr{T},\rho)$ and assume* diam($\mathscr{T}$) ≤ 1. *Then*

$$\mathbb{E}\left[\sup_{t\in\mathscr{T}} X_t\right] \leq C\sum_{k=0}^{+\infty} 2^{-k}\sqrt{\log\mathcal{N}(\mathscr{T},\rho,2^{-k})}$$

for some constant $0 \leq C < +\infty$.

Proof Recall that we assume that $\mathscr{T}$ is countable. Let $\mathscr{T}_j \subseteq \mathscr{T}, j \geq 1$, be a sequence of
finite sets such that $\mathscr{T}_j \uparrow \mathscr{T}$. By monotone convergence (Proposition B.4.14),

$$\mathbb{E}\left[\sup_{t\in\mathscr{T}} X_t\right] = \sup_{j\geq 1}\mathbb{E}\left[\sup_{t\in\mathscr{T}_j} X_t\right].$$

Moreover, $\mathcal{N}(\mathscr{T}_j,\rho,\varepsilon) \leq \mathcal{N}(\mathscr{T},\rho,\varepsilon)$ for any $\varepsilon > 0$ since $\mathscr{T}_j \subseteq \mathscr{T}$. Hence, it suffices to
handle the case $|\mathscr{T}| < +\infty$.
ε-nets at all scales. For each $k \geq 0$, let N_k be an 2^{-k}-net of $\mathscr{T}$ with $|N_k| = \mathcal{N}(\mathscr{T},\rho,2^{-k})$
and projection map π_k. Because diam($\mathscr{T}$) ≤ 1, $N_0 = \{t_0\}$, where $t_0 \in \mathscr{T}$ can be taken
arbitrarily. Moreover, because $\mathscr{T}$ is finite, there is $1 \leq \kappa < +\infty$ such that $N_k = \mathscr{T}$ for all
$k \geq \kappa$. In particular, $\pi_\kappa(t) = t$ for all $t \in \mathscr{T}$. By a telescoping argument,

$$X_t = X_{t_0} + \sum_{k=0}^{\kappa-1}\left(X_{\pi_{k+1}(t)} - X_{\pi_k(t)}\right).$$

Taking a supremum and then an expectation gives

$$\mathbb{E}\left[\sup_{t\in\mathscr{T}} X_t\right] \leq \sum_{k=0}^{\kappa-1}\mathbb{E}\left[\sup_{t\in\mathscr{T}}\left(X_{\pi_{k+1}(t)} - X_{\pi_k(t)}\right)\right], \tag{2.4.36}$$

where we used $\mathbb{E}[X_{t_0}] = 0$.
Sub-Gaussian bound. We use the maximal inequality (Theorem 2.4.21) to bound the expectation in (2.4.36). For each k, the number of distinct elements in the supremum is at most

$$\begin{aligned}
\left|\{(\pi_k(t),\pi_{k+1}(t))\colon t\in\mathscr{T}\}\right| &\leq |N_k \times N_{k+1}| \\
&= |N_k| \times |N_{k+1}| \\
&\leq (\mathcal{N}(\mathscr{T},\rho,2^{-k-1}))^2.
\end{aligned}$$

For any $t \in \mathcal{T}$, by the triangle inequality,

$$\rho(\pi_k(t), \pi_{k+1}(t)) \leq \rho(\pi_k(t), t) + \rho(t, \pi_{k+1}(t)) \leq 2^{-k} + 2^{-k-1} \leq 2^{-k+1},$$

so that

$$X_{\pi_{k+1}(t)} - X_{\pi_k(t)} \in s\mathcal{G}(\mathcal{K}^2 2^{-2k+2})$$

for some $0 < \mathcal{K} < +\infty$ by the sub-Gaussian increments assumption. We can therefore apply Theorem 2.4.21 to get

$$\mathbb{E}\left[\sup_{t \in \mathcal{T}}\left(X_{\pi_{k+1}(t)} - X_{\pi_k(t)}\right)\right] \leq \sqrt{2\mathcal{K}^2 2^{-2k+2} \log(\mathcal{N}(\mathcal{T}, \rho, 2^{-k-1})^2)}$$

$$\leq C 2^{-k-1} \sqrt{\log \mathcal{N}(\mathcal{T}, \rho, 2^{-k-1})}$$

for some constant $0 \leq C < +\infty$.

To finish the argument, we plug back into (2.4.36),

$$\mathbb{E}\left[\sup_{t \in \mathcal{T}} X_t\right] \leq \sum_{k=0}^{\kappa-1} C 2^{-k-1} \sqrt{\log \mathcal{N}(\mathcal{T}, \rho, 2^{-k-1})},$$

which implies the claim.   ∎

Using a similar argument, one can derive a tail inequality.

Theorem 2.4.32 (Chaining tail inequality). *Let $\{X_t\}_{t \in \mathcal{T}}$ be a zero-mean stochastic process with sub-Gaussian increments on $(\mathcal{T}, \rho)$ and assume* $\mathrm{diam}(\mathcal{T}) \leq 1$. *Then, for all $t_0 \in \mathcal{T}$ and $\beta > 0$,*

$$\mathbb{P}\left[\sup_{t \in \mathcal{T}}(X_t - X_{t_0}) \geq C \sum_{k=0}^{+\infty} 2^{-k} \sqrt{\log \mathcal{N}(\mathcal{T}, \rho, 2^{-k})} + \beta\right] \leq C \exp\left(-\frac{\beta^2}{C}\right)$$

for some constant $0 \leq C < +\infty$.

We give an application of the discrete Dudley inequality in Section 2.4.6.

2.4.5 ▷ *Data Science: Johnson–Lindenstrauss Lemma and Application to Compressed Sensing*

In this section, we discuss an application of the Chernoff–Cramér method (Section 2.4.1) to dimension reduction in data science. We use once again an ε-net argument (Section 2.4.4).

Johnson–Lindenstrauss lemma

The Johnson–Lindenstrauss lemma states roughly that, for any collection of points in a high-dimensional Euclidean space, one can find an embedding of much lower dimension that roughly preserves the metric relationships of the points, that is, their distances. Remarkably, no structure is assumed on the original points and the result is *independent of the input dimension*. The method of proof simply involves performing a random projection.

Lemma 2.4.33 (Johnson–Lindenstrauss lemma). *For any set of points $\boldsymbol{x}^{(1)}, \ldots, \boldsymbol{x}^{(m)}$ in $\mathbb{R}^n$ and $\theta \in (0, 1)$, there exists a mapping $f : \mathbb{R}^n \to \mathbb{R}^d$ with $d = \Theta(\theta^{-2} \log m)$ such that the following hold: for all i, j,*

$$(1 - \theta)\|\boldsymbol{x}^{(i)} - \boldsymbol{x}^{(j)}\|_2 \leq \|f(\boldsymbol{x}^{(i)}) - f(\boldsymbol{x}^{(j)})\|_2 \leq (1 + \theta)\|\boldsymbol{x}^{(i)} - \boldsymbol{x}^{(j)}\|_2. \tag{2.4.37}$$

We use the probabilistic method: we derive a "distributional" version of the result that, in turn, implies Lemma 2.4.33 by showing that a mapping with the desired properties exists with positive probability. Before stating this claim formally, we define the explicit random linear mapping we will employ. Let A be a $d \times n$ matrix whose entries are independent $N(0, 1)$. Note that, for any fixed $\boldsymbol{z} \in \mathbb{R}^n$,

$$\mathbb{E}\,\|A\boldsymbol{z}\|_2^2 = \mathbb{E}\left[\sum_{i=1}^{d}\left(\sum_{j=1}^{n} A_{ij}z_j\right)^2\right] = d\,\mathrm{Var}\left[\sum_{j=1}^{n} A_{1j}z_j\right] = d\|\boldsymbol{z}\|_2^2, \tag{2.4.38}$$

where we used the independence of the A_{ij}s (and, in particular, of the rows of A) and the fact that

$$\mathbb{E}\left[\sum_{j=1}^{n} A_{ij}z_j\right] = 0. \tag{2.4.39}$$

Hence, the normalized mapping

$$L = \frac{1}{\sqrt{d}}A$$

preserves the squared Euclidean norm "on average," that is, $\mathbb{E}\,\|L\boldsymbol{z}\|_2^2 = \|\boldsymbol{z}\|_2^2$. We use the Chernoff–Cramér method to prove a high-probability result.

Lemma 2.4.34 *Fix $\delta, \theta \in (0, 1)$. Then the random linear mapping L above with $d = \Theta(\theta^{-2}\log \delta^{-1})$ is such that for any $\boldsymbol{z} \in \mathbb{R}^n$ with $\|\boldsymbol{z}\|_2 = 1$,*

$$\mathbb{P}\,[|\,\|L\boldsymbol{z}\|_2 - 1| \geq \theta] \leq \delta. \tag{2.4.40}$$

Before proving Lemma 2.4.34, we argue that it implies the Johnson–Lindenstrauss lemma (Lemma 2.4.33). Simply take $\delta = 1/(2\binom{m}{2})$, apply the previous lemma to each normalized pairwise difference $\boldsymbol{z} = (\boldsymbol{x}^{(i)} - \boldsymbol{x}^{(j)})/\|\boldsymbol{x}^{(i)} - \boldsymbol{x}^{(j)}\|_2$, and use a union bound over all $\binom{m}{2}$ such pairs. The probability that any of the inequalities (2.4.37) is not satisfied by the linear mapping $f(\boldsymbol{z}) = L\boldsymbol{z}$ is then at most $1/2$. Hence, a mapping with the desired properties exists for $d = \Theta(\theta^{-2} \log m)$.

Proof of Lemma 2.4.34 We prove one direction. Specifically, we establish

$$\mathbb{P}\,[\|L\boldsymbol{z}\|_2 \geq 1 + \theta] \leq \exp\left(-\frac{3}{4}d\theta^2\right). \tag{2.4.41}$$

Note that the right-hand side is $\leq \delta$ for $d = \Theta(\theta^{-2}\log\delta^{-1})$. An inequality in the other direction can be proved similarly by working with $-W$ below.

Recall that a sum of independent Gaussians is Gaussian (just compute the convolution and complete the squares). So

$$(A\boldsymbol{z})_k \sim N(0, \|\boldsymbol{z}\|_2^2) = N(0, 1) \quad \forall k,$$

where we argued as in (2.4.38) to compute the variance. Hence,

$$W = \|Az\|_2^2 = \sum_{k=1}^{d} (Az)_k^2$$

is a sum of squares of independent Gaussians, that is, χ^2-distributed random variables. By (2.4.18) and independence,

$$M_W(s) = \frac{1}{(1 - 2s)^{d/2}}.$$

Applying the Chernoff–Cramér bound (2.4.2) with $s = \frac{1}{2}(1 - d/\beta)$ gives

$$\mathbb{P}[W \geq \beta] \leq \frac{M_W(s)}{e^{s\beta}} = \frac{1}{e^{s\beta}(1 - 2s)^{d/2}} = e^{(d-\beta)/2} \left(\frac{\beta}{d} \right)^{d/2}.$$

Finally, take $\beta = d(1 + \theta)^2$. Rearranging we get

$$\begin{aligned}
\mathbb{P}[\|Lz\|_2 \geq 1 + \theta] &= \mathbb{P}[\|Az\|_2^2 \geq d(1 + \theta)^2] \\
&= \mathbb{P}[W \geq \beta] \\
&\leq e^{d[1-(1+\theta)^2]/2} \left[(1 + \theta)^2 \right]^{d/2} \\
&= \exp\left(-d(\theta + \theta^2/2 - \log(1 + \theta)) \right) \\
&\leq \exp\left(-\frac{3}{4} d\theta^2 \right),
\end{aligned}$$

where we used $\log(1 + x) \leq x - x^2/4$ on $[0, 1]$ (see Exercise 1.16).   ∎

Remark 2.4.35 *The Johnson–Lindenstrauss lemma is essentially optimal [Alo03, section 9]: any set of n points with all pairwise distances in $[1 - \theta, 1 + \theta]$ requires at least $\Omega(\log n/ (\theta^2 \log \theta^{-1}))$ dimensions. Note, however, that it relies crucially on the use of the Euclidean norm [BC03].*

To give some further geometric insights into the proof, we make a series of observations:

1. The d rows of $\frac{1}{\sqrt{n}} A$ are "on average" orthonormal. Indeed, note that for $i \neq j$,

$$\mathbb{E}\left[\frac{1}{n} \sum_{k=1}^{n} A_{ik} A_{jk} \right] = \mathbb{E}[A_{i1}] \, \mathbb{E}[A_{j1}] = 0$$

by independence and

$$\mathbb{E}\left[\frac{1}{n} \sum_{k=1}^{n} A_{ik}^2 \right] = \mathbb{E}[A_{i1}^2] = 1$$

since the A_{ik}s have mean 0 and variance 1. When n is large, those two quantities are concentrated around their mean. Fix a unit vector z. Then $\frac{1}{\sqrt{n}} Az$ corresponds approximately to an orthogonal projection of z onto a uniformly chosen random subspace of dimension d.

2. Now observe that projecting z on a uniform random subspace of dimension d can be done in the following way: first apply a uniformly chosen random rotation to z; and then project the resulting vector on the first d dimensions. In other words, $\frac{1}{\sqrt{n}}\|Az\|_2$ is approximately distributed as the norm of the first d components of a uniform unit vector in $\mathbb{R}^n$. To analyze this quantity, note that a vector in $\mathbb{R}^n$ whose components are independent $N(0, 1)$, when divided by its norm, produces a uniform vector in $\mathbb{R}^n$. When d is large, the norm of the first d components of that vector is therefore a ratio whose numerator is concentrated around $\sqrt{d}$ and whose denominator is concentrated around $\sqrt{n}$ (by calculations similar to those in the first point).

3. Hence, $\|Lz\|_2 = \sqrt{\frac{n}{d}} \times \frac{1}{\sqrt{n}}\|Az\|_2$ should be concentrated around 1.

The Johnson–Lindenstrauss lemma makes it possible to solve certain computational problems (e.g., finding the nearest point to a query) more efficiently by working in a smaller dimension. We discuss a different type of application next.

Compressed sensing

In the *compressed sensing* problem, one seeks to recover a signal $x \in \mathbb{R}^n$ from a small number of linear measurements $(Lx)_i$, $i = 1, \ldots, d$. In complete generality, one needs n such measurements to recover *any* unknown $x \in \mathbb{R}^n$ as the *sensing matrix L* must be invertible (or, SENSING more precisely, injective). However, by imposing extra structure on the signal and choosing MATRIX the sensing matrix appropriately, much better results can be obtained. Compressed sensing relies on sparsity.

Definition 2.4.36 (Sparse vectors). *We say that a vector $z \in \mathbb{R}^n$ is k-sparse if it has at most* K-SPARSE *k non-zero entries. We let $\mathscr{S}_k^n$ be the set of k-sparse vectors in $\mathbb{R}^n$. Note that $\mathscr{S}_k^n$ is a union* VECTOR *of $\binom{n}{k}$ linear subspaces, one for each support of the non-zero entries.*

To solve the compressed sensing problem over k-sparse vectors, it suffices to find a sensing matrix L satisfying that all subsets of $2k$ columns are linearly independent. Indeed, if $x, x' \in \mathscr{S}_k^n$, then $x - x'$ has at most $2k$ non-zero entries. Hence, in order to have $L(x - x') = 0$, it must be that $x - x' = 0$ under the previous condition on L. That implies the required injectivity. The implication goes in the other direction as well. Observe for instance that the matrix used in the proof of the Johnson–Lindenstrauss lemma satisfies this property as long as $d \geq 2k$: because of the continuous density of its entries, the probability that $2k$ of its columns are linearly dependent is 0 when $d \geq 2k$. For practical applications, however, other requirements must be met, in particular, computational efficiency. We describe such a computationally efficient approach.

The following definition will play a key role. Roughly speaking, a restricted isometry preserves enough of the metric structure of $\mathscr{S}_k^n$ to be invertible on its image.

Definition 2.4.37 (Restricted isometry property). *A $d \times n$ linear mapping L satisfies the* RESTRICTED *(k, θ)-restricted isometry property (RIP) if for all $z \in \mathscr{S}_k^n$,* ISOMETRY
PROPERTY

$$(1 - \theta)\|z\|_2 \leq \|Lz\|_2 \leq (1 + \theta)\|z\|_2. \tag{2.4.42}$$

We say that L is (k, θ)-RIP.

Given a (k, θ)-RIP matrix L, can we recover $z \in \mathscr{S}_k^n$ from Lz? And how small can d be? The next two claims answer these questions.

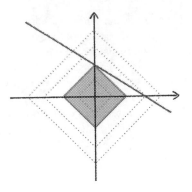

Figure 2.8 Because ℓ^1 balls (square) have corners, minimizing the ℓ^1 norm over a linear subspace (line) tends to produce sparse solutions.

Lemma 2.4.38 (Sensing matrix). *Let A be a $d \times n$ matrix whose entries are i.i.d. $N(0,1)$ and let $L = \frac{1}{\sqrt{d}}A$. There is a constant $0 < C < +\infty$ such that if $d \geq Ck \log n$, then L is $(10k, 1/3)$-RIP with probability at least $1 - 1/n$.*

Lemma 2.4.39 (Sparse signal recovery). *Let L be $(10k, 1/3)$-RIP. Then for any $x \in \mathscr{S}_k^n$, the unique solution to the following minimization problem*

$$\min_{z \in \mathbb{R}^n} \|z\|_1 \quad subject\ to \quad Lz = Lx \tag{2.4.43}$$

is $z^* = x$.

It may seem that a more natural alternative approach to (2.4.43) is to instead minimize the *number of non-zero entries* in z, that is, $\|z\|_0$. However, the advantage of the ℓ^1 norm is that the problem can then be formulated as a linear program, that is, the minimization of a linear objective subject to linear inequalities (see Exercise 2.13). This permits much faster computation of the solution using standard techniques – while still leading to a sparse solution. See Figure 2.8 for some insights as to why ℓ^1 indeed promotes sparsity.

Putting the two lemmas together shows we obtain the next claim:

Claim 2.4.40 *Let L be as in Lemma 2.4.38 with $d = \Theta(k \log n)$. With probability $1 - o(1)$, any $x \in \mathscr{S}_k^n$ can be recovered from the input Lx by solving (2.4.43).*

Note that d can in general be much smaller than n and not far from the $2k$ bound we derived above.

ε-**net argument** We start with the proof of Lemma 2.4.38. The claim does *not* follow immediately from the (distributional) Johnson–Lindenstrauss lemma (i.e., Lemma 2.4.34). Indeed, that lemma implies that a (normalized) matrix with i.i.d. standard Gaussian entries is an approximate isometry *on a finite set of points*. Here we need a linear mapping that is an approximate isometry for *all* vectors in $\mathscr{S}_k^n$, an uncountable space.

For a subset of indices $J \subseteq [n]$ and a vector $y \in \mathbb{R}^n$, we let y_J be the vector y restricted to the entries in J, that is, the subvector $(y_j)_{j \in J}$. Fix a subset of indices $I \subseteq [n]$ of size $10k$.

We need the RIP condition (Definition 2.4.37) to hold for all $z \in \mathbb{R}^n$ with non-zero entries in I (and all such I). The way to achieve this is to use an ε-net argument, as described in Section 2.4.4. Indeed, notice that, for $z \neq \mathbf{0}$, the function $\|Lz\|_2 / \|z\|_2$

1. does not depend on the norm of z, so that we can restrict ourselves to the compact set $\partial B_I := \{z : z_{[n] \setminus I} = \mathbf{0}, \|z\|_2 = 1\}$, and

2. is continuous on ∂B_I, so that it suffices to construct a fine enough covering of ∂B_I by a finite collection of balls (i.e., an ε-net) and apply Lemma 2.4.34 to the centers of those balls.

Proof of Lemma 2.4.38 Let $I \subseteq [n]$ be a subset of indices of size $k' := 10k$. There are $\binom{n}{k'} \leq n^{k'} = \exp(k' \log n)$ such subsets and we denote their collection by $\mathcal{I}(k', n)$. We let N_I be an ε-net of ∂B_I. By Claim 2.4.26, we can choose one of size at most $(3/\varepsilon)^{k'}$. We take

$$\varepsilon = \frac{1}{C' \sqrt{6n \log n}}$$

for a constant C' that will be determined below. The reason for this choice will become clear when we set C'. The union of all ε-nets has size

$$\left| \cup_{I \in \mathcal{I}(k', n)} N_I \right| \leq n^{k'} \left(\frac{3}{\varepsilon} \right)^{k'} \leq \exp(C'' k' \log n)$$

for some $C'' > 0$. Our goal is to show that

$$\sup_{z \in \cup_{I \in \mathcal{I}(k', n)} \partial B_I} |\|Lz\|_2 - 1| \leq \frac{1}{3}. \tag{2.4.44}$$

We seek to apply the inequality (2.4.32).

Applying Johnson–Lindenstrauss to the ε-nets: The first step is to control the supremum in (2.4.44) – restricted to the ε-nets. Lemma 2.4.34 is exactly what we need for this. Take $\theta = 1/6$, $\delta = 1/(2n| \cup_I N_I|)$, and

$$d = \Theta \left(\theta^{-2} \log(2n| \cup_I N_I|) \right) = \Theta(k' \log n),$$

as required by the lemma. Then, by a union bound over the N_Is, with probability $1 - 1/(2n)$, we have

$$\sup_{z \in \cup_I N_I} |\|Lz\|_2 - 1| \leq \frac{1}{6}. \tag{2.4.45}$$

Lipschitz continuity: The next step is to establish Lipschitz continuity of $|\|Lz\|_2 - 1|$. For vectors $y, z \in \mathbb{R}^n$, by repeated applications of the triangle inequality, we have

$$\left| |\|Lz\|_2 - 1| - |\|Ly\|_2 - 1| \right| \leq |\|Lz\|_2 - \|Ly\|_2| \leq \|L(z - y)\|_2.$$

To bound the rightmost expression, we let A_* be the largest entry of A in absolute value and note that

$$\|L(z - y)\|_2^2 = \sum_{i=1}^{d} \left(\sum_{j=1}^{n} L_{ij}(z_j - y_j) \right)^2$$

$$\leq \sum_{i=1}^{d} \left(\sum_{j=1}^{n} L_{ij}^2 \right) \left(\sum_{j=1}^{n} (z_j - y_j)^2 \right)$$

$$\leq dn \left(\frac{1}{\sqrt{d}} A_* \right)^2 \|z - y\|_2^2$$

$$\leq n A_*^2 \|z - y\|_2^2,$$

where we used Cauchy–Schwarz (Theorem B.4.8) on the second line. Taking the square root, we see that the (random) Lipschitz constant of $|\|Lz\|_2 - 1|$ (with respect to the Euclidean metric) is at most $K := \sqrt{n} A_*$.

Controlling the Lipschitz constant: So it remains to control A_*. For this we use the Chernoff–Cramér bound for Gaussians (see (2.4.4)), which implies by a union bound over the entries of A that

$$\mathbb{P}[A_* \geq C' \sqrt{\log n}] \leq \mathbb{P}\left[\exists i, j, \ |A_{ij}| \geq C' \sqrt{\log n} \right]$$

$$\leq n^2 \exp \left(-\frac{(C' \sqrt{\log n})^2}{2} \right)$$

$$\leq \frac{1}{2n}$$

for a $C' > 0$ large enough. Hence, with probability $1 - 1/(2n)$, we have $A_* < C' \sqrt{\log n}$ and

$$K\varepsilon \leq \frac{1}{6} \tag{2.4.46}$$

by the choice of ε made previously.

Putting everything together: We apply (2.4.32). Combining (2.4.45) and (2.4.46), with probability $1 - 1/n$, the claim (2.4.44) holds. That concludes the proof. ∎

ℓ^1 **minimization** Finally we prove Lemma 2.4.39 (which can be skipped).

Proof of Lemma 2.4.39 Let z^* be a solution to (2.4.43) and note that such a solution exists because $z = x$ satisfies the constraint. Without loss of generality assume that only the first k entries of x are non-zero, that is, $x_{[n]\setminus[k]} = \mathbf{0}$. Moreover, order the remaining entries of x so that the residual $r = z^* - x$ has its entries $r_{[n]\setminus[k]}$ in non-increasing order in absolute value. Our goal is to show that $\|r\|_2 = 0$.

In order to leverage the RIP condition, we break up the vector r into $9k$-long subvectors. Let

$$I_0 = [k], \quad I_i = \{(9(i-1)+1)k+1, \ldots, (9i+1)k\} \ \forall i \geq 1,$$

and $\bar{I}_i = \bigcup_{j>i} I_j$. We will also need $I_{01} = I_0 \cup I_1$ and $\bar{I}_{01} = \bar{I}_1$.

We first use the optimality of z^*. Note that $x_{\bar{I}_0} = 0$ implies that

$$\|z^*\|_1 = \|z_{I_0}^*\|_1 + \|z_{\bar{I}_0}^*\|_1 = \|z_{I_0}^*\|_1 + \|r_{\bar{I}_0}\|_1$$

and

$$\|\boldsymbol{x}\|_1 = \|\boldsymbol{x}_{I_0}\|_1 \leq \|\boldsymbol{z}_{I_0}^*\|_1 + \|\boldsymbol{r}_{I_0}\|_1$$

by the triangle inequality. Since $\|\boldsymbol{z}^*\|_1 \leq \|\boldsymbol{x}\|_1$ by optimality (and the fact that $\boldsymbol{x}$ satisfies the constraint), we then have

$$\|\boldsymbol{r}_{\bar{I}_0}\|_1 \leq \|\boldsymbol{r}_{I_0}\|_1. \tag{2.4.47}$$

On the other hand, the RIP condition gives a similar inequality in the other direction. Indeed, notice that $L\boldsymbol{r} = \boldsymbol{0}$ by the constraint in (2.4.43) or, put differently, $L\boldsymbol{r}_{I_{01}} = -\sum_{i \geq 2} L\boldsymbol{r}_{I_i}$. Then, by the RIP condition and the triangle inequality, we have

$$\frac{2}{3}\|\boldsymbol{r}_{I_{01}}\|_2 \leq \|L\boldsymbol{r}_{I_{01}}\|_2 \leq \sum_{i \geq 2} \|L\boldsymbol{r}_{I_i}\|_2 \leq \frac{4}{3} \sum_{i \geq 2} \|\boldsymbol{r}_{I_i}\|_2, \tag{2.4.48}$$

where we used the fact that by construction $\boldsymbol{r}_{I_{01}}$ is $10k$-sparse and each $\boldsymbol{r}_{I_i}$ is $9k$-sparse.

We note that by the ordering of the entries of $\boldsymbol{x}$,

$$\|\boldsymbol{r}_{I_{i+1}}\|_2^2 \leq 9k \left(\frac{\|\boldsymbol{r}_{I_i}\|_1}{9k}\right)^2 = \frac{\|\boldsymbol{r}_{I_i}\|_1^2}{9k}, \tag{2.4.49}$$

where we bounded $\boldsymbol{r}_{I_{i+1}}$ entrywise by the expression in parenthesis. Combining (2.4.47) and (2.4.49), and using that $\|\boldsymbol{r}_{I_0}\|_1 \leq \sqrt{k}\|\boldsymbol{r}_{I_0}\|_2$ by Cauchy–Schwarz, we have

$$\sum_{i \geq 2} \|\boldsymbol{r}_{I_i}\|_2 \leq \sum_{j \geq 1} \frac{\|\boldsymbol{r}_{I_j}\|_1}{\sqrt{9k}} = \frac{\|\boldsymbol{r}_{\bar{I}_0}\|_1}{3\sqrt{k}} \leq \frac{\|\boldsymbol{r}_{I_0}\|_1}{3\sqrt{k}} \leq \frac{\|\boldsymbol{r}_{I_0}\|_2}{3} \leq \frac{\|\boldsymbol{r}_{I_{01}}\|_2}{3}.$$

Plugging this back into (2.4.48) gives

$$\|\boldsymbol{r}_{I_{01}}\|_2 \leq 2 \sum_{i \geq 2} \|\boldsymbol{r}_{I_i}\|_2 \leq \frac{2}{3}\|\boldsymbol{r}_{I_{01}}\|_2,$$

which implies $\boldsymbol{r}_{I_{01}} = \boldsymbol{0}$. In particular, $\boldsymbol{r}_{I_0} = \boldsymbol{0}$ and, by (2.4.47), $\boldsymbol{r}_{\bar{I}_0} = \boldsymbol{0}$ as well. We have shown that $\boldsymbol{r} = \boldsymbol{0}$. Or, in other words, $\boldsymbol{z}^* = \boldsymbol{x}$. ∎

Remark 2.4.41 *Lemma 2.4.39 can be extended to noisy measurements using a modification of (2.4.43). This provides some robustness to noise which is important in applications. See [CRT06b].*

2.4.6 ▷ Data Science: Classification, Empirical Risk Minimization, and VC Dimension

In the *binary classification* problem, one is given samples $\mathcal{S}_n = \{(X_i, C(X_i))\}_{i=1}^n$ where $X_i \in$ BINARY $\mathbb{R}^d$ is a feature vector and $C(X_i) \in \{0, 1\}$ is a label. The feature vectors are assumed to be CLASSIFICATION independent samples from an unknown probability measure μ, and $C \colon \mathbb{R}^d \to \{0, 1\}$ is a measurable Boolean function. For instance, the feature vector might be an image (encoded as a vector) and the label might indicate "cat" (label 0) or "dog" (label 1). Our goal is to learn the function (or concept) C from the samples.

More precisely, we seek to construct a *hypothesis* $h \colon \mathbb{R}^d \to \{0, 1\}$ that is a good approx- HYPOTHESIS imation to C in the sense that it predicts the label well on a new sample (from the same distribution). Formally, we want h to have small *true risk* (or *generalization error*): TRUE RISK

$$R(h) = \mathbb{P}[h(X) \neq C(X)],$$

where $X \sim \mu$. Because we only have access to the distribution μ through the samples, it is natural to estimate the true risk of the hypothesis h using the samples as

$$R_n(h) = \frac{1}{n} \sum_{i=1}^{n} \mathbf{1}\{h(X_i) \neq C(X_i)\},$$

EMPIRICAL RISK

EMPIRICAL RISK MINIMIZER

which is called the *empirical risk*. Indeed, observe that $\mathbb{E}R_n(h) = R(h)$ and, by the law of large numbers, $R_n(h) \to R(h)$ almost surely as $n \to +\infty$. Ignoring computational considerations, one can then formally define an *empirical risk minimizer*

$$h^* \in \text{ERM}_{\mathcal{H}}(\mathcal{S}_n) = \{h \in \mathcal{H}: R_n(h) \leq R_n(h'), \forall h' \in \mathcal{H}\},$$

HYPOTHESIS CLASS

where $\mathcal{H}$, the *hypothesis class*, is a given collection of Boolean functions over $\mathbb{R}^d$. We assume further that h^* can be defined as a measurable function of the samples.

Overfitting Why restrict the hypothesis class? It turns out that minimizing the empirical risk over *all* Boolean functions makes it impossible to achieve an arbitrarily small risk. Intuitively considering too rich a class of functions, that is, functions that too intricately follow the data, leads to overfitting: the learned hypothesis will fit the sampled data, but it may not generalize well to unseen examples. A *learner* $\mathcal{A}$ is a map from samples to measurable Boolean functions over $\mathbb{R}^d$, that is, for any n and any $\mathcal{S}_n \in (\mathbb{R}^d \times \{0,1\})^n$, the learner outputs a function $\mathcal{A}(\cdot, \mathcal{S}_n): \mathbb{R}^d \to \{0,1\}$. The following theorem shows that any learner has fundamental limitations if all concepts are possible.

LEARNER

Theorem 2.4.42 (No free lunch). *For any learner $\mathcal{A}$ and any finite $\mathcal{X} \subseteq \mathbb{R}^d$ of even size $|\mathcal{X}| =: 2m > 4$, there exist a concept $C: \mathcal{X} \to \{0,1\}$ and a distribution μ over $\mathcal{X}$ such that*

$$\mathbb{P}[R(\mathcal{A}(\cdot, \mathcal{S}_m)) \geq 1/8] \geq 1/8, \qquad (2.4.50)$$

where $\mathcal{S}_m = \{(X_i, C(X_i))\}_{i=1}^{m}$ with independent $X_i \sim \mu$.

The gist of the proof is intuitive. In essence, if the target concept is arbitrary and we only get to see half of the possible instances, then we have learned nothing about the other half and cannot expect low generalization error.

Proof of Theorem 2.4.42   We let μ be uniform over $\mathcal{X}$. To prove the existence of a concept satisfying (2.4.50), we use the probabilistic method (Section 2.2.1) and pick C at random. For each $x \in \mathcal{X}$, we set $C(x) := Y_x$, where the Y_xs are i.i.d. uniform in $\{0,1\}$.

We first bound $\mathbb{E}[R(\mathcal{A}(\cdot, \mathcal{S}_m))]$, where the expectation runs over both random labels $\{Y_x\}_{x \in \mathcal{X}}$ and the samples $\mathcal{S}_m = \{(X_i, C(X_i))\}_{i=1}^{m}$. For an additional independent sample $X \sim \mu$, we will need the event that the learner, given samples $\mathcal{S}_m$, makes an incorrect prediction on X

$$B = \{\mathcal{A}(X, \mathcal{S}_m)) \neq Y_X\},$$

and the event that X is observed in the samples $\mathcal{S}_m$

$$O = \{X \in \{X_1, \ldots, X_m\}\}.$$

By the tower property (Lemma B.6.16),

$$
\begin{aligned}
\mathbb{E}[R(\mathcal{A}(\,\cdot\,,\mathcal{S}_m))] &= \mathbb{P}[B] \\
&= \mathbb{E}[\mathbb{P}[B \mid \mathcal{S}_m]] \\
&= \mathbb{E}\left[\mathbb{P}[B \mid O,\mathcal{S}_m]\mathbb{P}[O \mid \mathcal{S}_m] + \mathbb{P}[B \mid O^c,\mathcal{S}_m]\mathbb{P}[O^c \mid \mathcal{S}_m]\right] \\
&\geq \mathbb{E}\left[\mathbb{P}[B \mid O^c,\mathcal{S}_m]\mathbb{P}[O^c \mid \mathcal{S}_m]\right] \\
&\geq \frac{1}{2} \times \frac{1}{2},
\end{aligned}
$$

where we used that

- $\mathbb{P}[O^c \mid \mathcal{S}_m] \geq 1/2$ because $|\mathcal{X}| = 2m$ and μ is uniform, and
- $\mathbb{P}[B \mid O^c,\mathcal{S}_m] = 1/2$ because for any $x \notin \{X_1,\ldots,X_m\}$ the prediction $\mathcal{A}(x,\mathcal{S}_m) \in \{0,1\}$ is independent of Y_x and the latter is uniform.

Conditioning over the concept, we have proved that

$$
\mathbb{E}\left[\mathbb{E}[R(\mathcal{A}(\,\cdot\,,\mathcal{S}_m)) \mid \{Y_x\}_{x\in\mathcal{X}}]\right] \geq \frac{1}{4}.
$$

Hence, by the first moment principle (Theorem 2.2.1),

$$
\mathbb{P}[\mathbb{E}[R(\mathcal{A}(\,\cdot\,,\mathcal{S}_m)) \mid \{Y_x\}_{x\in\mathcal{X}}] \geq 1/4] > 0,
$$

where the probability is taken over $\{Y_x\}_{x\in\mathcal{X}}$. That is, there exists a choice $\{y_x\}_{x\in\mathcal{X}} \in \{0,1\}^{\mathcal{X}}$ such that

$$
\mathbb{E}[R(\mathcal{A}(\,\cdot\,,\mathcal{S}_m)) \mid \{Y_x = y_x\}_{x\in\mathcal{X}}] \geq 1/4. \tag{2.4.51}
$$

Finally, to prove (2.4.50), we use a variation on Markov's inequality (Theorem 2.1.1) for $[0,1]$-valued random variables. If $Z \in [0,1]$ is a random variable with $\mathbb{E}[Z] = \mu$ and $\alpha \in [0,1]$, then

$$
\mathbb{E}[Z] \leq \alpha \times \mathbb{P}[Z < \alpha] + 1 \times \mathbb{P}[Z \geq \alpha] \leq \mathbb{P}[Z \geq \alpha] + \alpha.
$$

Taking $\alpha = \mu/2$ gives

$$
\mathbb{P}[Z \geq \mu/2] \geq \mu/2.
$$

Going back to (2.4.51), we obtain

$$
\mathbb{P}\left[R(\mathcal{A}(\,\cdot\,,\mathcal{S}_m)) \geq \frac{1}{8} \,\middle|\, \{Y_x = y_x\}_{x\in\mathcal{X}}\right] \geq \frac{1}{8},
$$

establishing the claim. ∎

The way out is to "limit the complexity" of the hypotheses. For instance, we could restrict ourselves to half-spaces

$$
\mathcal{H}_{\mathrm{H}} = \left\{h(x) = \mathbf{1}\{x^T u \geq \alpha\} : u \in \mathbb{R}^d, \alpha \in \mathbb{R}\right\},
$$

or axis-aligned boxes

$$
\mathcal{H}_{\mathrm{B}} = \{h(x) = \mathbf{1}\{x_i \in [\alpha_i,\beta_i],\ \forall i\} : -\infty \leq \alpha_i \leq \beta_i \leq \infty,\ \forall i\}.
$$

In order for the empirical risk minimizer h^* to have a generalization error close to the best achievable error, we need the empirical risk of the learned hypothesis $R_n(h^*)$ to be close to its expectation $R(h^*)$, which is guaranteed by the law of large numbers for sufficiently large n. But that is not enough, we also need that same property to hold *for all hypotheses in $\mathcal{H}$ simultaneously*. Otherwise we could be fooled by a poorly performing hypothesis with unusually good empirical risk on the samples. The hypothesis class is typically infinite and, therefore, controlling empirical risk deviations from their expectations uniformly over $\mathcal{H}$ is not straightforward.

Uniform deviations Our goal in this section is to show how to bound

$$\mathbb{E}\left[\sup_{h\in\mathcal{H}}\{R_n(h) - R(h)\}\right] = \mathbb{E}\left[\sup_{h\in\mathcal{H}}\left\{\frac{1}{n}\sum_{i=1}^{n}\ell(h,X_i) - \mathbb{E}[\ell(h,X)]\right\}\right] \qquad (2.4.52)$$

in terms of a measure of complexity of the class $\mathcal{H}$, where we defined the loss $\ell(h,x) = \mathbf{1}\{h(x) \neq C(x)\}$ to simplify the notation. We assume that $\mathcal{H}$ is countable. (Observe for instance that, for $\mathcal{H}_\mathrm{H}$ and $\mathcal{H}_\mathrm{B}$, nothing is lost by assuming that the parameters defining the hypotheses are rational-valued.)

Controlling deviations uniformly over $\mathcal{H}$ as in (2.4.52) allows one to provide guarantees on the empirical risk minimizer. Indeed, for any $h' \in \mathcal{H}$,

$$\begin{aligned}
R(h^*) &= R_n(h^*) + \{R(h^*) - R_n(h^*)\} \\
&\leq R_n(h^*) + \sup_{h\in\mathcal{H}}\{R(h) - R_n(h)\} \\
&\leq R_n(h') + \sup_{h\in\mathcal{H}}\{R(h) - R_n(h)\} \\
&= R(h') + \{R_n(h') - R(h')\} + \sup_{h\in\mathcal{H}}\{R(h) - R_n(h)\} \\
&\leq R(h') + \sup_{h\in\mathcal{H}}\{R_n(h) - R(h)\} + \sup_{h\in\mathcal{H}}\{R(h) - R_n(h)\},
\end{aligned}$$

where, on the third line, we used the definition of the empirical risk minimizer. Taking an infimum over h', then an expectation over the samples, and rearranging gives

$$\begin{aligned}
\mathbb{E}[R(h^*)] &- \inf_{h'\in\mathcal{H}} R(h') \\
&\leq \mathbb{E}\left[\sup_{h\in\mathcal{H}}\{R_n(h) - R(h)\}\right] + \mathbb{E}\left[\sup_{h\in\mathcal{H}}\{R(h) - R_n(h)\}\right]. \qquad (2.4.53)
\end{aligned}$$

This inequality allows us to relate two quantities of interest: the expected true risk of the empirical risk minimizer (i.e., $\mathbb{E}[R(h^*)]$) and the best possible true risk (i.e., $\inf_{h'\in\mathcal{H}} R(h')$). The first term on the right-hand side is (2.4.52) and the second one can be bounded in a similar fashion as we argue below. Observe that the suprema are inside the expectations and that the random variables $R_n(h) - R(h)$ are highly correlated. Indeed, two similar hypotheses will produce similar predictions. While the absence of independence in some sense makes bounding this type expectation harder, the correlation is ultimately what allows us to tackle infinite classes $\mathcal{H}$.

To bound (2.4.52), we use the methods of Section 2.4.4. As a first step, we apply the symmetrization trick, which we introduced in Section 2.4.2 to give a proof of Hoeffding's lemma (Lemma 2.4.12). Let $(\varepsilon_i)_{i=1}^{n}$ be i.i.d. uniform random variables in $\{-1, +1\}$ (i.e., Rademacher variables) and let $(X_i')_{i=1}^{n}$ be an independent copy of $(X_i)_{i=1}^{n}$. Then,

$$\mathbb{E}\left[\sup_{h\in\mathcal{H}}\{R_n(h)-R(h)\}\right]$$

$$=\mathbb{E}\left[\sup_{h\in\mathcal{H}}\left\{\frac{1}{n}\sum_{i=1}^{n}\ell(h,X_i)-\mathbb{E}[\ell(h,X)]\right\}\right]$$

$$=\mathbb{E}\left[\sup_{h\in\mathcal{H}}\left\{\frac{1}{n}\sum_{i=1}^{n}[\ell(h,X_i)-\mathbb{E}[\ell(h,X_i')\,|\,(X_j)_{j=1}^{n}]]\right\}\right]$$

$$=\mathbb{E}\left[\sup_{h\in\mathcal{H}}\mathbb{E}\left[\frac{1}{n}\sum_{i=1}^{n}[\ell(h,X_i)-\ell(h,X_i')]\,\Bigg|\,(X_j)_{j=1}^{n}\right]\right]$$

$$\leq\mathbb{E}\left[\sup_{h\in\mathcal{H}}\left\{\frac{1}{n}\sum_{i=1}^{n}[\ell(h,X_i)-\ell(h,X_i')]\right\}\right],$$

where on the fourth line we used taking it out what is known (Lemma B.6.13) and on the fifth line we used $\sup_h \mathbb{E}Y_h \leq \mathbb{E}[\sup_h Y_h]$ and the tower property. Next we note that $\ell(h,X_i) - \ell(h,X_i')$ is symmetric and independent of ε_i (which is also symmetric) to deduce that the last line above is

$$=\mathbb{E}\left[\sup_{h\in\mathcal{H}}\left\{\frac{1}{n}\sum_{i=1}^{n}\varepsilon_i[\ell(h,X_i)-\ell(h,X_i')]\right\}\right]$$

$$\leq\mathbb{E}\left[\sup_{h\in\mathcal{H}}\frac{1}{n}\sum_{i=1}^{n}\varepsilon_i\ell(h,X_i)+\sup_{h\in\mathcal{H}}\frac{1}{n}\sum_{i=1}^{n}(-\varepsilon_i)\ell(h,X_i')\right]$$

$$=2\,\mathbb{E}\left[\sup_{h\in\mathcal{H}}\frac{1}{n}\sum_{i=1}^{n}\varepsilon_i\ell(h,X_i)\right].$$

The exact same argument also applies to the second term on the right-hand side of (2.4.53), so

$$\mathbb{E}[R(h^*)]-\inf_{h'\in\mathcal{H}}R(h')\leq 4\,\mathbb{E}\left[\sup_{h\in\mathcal{H}}\frac{1}{n}\sum_{i=1}^{n}\varepsilon_i\ell(h,X_i)\right]. \tag{2.4.54}$$

Changing the normalization slightly, we define the process

$$Z_n(h)=\frac{1}{\sqrt{n}}\sum_{i=1}^{n}\varepsilon_i\ell(h,X_i),\quad h\in\mathcal{H}. \tag{2.4.55}$$

Our task reduces to upper bounding

$$\mathbb{E}\left[\sup_{h\in\mathcal{H}}Z_n(h)\right]. \tag{2.4.56}$$

Note that we will not compute the best possible true risk (which in general could be "bad," i.e., large) – only how close the empirical risk minimizer gets to it.

VC dimension We make two observations about $Z_n(h)$.

1. It is centered. Also, as a weighted sum of independent random variables in $[-1,1]$, it is sub-Gaussian with variance factor 1 by the general Hoeffding inequality (Theorem 2.4.9) and Hoeffding's lemma (Lemma 2.4.12).

2. It depends only on the values of the hypothesis h at a finite number of points, $X_1, \ldots, X_n$. Hence, while the supremum in (2.4.56) is over a potentially infinite class of functions $\mathcal{H}$, it is in effect a supremum over at most 2^n functions, that is, all the possible restrictions of the hs to $(X_i)_{i=1}^n$.

A naive application of the maximal inequality in Lemma 2.4.21, together with the two observations above, gives

$$\mathbb{E}\left[\sup_{h \in \mathcal{H}} Z_n(h)\right] \leq \sqrt{2 \log 2^n} = \sqrt{2n \log 2}.$$

Unfortunately, plugging this back into (2.4.54) gives an upper bound, which fails to converge to 0 as $n \to +\infty$.

To obtain a better bound, we show that in general the number of distinct restrictions of $\mathcal{H}$ to n points can grow much slower than 2^n.

Definition 2.4.43 (Shattering). *Let $\Lambda = \{\ell_1, \ldots, \ell_n\} \subseteq \mathbb{R}^d$ be a finite set and let $\mathcal{H}$ be a class of Boolean functions on $\mathbb{R}^d$. The restriction of $\mathcal{H}$ to Λ is*

$$\mathcal{H}_\Lambda = \{(h(\ell_1), \ldots, h(\ell_n)) : h \in \mathcal{H}\}.$$

We say that Λ is shattered *by $\mathcal{H}$ if $|\mathcal{H}_\Lambda| = 2^{|\Lambda|}$, that is, if all Boolean functions over Λ can be obtained by restricting a function in $\mathcal{H}$ to the points in Λ.*

Definition 2.4.44 (VC dimension). *Let $\mathcal{H}$ be a class of Boolean functions on $\mathbb{R}^d$. The* VC *dimension of $\mathcal{H}$, denoted $\mathrm{vc}(\mathcal{H})$, is the maximum cardinality of a set shattered by $\mathcal{H}$.*

We prove the following combinatorial lemma at the end of this section.

Lemma 2.4.45 (Sauer's lemma). *Let $\mathcal{H}$ be a class of Boolean functions on $\mathbb{R}^d$. For any finite set $\Lambda = \{\ell_1, \ldots, \ell_n\} \subseteq \mathbb{R}^d$,*

$$|\mathcal{H}_\Lambda| \leq \left(\frac{en}{\mathrm{vc}(\mathcal{H})}\right)^{\mathrm{vc}(\mathcal{H})}.$$

That is, the number of distinct restrictions of $\mathcal{H}$ to any n points grows at most as $\propto n^{\mathrm{vc}(\mathcal{H})}$.

Returning to $\mathbb{E}[\sup_{h \in \mathcal{H}} Z_n(h)]$, we get the following inequality.

Lemma 2.4.46 *There exists a constant $0 < C < +\infty$ such that, for any countable class of measurable Boolean functions $\mathcal{H}$ over $\mathbb{R}^d$,*

$$\mathbb{E}\left[\sup_{h \in \mathcal{H}} Z_n(h)\right] \leq C\sqrt{\mathrm{vc}(\mathcal{H}) \log n}. \tag{2.4.57}$$

Proof Recall that $Z_n(h) \in s\mathcal{G}(1)$. Since the supremum over $\mathcal{H}$, when seen as restricted to $\{X_1, \ldots, X_n\}$, is in fact a supremum over at most $\left(\frac{en}{\mathrm{vc}(\mathcal{H})}\right)^{\mathrm{vc}(\mathcal{H})}$ functions by Sauer's lemma (Lemma 2.4.45), we have by Lemma 2.4.21,

$$\mathbb{E}\left[\sup_{h \in \mathcal{H}} Z_n(h)\right] \leq \sqrt{2 \log\left[\left(\frac{en}{\mathrm{vc}(\mathcal{H})}\right)^{\mathrm{vc}(\mathcal{H})}\right]}.$$

That proves the claim. ∎

Returning to (2.4.54), the previous lemma finally implies

$$\mathbb{E}[R(h^*)] - \inf_{h' \in \mathcal{H}} R(h') \leq 4C\sqrt{\frac{\text{vc}(\mathcal{H}) \log n}{n}}.$$

For hypothesis classes with finite VC dimension, the bound goes to 0 as $n \to +\infty$.
We give some examples.

Example 2.4.47 (VC dimension of half-spaces). Consider the class of half-spaces.

Claim 2.4.48

$$\text{vc}(\mathcal{H}_{\text{H}}) = d + 1.$$

We only prove the case $d = 1$, where $\mathcal{H}_{\text{H}}$ reduces to half-lines $(-\infty, \gamma]$ or $[\gamma, +\infty)$. Clearly, any set $\Lambda = \{\ell_1, \ell_2\} \subseteq \mathbb{R}$ with elements is shattered by $\mathcal{H}_{\text{H}}$. On the other hand, for any $\Lambda = \{\ell_1, \ell_2, \ell_3\}$ with $\ell_1 < \ell_2 < \ell_3$, any half-line containing ℓ_1 and ℓ_3 necessarily includes ℓ_2 as well. Hence, no set of size 3 is shattered by $\mathcal{H}_{\text{H}}$. ◄

Example 2.4.49 (VC dimension of boxes). Consider the class of axis-aligned boxes.

Claim 2.4.50

$$\text{vc}(\mathcal{H}_{\text{B}}) = 2d.$$

We only prove the case $d = 2$, where $\mathcal{H}_{\text{B}}$ reduces to rectangles. The four-point set $\Lambda = \{(-1, 0), (1, 0), (0, -1), (0, 1)\}$ is shattered by $\mathcal{H}_{\text{B}}$. Indeed, the rectangle $[-1, 1] \times [-1, 1]$ contains Λ, with each side of the rectangle containing one of the points. Moving any side inward by $\varepsilon < 1$ removes the corresponding point from the rectangle without affecting the other ones. Hence, any subset of Λ can be obtained by this procedure.

On the other hand, let $\Lambda = \{\ell_1, \ldots, \ell_5\} \subseteq \mathbb{R}^2$ be any set of five distinct points. If the points all lie on the same axis-aligned line, then an argument similar to the half-line case in Claim 2.4.48 shows that Λ is not shattered. Otherwise consider the axis-aligned rectangle with smallest area containing Λ. For each side of the rectangle, choose one point of Λ that lies on it. These necessarily exist (otherwise the rectangle could be made even smaller) and denote them by x_{N} for the highest, x_{E} for the rightmost, x_{S} for the lowest, and x_{W} for the leftmost. Note that they may not be distinct, but in any case at least one point in Λ, say ℓ_5 without loss of generality, is not in the list. Now observe that any axis-aligned rectangle containing $x_{\text{N}}, x_{\text{E}}, x_{\text{S}}, x_{\text{W}}$ must also contain ℓ_5 since its coordinates are sandwiched between the bounds defined by those points. Hence, no set of size 5 is shattered. That proves the claim. ◄

These two examples also provide insights into Sauer's lemma. Consider the case of rectangles for instance. Over a collection of n sample points, a rectangle defines the same $\{0, 1\}$-labeling as the *minimal-area rectangle containing the same points*. Because each side of a minimal-area rectangle must touch at least one point in the sample, there are at most n^4 such rectangles, and hence there are at most $n^4 \ll 2^n$ restrictions of $\mathcal{H}_{\text{B}}$ to these sample points.

Application of chaining It turns out that the $\sqrt{\log n}$ factor in (2.4.57) is not optimal. We use chaining (Section 2.4.4) to improve the bound.

We claim that the process $\{Z_n(h)\}_{h \in \mathcal{H}}$ has sub-Gaussian increments under an appropriately defined pseudometric. Indeed, conditioning on $(X_i)_{i=1}^n$, by the general Hoeffding inequality

(Theorem 2.4.9) and Hoeffding's lemma (Lemma 2.4.12), we have that the increment (as a function of the ε_is which have variance factor 1)

$$Z_n(g) - Z_n(h) = \sum_{i=1}^{n} \varepsilon_i \frac{\ell(g, X_i) - \ell(h, X_i)}{\sqrt{n}}$$

is sub-Gaussian with variance factor

$$\sum_{i=1}^{n} \left(\frac{\ell(g, X_i) - \ell(h, X_i)}{\sqrt{n}} \right)^2 \times 1 = \frac{1}{n} \sum_{i=1}^{n} [\ell(g, X_i) - \ell(h, X_i)]^2.$$

Define the pseudometric

$$\rho_n(g, h) = \left[\frac{1}{n} \sum_{i=1}^{n} [\ell(g, X_i) - \ell(h, X_i)]^2 \right]^{1/2} = \left[\frac{1}{n} \sum_{i=1}^{n} [g(X_i) - h(X_i)]^2 \right]^{1/2},$$

where we used that $\ell(h, x) = \mathbf{1}\{h(x) \neq C(x)\}$ by definition. It satisfies the triangle inequality since it can be expressed as a Euclidean norm. In fact, it will be useful to recast it in a more general setting. For a probability measure η over $\mathbb{R}^d$, define

$$\|g - h\|_{L^2(\eta)}^2 = \int_{\mathbb{R}^d} (f(x) - g(x))^2 d\eta(x).$$

Let μ_n be the empirical measure

$$\mu_n = \mu_{(X_i)_{i=1}^n} := \frac{1}{n} \sum_{i=1}^{n} \delta_{X_i}, \qquad (2.4.58)$$

where δ_x is the probability measure that puts mass 1 on x. Then, we can rewrite

$$\rho_n(g, h) = \|g - h\|_{L^2(\mu_n)}.$$

Hence we have shown that, conditioned on the samples, the process $\{Z_n(h)\}_{h \in \mathcal{H}}$ has sub-Gaussian increments with respect to $\| \cdot \|_{L^2(\mu_n)}$. Note that the pseudometric here is *random* as it depends on the samples. Though, by the law of large numbers, $\|g - h\|_{L^2(\mu_n)}$ approaches its expectation, $\|g - h\|_{L^2(\mu)}$, as $n \to +\infty$.

Applying the discrete Dudley inequality (Theorem 2.4.31), we obtain the following bound.

Lemma 2.4.51 *There exists a constant $0 < C < +\infty$ such that, for any countable class of measurable Boolean functions $\mathcal{H}$ over $\mathbb{R}^d$,*

$$\mathbb{E} \left[\sup_{h \in \mathcal{H}} Z_n(h) \right] \leq C \mathbb{E} \left[\sum_{k=0}^{+\infty} 2^{-k} \sqrt{\log \mathcal{N}(\mathcal{H}, \| \cdot \|_{L^2(\mu_n)}, 2^{-k})} \right],$$

where μ_n is the empirical measure over the samples $(X_i)_{i=1}^n$.

Proof Because $\mathcal{H}$ comprises only Boolean functions, it follows that under the pseudometric $\| \cdot \|_{L^2(\mu_n)}$ the diameter is bounded by 1. We apply the discrete Dudley inequality conditioned on $(X_i)_{i=1}^n$. Then we take an expectation over the samples. ∎

Our use of the symmetrization trick is more intuitive than it may have appeared at first. The central limit theorem indicates that the fluctuations of centered averages such as

$$(R_n(g) - R(g)) - (R_n(h) - R(h))$$

tend to cancel out and that, in the limit, the variance alone characterizes the overall behavior. The ε_is in some sense explicitly capture the canceling part of this phenomenon, while ρ_n captures the scale of the resulting global fluctuations in the increments.

Our final task is to bound the covering numbers $\mathcal{N}(\mathcal{H}, \| \cdot \|_{L^2(\mu_n)}, 2^{-k})$.

Theorem 2.4.52 (Covering numbers via VC dimension). *There exists a constant $0 < C < +\infty$ such that, for any class of measurable Boolean functions $\mathcal{H}$ over $\mathbb{R}^d$, any probability measure η over $\mathbb{R}^d$, and any $\varepsilon \in (0, 1)$,*

$$\mathcal{N}(\mathcal{H}, \| \cdot \|_{L^2(\eta)}, \varepsilon) \leq \left(\frac{2}{\varepsilon} \right)^{C \, \mathrm{vc}(\mathcal{H})}.$$

Before proving Theorem 2.4.52, we derive its implications for uniform deviations. Compare the following bound to Lemma 2.4.46.

Lemma 2.4.53 *There exists a constant $0 < C < +\infty$ such that, for any countable class of measurable Boolean functions $\mathcal{H}$ over $\mathbb{R}^d$,*

$$\mathbb{E} \left[\sup_{h \in \mathcal{H}} Z_n(h) \right] \leq C \sqrt{\mathrm{vc}(\mathcal{H})}.$$

Proof By Lemma 2.4.51,

$$\mathbb{E} \left[\sup_{h \in \mathcal{H}} Z_n(h) \right] \leq C \, \mathbb{E} \left[\sum_{k=0}^{+\infty} 2^{-k} \sqrt{\log \mathcal{N}(\mathcal{H}, \| \cdot \|_{L^2(\mu_n)}, 2^{-k})} \right]$$

$$\leq C \, \mathbb{E} \left[\sum_{k=0}^{+\infty} 2^{-k} \sqrt{\log \left(\frac{2}{2^{-k}} \right)^{C' \, \mathrm{vc}(\mathcal{H})}} \right]$$

$$= C \sqrt{\mathrm{vc}(\mathcal{H})} \, \mathbb{E} \left[\sum_{k=0}^{+\infty} 2^{-k} \sqrt{k+1} \sqrt{C' \log 2} \right]$$

$$\leq C'' \sqrt{\mathrm{vc}(\mathcal{H})}$$

for some $0 < C'' < +\infty$. ∎

It remains to prove Theorem 2.4.52.

Proof of Theorem 2.4.52 Let $\mathcal{G} = \{g_1, \ldots, g_N\} \subseteq \mathcal{H}$ be a maximal ε-packing of $\mathcal{H}$ with $N \geq \mathcal{N}(\mathcal{H}, \| \cdot \|_{L^2(\eta)}, \varepsilon)$, which exists by Lemma 2.4.24. We use the probabilistic method (Section 2.2) and Hoeffding's inequality for bounded variables (Theorem 2.4.10) to show that there exists a small number of points $\{x_1, \ldots, x_m\}$ such that $\mathcal{G}$ is still a good packing when $\mathcal{H}$ is restricted to the x_is. Then we use Sauer's lemma (Lemma 2.4.45) to conclude.

1. **Restriction.** By construction, the collection $\mathcal{G}$ satisfies

$$\|g_i - g_j\|_{L^2(\eta)} > \varepsilon, \quad \forall i \neq j.$$

For an integer m that we will choose as small as possible below, let $\mathbb{X} = \{X_1, \ldots, X_m\}$ be i.i.d. samples from η and let $\mu_{\mathbb{X}}$ be the corresponding empirical measure (as defined in (2.4.58)). Observe that, for any $i \neq j$,

$$\mathbb{E}\left[\|g_i - g_j\|_{L^2(\mu_{\mathbb{X}})}^2\right] = \mathbb{E}\left[\frac{1}{m}\sum_{k=1}^m [g_i(X_k) - g_j(X_k)]^2\right] = \|g_i - g_j\|_{L^2(\eta)}^2.$$

Moreover, $[g_i(X_k) - g_j(X_k)]^2 \in [0, 1]$. Hence, by Hoeffding's inequality there exists a constant $0 < C < +\infty$ and an $m \leq C\varepsilon^{-4}\log N$ such that

$$\mathbb{P}\left[\|g_i - g_j\|_{L^2(\eta)}^2 - \|g_i - g_j\|_{L^2(\mu_{\mathbb{X}})}^2 \geq \frac{3\varepsilon^2}{4}\right]$$

$$= \mathbb{P}\left[m\|g_i - g_j\|_{L^2(\eta)}^2 - \sum_{k=1}^m [g_i(X_k) - g_j(X_k)]^2 \geq m\frac{3\varepsilon^2}{4}\right]$$

$$\leq \exp\left(-\frac{2(m \cdot 3\varepsilon^2/4)^2}{m}\right)$$

$$= \exp\left(-\frac{9}{8}m\varepsilon^4\right)$$

$$< \frac{1}{N^2}.$$

This implies that, for this choice of m,

$$\mathbb{P}\left[\|g_i - g_j\|_{L^2(\mu_{\mathbb{X}})} > \frac{\varepsilon}{2} \;\forall i \neq j\right] > 0,$$

where the probability is over the samples. Therefore, there must be a set $\mathcal{X} = \{x_1, \ldots, x_m\} \subseteq \mathbb{R}^d$ such that

$$\|g_i - g_j\|_{L^2(\mu_{\mathcal{X}})} > \frac{\varepsilon}{2} \;\forall i \neq j. \tag{2.4.59}$$

2. **VC bound.** In particular, by (2.4.59), the functions in $\mathcal{G}$ restricted to $\mathcal{X}$ are distinct. By Sauer's lemma (Lemma 2.4.45),

$$N = |\mathcal{G}_{\mathcal{X}}| \leq |\mathcal{H}_{\mathcal{X}}| \leq \left(\frac{em}{\text{vc}(\mathcal{H})}\right)^{\text{vc}(\mathcal{H})} \leq \left(\frac{eC\varepsilon^{-4}\log N}{\text{vc}(\mathcal{H})}\right)^{\text{vc}(\mathcal{H})}. \tag{2.4.60}$$

Using that $\frac{1}{2D}\log N = \log N^{1/2D} \leq N^{1/2D}$, where $D = \text{vc}(\mathcal{H})$, we get

$$\left(\frac{eC\varepsilon^{-4}\log N}{\text{vc}(\mathcal{H})}\right)^{\text{vc}(\mathcal{H})} \leq \left(C'\varepsilon^{-4}\right)^{\text{vc}(\mathcal{H})} N^{1/2}, \tag{2.4.61}$$

where $C' = 2eC$. Plugging (2.4.61) back into (2.4.60) and rearranging gives

$$N \leq \left(C'\varepsilon^{-4}\right)^{2\,\text{vc}(\mathcal{H})}.$$

That concludes the proof.   ∎

Proof of Sauer's lemma Recall from Appendix A (see also Exercise 1.4) that for integers $0 < d \leq n$,

$$\sum_{k=0}^{d} \binom{n}{k} \leq \left(\frac{en}{d}\right)^d. \tag{2.4.62}$$

Sauer's lemma (Lemma 2.4.45) follows from the following claim.

Lemma 2.4.54 (Pajor). *Let $\mathcal{H}$ be a class of Boolean functions on $\mathbb{R}^d$ and let $\Lambda = \{\ell_1, \ldots, \ell_n\}$ $\subseteq \mathbb{R}^d$ be any finite subset. Then*

PAJOR'S
LEMMA

$$|\mathcal{H}_\Lambda| \leq |\{S \subseteq \Lambda : S \text{ is shattered by } \mathcal{H}\}|,$$

where the right-hand side includes the empty set.

Going back to Sauer's lemma, by Lemma 2.4.54 we have the upper bound

$$|\mathcal{H}_\Lambda| \leq |\{S \subseteq \Lambda : S \text{ is shattered by } \mathcal{H}\}|.$$

By definition of the VC-dimension (Definition 2.4.44), the subsets $S \subseteq \Lambda$ that are shattered by $\mathcal{H}$ have size at most $\text{vc}(\mathcal{H})$. So the right-hand side is bounded above by the total number of subsets of size at most $d = \text{vc}(\mathcal{H})$ of a set of size n. By (2.4.62), this gives

$$|\mathcal{H}_\Lambda| \leq \left(\frac{en}{\text{vc}(\mathcal{H})}\right)^{\text{vc}(\mathcal{H})},$$

which establishes Sauer's lemma.

So it remains to prove Lemma 2.4.54.

Proof of Lemma 2.4.54 We prove the claim by induction on the size n of Λ. The result is trivial for $n = 1$. Assume the result is true for any $\mathcal{H}$ and any subset of size $n - 1$. To apply induction, for $\iota = 0, 1$ we let

$$\mathcal{H}^\iota = \{h \in \mathcal{H} : h(\ell_n) = \iota\},$$

and we set

$$\Lambda' = \{\ell_1, \ldots, \ell_{n-1}\}.$$

It will be convenient to introduce the following notation:

$$\mathcal{S}(\Lambda; \mathcal{H}) = |\{S \subseteq \Lambda : S \text{ is shattered by } \mathcal{H}\}|.$$

Because $|\mathcal{H}_\Lambda| = |\mathcal{H}^0_{\Lambda'}| + |\mathcal{H}^1_{\Lambda'}|$ and the induction hypothesis implies $\mathcal{S}(\Lambda'; \mathcal{H}^\iota) \geq |\mathcal{H}^\iota_{\Lambda'}|$ for $\iota = 0, 1$, it suffices to show that

$$\mathcal{S}(\Lambda; \mathcal{H}) \geq \mathcal{S}(\Lambda'; \mathcal{H}^0) + \mathcal{S}(\Lambda'; \mathcal{H}^1). \tag{2.4.63}$$

There are two types of sets that contribute to the right-hand side.

- *One but not both.* Let $S \subseteq \Lambda'$ be a set that contributes to one of $\mathcal{S}(\Lambda'; \mathcal{H}^0)$ or $\mathcal{S}(\Lambda'; \mathcal{H}^1)$ but not both. Then, S is a subset of the larger set Λ and it is certainly shattered by the larger collection $\mathcal{H}$. Hence, it also contributes to the left-hand side of (2.4.63).

- *Both.* Let $S \subseteq \Lambda'$ be a set that contributes to both $\mathcal{S}(\Lambda'; \mathcal{H}^0)$ and $\mathcal{S}(\Lambda'; \mathcal{H}^1)$. Hence, it contributes two to the right-hand side of (2.4.63). As in the previous point, it is also included in $\mathcal{S}(\Lambda; \mathcal{H})$, *but it only contributes one to the left-hand side of* (2.4.63). It turns out that there is another set that contributes one to the left-hand side but zero to the right-hand side: the subset $S \cup \{\ell_n\}$. Indeed, by definition of $\mathcal{H}^\iota$, the subset $S \cup \{\ell_n\}$ cannot be shattered by it since all functions in it take the same value on ℓ_n. On the other hand, any Boolean function h on $S \cup \{\ell_n\}$ with $h(\ell_n) = \iota$ is realized in $\mathcal{H}^\iota$ since S itself is shattered by $\mathcal{H}^\iota$.

That concludes the proof.   ∎

Exercises

Exercise 2.1 (Moments of non-negative random variables). Prove (B.5.1). (Hint: Use Fubini's Theorem to compute the integral.)

Exercise 2.2 (Bonferroni inequalities). Let $A_1, \ldots, A_n$ be events and $B_n := \cup_i A_i$. Define

$$S^{(r)} := \sum_{1 \leq i_1 < \cdots < i_r \leq n} \mathbb{P}[A_{i_1} \cap \cdots \cap A_{i_r}]$$

and

$$X_n := \sum_{i=1}^{n} \mathbf{1}_{A_i}.$$

(i) Let $x_0 \leq x_1 \leq \cdots \leq x_s \geq x_{s+1} \geq \cdots \geq x_m$ be a *unimodal* sequence of non-negative reals such that $\sum_{j=0}^{m} (-1)^j x_j = 0$. Show that $\sum_{j=0}^{\ell} (-1)^j x_j$ is ≥ 0 for even ℓ and ≤ 0 for odd ℓ.

(ii) Show that, for all r,

$$\sum_{1 \leq i_1 < \cdots < i_r \leq n} \mathbf{1}_{A_{i_1}} \mathbf{1}_{A_{i_2}} \cdots \mathbf{1}_{A_{i_r}} = \binom{X_n}{r}.$$

(iii) Use (i) and (ii) to show that when $\ell \in [n]$ is odd

$$\mathbb{P}[B_n] \leq \sum_{r=1}^{\ell} (-1)^{r-1} S^{(r)}$$

and when $\ell \in [n]$ is even

$$\mathbb{P}[B_n] \geq \sum_{r=1}^{\ell} (-1)^{r-1} S^{(r)}.$$

These inequalities are called *Bonferroni inequalities*. The case $\ell = 1$ is Boole's inequality.

Exercise 2.3 (Percolation on $\mathbb{Z}^2$: a better bound). Let E_1 be the event that all edges are open in $[-N, N] \times [-N, N]$ and E_2 be the event that there is no closed self-avoiding dual cycle surrounding $[-N, N]^2$. By looking at $E_1 \cap E_2$, show that $\theta(p) > 0$ for $p > 2/3$.

Exercise 2.4 (Percolation on $\mathbb{Z}^d$: existence of critical threshold). Consider bond percolation on $\mathbb{L}^d$.

(i) Show that $p_c(\mathbb{L}^d) > 0$. (Hint: Count self-avoiding paths.)
(ii) Show that $p_c(\mathbb{L}^d) < 1$. (Hint: Use the result for $\mathbb{L}^2$.)

Exercise 2.5 (Sums of uncorrelated variables). Centered random variables $X_1, X_2, \ldots$ are *uncorrelated* if

$$\mathbb{E}[X_r X_s] = 0 \qquad \forall r \neq s.$$

(i) Assume further that $\mathrm{Var}[X_r] \leq C < +\infty$ for all r. Show that

$$\mathbb{P}\left[\frac{1}{n}\sum_{r \leq n} X_r \geq \beta\right] \leq \frac{C^2}{\beta^2 n}.$$

(ii) Use (i) to prove Theorem 2.1.6.

Exercise 2.6 (Pairwise independence: lack of concentration). Let $U = (U_1, \ldots, U_\ell)$ be uniformly distributed over $\{0, 1\}^\ell$. Let $n = 2^\ell - 1$. For all $v \in \{0, 1\}^\ell \backslash \mathbf{0}$, define

$$X_v = (U \cdot v) \mod 2.$$

(i) Show that the random variables X_v, $v \in \{0, 1\}^\ell \backslash \mathbf{0}$, are uniformly distributed in $\{0, 1\}$ and pairwise independent.
(ii) Show that for any event A measurable with respect to $\sigma(X_v, v \in \{0, 1\}^\ell \backslash \mathbf{0})$, $\mathbb{P}[A]$ is either 0 or $\geq 1/(n + 1)$.

Exercise 2.5 shows that pairwise independence implies "polynomial concentration" of the average of square-integrable X_vs. On the other hand, the current exercise suggests that in general pairwise independence cannot imply "exponential concentration."

Exercise 2.7 (Chernoff bound for Poisson trials). Using the Chernoff–Cramér method, prove part (i) of Theorem 2.4.7. Show that part (ii) follows from part (i).

Exercise 2.8 (Stochastic knapsack: some details). Consider the stochastic fractional knapsack problem in Section 2.4.3.

(i) Prove that the greedy algorithm described there gives an optimal solution to problem (2.4.21).
(ii) Prove Claim 2.4.20 for $\tau \in (0, 1/6)$.

Exercise 2.9 (Stochastic knapsack: 0-1 version). Consider the stochastic fractional knapsack problem in Section 2.4.3.

(i) Adapt the greedy algorithm for the 0-1 knapsack problem and show that it is not optimal in general. (Hint: Construct a counter-example with two items.)
(ii) Prove Claim 2.4.20 for the greedy solution of (i).

Exercise 2.10 (A proof of Pólya's theorem). Let (S_t) be simple random walk on $\mathbb{L}^d$ started at the origin 0.

(i) For $d = 1$, use Stirling's formula (see Appendix A) to show that $\mathbb{P}[S_{2n} = 0] = \Theta(n^{-1/2})$.

(ii) For $j = 1, \ldots, d$, let $N_t^{(j)}$ be the number of steps in the jth coordinate by time t. Show that

$$\mathbb{P}\left[N_n^{(j)} \in \left[\frac{n}{2d}, \frac{3n}{2d}\right], \forall j\right] \geq 1 - \exp(-\kappa_d n)$$

for some constant $\kappa_d > 0$.

(iii) Use (i) and (ii) to show that, for any $d \geq 3$, $\mathbb{P}[S_{2n} = 0] = O(n^{-d/2})$.

Exercise 2.11 (Maximum degree). Let $G_n = (V_n, E_n) \sim \mathbb{G}_{n,p_n}$ be an Erdős–Rényi graph with n vertices and density p_n. Suppose $np_n = C \log n$ for some $C > 0$. Let D_n be the maximum degree of G_n. Use Bernstein's inequality to show that for any $\varepsilon > 0$,

$$\mathbb{P}[D_n \geq (n-1)p_n + \max\{C, 4(1+\varepsilon)\}\log n] \to 0,$$

as $n \to +\infty$.

Exercise 2.12 (RIP versus orthogonality). Show that a $(k, 0)$-RIP matrix with $k \geq 2$ is orthogonal, that is, its columns are orthonormal.

Exercise 2.13 (Compressed sensing: linear programming formulation). Formulate (2.4.43) as a linear program, that is, the minimization of a linear objective subject to linear inequalities.

Exercise 2.14 (Compressed sensing: almost sparse case). By adapting the proof of Lemma 2.4.39, show the following "almost sparse" version. Let L be $(10k, 1/3)$-RIP. Then, for any $x \in \mathbb{R}^n$, the solution to (2.4.43) satisfies $\|z^* - x\|_2 = O(\eta(x)/\sqrt{k})$, where $\eta(x) := \min_{x' \in \mathscr{S}_k^n} \|x - x'\|_1$.

Exercise 2.15 (Spectral norm without independence). Give an example of a random matrix $A \in \mathbb{R}^{n \times n}$ whose entries are bounded, but not independent, such that the spectral norm is $\Omega(n)$ with high probability.

Exercise 2.16 (Spectral norm: symmetric matrix). Let $A \in \mathbb{R}^{n \times n}$ be a symmetric random matrix. We assume that entries on and above the diagonal $A_{i,j}, i \leq j$, are centered, independent, and sub-Gaussian with variance factor v. Each entry below the diagonal is equal to the corresponding entry above it. Prove an analogue of Theorem 2.4.28 for A. (Hint: Mimic the proof of Theorem 2.4.28.)

Exercise 2.17 (Chaining tail inequality). Prove Theorem 2.4.32.

Exercise 2.18 (Poisson convergence: method of moments). Let $A_1, \ldots, A_n$ be events and $A := \cup_i A_i$. Define

$$S^{(r)} := \sum_{1 \leq i_1 < \cdots < i_r \leq n} \mathbb{P}[A_{i_1} \cap \cdots \cap A_{i_r}]$$

and

$$X_n := \sum_{i=1}^n A_i.$$

Assume that there is $\mu > 0$ such that, for all r,

$$S^{(r)} \to \frac{\mu^r}{r!}.$$

Use Exercise 2.2 and a Taylor expansion of $e^{-\mu}$ to show that

$$\mathbb{P}[X_n = 0] \to e^{-\mu}.$$

In fact, $X_n \xrightarrow{\mathrm{d}} \mathrm{Poi}(\mu)$ (no need to prove this). This is a special case of the *method of moments*.

Exercise 2.19 (Connectivity: critical window). Using Exercise 2.18 show that, when $p_n = \frac{\log n + s}{n}$, the probability that an Erdős–Rényi graph $G_n \sim \mathbb{G}_{n,p_n}$ contains no isolated vertex converges to $e^{-e^{-s}}$.

Bibliographic Remarks

Section 2.1 For more on moment-generating functions, see [Bil12, section 21].

Section 2.2 The examples in Section 2.2.1 are taken from [AS11, sections 2.4, 3.2]. A fascinating account of the longest increasing subsequence problem is given in [Rom15], from which the material in Section 2.2.3 is taken. The contour lemma, Lemma 2.2.14, is attributed to Whitney [Whi32] and is usually proved "by picture" [Gri10a, Figure 3.1]. A formal proof of the lemma can be found in [Kes82, Appendix A]. For much more on percolation, see [Gri10b]. A gentler introduction is provided in [Ste].

Section 2.3 The presentation in Section 2.3.2 follows [AS11, section 4.4] and [JLR11, section 3.1]. The result for general subgraphs is due to Bollobás [Bol81]. A special case (including cliques) was proved by Erdős and Rényi [ER60]. For variants of the small subgraph containment problem involving copies that are induced, disjoint, isolated, and so on, see, for example, [JLR11, chapter 3]. For corresponding results for larger subgraphs, such as cycles or matchings, see, for example, [Bol01]. The connectivity threshold in Section 2.3.2 is also due to the same authors [ER59]. The presentation here follows [vdH17, section 5.2]. For more on the method of moments, see, for example, [Dur10, section 3.3.5] or [JLR11, section 6.1]. Claim 2.3.11 is due to R. Lyons [Lyo90].

Section 2.4 The use of the moment-generating function to derive tail bounds for sums of independent random variables was pioneered by Cramér [Cra38], Bernstein [Ber46], and Chernoff [Che52]. For much more on concentration inequalities, see, for example, [BLM13]. The basics of large deviation theory are covered in [Dur10, section 2.6]. See also [RAS15] and [DZ10]. Section 2.4.2 is based partly on [Ver18] and [Lug, section 3.2]. Section 2.4.3 is based on [FR98, section 5.3]. Very insightful, and much deeper, treatment of the material in Section 2.4.4 can be found in [Ver18, vH16]. The presentation in Section 2.4.5 is inspired by [Har, Lectures 6 and 8] and [Tao]. The Johnson–Lindenstrauss lemma was first proved by Johnson and Lindenstrauss using non-probabilistic arguments [JL84]. The idea of using random projections to simplify the proof was introduced by Frankl and Maehara [FM88] and the proof presented here based on Gaussian projections is due to Indyk and

Motwani [IM98]. See [Ach03] for an overview of the various proofs known. For more on the random projection method, see [Vem04]. For algorithmic applications of the Johnson–Lindenstrauss lemma, see, for example, [Har, Lecture 7]. Compressed sensing emerged in the works of Donoho [Don06] and Candès, Romberg, and Tao [CRT06a, CRT06b]. The restricted isometry property was introduced by Candès and Tao [CT05]. Lemma 2.4.39 is due to Candés, Romberg, and Tao [CRT06b]. The proof of Lemma 2.4.38 presented here is due to Baraniuk et al. [BDDW08]. A survey of compressed sensing can be found in [CW08]. A thorough mathematical introduction to compressed sensing can be found in [FR13]. The material in Section 2.4.2 can be found in [BLM13, chapter 2]. Hoeffding's lemma and inequality are due to Hoeffding [Hoe63]. Section 2.4.6 borrows from [Ver18, vH16, SSBD14, Haz16]. The proof of Sauer's lemma follows [Ver18, section 8.3.3]. For a proof of Claim 2.4.48 in general dimension d, see, for example, [SSBD14, section 9.1.3].

Martingales and Potentials

In this chapter we turn to *martingales*, which play a central role in probability theory. We illustrate their use in a number of applications to the analysis of discrete stochastic processes. After some background on stopping times and a brief review of basic martingale properties and results in Section 3.1, we develop two major directions. In Section 3.2, we show how martingales can be used to derive a substantial generalization of our previous concentration inequalities – from the *sums* of independent random variables we focused on in Chapter 2 to *nonlinear functions* with Lipschitz properties. In particular, we give several applications of the method of bounded differences to random graphs. We also discuss bandit problems in machine learning. In the second thread in Section 3.3, we give an introduction to *potential theory* and *electrical network theory* for Markov chains. This toolkit in particular provides bounds on hitting times for random walks on networks, with important implications in the study of recurrence among other applications. We also introduce Wilson's remarkable method for generating uniform spanning trees.

3.1 Background

We begin with a quick review of stopping times and martingales. Along the way, we prove a few useful results. In particular, we derive some bounds on hitting times and cover times of Markov chains.

Throughout, $(\Omega, \mathcal{F}, (\mathcal{F}_t)_{t \in \mathbb{Z}_+}, \mathbb{P})$ is a filtered space. See Appendix B for a formal definition. Recall that, intuitively, the σ-algebra $\mathcal{F}_t$ in the filtration $(\mathcal{F}_t)_t$ represents "the information known at time t." All time indices are discrete (in $\mathbb{Z}_+$ unless stated otherwise). We will also use the notation $\overline{\mathbb{Z}}_+ := \{0, 1, \ldots, +\infty\}$ to allow time $+\infty$.

3.1.1 Stopping Times

Definitions Roughly speaking, a stopping time is a random time whose value is determined by a rule not depending on the future. Formally:

Definition 3.1.1 (Stopping time). *A random variable* $\tau : \Omega \to \overline{\mathbb{Z}}_+$ *is called a* stopping
time *if*

$$\{\tau \leq t\} \in \mathcal{F}_t, \ \forall t \in \overline{\mathbb{Z}}_+,$$

or, equivalently,

$$\{\tau = t\} \in \mathcal{F}_t, \ \forall t \in \overline{\mathbb{Z}}_+.$$

To see this equivalence, note that $\{\tau = t\} = \{\tau \leq t\} \setminus \{\tau \leq t-1\}$, and $\{\tau \leq t\} = \cup_{i \leq t} \{\tau = i\}$.

Example 3.1.2 (Hitting time). Let $(A_t)_{t \in \mathbb{Z}_+}$, with values in $(E, \mathcal{E})$, be adapted and $B \in \mathcal{E}$. Then,

$$\tau = \inf\{t \geq 0 \colon A_t \in B\}$$

HITTING TIME   is a stopping time known as a *hitting time*. In contrast, the last visit to a set is typically not a stopping time.   ◀

Let τ be a stopping time. Denote by $\mathcal{F}_\tau$ the set of all events F such that, $\forall t \in \overline{\mathbb{Z}}_+$, $F \cap \{\tau = t\} \in \mathcal{F}_t$. Intuitively, the σ-algebra $\mathcal{F}_\tau$ captures the information up to time τ. The following lemmas help clarify the definition of $\mathcal{F}_\tau$.

Lemma 3.1.3 $\mathcal{F}_\tau = \mathcal{F}_s$ *if* $\tau := s$, $\mathcal{F}_\tau = \mathcal{F}_\infty = \sigma(\cup_t \mathcal{F}_t)$ *if* $\tau := +\infty$ *and* $\mathcal{F}_\tau \subseteq \mathcal{F}_\infty$ *for any* τ.

Proof In the first case, note that $F \cap \{\tau = t\}$ is empty if $t \neq s$ and is F if $t = s$. So if $F \in \mathcal{F}_\tau$ then $F = F \cap \{\tau = s\} \in \mathcal{F}_s$ by definition of $\mathcal{F}_\tau$, and if $F \in \mathcal{F}_s$ then $F = F \cap \{\tau = t\} \in \mathcal{F}_t$ for all t by definition of τ. So we have proved both inclusions. This works also for $t = +\infty$. For the third claim note that, for any $F \in \mathcal{F}_\tau$,

$$F = \cup_{t \in \overline{\mathbb{Z}}_+} F \cap \{\tau = t\} \in \mathcal{F}_\infty,$$

again by definition of $\mathcal{F}_\tau$.   ∎

Lemma 3.1.4 *If* (X_t) *is adapted and* τ *is a stopping time then* $X_\tau \in \mathcal{F}_\tau$ *(where we assume that* $X_\infty \in \mathcal{F}_\infty$, *for example, by setting* $X_\infty := \liminf X_n$).

Proof For $B \in \mathcal{E}$,

$$\{X_\tau \in B\} \cap \{\tau = t\} = \{X_t \in B\} \cap \{\tau = t\} \in \mathcal{F}_t$$

by definition of τ. That shows X_τ is measurable with respect to $\mathcal{F}_\tau$ as claimed.   ∎

Lemma 3.1.5 *If* σ, τ *are stopping times, then* $\mathcal{F}_{\sigma \wedge \tau} \subseteq \mathcal{F}_\tau$.

Proof Let $F \in \mathcal{F}_{\sigma \wedge \tau}$. Note that

$$F \cap \{\tau = t\} = \cup_{s \leq t}[(F \cap \{\sigma \wedge \tau = s\}) \cap \{\tau = t\}] \in \mathcal{F}_t.$$

Indeed, the expression in parenthesis is in $\mathcal{F}_s \subseteq \mathcal{F}_t$ by definition of $\mathcal{F}_{\sigma \wedge \tau}$ and $\{\tau = t\} \in \mathcal{F}_t$.   ∎

Let (X_t) be a Markov chain on a countable space V. The following two examples of stopping times will play an important role.

FIRST RETURN   **Definition 3.1.6** (First visit and return). *The* first visit time *and* first return time *to* $x \in V$ *are*

$$\tau_x := \inf\{t \geq 0 \colon X_t = x\} \quad and \quad \tau_x^+ := \inf\{t \geq 1 \colon X_t = x\}.$$

Similarly, τ_B *and* τ_B^+ *are the first visit time and first return time to* $B \subseteq V$.

COVER TIME   **Definition 3.1.7** (Cover time). *Assume* V *is finite. The* cover time *of* (X_t) *is the first time that all states have been visited, that is,*

$$\tau_{\mathrm{cov}} := \inf\{t \geq 0 \colon \{X_0, \ldots, X_t\} = V\}.$$

Strong Markov property Let (X_t) be a Markov chain with transition matrix P and initial distribution μ. Let $\mathcal{F}_t = \sigma(X_0, \ldots, X_t)$. Recall that the Markov property (Theorem 1.1.18) says that, given the present, the future is independent of the past. The Markov property naturally extends to stopping times. Let τ be a stopping time with $\mathbb{P}[\tau < +\infty] > 0$. In its simplest form we have:

$$\mathbb{P}[X_{\tau+1} = y \mid \mathcal{F}_\tau] = \mathbb{P}_{X_\tau}[X_{\tau+1} = y] = P(X_\tau, y).$$

In other words, the chain "starts afresh" at a stopping time with the state at that time as a starting point. More generally:

Theorem 3.1.8 (Strong Markov property). *Let $f_t \colon V^\infty \to \mathbb{R}$ be a sequence of measurable functions, uniformly bounded in t and let $F_t(x) := \mathbb{E}_x[f_t((X_s)_{s \geq 0})]$. On $\{\tau < +\infty\}$,*

$$\mathbb{E}[f_\tau((X_{\tau+t})_{t \geq 0}) \mid \mathcal{F}_\tau] = F_\tau(X_\tau).$$

Throughout, when we say that two random variables Y, Z are equal on an event B, we mean formally that $Y\mathbf{1}_B = Z\mathbf{1}_B$ almost surely.

Proof of Theorem 3.1.8 We use that

$$\mathbb{E}[f_\tau((X_{\tau+t})_{t \geq 0}) \mid \mathcal{F}_\tau]\mathbf{1}_{\tau < +\infty} = \mathbb{E}[f_\tau((X_{\tau+t})_{t \geq 0})\mathbf{1}_{\tau < +\infty} \mid \mathcal{F}_\tau].$$

Let $A \in \mathcal{F}_\tau$. Summing over the possible values of τ, using the tower property (Lemma B.6.16) and then the Markov property

$$\begin{aligned}
&\mathbb{E}[f_\tau((X_{\tau+t})_{t \geq 0})\mathbf{1}_{\tau < +\infty}; A] \\
&= \mathbb{E}[f_\tau((X_{\tau+t})_{t \geq 0}); A \cap \{\tau < +\infty\}] \\
&= \sum_{s \geq 0} \mathbb{E}[f_s((X_{s+t})_{t \geq 0}); A \cap \{\tau = s\}] \\
&= \sum_{s \geq 0} \mathbb{E}[\mathbb{E}[f_s((X_{s+t})_{t \geq 0}); A \cap \{\tau = s\} \mid \mathcal{F}_s]] \\
&= \sum_{s \geq 0} \mathbb{E}[\mathbf{1}_{A \cap \{\tau = s\}}\mathbb{E}[f_s((X_{s+t})_{t \geq 0}) \mid \mathcal{F}_s]] \\
&= \sum_{s \geq 0} \mathbb{E}[\mathbf{1}_{A \cap \{\tau = s\}}F_s(X_s)] \\
&= \sum_{s \geq 0} \mathbb{E}[F_s(X_s); A \cap \{\tau = s\}] \\
&= \mathbb{E}[F_\tau(X_\tau); A \cap \{\tau < +\infty\}] \\
&= \mathbb{E}[F_\tau(X_\tau)\mathbf{1}_{\tau < +\infty}; A],
\end{aligned}$$

where, on the fifth line, we used that $A \cap \{\tau = s\} \in \mathcal{F}_s$ by definition of $\mathcal{F}_\tau$ and taking out what is known (Lemma B.6.13). The definition of the conditional expectation (Theorem B.6.1) concludes the proof. ∎

The following typical application of the strong Markov property (Theorem 3.1.8) is useful.

Theorem 3.1.9 (Reflection principle). *Let $X_1, X_2, \ldots$ be i.i.d. with a distribution symmetric about 0 and let $S_t = \sum_{i \le t} X_i$. Then, for $b > 0$,*

$$\mathbb{P}\left[\sup_{i \le t} S_i \ge b\right] \le 2\,\mathbb{P}[S_t \ge b].$$

Proof Let $\tau := \inf\{i \le t : S_i \ge b\}$. By the strong Markov property, on $\{\tau < t\}$, $S_t - S_\tau$ is independent of $\mathcal{F}_\tau$ and is symmetric about 0. In particular, it has probability at least $1/2$ of being greater or equal to 0 by the first moment principle (Theorem 2.2.1), an event which implies that S_t is greater than or equal to b. Hence,

$$\mathbb{P}[S_t \ge b] \ge \mathbb{P}[\tau = t] + \frac{1}{2}\mathbb{P}[\tau < t] \ge \frac{1}{2}\mathbb{P}[\tau \le t].$$

(Exercise 3.1 asks for a more formal proof.)   ∎

In the case of simple random walk on $\mathbb{Z}$, we get a stronger statement.

Theorem 3.1.10 (Reflection principle: simple random walk). *Let (S_t) be simple random walk on $\mathbb{Z}$ started at 0. Then, $\forall a, b, t > 0$,*

$$\mathbb{P}[S_t = b + a] = \mathbb{P}\left[S_t = b - a,\ \sup_{i \le t} S_i \ge b\right],$$

and

$$\mathbb{P}\left[\sup_{i \le t} S_i \ge b\right] = \mathbb{P}[S_t = b] + 2\,\mathbb{P}[S_t > b].$$

Proof Reflect the sub-path after the first visit to b across the line $y = b$. Summing over $a > 0$ and rearranging gives the second claim.   ∎

We record another related result that will be useful later.

Theorem 3.1.11 (Ballot theorem). *In an election with n voters, candidate A gets α votes and candidate B gets $\beta < \alpha$ votes. The probability that A leads B throughout the counting is $\frac{\alpha - \beta}{n}$.*

kTH RETURN **Recurrence** Let (X_t) be a Markov chain on a countable state space V. The *time of the kth return to y* is (letting $\tau_y^0 := 0$)

$$\tau_y^k := \inf\{t > \tau_y^{k-1} : X_t = y\}.$$

In particular, $\tau_y^1 = \tau_y^+$. Define $\rho_{xy} := \mathbb{P}_x[\tau_y^+ < +\infty]$. Then by the strong Markov property (and induction)

$$\mathbb{P}_x[\tau_y^k < +\infty] = \rho_{xy}\rho_{yy}^{k-1}. \tag{3.1.1}$$

(Exercise 3.2 asks for a more formal proof.) Letting

$$N_y := \sum_{t > 0} \mathbf{1}_{\{X_t = y\}} = \sum_{k \ge 1} \mathbf{1}_{\{\tau_y^k < +\infty\}}$$

be the number of visits to y after time 0, by linearity

$$\mathbb{E}_x[N_y] = \frac{\rho_{xy}}{1 - \rho_{yy}}. \tag{3.1.2}$$

When $\rho_{yy} < 1$, we have $\mathbb{E}_y[N_y] < +\infty$ by (3.1.2), and in particular $\tau_y^k = +\infty$ for some k. Or $\rho_{yy} = 1$ and, starting at $x = y$, we have $\tau_y^k < +\infty$ almost surely for all k by (3.1.1). That leads us to the following dichotomy.

Definition 3.1.12 (Recurrence). *A state x is* recurrent *if $\rho_{xx} = 1$. Otherwise it is* transient. *We* RECURRENT *refer to the recurrence or transience of a state as its* type. *Let x be recurrent. If in addition $\mathbb{E}_x[\tau_x^+] < +\infty$, we say that x is* positive recurrent; *otherwise we say that it is* null recurrent. *A chain is* recurrent *(or* transient, *or* positive recurrent, *or* null recurrent*) if all its states are.*

Recurrence is "contagious" in the following sense.

Lemma 3.1.13 *If x is recurrent and $\rho_{xy} > 0$, then y is recurrent and $\rho_{yx} = \rho_{xy} = 1$.*

A subset $C \subseteq V$ is *closed* if $x \in C$ and $\rho_{xy} > 0$ implies $y \in C$. A subset $D \subseteq V$ is *irreducible* if $x, y \in D$ implies $\rho_{xy} > 0$. This definition is consistent with (and generalizes to sets) the one we gave in Section 1.1.2. Recall that we have the following decomposition theorem.

Theorem 3.1.14 (Decomposition theorem). *Let $R := \{x: \rho_{xx} = 1\}$ be the recurrent states of the chain. Then R can be written as a disjoint union $\cup_j R_j$, where each R_j is closed and irreducible.*

Example 3.1.15 (Simple random walk on $\mathbb{Z}$). Consider simple random walk (S_t) on $\mathbb{Z}$ started at 0. The chain is clearly irreducible so it suffices to check the type of state 0 by Lemma 3.1.13. First note the periodicity of this chain. So we look at S_{2t}. Then by Stirling's formula (see Appendix A),

$$\mathbb{P}[S_{2t} = 0] = \binom{2t}{t} 2^{-2t} \sim 2^{-2t} \frac{(2t)^{2t}}{(t^t)^2} \frac{\sqrt{2t}}{\sqrt{2\pi t}} \sim \frac{1}{\sqrt{\pi t}}.$$

Thus,

$$\mathbb{E}[N_0] = \sum_{t>0} \mathbb{P}[S_t = 0] = +\infty,$$

and the chain is recurrent. ◀

Return times are closely related to stationary measures. We recall the following standard results without proof. We gave an alternative proof of the existence of a unique stationary distribution in the finite, irreducible case in Theorem 1.1.24.

Theorem 3.1.16 *Let x be a recurrent state. Then the following defines a stationary measure:*

$$\mu_x(y) := \mathbb{E}_x \left[\sum_{0 \le t < \tau_x^+} \mathbf{1}_{\{X_t = y\}} \right].$$

Theorem 3.1.17 *If (X_t) is irreducible and recurrent, then the stationary measure is unique up to a constant multiple.*

Theorem 3.1.18 *If there is a stationary distribution π, then all states y that have $\pi(y) > 0$ are recurrent.*

Theorem 3.1.19 *If (X_t) is irreducible and has a stationary distribution π, then*

$$\pi(x) = \frac{1}{\mathbb{E}_x \tau_x^+}.$$

Theorem 3.1.20 *If (X_t) is irreducible, then the following are equivalent.*

(i) *There is a stationary distribution.*
(ii) *All states are positive recurrent.*
(iii) *There is a positive recurrent state.*

We have seen previously that, in the irreducible, positive recurrent, aperiodic case, there is convergence to stationarity (see Theorem 1.1.33). In the transient and null recurrent cases, there is no stationary distribution to converge to by Theorem 3.1.20. Instead, we have the following.

Theorem 3.1.21 (Convergence of P^t: transient and null recurrent cases). *If P is an irreducible chain which is either transient or null recurrent, we have for all x, y that*

$$\lim_t P^t(x, y) = 0.$$

Proof We only prove the transient case. In that case, we showed in (3.1.2) that

$$\sum_t P^t(x, y) = \mathbb{E}_x \left[\sum_t \mathbf{1}_{\{X_t = y\}} \right] = \mathbb{E}_x[N_y] < +\infty.$$

Hence, $P^t(x, y) \to 0$. ∎

A useful identity A slight generalization of the "cycle trick" used in the proof of Theorem 3.1.16 gives a useful identity.

GREEN
FUNCTION

Definition 3.1.22 (Green function). *Let σ be a stopping time for a Markov chain (X_t). The Green function of the chain stopped at σ is given by*

$$\mathcal{G}_\sigma(x, y) = \mathbb{E}_x \left[\sum_{0 \leq t < \sigma} \mathbf{1}_{\{X_t = y\}} \right], \qquad x, y \in V, \tag{3.1.3}$$

that is, it is the expected number of visits to y before σ when started at x.

Lemma 3.1.23 (Occupation measure identity). *Consider an irreducible, positive recurrent Markov chain $(X_t)_{t \geq 0}$ with transition matrix P and stationary distribution π. Let x be a state and σ be a stopping time such that $\mathbb{E}_x[\sigma] < +\infty$ and $\mathbb{P}_x[X_\sigma = x] = 1$. For any y,*

$$\mathcal{G}_\sigma(x, y) = \pi_y \, \mathbb{E}_x[\sigma].$$

Proof By the uniqueness of the stationary measure up to constant multiple (Theorem 3.1.17), it suffices to show that $\mathcal{G}_\sigma(x, y)$ satisfies the system for a stationary measure as a function of y

$$\sum_y \mathcal{G}_\sigma(x, y) P(y, z) = \mathcal{G}_\sigma(x, z) \quad \forall z, \tag{3.1.4}$$

and use the fact that

$$\sum_y \mathscr{G}_\sigma(x,y) = \sum_y \mathbb{E}_x \left[\sum_{0 \le t < \sigma} \mathbf{1}_{\{X_t=y\}} \right] = \mathbb{E}_x[\sigma].$$

To check (3.1.4), because $X_\sigma = X_0$ almost surely, observe that

$$\mathscr{G}_\sigma(x,z) = \mathbb{E}_x \left[\sum_{0 \le t < \sigma} \mathbf{1}_{X_t=z} \right]$$

$$= \mathbb{E}_x \left[\sum_{0 \le t < \sigma} \mathbf{1}_{X_{t+1}=z} \right]$$

$$= \sum_{t \ge 0} \mathbb{P}_x[X_{t+1} = z, \sigma > t].$$

Since $\{\sigma > t\} \in \mathcal{F}_t$, applying the Markov property we get

$$\mathscr{G}_\sigma(x,z) = \sum_{t \ge 0} \sum_y \mathbb{P}_x[X_t = y, X_{t+1} = z, \sigma > t]$$

$$= \sum_{t \ge 0} \sum_y \mathbb{P}_x[X_{t+1} = z \mid X_t = y, \sigma > t] \, \mathbb{P}_x[X_t = y, \sigma > t]$$

$$= \sum_{t \ge 0} \sum_y P(y,z) \, \mathbb{P}_x[X_t = y, \sigma > t]$$

$$= \sum_y \mathscr{G}_\sigma(x,y) P(y,z),$$

which establishes (3.1.4) and proves the claim. ∎

Here is a typical application of this lemma.

Corollary 3.1.24 *In the setting of Lemma 3.1.23, for all $x \ne y$,*

$$\mathbb{P}_x[\tau_y < \tau_x^+] = \frac{1}{\pi_x(\mathbb{E}_x[\tau_y] + \mathbb{E}_y[\tau_x])}.$$

Proof Let σ be the time of the first visit to x after the first visit to y. Then, $\mathbb{E}_x[\sigma] = \mathbb{E}_x[\tau_y] + \mathbb{E}_y[\tau_x] < +\infty$, where we used that the chain is irreducible and positive recurrent. By the strong Markov property, the number of visits to x before the first visit to y is geometric with success probability $\mathbb{P}_x[\tau_y < \tau_x^+]$. Moreover, the number of visits to x after the first visit to y but before σ is 0 by definition. Hence, $\mathscr{G}_\sigma(x,y)$ is the mean of the geometric distribution, namely, $1/\mathbb{P}_x[\tau_y < \tau_x^+]$. Applying the occupation measure identity gives the result. ∎

3.1.2 ▷ *Markov Chains: Exponential Tail of Hitting Times and Some Cover Time Bounds*

Tail of a hitting time On a finite state space, the tail of any hitting time converges to 0 exponentially fast.

Lemma 3.1.25 *Let (X_t) be a finite, irreducible Markov chain with state space V. For any subset of states $A \subseteq V$ and initial distribution μ:*

(i) *It holds that $\mathbb{E}_\mu[\tau_A] < +\infty$ (and, in particular, $\tau_A < +\infty$ a.s.).*

(ii) *Letting $\bar{t}_A := \max_x \mathbb{E}_x[\tau_A]$, we have the tail bound*

$$\mathbb{P}_\mu[\tau_A > t] \le \exp\left(-\left\lfloor \frac{t}{\lceil e\bar{t}_A \rceil} \right\rfloor\right).$$

Proof For any integer m, for some distribution θ over the state space V, by the strong Markov property (Theorem 3.1.8),

$$\mathbb{P}_\mu[\tau_A > ms \mid \tau_A > (m-1)s] = \mathbb{P}_\theta[\tau_A > s] \le \max_x \mathbb{P}_x[\tau_A > s] =: \alpha_s.$$

Choose s large enough that, from any x, there is a path to A of length at most s of positive probability. Such an s exists by irreducibility. In particular, $\alpha_s < 1$.

By induction, $\mathbb{P}_\mu[\tau_A > ms] \le \alpha_s^m$, or put differently,

$$\mathbb{P}_\mu[\tau_A > t] \le \alpha_s^{\lfloor \frac{t}{s} \rfloor}. \tag{3.1.5}$$

The result for the expectation follows from

$$\mathbb{E}_\mu[\tau_A] = \sum_{t \ge 0} \mathbb{P}_\mu[\tau_A > t] \le \sum_t \alpha_s^{\lfloor \frac{t}{s} \rfloor} < +\infty$$

since $\alpha_s < 1$.

By Markov's inequality (Theorem 2.1.1),

$$\alpha_s = \max_x \mathbb{P}_x[\tau_A > s] \le \frac{\bar{t}_A}{s}.$$

Plugging back into (3.1.5) gives $\mathbb{P}_\mu[\tau_A > t] \le \left(\frac{\bar{t}_A}{s}\right)^{\lfloor \frac{t}{s} \rfloor}$. By differentiating with respect to s, it can be checked that a good choice for s is $\lceil e\bar{t}_A \rceil$. Simplifying gives the second claim. ∎

Application to cover times We give an application of the previous bound to cover times. Let (X_t) be a finite, irreducible Markov chain on V with $n := |V| > 1$. Recall that the cover time is $\tau_{\text{cov}} := \max_y \tau_y$. We bound the mean cover time in terms of

$$\bar{t}_{\text{hit}} := \max_{x \ne y} \mathbb{E}_x \tau_y.$$

Claim 3.1.26

$$\max_x \mathbb{E}_x[\tau_{\text{cov}}] \le (3 + \log n)\lceil e\,\bar{t}_{\text{hit}} \rceil.$$

Proof By a union bound over all states to be visited and Lemma 3.1.25,

$$\max_x \mathbb{P}_x[\tau_{\text{cov}} > t] \le \min\left\{1, n \exp\left(-\left\lfloor \frac{t}{\lceil e\,\bar{t}_{\text{hit}} \rceil} \right\rfloor\right)\right\}.$$

Summing over $t \in \mathbb{Z}_+$ and appealing to the sum of a geometric series,

$$\max_x \mathbb{E}_x[\tau_{\text{cov}}] \le (\log n + 1)\lceil e\,\bar{t}_{\text{hit}} \rceil + \frac{1}{1 - e^{-1}}\lceil e\,\bar{t}_{\text{hit}} \rceil,$$

where the first term on the right-hand side comes from the fact that until $t \geq (\log n + 1) \lceil e \, \bar{t}_{\text{hit}} \rceil$ the upper bound above is 1. The factor $\lceil e \, \bar{t}_{\text{hit}} \rceil$ in the second term on the right-hand side comes from the fact that we must break up the series into blocks of size $\lceil e \, \bar{t}_{\text{hit}} \rceil$. Simplifying gives the claim. $\blacksquare$

A clever argument gives a better constant factor as well as a lower bound.

Theorem 3.1.27 (Matthews' cover time bounds). *Let*

$$\underline{t}_{\text{hit}}^A := \min_{x,y \in A, \, x \neq y} \mathbb{E}_x \tau_y$$

and $h_n := \sum_{m=1}^{n} \frac{1}{m}$.
  Then,

$$\max_x \mathbb{E}_x[\tau_{\text{cov}}] \leq h_n \, \bar{t}_{\text{hit}} \tag{3.1.6}$$

and

$$\min_x \mathbb{E}_x[\tau_{\text{cov}}] \geq \max_{A \subseteq V} h_{|A|-1} \, \underline{t}_{\text{hit}}^A. \tag{3.1.7}$$

Clearly, $\max_{x \neq y} \underline{t}_{\text{hit}}^{\{x,y\}}$ is a lower bound on the worst expected cover time. Lower bound (3.1.7) says that a tighter bound is obtained by finding a larger subset of states A that are "far away" from each other.
  We sketch the proof of the lower bound for $A = V$, which we assume is $[n]$ without loss of generality. The other cases are similar. Let $(J_1, \ldots, J_n)$ be a uniform random ordering of V, let $C_m := \max_{i \leq J_m} \tau_i$, and let L_m be the last state visited among $J_1, \ldots, J_m$. Then, for $m \geq 2$,

$$\mathbb{E}_x[C_m - C_{m-1} \mid J_1, \ldots, J_m, \{X_t, t \leq C_{m-1}\}] \geq \underline{t}_{\text{hit}}^V \mathbf{1}_{\{L_m = J_m\}}.$$

By symmetry, $\mathbb{P}[L_m = J_m] = \frac{1}{m}$. To see this, first pick the set of vertices corresponding to $\{J_1, \ldots, J_m\}$, wait for all of those vertices to be visited, then pick the ordering. Moreover, observe that $\mathbb{E}_x C_1 \geq (1 - \frac{1}{n}) \underline{t}_{\text{hit}}^V$, where the factor of $(1 - \frac{1}{n})$ accounts for the probability that $J_1 \neq x$. Taking expectations in the previous display and summing over m gives the result.
  Exercise 3.3 asks for a proof that the bounds above cannot in general be improved up to smaller order terms.

3.1.3 Martingales

Definition Martingales are an important class of stochastic processes that correspond intuitively to the "probabilistic version of a monotone sequence." They hide behind many processes and have properties that make them powerful tools in the analysis of processes where they have been identified. Formally:

Definition 3.1.28 (Martingale). *An adapted process $(M_t)_{t \geq 0}$ with $\mathbb{E}|M_t| < +\infty$ for all t is a* martingale *if*

$$\mathbb{E}[M_{t+1} \mid \mathcal{F}_t] = M_t \qquad \forall t \geq 0.$$

If equality is replaced with $\leq$ or $\geq$, we get a supermartingale or a submartingale, respectively. We say that a martingale is bounded in L^p *if $\sup_t \mathbb{E}[|X_t|^p] < +\infty$.*

Recall that adapted (Definition B.7.5) simply means that $M_t \in \mathcal{F}_t$, that is, roughly speaking M_t is "known at time t." Note that for a martingale, by the tower property (Lemma B.6.16), we have $\mathbb{E}[M_t \mid \mathcal{F}_s] = M_s$ for all $t > s$, and similarly (with inequalities) for supermartingales and submartingales.

We start with a straightforward example.

Example 3.1.29 (Sums of i.i.d. random variables with mean 0). Let $X_0, X_1, \ldots$ be i.i.d. integrable, centered random variables, $\mathcal{F}_t = \sigma(X_0, \ldots, X_t)$, $S_0 = 0$, and $S_t = \sum_{1 \le i \le t} X_i$. Note that $\mathbb{E}|S_t| < \infty$ by the triangle inequality. By taking out what is known and the role of independence lemma (Lemma B.6.14), we obtain

$$\mathbb{E}[S_t \mid \mathcal{F}_{t-1}] = \mathbb{E}[S_{t-1} + X_t \mid \mathcal{F}_{t-1}] = S_{t-1} + \mathbb{E}[X_t] = S_{t-1},$$

which proves that (S_t) is a martingale. ◄

Martingales however are richer than random walks with centered steps. For instance, mixtures of such random walks are also martingales.

Example 3.1.30 (Mixtures of random walks). Consider again the setting of Example 3.1.29. This time assume that X_0 is uniformly distributed in $\{1, 2\}$ and define

$$R_t = X_0 S_t, \quad t \ge 0.$$

Then, because (S_t) is a martingale,

$$\mathbb{E}[R_t \mid \mathcal{F}_{t-1}] = X_0 \mathbb{E}[S_t \mid \mathcal{F}_{t-1}] = X_0 S_{t-1} = R_{t-1},$$

so (R_t) is also a martingale.

Further examples Martingales can also be a little more hidden. Here are two examples.

Example 3.1.31 (Variance of a sum of i.i.d. random variables). Consider again the setting of Example 3.1.29 with $\sigma^2 := \mathrm{Var}[X_1] < \infty$. Define

$$M_t = S_t^2 - t\sigma^2.$$

Note that by the triangle inequality and the fact that S_t has mean zero and is a sum of independent random variables,

$$\mathbb{E}|M_t| \le \sum_{1 \le i \le t} \mathrm{Var}[X_i] + t\sigma^2 \le 2t\sigma^2 < +\infty.$$

Moreover, arguing similarly to the previous example, and using the fact that both X_t and S_{t-1} are square integrable,

$$\begin{aligned}
\mathbb{E}[M_t \mid \mathcal{F}_{t-1}] &= \mathbb{E}[(X_t + S_{t-1})^2 - t\sigma^2 \mid \mathcal{F}_{t-1}] \\
&= \mathbb{E}[X_t^2 + 2X_t S_{t-1} + S_{t-1}^2 - t\sigma^2 \mid \mathcal{F}_{t-1}] \\
&= \sigma^2 + 0 + S_{t-1}^2 - t\sigma^2 \\
&= M_{t-1},
\end{aligned}$$

which proves that (M_t) is a martingale. ◄

Example 3.1.32 (Eigenvectors of a transition matrix). Let $(X_t)_{t \geq 0}$ be a finite Markov chain with state space E and transition matrix P, and let $(\mathcal{F}_t)_{t \geq 0}$ be the corresponding filtration. Suppose $f : E \to \mathbb{R}$ is such that

$$\sum_j P(i,j) f(j) = \lambda f(i) \; \forall i \in S.$$

In other words, f is a (right) eigenvector of P with eigenvalue λ. Define

$$M_t = \lambda^{-t} f(X_t).$$

Note that by the finiteness of the state space

$$\mathbb{E}|M_t| < +\infty,$$

and that further by the Markov property

$$
\begin{aligned}
\mathbb{E}[M_t \mid \mathcal{F}_{t-1}] &= \lambda^{-t} \mathbb{E}[f(X_t) \mid \mathcal{F}_{t-1}] \\
&= \lambda^{-t} \sum_j P(X_{t-1}, j) f(j) \\
&= \lambda^{-t} \cdot \lambda f(X_{t-1}) \\
&= M_{t-1}.
\end{aligned}
$$

That is, (M_t) is a martingale. ◀

Or we can create martingales out of thin air. We give two important examples that will appear later.

Example 3.1.33 (Doob martingale: accumulating data). Let X with $\mathbb{E}|X| < +\infty$. Define $M_t = \mathbb{E}[X \mid \mathcal{F}_t]$. Note that $\mathbb{E}|M_t| \leq \mathbb{E}|X| < +\infty$ by Jensen's inequality, and

$$\mathbb{E}[M_t \mid \mathcal{F}_{t-1}] = \mathbb{E}[X \mid \mathcal{F}_{t-1}] = M_{t-1}$$

by the tower property. This is known as a *Doob martingale*. Intuitively, this process tracks our expectation of the unobserved X as "more information becomes available." See the exposure martingales in Section 3.2.3 for a concrete illustration of this idea. ◀

DOOB
MARTINGALE

Example 3.1.34 (Martingale transform). Let $(X_t)_{t \geq 1}$ be an integrable, adapted process and let $(C_t)_{t \geq 1}$ be a bounded, predictable process. Recall that predictable (Definition B.7.6) means $C_t \in \mathcal{F}_{t-1}$ for all t, that is, roughly speaking C_t is "known at time $t - 1$." Define

$$N_t = \sum_{i \leq t} (X_i - \mathbb{E}[X_i \mid \mathcal{F}_{i-1}]) C_i.$$

Then,

$$\mathbb{E}|N_t| \leq \sum_{i \leq t} 2 \mathbb{E}|X_t| K < +\infty,$$

where $|C_t| < K$ for all $t \geq 1$, and

$$
\begin{aligned}
\mathbb{E}[N_t - N_{t-1} \mid \mathcal{F}_{t-1}] &= \mathbb{E}[(X_t - \mathbb{E}[X_t \mid \mathcal{F}_{t-1}]) C_t \mid \mathcal{F}_{t-1}] \\
&= C_t(\mathbb{E}[X_t \mid \mathcal{F}_{t-1}] - \mathbb{E}[X_t \mid \mathcal{F}_{t-1}]) \\
&= 0,
\end{aligned}
$$

by taking out what is known. So (N_t) is a martingale.

When (X_t) is itself a martingale (in which case $\mathbb{E}[X_i \mid \mathcal{F}_{i-1}] = X_{i-1}$ in the definition of N_t), this is a sort of "stochastic (Stieltjes) integral." When, instead, (X_t) is a supermartingale (respectively submartingale) and (C_t) is non-negative and bounded, then the same computation shows that

$$N_t = \sum_{i \leq t} (X_i - X_{i-1}) C_i$$

defines a supermartingale (respectively submartingale).   ◀

As implied by the next lemma, an immediate consequence of Jensen's inequality (in its conditional version of Lemma B.6.12), submartingales naturally arise as convex functions of martingales.

Lemma 3.1.35 *If $(M_t)_{t \geq 0}$ is a martingale and ϕ is a convex function such that $\mathbb{E}|\phi(M_t)| < +\infty$ for all t, then $(\phi(M_t))_{t \geq 0}$ is a submartingale. Moreover, if $(M_t)_{t \geq 0}$ is a submartingale and ϕ is an increasing convex function with $\mathbb{E}|\phi(M_t)| < +\infty$ for all t, then $(\phi(M_t))_{t \geq 0}$ is a submartingale.*

Martingales and stopping times A fundamental reason explaining the utility of martingales in analyzing a variety of stochastic processes is that they play nicely with stopping times, in particular, through what is known as the *optional stopping theorem* (in its various forms). We will encounter many applications of this important result. First a definition:

Definition 3.1.36 *Let (M_t) be an adapted process and σ be a stopping time. Then,*

$$M_t^\sigma(\omega) := M_{\sigma(\omega) \wedge t}(\omega)$$

STOPPED
PROCESS
is M_t stopped at σ.

Lemma 3.1.37 *Let (M_t) be a supermartingale and σ be a stopping time. Then the stopped process (M_t^σ) is a supermartingale and in particular*

$$\mathbb{E}[M_t] \leq \mathbb{E}[M_{\sigma \wedge t}] \leq \mathbb{E}[M_0].$$

The same result holds with equalities if (M_t) is a martingale, and with inequalities in the opposite direction if (M_t) is a submartingale.

Proof Note that

$$M_t^\sigma - M_0 = \sum_{i \leq t} C_i (X_i - X_{i-1}),$$

with $C_i = \mathbf{1}\{i \leq \sigma\} \in \mathcal{F}_{i-1}$ (which is non-negative and bounded) and $X_i = M_i$ for all i, and use Example 3.1.34 to conclude that $\mathbb{E}[M_{\sigma \wedge t}] \leq \mathbb{E}[M_0]$.

On the other hand,

$$M_t - M_t^\sigma = \sum_{i \leq t} (1 - C_i)(X_i - X_{i-1}).$$

So the other inequality follows from the same argument.   ■

Theorem 3.1.38 (Doob's optional stopping theorem). *Let (M_t) be a supermartingale and σ be a stopping time. Then, M_σ is integrable and*

$$\mathbb{E}[M_\sigma] \leq \mathbb{E}[M_0]$$

if any of the following conditions hold:

(i) σ is bounded;

(ii) (M_t) is uniformly bounded and σ is almost surely finite;

(iii) $\mathbb{E}[\sigma] < +\infty$ and (M_t) has bounded increments (i.e., there is $c > 0$ such that $|M_t - M_{t-1}| \leq c$ a.s. for all t);

(iv) (M_t) is non-negative and σ is almost surely finite.

The first three imply equality if (M_t) is a martingale.

Proof Case (iv) is Fatou's lemma (Proposition B.4.14). We prove (iii). We leave the proof of the other claims as an exercise (see Exercise 3.5).

From Lemma 3.1.37, we have

$$\mathbb{E}[M_{\sigma \wedge t} - M_0] \leq 0. \tag{3.1.8}$$

Furthermore, the assumption that $\mathbb{E}[\sigma] < +\infty$ implies that $\sigma < +\infty$ almost surely. Hence, we seek to take a limit as $t \to +\infty$ *inside the expectation*. To justify swapping limit and expectation, note that by a telescoping sum

$$|M_{\sigma \wedge t} - M_0| \leq \left| \sum_{s \leq \sigma \wedge t} (M_s - M_{s-1}) \right|$$

$$\leq \sum_{s \leq \sigma} |M_s - M_{s-1}|$$

$$\leq c\sigma.$$

The claim now follows from dominated convergence (Proposition B.4.14). Equality holds if (M_t) is a martingale. ∎

Although the optional stopping theorem (Theorem 3.1.38) is useful, one often works directly with Lemma 3.1.37 and applies suitable limit theorems (see Proposition B.4.14). The following martingale-based proof of Wald's first identity provides an illustration.

Theorem 3.1.39 (Wald's first identity). *Let $X_1, X_2, \ldots \in L^1$ be i.i.d. with $\mathbb{E}[X_1] = \mu$ and let $\tau \in L^1$ be a stopping time. Let $S_t = \sum_{s=1}^{t} X_s$. Then,*

$$\mathbb{E}[S_\tau] = \mu \, \mathbb{E}[\tau].$$

Proof We first prove the result for non-negative X_is. By Example 3.1.29, $S_t - t\mu$ is a martingale and Lemma 3.1.37 implies that $\mathbb{E}[S_{\tau \wedge t} - \mu(\tau \wedge t)] = 0$, or

$$\mathbb{E}[S_{\tau \wedge t}] = \mu \, \mathbb{E}[\tau \wedge t].$$

Note that, in the non-negative case, we have $S_{\tau \wedge t} \uparrow S_\tau$ and $\tau \wedge t \uparrow \tau$. Thus, by monotone convergence (Proposition B.4.14), the claim $\mathbb{E}[S_\tau] = \mu \, \mathbb{E}[\tau]$ follows in that case.

Consider now the general case. Again, $\mathbb{E}[S_{\tau \wedge t}] = \mu \, \mathbb{E}[\tau \wedge t]$ and $\mathbb{E}[\tau \wedge t] \uparrow \mathbb{E}[\tau]$. Applying the previous argument to the sum of non-negative random variables $R_t = \sum_{s=1}^{t} |X_s|$ shows that $\mathbb{E}[R_\tau] = \mathbb{E}|X_1|\,\mathbb{E}[\tau] < +\infty$ by assumption. Since $|S_{\tau \wedge t}| \leq R_\tau$ for all t by the triangle inequality, dominated convergence (Proposition B.4.14) implies $\mathbb{E}[S_{\tau \wedge t}] \to \mathbb{E}[S_\tau]$ and we are done. ∎

We also recall Wald's second identity. The proof, which we omit, uses the martingale in Example 3.1.31.

Theorem 3.1.40 (Wald's second identity). *Let $X_1, X_2, \ldots \in L^2$ be i.i.d. with $\mathbb{E}[X_1] = 0$ and $\mathrm{Var}[X_1] = \sigma^2$ and let $\tau \in L^1$ be a stopping time. Then,*

$$\mathbb{E}[S_\tau^2] = \sigma^2 \mathbb{E}[\tau].$$

GAMBLER'S
RUIN

We illustrate Wald's identities on the *gambler's ruin* problem that is characteristic of applications of stopping times in Markov chains. We consider the "unbiased" and "biased" cases separately.

Example 3.1.41 (Gambler's ruin: unbiased case). Let (S_t) be simple random walk on $\mathbb{Z}$ started at 0 and let $\tau = \tau_a \wedge \tau_b$, where $-\infty < a < 0 < b < +\infty$, where the first visit time τ_x was defined formally in Definition 3.1.6.

Claim 3.1.42 *We have:*

(i) $\tau < +\infty$ *almost surely;*
(ii) $\mathbb{P}[\tau_a < \tau_b] = \frac{b}{b-a}$;
(iii) $\mathbb{E}[\tau] = -ab$;
(iv) $\tau_a < +\infty$ *almost surely but $\mathbb{E}[\tau_a] = +\infty$.*

Proof We prove the claims in order.

(i) We argue that in fact $\mathbb{E}[\tau] < \infty$. That follows immediately from the exponential tail of hitting times in Lemma 3.1.25 for the chain $(S_{\tau \wedge t})$ whose (effective) state space, $\{a, a+1, \ldots, b\}$, is finite.

(ii) By Wald's first identity (Theorem 3.1.39) and (i), we have $\mathbb{E}[S_\tau] = 0$ or

$$a\,\mathbb{P}[S_\tau = a] + b\,\mathbb{P}[S_\tau = b] = 0,$$

that is, using $\mathbb{P}[S_\tau = a] = 1 - \mathbb{P}[S_\tau = b] = \mathbb{P}[\tau_a < \tau_b]$,

$$\mathbb{P}[\tau_a < \tau_b] = \frac{b}{b-a} \qquad \text{and} \qquad \mathbb{P}[\tau_a < +\infty] \geq \mathbb{P}[\tau_a < \tau_b] \to 1,$$

where we took $b \to \infty$ in the first expression to obtain the second one.

(iii) Because $\sigma^2 = 1$, Wald's second identity (Theorem 3.1.40) says that $\mathbb{E}[S_\tau^2] = \mathbb{E}[\tau]$. Furthermore, we have by (ii),

$$\mathbb{E}[S_\tau^2] = \frac{b}{b-a}a^2 + \frac{-a}{b-a}b^2 = -ab.$$

Thus, $\mathbb{E}[\tau] = -ab$.

(iv) The first claim was proved in (ii). When $b \to +\infty$, $\tau = \tau_a \wedge \tau_b \uparrow \tau_a$ and monotone convergence applied to (iii) gives that $\mathbb{E}[\tau_a] = +\infty$.

That concludes the proof. ∎

Note that (iv) shows that the L^1 condition on the stopping time in Wald's second identity (Theorem 3.1.40) is necessary. Indeed, we have shown $a^2 = \mathbb{E}[S_{\tau_a}^2] \neq \sigma^2 \mathbb{E}[\tau_a] = +\infty$. ◄

Example 3.1.43 (Gambler's ruin: biased case). The *biased random walk on* $\mathbb{Z}$ with parameter $1/2 < p < 1$ is the process (S_t) with $S_0 = 0$ and $S_t = \sum_{s \leq t} X_s$, where the X_ss are i.i.d. in $\{-1, +1\}$ with $\mathbb{P}[X_1 = 1] = p$. Let again $\tau := \tau_a \wedge \tau_b$, where $a < 0 < b$. Define $q := 1 - p$, $\delta := p - q > 0$, and $\phi(x) := (q/p)^x$.

Claim 3.1.44 *We have:*

(i) $\tau < +\infty$ *almost surely;*
(ii) $\mathbb{P}[\tau_a < \tau_b] = \frac{\phi(b)-\phi(0)}{\phi(b)-\phi(a)}$;
(iii) $\mathbb{E}[\tau_b] = \frac{b}{2p-1}$;
(iv) $\tau_a = -\infty$ *with positive probability.*

Proof Let $\psi_t(x) := x - \delta t$. We use *two* martingales: $(\phi(S_t))$ and $(\psi_t(S_t))$. Observe that indeed both processes are clearly integrable and

$$\mathbb{E}[\phi(S_t) \mid \mathcal{F}_{t-1}] = p(q/p)^{S_{t-1}+1} + q(q/p)^{S_{t-1}-1} = \phi(S_{t-1})$$

and

$$\mathbb{E}[\psi_t(S_t) \mid \mathcal{F}_{t-1}] = p[S_{t-1} + 1 - \delta t] + q[S_{t-1} - 1 - \delta t] = \psi_{t-1}(S_{t-1}).$$

(i) This claim follows by the same argument as in the unbiased case.
(ii) Note that $(\phi(S_t))$ is a non-negative, bounded martingale since $q < p$ by assumption. By Lemma 3.1.37 and dominated convergence (Proposition B.4.14),

$$\phi(0) = \mathbb{E}[\phi(S_\tau)] = \mathbb{P}[\tau_a < \tau_b]\phi(a) + \mathbb{P}[\tau_a > \tau_b]\phi(b),$$

or, rearranging, $\mathbb{P}[\tau_a < \tau_b] = \frac{\phi(b)-\phi(0)}{\phi(b)-\phi(a)}$. Taking $b \to +\infty$, by monotonicity

$$\mathbb{P}[\tau_a < +\infty] = \frac{1}{\phi(a)} < 1, \tag{3.1.9}$$

so that $\tau_a = +\infty$ with positive probability. On the other hand, $\mathbb{P}[\tau_b < \tau_a] = 1 - \mathbb{P}[\tau_a < \tau_b] = \frac{\phi(0)-\phi(a)}{\phi(b)-\phi(a)}$, and taking $a \to -\infty$,

$$P[\tau_b < +\infty] = 1.$$

(iii) By Lemma 3.1.37 applied to $(\psi_t(S_t))$,

$$0 = \mathbb{E}[S_{\tau_b \wedge t} - \delta(\tau_b \wedge t)]. \tag{3.1.10}$$

By monotone convergence (Proposition B.4.14), $\mathbb{E}[\tau_b \wedge t] \uparrow \mathbb{E}[\tau_b]$. Furthermore, observe that $-\inf_t S_t \geq 0$ almost surely since $S_0 = 0$. Moreover, for $x \geq 0$, by (3.1.9),

$$\mathbb{P}[-\inf_t S_t \geq x] = \mathbb{P}[\tau_{-x} < +\infty] = \left(\frac{q}{p}\right)^x$$

so that $\mathbb{E}[-\inf_t S_t] = \sum_{x \geq 1} \mathbb{P}[-\inf_t S_t \geq x] < +\infty$. Hence, in (3.1.10), we can use dominated convergence (Proposition B.4.14) with

$$|S_{\tau_b \wedge t}| \leq \max\{b, -\inf_t S_t\},$$

and the fact that $\tau_b < +\infty$ almost surely from (ii) to deduce that $\mathbb{E}[\tau_b] = \frac{\mathbb{E}[S_{\tau_b}]}{p-q} = \frac{b}{2p-1}$.

(iv) That claim was proved in (ii).

That concludes the proof.   ∎

Note that, in (iii) above, in order to apply Wald's first identity directly we would have had to prove that $\tau_b \in L^1$ first.   ◀

We also obtain the following maximal version of Markov's inequality (Theorem 2.1.1).

Theorem 3.1.45 (Doob's submartingale inequality). *Let (M_t) be a non-negative submartingale. Then, for $b > 0$,*

$$\mathbb{P}\left[\sup_{0 \leq s \leq t} M_s \geq b\right] \leq \frac{\mathbb{E}[M_t]}{b}.$$

Observe that a naive application of Markov's inequality implies only that

$$\sup_{0 \leq s \leq t} \mathbb{P}[M_s \geq b] \leq \frac{\mathbb{E}[M_t]}{b},$$

where we used that $\mathbb{E}[M_s] \leq \mathbb{E}[M_t]$ for all $0 \leq s \leq t$ for a submartingale. Introducing an appropriate stopping time immediately gives something stronger. (Exercise 3.6 asks for the supermartingale version of this.)

Proof Let σ be the first time that $M_t \geq b$. Then the event of interest can be characterized as

$$\left\{\sup_{0 \leq s \leq t} M_s \geq b\right\} = \{M_{\sigma \wedge t} \geq b\}.$$

By Markov's inequality,

$$\mathbb{P}[M_{\sigma \wedge t} \geq b] \leq \frac{\mathbb{E}[M_{\sigma \wedge t}]}{b}.$$

Lemma 3.1.37 implies that $\mathbb{E}[M_{\sigma \wedge t}] \leq \mathbb{E}[M_t]$, which concludes the proof.   ∎

One consequence of the previous bound is a strengthening of Chebyshev's inequality (Theorem 2.1.2) for sums of independent random variables.

Corollary 3.1.46 (Kolmogorov's maximal inequality). *Let $X_1, X_2, \ldots$ be independent random variables with $\mathbb{E}[X_i] = 0$ and $\mathrm{Var}[X_i] < +\infty$. Define $S_t = \sum_{i \leq t} X_i$. Then, for $\beta > 0$,*

$$\mathbb{P}\left[\max_{i \leq t} |S_i| \geq \beta\right] \leq \frac{\mathrm{Var}[S_t]}{\beta^2}.$$

Proof By Example 3.1.29, (S_t) is a martingale. By Lemma 3.1.35, (S_t^2) is hence a (non-negative) submartingale. The result follows from Doob's submartingale inequality (Theorem 3.1.45).   ∎

Convergence Finally another fundamental result about martingales is the following convergence theorem, which we state without proof. We give a quick application in Example 3.1.49.

Theorem 3.1.47 (Martingale convergence theorem). *Let (M_t) be a supermartingale bounded in L^1. Then, (M_t) converges almost surely to a finite limit M_∞. Moreover, letting $M_\infty :=$ $\limsup_t M_t$, then $M_\infty \in \mathcal{F}_\infty$ and $\mathbb{E}|M_\infty| < +\infty$.*

Corollary 3.1.48 (Convergence of non-negative supermartingales). *If (M_t) is a non-negative supermartingale, then M_t converges almost surely to a finite limit M_∞ with $\mathbb{E}[M_\infty] \le \mathbb{E}[M_0]$.*

Proof By the supermartingale property, (M_t) is bounded in L^1 since

$$\mathbb{E}|M_t| = \mathbb{E}[M_t] \le \mathbb{E}[M_0], \; \forall t.$$

Then we use the martingale convergence theorem (Theorem 3.1.47) and Fatou's lemma (Proposition B.4.14). ∎

Example 3.1.49 (Pólya's urn). An urn contains one red ball and one green ball. At each time, we pick one ball and put it back with an extra ball of the same color. This process is known as *Pólya's urn*. Let R_t (respectively G_t) be the number of red balls (respectively green balls) after the tth draw. Let

$$\mathcal{F}_t := \sigma(R_0, G_0, R_1, G_1, \dots, R_t, G_t).$$

Define M_t to be the fraction of green balls after the tth draw. Then,

$$\begin{aligned}
\mathbb{E}[M_t \mid \mathcal{F}_{t-1}] &= \frac{R_{t-1}}{G_{t-1} + R_{t-1}} \frac{G_{t-1}}{G_{t-1} + R_{t-1} + 1} \\
&\quad + \frac{G_{t-1}}{G_{t-1} + R_{t-1}} \frac{G_{t-1} + 1}{G_{t-1} + R_{t-1} + 1} \\
&= \frac{G_{t-1}}{G_{t-1} + R_{t-1}} \\
&= M_{t-1}.
\end{aligned}$$

Since $M_t \ge 0$ and is a martingale, we have $M_t \to M_\infty$ almost surely. In fact, Exercise 3.4 asks for a proof that

$$\mathbb{P}[G_t = m + 1] = \binom{t}{m} \frac{m!(t-m)!}{(t+1)!} = \frac{1}{t+1}.$$

So taking a limit as $t \to +\infty$,

$$\mathbb{P}[M_t \le x] = \frac{\lfloor x(t+2) - 1 \rfloor}{t+1} \to x.$$

That is, (M_t) converges in distribution to a uniform random variable on $[0, 1]$. ◄

Convergence of the expectation in general requires stronger conditions. A simple case is boundedness in L^2. Before stating the result, we derive a key property of martingales in L^2 which will be useful later.

Lemma 3.1.50 (Orthogonality of increments). *Let (M_t) be a martingale with $M_t \in L^2$. Let $s \leq t \leq u \leq v$. Then,*

$$\langle M_t - M_s, M_v - M_u \rangle = 0,$$

where $\langle X, Y \rangle = \mathbb{E}[XY]$.

Proof Use $M_u = \mathbb{E}[M_v \mid \mathcal{F}_u]$ and $M_t - M_s \in \mathcal{F}_u$, and apply the L^2 characterization of the conditional expectation (Theorem B.6.2). ∎

In other words, martingale increments over disjoint time intervals are uncorrelated (provided the second moment exists). Note that this is weaker than the independence of increments of random walks. (See Section 3.2.1 for more discussion on this.)

Theorem 3.1.51 (Convergence of martingales bounded in L^2). *Let (M_t) be a martingale with $M_t \in L^2$. Then, (M_t) is bounded in L^2 if and only if*

$$\sum_{k \geq 1} \mathbb{E}[(M_t - M_{t-1})^2] < +\infty.$$

When this is the case, M_t converges almost surely and in L^2 to a finite limit M_∞, and furthermore

$$\mathbb{E}[M_t] \to \mathbb{E}[M_\infty] < +\infty$$

as $t \to +\infty$.

Proof Writing M_t as a telescoping sum of increments, the orthogonality of increments (Lemma 3.1.50) implies

$$\mathbb{E}[M_t^2] = \mathbb{E}\left[\left(M_0 + \sum_{s=1}^{t} (M_s - M_{s-1}) \right)^2 \right]$$

$$= \mathbb{E}[M_0^2] + \sum_{s=1}^{t} \mathbb{E}[(M_s - M_{s-1})^2],$$

proving the first claim.

By the monotonicity of norms (Lemma B.4.16), (M_t) bounded in L^2 implies that (M_t) is bounded in L^1, which, in turn, implies that M_t converges almost surely to a finite limit M_∞ with $\mathbb{E}|M_\infty| < +\infty$ by Theorem 3.1.47. Then, using Fatou's lemma (Proposition B.4.14) in

$$\mathbb{E}[(M_{t+s} - M_t)^2] = \sum_{t+1 \leq i \leq t+s} \mathbb{E}[(M_i - M_{i-1})^2]$$

gives

$$\mathbb{E}[(M_\infty - M_t)^2] \leq \sum_{t+1 \leq i} \mathbb{E}[(M_i - M_{i-1})^2].$$

The right-hand side goes to 0 since the series is finite, which proves the second claim.

The last claim follows from Lemmas B.4.16 and B.4.17. ∎

3.1.4 ▷ *Percolation: Critical Regime on Infinite d-Regular Tree*

Consider bond percolation (see Definition 1.2.1) on the infinite d-regular tree $\mathbb{T}_d$ rooted at a vertex 0. In Section 2.3.3, we showed that

$$p_c(\mathbb{T}_d) = \sup\{p \in [0,1] \colon \mathbb{P}_p[|\mathcal{C}_0| = +\infty] = 0\} = \frac{1}{d-1},$$

where recall that $\mathcal{C}_0$ is the open cluster of the root. Here we consider the critical case, that is, we set density $p = \frac{1}{d-1}$. (The same results apply to the infinite b-ary tree $\widehat{\mathbb{T}}_b$ with $d = b+1$.) Assume $d \geq 3$ (since $d = 2$ is simply a path).

First:

Claim 3.1.52 $|\mathcal{C}_0| < +\infty$ *almost surely.*

Let $X_n := |\partial_n \cap \mathcal{C}_0|$, where ∂_n are the nth level vertices. In Section 2.3.3, we proved the same claim in the subcritical case using the first moment method. It does not work here because

$$\mathbb{E}X_n = d(d-1)^{n-1}p^n = \frac{d}{d-1} \nrightarrow 0.$$

Instead, we use a martingale argument which will be generalized when we discuss branching processes in Section 6.1.

Proof of Claim 3.1.52 Let $b := d - 1$ be the branching ratio. Because the root has a different number of children, we consider the descendants of its children. Let Z_n be the number of vertices in the open cluster of the first child of the root n levels below it and let $\mathcal{F}_n = \sigma(Z_0, \ldots, Z_n)$. Then, $Z_0 = 1$ and

$$\mathbb{E}[Z_n \mid \mathcal{F}_{n-1}] = bpZ_{n-1} = Z_{n-1}.$$

So (Z_n) is a non-negative, integer-valued martingale and it converges almost surely to a finite limit by Corollary 3.1.48. (In particular, $\mathbb{E}[Z_n] = 1$, which will be useful below.) But, clearly, for any integer $k > 0$ and $N \geq 0$,

$$\mathbb{P}[Z_n = k \ \forall n \geq N] = 0,$$

so it must be that the limit is 0 almost surely. In other words, Z_n is eventually 0 for all n large enough. This is true for every child of the root. Hence, the open cluster of the root is finite almost surely. ∎

On the other hand:

Claim 3.1.53

$$\mathbb{E}|\mathcal{C}_0| = +\infty.$$

Proof Consider the descendant subtree, T_1, of the first child of the root, which we denote by 1. Let $\widetilde{\mathcal{C}}_1$ be the open cluster of 1 in T_1. As we showed in the previous claim, the expected number of vertices on any level of T_1 is 1. So $\mathbb{E}|\widetilde{\mathcal{C}}_1| = +\infty$ by summing over the levels. ∎

3.2 Concentration for Martingales and Applications

The Chernoff–Cramér method extends naturally to martingales. This observation leads to powerful new tail bounds that hold far *beyond the case of sums of independent variables*. In particular, it will allow us to prove one version of the more general concentration phenomenon, which can be stated informally as: a function $f(X_1, \ldots, X_n)$ of many independent random variables that is not too sensitive to any of its coordinates tends to be close to its mean.

3.2.1 Azuma–Hoeffding Inequality

The main result of this section is the following generalization of Hoeffding's inequality (Theorem 2.4.10).

Theorem 3.2.1 (Maximal Azuma–Hoeffding inequality). *Let* $(Z_t)_{t \in \mathbb{Z}_+}$ *be a martingale with respect to the filtration* $(\mathcal{F}_t)_{t \in \mathbb{Z}_+}$. *Assume that there are predictable processes* (A_t) *and* (B_t) *(i.e.,* $A_t, B_t \in \mathcal{F}_{t-1}$*) and constants* $0 < c_t < +\infty$ *such that for all* $t \geq 1$, *almost surely,*

$$A_t \leq Z_t - Z_{t-1} \leq B_t \qquad and \qquad B_t - A_t \leq c_t.$$

Then, for all $\beta > 0$,

$$\mathbb{P}\left[\sup_{0 \leq i \leq t} (Z_i - Z_0) \geq \beta \right] \leq \exp\left(-\frac{2\beta^2}{\sum_{i \leq t} c_i^2} \right).$$

Applying this inequality to $(-Z_t)$ gives a tail bound in the other direction.

Proof of Theorem 3.2.1   As in the Chernoff–Cramér method, we start by applying Markov's inequality (Theorem 2.1.1). Here we use the maximal version for submartingales, Doob's submartingale inequality (Theorem 3.1.45). First notice that e^{sx} is increasing and convex for $s > 0$, so that by Lemma 3.1.35 the process $(e^{s(Z_t - Z_0)})_t$ is a submartingale. Hence, for $s > 0$, by Theorem 3.1.45

$$\begin{aligned}
\mathbb{P}\left[\sup_{0 \leq i \leq t} (Z_i - Z_0) \geq \beta \right] &= \mathbb{P}\left[\sup_{0 \leq i \leq t} e^{s(Z_i - Z_0)} \geq e^{s\beta} \right] \\
&\leq \frac{\mathbb{E}\left[e^{s(Z_t - Z_0)} \right]}{e^{s\beta}} \\
&= \frac{\mathbb{E}\left[e^{s \sum_{r=1}^{t} (Z_r - Z_{r-1})} \right]}{e^{s\beta}}.
\end{aligned} \qquad (3.2.1)$$

Unlike the Chernoff–Cramér case, however, the terms in the exponent are not independent. Instead, to exploit the martingale property, we condition on the filtration. By taking out what is known (Lemma B.6.13),

$$\mathbb{E}\left[\mathbb{E}\left[e^{s \sum_{r=1}^{t} (Z_r - Z_{r-1})} \,\Big|\, \mathcal{F}_{t-1} \right] \right] = \mathbb{E}\left[e^{s \sum_{r=1}^{t-1} (Z_r - Z_{r-1})} \, \mathbb{E}\left[e^{s(Z_t - Z_{t-1})} \,\Big|\, \mathcal{F}_{t-1} \right] \right].$$

The martingale property and the assumption in the statement imply that, conditioned on $\mathcal{F}_{t-1}$, the random variable $Z_t - Z_{t-1}$ is centered and lies in an interval of length c_t. Hence, by Hoeffding's lemma (Lemma 2.4.12), it holds almost surely that

$$\mathbb{E}\left[e^{s(Z_t - Z_{t-1})} \,\big|\, \mathcal{F}_{t-1}\right] \leq \exp\left(\frac{s^2 c_t^2/4}{2}\right) = \exp\left(\frac{c_t^2 s^2}{8}\right). \tag{3.2.2}$$

Using the tower property (Lemma B.6.16) and arguing by induction, we obtain

$$\mathbb{E}\left[e^{s(Z_t - Z_0)}\right] \leq \exp\left(\frac{s^2 \sum_{r \leq t} c_r^2}{8}\right).$$

Put differently, we have proved that $Z_t - Z_0$ is sub-Gaussian with variance factor $\frac{1}{4}\sum_{r \leq t} c_r^2$. By (2.4.16) (or, equivalently, by choosing $s = \beta / \frac{1}{4}\sum_{r \leq t} c_r^2$ in (3.2.1)), we get the result. $\blacksquare$

In Theorem 3.2.1, the *martingale difference* sequence (X_t), where $X_t := Z_t - Z_{t-1}$, is not only "pairwise uncorrelated" by Lemma 3.1.50, that is,

$$\mathbb{E}[X_s X_r] = 0 \qquad \forall r \neq s,$$

but it is in fact "mutually uncorrelated," that is,

$$\mathbb{E}\left[X_{j_1} \cdots X_{j_k}\right] = 0 \qquad \forall k \geq 1,\ \forall\, 1 \leq j_1 < \cdots < j_k.$$

This stronger property helps explain why $\sum_{r \leq t} X_r$ is highly concentrated. This point is the subject of Exercise 3.7, which guides the reader through a slightly different proof of the Azuma–Hoeffding inequality. Compare with Exercises 2.5 and 2.6.

MARTINGALE
DIFFERENCE

3.2.2 Method of Bounded Differences

The power of the maximal Azuma–Hoeffding inequality (Theorem 3.2.1) is that it produces tail inequalities for quantities other than sums of independent variables. The setting is the following. Let $X_1, \ldots, X_n$ be independent random variables where X_i is $\mathcal{X}_i$-valued for all i and let $X = (X_1, \ldots, X_n)$. Assume that $f : \mathcal{X}_1 \times \cdots \times \mathcal{X}_n \to \mathbb{R}$ is a measurable function. Our goal is to characterize the concentration properties of $f(X)$ around its expectation in terms of its "discrete derivatives":

$$D_i f(x) := \sup_{y \in \mathcal{X}_i} f(x_1, \ldots, x_{i-1}, y, x_{i+1}, \ldots, x_n)$$

$$- \inf_{y' \in \mathcal{X}_i} f(x_1, \ldots, x_{i-1}, y', x_{i+1}, \ldots, x_n),$$

where $x = (x_1, \ldots, x_n) \in \mathcal{X}_1 \times \cdots \times \mathcal{X}_n$. We think of $D_i f(x)$ as a measure of the "sensitivity" of f to its ith coordinate.

High-level idea

We begin with two easier bounds that we will improve below. The trick to analyzing the concentration of $f(X)$ is to consider the Doob martingale (see Example 3.1.33)

$$Z_i = \mathbb{E}[f(X) \,|\, \mathcal{F}_i], \tag{3.2.3}$$

where $\mathcal{F}_i = \sigma(X_1, \ldots, X_i)$, which is well defined provided $\mathbb{E}|f(X)| < +\infty$. Note that

$$Z_n = \mathbb{E}[f(X) \mid \mathcal{F}_n] = f(X)$$

and

$$Z_0 = \mathbb{E}[f(X)],$$

so that we can write

$$f(X) - \mathbb{E}[f(X)] = \sum_{i=1}^{n} (Z_i - Z_{i-1}).$$

Intuitively, the martingale difference $Z_i - Z_{i-1}$ tracks the change in our expectation of $f(X)$ as X_i is revealed.

In fact, a clever probabilistic argument relates martingale differences directly to discrete derivatives. Let $X' = (X'_1, \ldots, X'_n)$ be an independent copy of X and let

$$X^{(i)} = (X_1, \ldots, X_{i-1}, X'_i, X_{i+1}, \ldots, X_n).$$

Then,

$$\begin{aligned}
Z_i - Z_{i-1} &= \mathbb{E}[f(X) \mid \mathcal{F}_i] - \mathbb{E}[f(X) \mid \mathcal{F}_{i-1}] \\
&= \mathbb{E}[f(X) \mid \mathcal{F}_i] - \mathbb{E}[f(X^{(i)}) \mid \mathcal{F}_{i-1}] \\
&= \mathbb{E}[f(X) \mid \mathcal{F}_i] - \mathbb{E}[f(X^{(i)}) \mid \mathcal{F}_i] \\
&= \mathbb{E}[f(X) - f(X^{(i)}) \mid \mathcal{F}_i].
\end{aligned}$$

Note that we crucially used the independence of the X_ks in the second and third lines. But then, by Jensen's inequality (Lemma B.6.12),

$$|Z_i - Z_{i-1}| \le \|D_i f\|_\infty. \tag{3.2.4}$$

By the orthogonality of increments of martingales in L^2 (Lemma 3.1.50), we immediately obtain a bound on the variance of f

$$\mathrm{Var}[f(X)] = \mathbb{E}[(Z_n - Z_0)^2] = \sum_{i=1}^{n} \mathbb{E}\left[(Z_i - Z_{i-1})^2\right] \le \sum_{i=1}^{n} \|D_i f\|_\infty^2. \tag{3.2.5}$$

By the maximal Azuma–Hoeffding inequality and the fact that

$$Z_i - Z_{i-1} \in [-\|D_i f\|_\infty, \|D_i f\|_\infty],$$

we also get a bound on the tail

$$\mathbb{P}[f(X) - \mathbb{E}[f(X)] \ge \beta] \le \exp\left(-\frac{\beta^2}{2 \sum_{i \le n} \|D_i f\|_\infty^2}\right). \tag{3.2.6}$$

A more careful analysis, which we detail below, leads to better bounds.

We emphasize that, although it may not be immediately obvious, independence plays a crucial role in the bound (3.2.4), as the next example shows.

Example 3.2.2 (A counterexample). Let $f(x_1, \ldots, x_n) = x_1 + \cdots + x_n$, where $x_i \in \{-1, 1\}$ for all i. Then,

$$\|D_1 f\|_\infty = \sup_{x_2, \ldots, x_n} [(1 + x_2 + \cdots + x_n) - (-1 + x_2 + \cdots + x_n)] = 2,$$

and similarly, $\|D_i f\|_\infty = 2$ for $i = 2, \ldots, n$. Let X_1 be a uniform random variable on $\{-1, 1\}$. First consider the case where we set $X_2, \ldots, X_n$ all equal to X_1. Then,

$$\mathbb{E}[f(X_1, \ldots, X_n)] = 0$$

and

$$\mathbb{E}[f(X_1, \ldots, X_n) \,|\, X_1] = nX_1,$$

so that

$$|\mathbb{E}[f(X_1, \ldots, X_n) \,|\, X_1] - \mathbb{E}[f(X_1, \ldots, X_n)]| = n > 2.$$

In particular, the corresponding Doob martingale does not have increments bounded by $\|D_i f\|_\infty = 2$.

For a less extreme example that has support over all of $\{-1, 1\}^n$, let

$$U_i = \begin{cases} 1 & \text{w.p. } 1 - \varepsilon, \\ -1 & \text{w.p. } \varepsilon, \end{cases}$$

for some $\varepsilon > 0$ independently for all $i = 1, \ldots, n - 1$. Let again X_1 be a uniform random variable on $\{-1, 1\}$ and, for $i = 2, \ldots, n$, define the random variable $X_i = U_{i-1} X_{i-1}$, that is, X_i is the same as X_{i-1} with probability $1 - \varepsilon$ and otherwise is flipped. Then,

$$\begin{aligned} \mathbb{E}[f(X_1, \ldots, X_n)] &= \mathbb{E}\left[X_1 + \cdots + X_n\right] \\ &= \mathbb{E}\left[X_1 \left(1 + \sum_{i=1}^{n-1} \prod_{j \leq i} U_j\right)\right] \\ &= \mathbb{E}[X_1] \, \mathbb{E}\left[1 + \sum_{i=1}^{n-1} \prod_{j \leq i} U_j\right] \\ &= 0, \end{aligned}$$

by the independence of X_1 and the U_is. Similarly,

$$\mathbb{E}[f(X_1, \ldots, X_n) \,|\, X_1] = X_1 \, \mathbb{E}\left[1 + \sum_{i=1}^{n-1} \prod_{j \leq i} U_j\right] = X_1 \left(\sum_{i=0}^{n-1} (1 - 2\varepsilon)^i\right),$$

so that

$$|\mathbb{E}[f(X_1, \ldots, X_n) \,|\, X_1] - \mathbb{E}[f(X_1, \ldots, X_n)]| = \left(\sum_{i=0}^{n-1} (1 - 2\varepsilon)^i\right) > 2,$$

for ε small enough and $n \geq 3$. In particular, the corresponding Doob martingale does not have increments bounded by $\|D_i f\|_\infty = 2$. ◄

Variance bounds

We give improved bounds on the variance. Our first bound explicitly decomposes the variance of $f(X)$ over the contributions of its individual entries.

Theorem 3.2.3 (Tensorization of the variance). *Let $X_1, \ldots, X_n$ be independent random variables where X_i is $\mathcal{X}_i$-valued for all i and let $X = (X_1, \ldots, X_n)$. Assume that $f : \mathcal{X}_1 \times \cdots \times \mathcal{X}_n \to \mathbb{R}$ is a measurable function with $\mathbb{E}[f(X)^2] < +\infty$. Define $\mathcal{F}_i = \sigma(X_1, \ldots, X_i)$, $\mathcal{G}_i = \sigma(X_1, \ldots, X_{i-1}, X_{i+1}, \ldots, X_n)$, and $Z_i = \mathbb{E}[f(X) \mid \mathcal{F}_i]$. Then, we have*

$$\mathrm{Var}[f(X)] \leq \sum_{i=1}^{n} \mathbb{E}\left[\mathrm{Var}\left[f(X) \mid \mathcal{G}_i\right]\right].$$

(It may be helpful to recall the formula $\mathrm{Var}[Y] = \mathbb{E}\left[\mathrm{Var}\left[Y \mid \mathcal{H}\right]\right] + \mathrm{Var}\left[\mathbb{E}\left[Y \mid \mathcal{H}\right]\right]$.)

Proof of Theorem 3.2.3   The key lemma is the following.

Lemma 3.2.4

$$\mathbb{E}\left[\mathbb{E}\left[f(X) \mid \mathcal{G}_i\right] \mid \mathcal{F}_i\right] = \mathbb{E}\left[f(X) \mid \mathcal{F}_{i-1}\right]$$

Proof   By the tower property (Lemma B.6.16),

$$\mathbb{E}\left[f(X) \mid \mathcal{F}_{i-1}\right] = \mathbb{E}\left[\mathbb{E}\left[f(X) \mid \mathcal{G}_i\right] \mid \mathcal{F}_{i-1}\right].$$

Moreover, $\sigma(X_i)$ is independent of $\sigma(\mathcal{G}_i, \mathcal{F}_{i-1})$, so by the role of independence (Lemma B.6.14), we have

$$\mathbb{E}\left[\mathbb{E}\left[f(X) \mid \mathcal{G}_i\right] \mid \mathcal{F}_{i-1}\right] = \mathbb{E}\left[\mathbb{E}\left[f(X) \mid \mathcal{G}_i\right] \mid \mathcal{F}_{i-1}, X_i\right] = \mathbb{E}\left[\mathbb{E}\left[f(X) \mid \mathcal{G}_i\right] \mid \mathcal{F}_i\right].$$

Combining the last two displays gives the result.   ■

Again, we take advantage of the orthogonality of increments to write

$$\mathrm{Var}[f(X)] = \sum_{i=1}^{n} \mathbb{E}\left[(Z_i - Z_{i-1})^2\right].$$

By Lemma 3.2.4,

$$\begin{aligned}
(Z_i - Z_{i-1})^2 &= (\mathbb{E}\left[f(X) \mid \mathcal{F}_i\right] - \mathbb{E}\left[f(X) \mid \mathcal{F}_{i-1}\right])^2 \\
&= (\mathbb{E}\left[f(X) \mid \mathcal{F}_i\right] - \mathbb{E}\left[\mathbb{E}\left[f(X) \mid \mathcal{G}_i\right] \mid \mathcal{F}_i\right])^2 \\
&= (\mathbb{E}\left[f(X) - \mathbb{E}\left[f(X) \mid \mathcal{G}_i\right] \mid \mathcal{F}_i\right])^2 \\
&\leq \mathbb{E}\left[(f(X) - \mathbb{E}\left[f(X) \mid \mathcal{G}_i\right])^2 \mid \mathcal{F}_i\right],
\end{aligned}$$

where we used Jensen's inequality on the last line. Taking expectations and using the tower property,

$$\mathrm{Var}[f(X)] = \sum_{i=1}^{n} \mathbb{E}\left[(Z_i - Z_{i-1})^2\right]$$

$$\leq \sum_{i=1}^{n} \mathbb{E}\left[\mathbb{E}\left[(f(X) - \mathbb{E}[f(X)\,|\,\mathcal{G}_i])^2\,|\,\mathcal{F}_i\right]\right]$$

$$= \sum_{i=1}^{n} \mathbb{E}\left[(f(X) - \mathbb{E}[f(X)\,|\,\mathcal{G}_i])^2\right]$$

$$= \sum_{i=1}^{n} \mathbb{E}\left[\mathbb{E}\left[(f(X) - \mathbb{E}[f(X)\,|\,\mathcal{G}_i])^2\,|\,\mathcal{G}_i\right]\right]$$

$$= \sum_{i=1}^{n} \mathbb{E}\left[\mathrm{Var}\left[f(X)\,|\,\mathcal{G}_i\right]\right].$$

That concludes the proof.   ∎

We derive two useful consequences of the tensorization property of the variance. The first one is the *Efron–Stein inequality*.

Theorem 3.2.5 (Efron–Stein inequality). *Let $X_1,\ldots,X_n$ be independent random variables where X_i is $\mathcal{X}_i$-valued for all i and let $X = (X_1,\ldots,X_n)$. Assume that $f : \mathcal{X}_1 \times \cdots \times \mathcal{X}_n \to \mathbb{R}$ is a measurable function with $\mathbb{E}[f(X)^2] < +\infty$. Let $X' = (X_1',\ldots,X_n')$ be an independent copy of X and*

$$X^{(i)} = (X_1,\ldots,X_{i-1},X_i',X_{i+1},\ldots,X_n).$$

Then,

$$\mathrm{Var}[f(X)] \leq \frac{1}{2} \sum_{i=1}^{n} \mathbb{E}[(f(X) - f(X^{(i)}))^2].$$

Proof   Observe that if Y' is an independent copy of $Y \in L^2$, then $\mathrm{Var}[Y] = \frac{1}{2}\mathbb{E}[(Y - Y')^2]$, which can be seen by adding and subtracting the mean, expanding and using independence. Hence,

$$\mathrm{Var}\left[f(X)\,|\,\mathcal{G}_i\right] = \frac{1}{2}\mathbb{E}[(f(X) - f(X^{(i)}))^2\,|\,\mathcal{G}_i],$$

where we used the independence of the X_is and X_i's. Plugging back into Theorem 3.2.3 gives the claim.   ∎

Our second consequence of Theorem 3.2.3 is a Poincaré-type inequality which relates the variance of a function to its *expected* "square gradient." Compare to the much weaker (3.2.5), which involves in each term a supremum rather than an expectation.

Theorem 3.2.6 (Bounded differences inequality). *Let $X_1,\ldots,X_n$ be independent random variables where X_i is $\mathcal{X}_i$-valued for all i and let $X = (X_1,\ldots,X_n)$. Assume that $f : \mathcal{X}_1 \times \cdots \times \mathcal{X}_n \to \mathbb{R}$ is a measurable function with $\mathbb{E}[f(X)^2] < +\infty$. Then,*

$$\mathrm{Var}[f(X)] \leq \frac{1}{4} \sum_{i=1}^{n} \mathbb{E}[D_i f(X)^2].$$

Proof   By Lemma 2.4.11,

$$\mathrm{Var}\,[f(X)\,|\,\mathcal{G}_i] \leq \frac{1}{4}D_i f(X)^2.$$

Plugging back into Theorem 3.2.3 gives the claim.   ∎

POINCARÉ
INEQUALITY **Remark 3.2.7** *For comparison, a version of the* Poincaré inequality *in one dimension asserts the following: let* $f \colon [0, T] \to \mathbb{R}$ *be continuously differentiable with* $f(0) = f(T) = 0$, $\int_0^T f(x)^2 + f'(x)^2 \mathrm{d}x < +\infty$ *and* $\int_0^T f(x)\mathrm{d}x = 0$, *then*

$$\int_0^T f(x)^2 \mathrm{d}x \leq C \int_0^T f'(x)^2 \mathrm{d}x, \tag{3.2.7}$$

where the best possible C *is* $T^2/4\pi^2$ *(see, for example, [SS03, chapter 3, Exercise 11]; this case is also known as* Wirtinger's inequality*). We give a quick proof for* $T = 1$ *with the suboptimal* $C = 1$. *Note that* $f(x) = \int_0^x f'(y)\mathrm{d}y$, *so by Cauchy–Schwarz (Theorem B.4.8),*

$$f(x)^2 \leq x \int_0^x f'(y)^2 \mathrm{d}y \leq \int_0^1 f'(y)^2 \mathrm{d}y.$$

The result follows by integration. Intuitively, for a function with mean 0 to have a large norm, it must have a large absolute derivative somewhere.

Example 3.2.8 (Longest common subsequence). Let $X_1, \ldots, X_{2n}$ be independent uniform random variables in $\{-1, +1\}$. Let Z be the length of the longest common subsequence in $(X_1, \ldots, X_n)$ and $(X_{n+1}, \ldots, X_{2n})$, that is,

$$
\begin{aligned}
Z = \max\big\{k \colon \exists 1 \leq i_1 < i_2 < \cdots < i_k \leq n \\
\text{and } n + 1 \leq j_1 < j_2 < \cdots < j_k \leq 2n \\
\text{such that } X_{i_1} = X_{j_1}, X_{i_2} = X_{j_2}, \ldots, X_{i_k} = X_{j_k}\big\}.
\end{aligned}
$$

Then, writing $Z = f(X_1, \ldots, X_{2n})$, it follows that $\|D_i f\|_\infty \leq 1$. Indeed, fix $\mathbf{x} = (x_1, \ldots, x_{2n})$ and let $\mathbf{x}^{i,+}$ (respectively $\mathbf{x}^{i,-}$) be $\mathbf{x}$, where the ith component is replaced with $+1$ (respectively -1). Assume without loss of generality that $f(\mathbf{x}^{i,-}) \leq f(\mathbf{x}^{i,+})$. Then, $|f(\mathbf{x}^{i,+}) - f(\mathbf{x}^{i,-})| \leq 1$ because removing the ith component (and its match) from a longest common subsequence when $x_i = +1$ (if present) decreases the length by 1. Since this is true for any $\mathbf{x}$, we have $\|D_i f\|_\infty \leq 1$. Finally, by the bounded differences inequality (Theorem 3.2.6),

$$\mathrm{Var}[Z] \leq \frac{1}{4} \sum_{i=1}^{2n} \|D_i f\|_\infty^2 \leq \frac{n}{2},$$

which is much better than the obvious $\mathrm{Var}[Z] \leq \mathbb{E}[Z^2] \leq n^2$. Note that we did not require any information about the expectation of Z.   ◀

McDiarmid's inequality

The following powerful consequence of the Azuma–Hoeffding inequality is commonly referred to as the *method of bounded differences*. Compare to (3.2.6).

Theorem 3.2.9 (McDiarmid's inequality). *Let $X_1, \ldots, X_n$ be independent random variables where X_i is $\mathcal{X}_i$-valued for all i, and let $X = (X_1, \ldots, X_n)$. Assume $f : \mathcal{X}_1 \times \cdots \times \mathcal{X}_n \to \mathbb{R}$ is a measurable function such that $\|D_i f\|_\infty < +\infty$ for all i. Then, for all $\beta > 0$,*

$$\mathbb{P}[f(X) - \mathbb{E}f(X) \geq \beta] \leq \exp\left(-\frac{2\beta^2}{\sum_{i \leq n} \|D_i f\|_\infty^2}\right).$$

Once again, applying the inequality to $-f$ gives a tail bound in the other direction.

Proof of Theorem 3.2.9   As before, we let

$$Z_i = \mathbb{E}[f(X) \mid \mathcal{F}_i],$$

where $\mathcal{F}_i = \sigma(X_1, \ldots, X_i)$, and we let $\mathcal{G}_i = \sigma(X_1, \ldots, X_{i-1}, X_{i+1}, \ldots, X_n)$. Then, it holds that $A_i \leq Z_i - Z_{i-1} \leq B_i$, where

$$B_i = \mathbb{E}\left[\sup_{y \in \mathcal{X}_i} f(X_1, \ldots, X_{i-1}, y, X_{i+1}, \ldots, X_n) - f(X) \,\middle|\, \mathcal{F}_{i-1}\right]$$

and

$$A_i = \mathbb{E}\left[\inf_{y \in \mathcal{X}_i} f(X_1, \ldots, X_{i-1}, y, X_{i+1}, \ldots, X_n) - f(X) \,\middle|\, \mathcal{F}_{i-1}\right].$$

Indeed, since $\sigma(X_i)$ is independent of $\mathcal{F}_{i-1}$ and $\mathcal{G}_i$, by the role of independence (Lemma B.6.14),

$$\begin{aligned}
Z_i &= \mathbb{E}\left[f(X) \mid \mathcal{F}_i\right] \\
&\leq \mathbb{E}\left[\sup_{y \in \mathcal{X}_i} f(X_1, \ldots, X_{i-1}, y, X_{i+1}, \ldots, X_n) \,\middle|\, \mathcal{F}_i\right] \\
&= \mathbb{E}\left[\sup_{y \in \mathcal{X}_i} f(X_1, \ldots, X_{i-1}, y, X_{i+1}, \ldots, X_n) \,\middle|\, \mathcal{F}_{i-1}, X_i\right] \\
&= \mathbb{E}\left[\sup_{y \in \mathcal{X}_i} f(X_1, \ldots, X_{i-1}, y, X_{i+1}, \ldots, X_n) \,\middle|\, \mathcal{F}_{i-1}\right],
\end{aligned}$$

and similarly for the other direction. Moreover, by definition, $B_i - A_i \leq \|D_i f\|_\infty := c_i$. The Azuma–Hoeffding inequality then gives the result. ∎

Examples

The moral of McDiarmid's inequality is that functions of *independent* variables that are *smooth*, in the sense that they do not depend too much on any one of their variables, are concentrated around their mean. Here are some straightforward applications.

Example 3.2.10 (Balls and bins: empty bins). Suppose we throw m balls into n bins independently, uniformly at random. The number of empty bins, $Z_{n,m}$, is centered at

$$\mathbb{E}Z_{n,m} = n\left(1 - \frac{1}{n}\right)^m.$$

Writing $Z_{n,m}$ as the sum of indicators $\sum_{i=1}^{n} \mathbf{1}_{B_i}$, where B_i is the event that bin i is empty, is a natural first attempt at proving concentration around the mean. However, there is a problem – the B_is are *not independent*. Indeed, because there is a fixed number of bins, the event B_i intuitively makes the other such events less likely. Instead, let X_j be the index of the bin in which ball j lands. The X_js are independent by construction and, moreover, letting $Z_{n,m} = f(X_1, \ldots, X_m)$ we have $\|D_i f\|_{\infty} \leq 1$. Indeed, moving a single ball changes the number of empty bins by at most 1 (if at all). Hence, by the method of bounded differences,

$$\mathbb{P}\left[\left|Z_{n,m} - n\left(1 - \frac{1}{n}\right)^m\right| \geq b\sqrt{m}\right] \leq 2e^{-2b^2}. \qquad \blacktriangleleft$$

Example 3.2.11 (Pattern matching). Let $X = (X_1, X_2, \ldots, X_n)$ be i.i.d. random variables taking values uniformly at random in a finite set S of size $s = |S|$. Let $a = (a_1, \ldots, a_k)$ be a fixed string of elements of S. We are interested in the number of occurrences of a as a (consecutive) substring in X, which we denote by N_n. Denote by E_i the event that the substring of X starting at i is a. Summing over the starting positions and using the linearity of expectation, the mean of N_n is

$$\mathbb{E}N_n = \mathbb{E}\left[\sum_{i=1}^{n-k+1} \mathbf{1}_{E_i}\right] = (n - k + 1)\left(\frac{1}{s}\right)^k.$$

However, the $\mathbf{1}_{E_i}$s are not independent. So we cannot use a Chernoff bound for Poisson trials (Theorem 2.4.7). Instead, we use the fact that $N_n = f(X)$, where $\|D_i f\|_{\infty} \leq k$, as each X_i appears in at most k substrings of length k. By the method of bounded differences, for all $b > 0$,

$$\mathbb{P}\left[\left|N_n - (n - k + 1)\left(\frac{1}{s}\right)^k\right| \geq bk\sqrt{n}\right] \leq 2e^{-2b^2}. \qquad \blacktriangleleft$$

The last two examples are perhaps not surprising in that they involve "sums of weakly independent" indicator variables. One might reasonably expect a sub-Gaussian-type inequality in that case. The next application is more striking and hints at connections to isoperimetric considerations (which we will not explore here).

Example 3.2.12 (Concentration of measure on the hypercube). For $A \subseteq \{0, 1\}^n$, a subset of the hypercube, and $r > 0$, we let

$$A_r = \left\{x \in \{0, 1\}^n : \inf_{a \in A} \|x - a\|_1 \leq r\right\}$$

be the points at ℓ^1 distance r from A. Fix $\varepsilon \in (0, 1/2)$ and assume that $|A| \geq \varepsilon 2^n$. Let λ_{ε} be such that $e^{-2\lambda_{\varepsilon}^2} = \varepsilon$. The following application of the method of bounded differences indicates that much of the uniform measure on the high-dimensional hypercube lies in a close neighborhood of any such set A. This is an example of the *concentration of measure phenomenon*.

Claim 3.2.13

$$r > 2\lambda_{\varepsilon}\sqrt{n} \implies |A_r| \geq (1 - \varepsilon)2^n.$$

Proof Let $X = (X_1, \ldots, X_n)$ be uniformly distributed in $\{0, 1\}^n$. Note that the coordinates are in fact independent. The function

$$f(\mathbf{x}) = \inf_{\mathbf{a} \in A} \|\mathbf{x} - \mathbf{a}\|_1$$

has $\|D_i f\|_\infty \leq 1$. Indeed, changing one coordinate of $\mathbf{x}$ can increase the ℓ^1 distance to the closest point to $\mathbf{x}$ by at most 1; in the other direction, if a one-coordinate change were to decrease f by more than 1, reversing it would produce an increase of that same amount – a contradiction. Hence, McDiarmid's inequality gives

$$\mathbb{P}\left[\mathbb{E}f(X) - f(X) \geq \beta\right] \leq \exp\left(-\frac{2\beta^2}{n}\right).$$

Choosing $\beta = \mathbb{E}f(X)$ and noting that $f(\mathbf{x}) \leq 0$ if and only if $\mathbf{x} \in A$ gives

$$\mathbb{P}[A] \leq \exp\left(-\frac{2(\mathbb{E}f(X))^2}{n}\right),$$

or, rearranging and using our assumption on A,

$$\mathbb{E}f(X) \leq \sqrt{\frac{1}{2}n \log \frac{1}{\mathbb{P}[A]}} \leq \sqrt{\frac{1}{2}n \log \frac{1}{\varepsilon}} = \lambda_\varepsilon \sqrt{n}.$$

By a second application of the method of bounded differences with $\beta = \lambda_\varepsilon \sqrt{n}$,

$$\mathbb{P}\left[f(X) \geq 2\lambda_\varepsilon \sqrt{n}\right] \leq \mathbb{P}\left[f(X) - \mathbb{E}f(X) \geq b\right] \leq \exp\left(-\frac{2\beta^2}{n}\right) = \varepsilon.$$

The result follows by observing that, with $r > 2\lambda_\varepsilon \sqrt{n}$,

$$\frac{|A_r|}{2^n} \geq \mathbb{P}\left[f(X) < 2\lambda_\varepsilon \sqrt{n}\right] \geq 1 - \varepsilon. \qquad \blacksquare$$

Claim 3.2.13 is striking for two reasons: (1) the radius $2\lambda_\varepsilon \sqrt{n}$ is much smaller than n, the diameter of $\{0, 1\}^n$; and (2) it applies to *any* A. The smallest r such that $|A_r| \geq (1 - \varepsilon)2^n$ in general depends on A. Here are two extremes.

For $\gamma > 0$, let

$$B(\gamma) := \left\{\mathbf{x} \in \{0, 1\}^n : \|\mathbf{x}\|_1 \leq \frac{n}{2} - \gamma \sqrt{\frac{n}{4}}\right\}.$$

Note that letting for $Y_n \sim B(n, \frac{1}{2})$,

$$\frac{1}{2^n}|B(\gamma)| = \sum_{\ell=0}^{\frac{n}{2} - \gamma\sqrt{\frac{n}{4}}} \binom{n}{\ell} 2^{-n} = \mathbb{P}\left[Y_n \leq \frac{n}{2} - \gamma\sqrt{\frac{n}{4}}\right]. \tag{3.2.8}$$

By the Berry–Esséen theorem (e.g., [Dur10, Theorem 3.4.9]), there is a $C > 0$ such that, after rearranging the final quantity in (3.2.8),

$$\left|\mathbb{P}\left[\frac{Y_n - n/2}{\sqrt{n/4}} \leq -\gamma\right] - \mathbb{P}[Z \leq -\gamma]\right| \leq \frac{C}{\sqrt{n}},$$

where $Z \sim N(0, 1)$. Let $\varepsilon < \varepsilon' < 1/2$ and let $\gamma_{\varepsilon'}$ be such that $\mathbb{P}[Z \leq -\gamma_{\varepsilon'}] = \varepsilon'$. Then, setting $A := B(\gamma_{\varepsilon'})$, for n large enough, we have $|A| \geq \varepsilon 2^n$ by (3.2.8). On the other hand, setting $r := \gamma_{\varepsilon'}\sqrt{n/4}$, we have $A_r \subseteq B(0)$, so that $|A_r| \leq \frac{1}{2}2^n < (1-\varepsilon)2^n$. We have shown that $r = \Omega(\sqrt{n})$ is in general required for Claim 3.2.13 to hold.

For an example at the other extreme, assume for simplicity that $N := \varepsilon 2^n$ is an integer. Let $A \subseteq \{0,1\}^n$ be constructed as follows: starting from the empty set, add points in $\{0,1\}^n$ to A independently, uniformly at random until $|A| = N$. Set $r := 2$. Each point selected in A has $\binom{n}{2}$ points within ℓ^1 distance 2. By a union bound, the probability that A_r does not cover all of $\{0,1\}^n$ is at most

$$\mathbb{P}[|\{0,1\}^n \backslash A_r| > 0] \leq \sum_{x \in \{0,1\}^n} \mathbb{P}[x \notin A_r] \leq 2^n \left(1 - \frac{\binom{n}{2}}{2^n}\right)^{\varepsilon 2^n} \leq 2^n e^{-\varepsilon\binom{n}{2}},$$

where, in the second inequality, we considered only the first N picks in the construction of A (possibly with repeats), and in the third inequality we used $1 - z \leq e^{-z}$ for all $z \in \mathbb{R}$ (see Exercise 1.16). In particular, as $n \to +\infty$,

$$\mathbb{P}[|\{0,1\}^n \backslash A_r| > 0] < 1.$$

So for n large enough there is a set A such that $A_r = \{0,1\}^n$, where $r = 2$. ◀

Remark 3.2.14 *In fact, it can be shown that sets of the form $\{x : \|x\|_1 \leq s\}$ have the smallest "expansion" among subsets of $\{0,1\}^n$ of the same size, a result known as Harper's vertex isoperimetric theorem. See, for example, [BLM13, Theorem 7.6 and Exercises 7.11–7.13].*

3.2.3 ▷ *Random Graphs: Exposure Martingale and Application to the Chromatic Number in Erdős–Rényi Model*

Exposure martingales In the context of the Erdős–Rényi graph model (Definition 1.2.2), a common way to apply the Azuma–Hoeffding inequality (Theorem 3.2.1) is to introduce an "exposure martingale." Let $G \sim \mathbb{G}_{n,p}$ and let F be any function on graphs such that $\mathbb{E}_{n,p}|F(G)| < +\infty$ for all n, p. Choose an arbitrary ordering of the vertices and, for $i = 1, \ldots, n$, denote by H_i the subgraph of G induced by the first i vertices. Then, the filtration $\mathcal{H}_i = \sigma(H_1, \ldots, H_i)$, $i = 1, \ldots, n$, corresponds to exposing the vertices of G one at a time. The Doob martingale

$$Z_i = \mathbb{E}_{n,p}[F(G) \mid \mathcal{H}_i], \qquad i = 1, \ldots, n,$$

VERTEX EXPOSURE MARTINGALE

is known as a *vertex exposure martingale*. An alternative way to define the filtration is to consider instead the random variables $X_i = (\mathbf{1}_{\{\{i,j\}\in G\}} : 1 \leq j \leq i)$ for $i = 2, \ldots, n$. In other words, X_i is a vector whose entries indicate the status (present or absent) of all potential edges incident to i and a vertex preceding it. Hence, $\mathcal{H}_i = \sigma(X_2, \ldots, X_i)$ for $i = 1, \ldots, n$ (and $\mathcal{H}_1$ is trivial as it corresponds to a graph with a single vertex and no edge). This representation has an important property: the X_is are *independent* as they pertain to disjoint subsets of edges. We are then in the setting of the method of bounded differences. Rewriting $F(G) = f(X_1, \ldots, X_n)$, the vertex exposure martingale coincides with the martingale (3.2.3) used in that context.

As an example, consider the chromatic number $\chi(G)$, that is, the smallest number of colors needed in a proper coloring of G. Define $f_\chi(X_1, \ldots, X_n) := \chi(G)$. We use the following combinatorial observation to bound $\|D_i f_\chi\|_\infty$.

Lemma 3.2.15 *Altering the status (absent or present) of edges incident to a fixed vertex v changes the chromatic number by at most 1.*

Proof Altering the status of edges incident to v increases the chromatic number by at most 1, since in the worst case one can simply use an extra color for v. On the other hand, if the chromatic number were to decrease by more than 1 after altering the status of edges incident to v, reversing the change and using the previous observation would produce a contradiction. ∎

A fortiori, since X_i depends on a *subset* of the edges incident to node i, Lemma 3.2.15 implies that $\|D_i f_\chi\|_\infty \leq 1$. Hence, for all $0 < p < 1$ and n, by an immediate application of McDiarmid's inequality (Theorem 3.2.9):

Claim 3.2.16

$$\mathbb{P}_{n,p}\left[|\chi(G) - \mathbb{E}_{n,p}[\chi(G)]| \geq b\sqrt{n-1}\right] \leq 2e^{-2b^2}.$$

Edge exposure martingales can be defined in a similar manner: reveal the edges one at a time in an arbitrary order. By Lemma 3.2.15, the corresponding function also satisfies the same ℓ^∞ bound. Observe however that, for the chromatic number, edge exposure results in a much weaker bound as the $\Theta(n^2)$ random variables produce only a *linear in n* deviation for the same tail probability. (The reader may want to ponder the apparent paradox: using a larger number of independent variables seemingly leads to weaker concentration in this case.) EDGE
EXPOSURE
MARTINGALE

Remark 3.2.17 *Note that Claim 3.2.16 tells us nothing about the expectation of $\chi(G)$. It turns out that, up to logarithmic factors, $\mathbb{E}_{n,p_n}[\chi(G)]$ is of order np_n when $p_n \sim n^{-\alpha}$ for some $0 < \alpha < 1$. We will not prove this result here. See the "Bibliographic remarks" at the end of this chapter for more on the chromatic number of Erdős–Rényi graphs.*

The chromatic number is concentrated on few values Much stronger concentration results can be obtained: when $p_n = n^{-\alpha}$ with $\alpha > \frac{1}{2}$, the chromatic number $\chi(G)$ is in fact concentrated on two values! We give a partial result along those lines which illustrates a less straightforward choice of martingale in the Azuma–Hoeffding inequality (Theorem 3.2.1).

Claim 3.2.18 *Let $p_n = n^{-\alpha}$ with $\alpha > \frac{5}{6}$ and let $G_n \sim \mathbb{G}_{n,p_n}$. Then, for any $\varepsilon > 0$ there is $\varphi_n := \varphi_n(\alpha, \varepsilon)$ such that*

$$\mathbb{P}_{n,p_n}[\varphi_n \leq \chi(G_n) \leq \varphi_n + 3] \geq 1 - \varepsilon$$

for all n large enough.

Proof We consider the following martingale. Let φ_n be the smallest integer such that

$$\mathbb{P}_{n,p_n}[\chi(G_n) \leq \varphi_n] > \frac{\varepsilon}{3}. \tag{3.2.9}$$

Let $F_n(G_n)$ be the minimal size of a set of vertices, U, in G_n such that $G_n \backslash U$ is φ_n-colorable. Let (Z_i) be the vertex exposure martingale associated to the quantity $F_n(G_n)$. The proof proceeds in two steps: we show that (1) all but $O(\sqrt{n})$ vertices can be φ_n-colored and (2) the remaining vertices can be colored using three *additional* colors. See Figure 3.1 for an illustration of the proof strategy.

We claim that (Z_i) has increments bounded by 1.

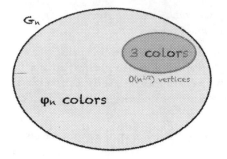

Figure 3.1 All but $O(\sqrt{n})$ vertices are colored using φ_n colors. The remaining vertices are colored using three additional colors.

Lemma 3.2.19 *Changing the edges incident to a single vertex can change F_n by at most 1.*

Proof Changing the edges incident to v can increase F_n by at most 1. Indeed, if F_n increases after such a change, it must be that $v \notin U$ since in the other case the edges incident to v would not affect the colorability of $G_n \setminus U$ – present or not. So we can add v to U and restore colorability. On the other hand, if F_n were to decrease by more than 1, reversing the change and using the previous observation would give a contradiction. ∎

Choose b_ε such that $e^{-b_\varepsilon^2/2} = \frac{\varepsilon}{3}$. Then, applying the Azuma–Hoeffding inequality to $(-Z_i)$,

$$\mathbb{P}_{n,p_n}\left[F_n(G_n) - \mathbb{E}_{n,p_n}[F_n(G_n)] \le -b_\varepsilon\sqrt{n-1}\right] \le \frac{\varepsilon}{3},$$

which, since $\mathbb{P}_{n,p_n}[F_n(G_n) = 0] = \mathbb{P}_{n,p_n}[\chi(G_n) \le \varphi_n] > \frac{\varepsilon}{3}$, implies that

$$\mathbb{E}_{n,p_n}[F_n(G_n)] \le b_\varepsilon\sqrt{n-1}.$$

Applying the Azuma–Hoeffding inequality to (Z_i) gives

$$\begin{aligned}
\mathbb{P}_{n,p_n}&\left[F_n(G_n) \ge 2b_\varepsilon\sqrt{n-1}\right]\\
&\le \mathbb{P}_{n,p_n}\left[F_n(G_n) - \mathbb{E}_{n,p_n}[F_n(G_n)] \ge b_\varepsilon\sqrt{n-1}\right]\\
&\le \frac{\varepsilon}{3}.
\end{aligned} \tag{3.2.10}$$

So with probability at least $1 - \frac{\varepsilon}{3}$, we can color all vertices but $2b_\varepsilon\sqrt{n-1}$ using φ_n colors. Let U be the remaining uncolored vertices.

We claim that, with high probability, we can color the vertices in U using at most three extra colors.

Lemma 3.2.20 *Fix $c > 0$, $\alpha > \frac{5}{6}$, and $\varepsilon > 0$. Let $G_n \sim \mathbb{G}_{n,p_n}$ with $p_n = n^{-\alpha}$. For all n large enough,*

$$\mathbb{P}_{n,p_n}\left[\text{every subset of } c\sqrt{n} \text{ vertices of } G_n \text{ can be 3-colored}\right] > 1 - \frac{\varepsilon}{3}. \tag{3.2.11}$$

Proof We use the first moment method (Theorem 2.2.6). We refer to a subset of vertices that is not 3-colorable but such that all of its subsets are as minimal, non-3-colorable. Let Y_n be the number of such subsets of size at most $c\sqrt{n}$ in G_n.

Any minimal, non-3-colorable subset W must have degree at least 3. Indeed, suppose that $w \in W$ has degree less than 3. Then, $W \backslash \{w\}$ is 3-colorable by definition. But, since w has fewer than three neighbors, it can also be properly colored without adding a new color – a contradiction. In particular, the subgraph of G_n induced by W must have at least $\frac{3}{2}|W|$ edges. Hence, the probability that a subset of vertices of G_n of size ℓ is minimal, non-3-colorable is at most

$$\binom{\binom{\ell}{2}}{\frac{3\ell}{2}} p_n^{\frac{3\ell}{2}},$$

by a union bound over all subsets of edges of size $\frac{3\ell}{2}$.

By the first moment method, by the binomial bounds $\binom{n}{\ell} \leq \left(\frac{en}{\ell}\right)^\ell$ (see Appendix A) and $\binom{\ell}{2} \leq \ell^2/2$, for some $c' \in (0, +\infty)$,

$$\mathbb{P}_{n,p_n}[Y_n > 0] \leq \mathbb{E}_{n,p_n} Y_n$$

$$\leq \sum_{\ell=4}^{c\sqrt{n}} \binom{n}{\ell} \binom{\binom{\ell}{2}}{\frac{3\ell}{2}} p_n^{\frac{3\ell}{2}}$$

$$\leq \sum_{\ell=4}^{c\sqrt{n}} \left(\frac{en}{\ell}\right)^\ell \left(\frac{e\ell}{3}\right)^{\frac{3\ell}{2}} n^{-\frac{3\ell\alpha}{2}}$$

$$\leq \sum_{\ell=4}^{c\sqrt{n}} \left(\frac{e^{\frac{5}{2}} n^{1-\frac{3\alpha}{2}} \ell^{\frac{1}{2}}}{3^{\frac{3}{2}}}\right)^\ell$$

$$\leq \sum_{\ell=4}^{c\sqrt{n}} \left(c' n^{\frac{5}{4}-\frac{3\alpha}{2}}\right)^\ell$$

$$\leq O\left(n^{\frac{5}{4}-\frac{3\alpha}{2}}\right)^4$$

$$\to 0,$$

as $n \to +\infty$, where we used that $\frac{5}{4} - \frac{3\alpha}{2} < \frac{5}{4} - \frac{5}{4} = 0$ when $\alpha > \frac{5}{6}$ so that the geometric series is dominated by its first term. Therefore, for n large enough, $\mathbb{P}_{n,p_n}[Y_n > 0] \leq \varepsilon/3$, concluding the proof. ∎

By the choice of φ_n in (3.2.9),

$$\mathbb{P}_{n,p_n}[\chi(G_n) < \varphi_n] \leq \frac{\varepsilon}{3}.$$

By (3.2.10) and (3.2.11) with $c = 2b_\varepsilon$,

$$\mathbb{P}_{n,p_n}[\chi(G_n) > \varphi_n + 3] \leq \frac{2\varepsilon}{3}.$$

So, overall,

$$\mathbb{P}_{n,p_n}[\varphi_n \leq \chi(G_n) \leq \varphi_n + 3] \geq 1 - \varepsilon.$$

That concludes the proof. ∎

3.2.4 ▷ *Random Graphs: Degree Sequence of Preferential Attachment Graphs*

Let $(G_t)_{t\geq 1} \sim \mathrm{PA}_1$ be a preferential attachment graph (Definition 1.2.3). A key feature of such graphs is a power-law degree sequence: the fraction of vertices with degree d behaves like $\propto d^{-\alpha}$ for some $\alpha > 0$, that is, it has a fat tail. Recall that we restrict ourselves to the tree case. In contrast, we will show in Section 4.1.4 that a (sparse) Erdős–Rényi random graph has an asymptotically Poisson-distributed degree sequence, and therefore a much thinner tail.

Power-law degree sequence Let $D_i(t)$ be the degree of the ith vertex in G_t, denoted v_i, and let

$$N_d(t) := \sum_{i=0}^{t} \mathbf{1}_{\{D_i(t)=d\}}$$

be the number of vertices of degree d in G_t. By construction $N_0(t) = 0$ for all t. Define the sequence

$$f_d := \frac{4}{d(d+1)(d+2)}, \qquad d \geq 1. \tag{3.2.12}$$

Our main claim is:

Claim 3.2.21

$$\frac{1}{t}N_d(t) \to_{\mathrm{p}} f_d \qquad \forall d \geq 1.$$

Proof The claim is immediately implied by the following lemmas.

Lemma 3.2.22 (Convergence of the mean).

$$\frac{1}{t}\mathbb{E}N_d(t) \to f_d \qquad \forall d \geq 1.$$

Lemma 3.2.23 (Concentration around the mean). *For any $\delta > 0$,*

$$\mathbb{P}\left[\left|\frac{1}{t}N_d(t) - \frac{1}{t}\mathbb{E}N_d(t)\right| \geq \sqrt{\frac{2\log\delta^{-1}}{t}}\right] \leq 2\delta \qquad \forall d \geq 1, \forall t.$$

An alternative representation of the process We start with the proof of Lemma 3.2.23, which is an application of the method of bounded differences.

Proof of Lemma 3.2.23 In our description of the preferential attachment process, the random choices made at each time depend in a seemingly complicated way on previous choices. In order to establish concentration of the process around its mean, we introduce a clever, alternative construction which has the advantage that it involves *independent* choices.

We start with a single vertex v_0. At time 1, we add a single vertex v_1 and an edge e_1 connecting v_0 and v_1. For bookkeeping, we orient edges away from the vertex of higher time index (but we ignore the orientations in the output). For a directed edge (i,j), we refer to i as its tail and j as its head. For all $s \geq 2$, let X_s be an independent, uniformly chosen edge extremity among the edges in G_{s-1}, that is, pick a uniform element in

$$\mathcal{X}_s := \{(1,\text{tail}), (1,\text{head}), \ldots, (s-1,\text{tail}), (s-1,\text{head})\}.$$

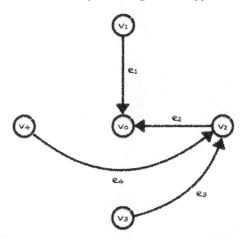

Figure 3.2 Graph obtained when $x_2 = (1, \text{head})$, $x_3 = (2, \text{tail})$, and $x_4 = (3, \text{head})$.

To form G_s, attach a new edge e_s to the vertex of G_{s-1} corresponding to X_s. A vertex of degree d' in G_{s-1} is selected with probability $\frac{d'}{2(s-1)}$, as it should. Note that X_s can be picked in advance independently of the sequence $(G_{s'})_{s' < s}$. For instance, if $x_2 = (1, \text{head})$, $x_3 = (2, \text{tail})$, and $x_4 = (3, \text{head})$, the graph obtained at time 4 is depicted in Figure 3.2.

We claim that $N_d(t) =: h(X_2, \ldots, X_t)$ as a function of $X_2, \ldots, X_t$ satisfies $\|D_i h\|_\infty \leq 2$. Indeed, let $(x_2, \ldots, x_t)$ be a realization of $(X_2, \ldots, X_t)$ and let $y \in \mathcal{X}_s$ with $y \neq x_s$. Replacing $x_s = (i, \text{end})$ with $y = (j, \text{end}')$, where $i, j \in \{1, \ldots, s-1\}$ and $\text{end}, \text{end}' \in \{\text{tail}, \text{head}\}$ has the effect of redirecting the head of edge e_s from the end of e_i to the end' of e_j. This redirection also brings along with it the heads of all other edges associated with the choice (s, head). But, crucially, those changes only affect the degrees of the vertices (i, end) and (j, end') in the original graph. Hence, the number of vertices with degree d changes by at most 2, as claimed. For instance, returning to the example of Figure 3.2. If we replace $x_3 = (2, \text{tail})$ with $y = (1, \text{tail})$, one obtains the graph in Figure 3.3. Note that only the degrees of vertices v_1 and v_2 are affected by this change.

By McDiarmid's inequality (Theorem 3.2.9), for all $\beta > 0$,

$$\mathbb{P}[|N_d(t) - \mathbb{E}N_d(t)| \geq \beta] \leq 2 \exp\left(-\frac{2\beta^2}{(2)^2(t-1)}\right),$$

which, choosing $\beta = \sqrt{2t \log \delta^{-1}}$, we can rewrite as

$$\mathbb{P}\left[\left|\frac{1}{t}N_d(t) - \frac{1}{t}\mathbb{E}N_d(t)\right| \geq \sqrt{\frac{2 \log \delta^{-1}}{t}}\right] \leq 2\delta.$$

That concludes the proof of the lemma. ∎

Dynamics of the mean Once again the method of bounded differences tells us nothing about the mean, which must be analyzed by other means. The proof of Lemma 3.2.22 does not rely on the Azuma–Hoeffding inequality but is given for completeness (and may be skipped).

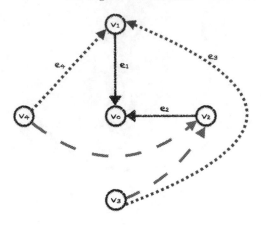

Figure 3.3 Substituting $x_3 = (2, \text{tail})$ with $y = (1, \text{tail})$ in the example of Figure 3.2 has the effect of replacing the dashed edges with the dotted edges. Note that only the degrees of vertices v_1 and v_2 are affected by this change.

Proof of Lemma 3.2.22   The idea of the proof is to derive a recursion for f_d by considering the evolution of $\mathbb{E}N_d(t)$ and taking a limit as $t \to +\infty$. Let $d \geq 1$. Observe that $\mathbb{E}N_d(t) = 0$ for $t \leq d-1$ since we need at least d edges to have a degree-d vertex. Moreover, by the description of the preferential attachment process, the following recursion holds for $t \geq d-1$:

$$\mathbb{E}N_d(t+1) - \mathbb{E}N_d(t) = \underbrace{\frac{d-1}{2t}\mathbb{E}N_{d-1}(t)}_{(a)} - \underbrace{\frac{d}{2t}\mathbb{E}N_d(t)}_{(b)} + \underbrace{\mathbf{1}_{\{d=1\}}}_{(c)} . \qquad (3.2.13)$$

Indeed, (a) $N_d(t)$ increases by 1 if a vertex of degree $d-1$ is picked, an event of probability $\frac{d-1}{2t}N_{d-1}(t)$ because the sum of degrees at time t is twice the number of edges (i.e., t); (b) $N_d(t)$ decreases by 1 if a vertex of degree d is picked, an event of probability $\frac{d}{2t}N_d(t)$; and (c) the last term comes from the fact that the new vertex always has degree 1. We rewrite (3.2.13) as

$$\begin{aligned}
\mathbb{E}N_d(t+1) &= \mathbb{E}N_d(t) + \frac{d-1}{2t}\mathbb{E}N_{d-1}(t) - \frac{d}{2t}\mathbb{E}N_d(t) + \mathbf{1}_{\{d=1\}} \\
&= \left(1 - \frac{d/2}{t}\right)\mathbb{E}N_d(t) + \left\{\frac{d-1}{2}\left[\frac{1}{t}\mathbb{E}N_{d-1}(t)\right] + \mathbf{1}_{\{d=1\}}\right\} \\
&=: \left(1 - \frac{d/2}{t}\right)\mathbb{E}N_d(t) + g_d(t), \qquad (3.2.14)
\end{aligned}$$

where $g_d(t)$ is defined as the expression in curly brackets on the second line. We will not solve this recursion explicitly. Instead, we seek to analyze its asymptotics, specifically we show that $\frac{1}{t}\mathbb{E}N_d(t) \to f_d$.

The key is to notice that the expression for $\mathbb{E}N_d(t+1)$ depends on $\frac{1}{t}\mathbb{E}N_{d-1}(t)$ – so we work by induction on d. Because of the form of the recursion, the following technical lemma is what we need to proceed.

Lemma 3.2.24 *Let f, g be non-negative functions of $t \in \mathbb{N}$ satisfying the following recursion*

$$f(t+1) = \left(1 - \frac{\alpha}{t}\right) f(t) + g(t) \qquad \forall t \geq t_0,$$

with $g(t) \to g \in [0, +\infty)$ as $t \to +\infty$, and where $\alpha > 0, t_0 \geq 2\alpha, f(t_0) \geq 0$ are constants. Then,

$$\frac{1}{t} f(t) \to \frac{g}{1 + \alpha},$$

as $t \to +\infty$.

The proof of this lemma is given after the proof of Claim 3.2.21. We first conclude the proof of Lemma 3.2.22. First let $d = 1$. In that case, $g_1(t) = g_1 := 1$, $\alpha := 1/2$, and $t_0 := 1$. By Lemma 3.2.24,

$$\frac{1}{t} \mathbb{E} N_1(t) \to \frac{1}{1 + 1/2} = \frac{2}{3} = f_1.$$

Assuming by induction that $\frac{1}{t} \mathbb{E} N_{d-1}(t) \to f_{d-1}$ we get

$$g_d(t) \to g_d := \frac{d-1}{2} f_{d-1},$$

as $t \to +\infty$. Using Lemma 3.2.24 with $\alpha := d/2$ and $t_0 := d - 1$, we obtain

$$\frac{1}{t} \mathbb{E} N_d(t) \to \frac{1}{1 + d/2} \left[\frac{d-1}{2} f_{d-1}\right] = \frac{d-1}{d+2} \cdot \frac{4}{(d-1)d(d+1)} = f_d.$$

That concludes the proof of Lemma 3.2.22. ∎

To prove Claim 3.2.21, we combine Lemmas 3.2.22 and 3.2.23. Fix any $d, \delta, \varepsilon > 0$. Choose t' large enough that for all $t \geq t'$,

$$\max\left\{\left|\frac{1}{t} \mathbb{E} N_d(t) - f_d\right|, \sqrt{\frac{2 \log \delta^{-1}}{t}}\right\} \leq \varepsilon.$$

Then,

$$\mathbb{P}\left[\left|\frac{1}{t} N_d(t) - f_d\right| \geq 2\varepsilon\right] \leq 2\delta$$

for all $t \geq t'$. That proves convergence in probability. ∎

Proof of the technical lemma It remains to prove Lemma 3.2.24.

Proof of Lemma 3.2.24 By induction on t, we have

$$\begin{aligned}
f(t+1) &= \left(1 - \frac{\alpha}{t}\right) f(t) + g(t) \\
&= \left(1 - \frac{\alpha}{t}\right)\left[\left(1 - \frac{\alpha}{t-1}\right) f(t-1) + g(t-1)\right] + g(t) \\
&= \left(1 - \frac{\alpha}{t}\right) g(t-1) + g(t) + \left(1 - \frac{\alpha}{t}\right)\left(1 - \frac{\alpha}{t-1}\right) f(t-1) \\
&= \cdots \\
&= \sum_{t=0}^{t-t_0} g(t-i) \prod_{j=0}^{i-1}\left(1 - \frac{\alpha}{t-j}\right) + f(t_0) \prod_{j=0}^{t-t_0}\left(1 - \frac{\alpha}{t-j}\right),
\end{aligned}$$

or

$$f(t+1) = \sum_{s=t_0}^{t} g(s) \prod_{r=s+1}^{t} \left(1 - \frac{\alpha}{r}\right) + f(t_0) \prod_{r=t_0}^{t} \left(1 - \frac{\alpha}{r}\right), \qquad (3.2.15)$$

where empty products are equal to 1. To guess the limit note that, for large s, $g(s)$ is roughly constant and that the product in the first term behaves like

$$\exp\left(-\sum_{r=s+1}^{t} \frac{\alpha}{r}\right) \approx \exp\left(-\alpha(\log t - \log s)\right) \approx \frac{s^\alpha}{t^\alpha}.$$

So approximating the sum by an integral we get that $f(t+1) \approx \frac{gt}{\alpha+1}$, which is indeed consistent with the claim.

Formally, we use that there is a constant $\gamma = 0.577\ldots$ such that (see, for example, [LL10, Lemma 12.1.3])

$$\sum_{\ell=1}^{m} \frac{1}{\ell} = \log m + \gamma + \Theta(m^{-1}),$$

and that by a Taylor expansion, for $|z| \le 1/2$,

$$\log(1-z) = -z + \Theta(z^2).$$

Fix $\eta > 0$ small and take t large enough that $\eta t > 2\alpha$ and $|g(s) - g| < \eta$ for all $s \ge \eta t$. Then, for $s + 1 \ge t_0$,

$$\sum_{r=s+1}^{t} \log\left(1 - \frac{\alpha}{r}\right) = -\sum_{r=s+1}^{t} \left\{\frac{\alpha}{r} + \Theta(r^{-2})\right\}$$

$$= -\alpha(\log t - \log s) + \Theta(s^{-1}),$$

so, taking exponentials,

$$\prod_{r=s+1}^{t} \left(1 - \frac{\alpha}{r}\right) = \frac{s^\alpha}{t^\alpha}(1 + \Theta(s^{-1})).$$

Hence,

$$\frac{1}{t} f(t_0) \prod_{r=t_0}^{t} \left(1 - \frac{\alpha}{r}\right) = \frac{t_0^\alpha}{t^{\alpha+1}}(1 + \Theta(t_0^{-1})) \to 0,$$

as $t \to +\infty$. Moreover,

$$\frac{1}{t} \sum_{s=\eta t}^{t} g(s) \prod_{r=s+1}^{t} \left(1 - \frac{\alpha}{r}\right) \le \frac{1}{t} \sum_{s=\eta t}^{t} (g + \eta) \frac{s^\alpha}{t^\alpha}(1 + \Theta(s^{-1}))$$

$$\le O(\eta) + (1 + \Theta(t^{-1})) \frac{g}{t^{\alpha+1}} \sum_{s=\eta t}^{t} s^\alpha$$

$$\le O(\eta) + (1 + \Theta(t^{-1})) \frac{g}{t^{\alpha+1}} \frac{(t+1)^{\alpha+1}}{\alpha+1}$$

$$\to O(\eta) + \frac{g}{\alpha+1},$$

where we bounded the sum on the second line by an integral. Similarly,

$$\frac{1}{t}\sum_{s=t_0}^{\eta t-1} g(s) \prod_{r=s+1}^{t} \left(1 - \frac{\alpha}{r}\right) \leq \frac{1}{t}\sum_{s=t_0}^{\eta t-1}(g + \eta)\frac{s^\alpha}{t^\alpha}(1 + \Theta(s^{-1}))$$

$$\leq \frac{\eta t}{t}(g + \eta)\frac{(\eta t)^\alpha}{t^\alpha}(1 + \Theta(t_0^{-1}))$$

$$\to O(\eta^{\alpha+1}).$$

Plugging these inequalities back into (3.2.15), we get

$$\limsup_t \frac{1}{t}f(t + 1) \leq \frac{g}{1 + \alpha} + O(\eta).$$

A similar inequality holds in the other direction. Letting $\eta \to 0$ concludes the proof. ∎

Remark 3.2.25 *A more quantitative result (uniform in t and d) can be derived. See, for example, [vdH17, sections 8.5, 8.6]. See the same reference for a generalization beyond trees.*

3.2.5 ▷ *Data Science: Stochastic Bandits and the Slicing Method*

In this section, we consider an application of the maximal Azuma–Hoeffding inequality (Theorem 3.2.1) to (multi-armed) bandit problems. These are meant as a simple model of sequential decision-making with limited information where a fundamental issue is trading off between exploitation of actions that have done well in the past and exploration of actions that might perform better in the future. A typical application is online advertising, where one must decide which advertisement to display to the next visitor to a website.

In the simplest version of the (two-arm) *stochastic bandit* problem, there are two *unknown* STOCHASTIC reward distributions ν_1, ν_2 over $[0, 1]$ with respective means $\mu_1 \neq \mu_2$. At each time $t = $ BANDIT $1, \ldots, n$, we request an independent sample from ν_{I_t}, where we are free to choose $I_t \in \{1, 2\}$ based on past choices and observed rewards $\{(I_s, Z_s)\}_{s<t}$. This will be referred to as pulling *arm* I_t. We then observe the reward $Z_t \sim \nu_{I_t}$. Letting $\mu^* := \mu_1 \vee \mu_2$, our goal is to minimize ARM

$$\overline{R}_n = n\mu^* - \mathbb{E}\left[\sum_{t=1}^{n} \mu_{I_t}\right], \tag{3.2.16}$$

which is known as the *pseudo-regret*. That is, we seek to make choices $(I_t)_{t=1}^n$ that minimize PSEUDO- the difference between the best achievable cumulative mean reward and the expected cumu- REGRET lative mean reward from our decisions. Note that the expectation in (3.2.16) is taken over the choices $(I_t)_{t=1}^n$, which themselves depend on the random rewards $(Z_s)_{t=1}^n$. Because ν_1 and ν_2 are unknown, there is a fundamental friction between exploiting the arm that has done best in the past and exploring further the other arm, which might perform better in the future.

One general approach that has proved effective in this type of problem is known as *optimism in the face of uncertainty*. Roughly speaking, we construct a set of plausible environments (in our case, the means of the reward distributions) that are consistent with observed data; then we make an optimal decision assuming that the true environment is the most favorable among them. A concrete implementation of this principle is the *Upper Confidence*

Martingales and Potentials

Bound (UCB) algorithm, which we now describe. In other words, we use a concentration
inequality to build a confidence interval for each reward mean, and then we pick the arm
with highest upper bound.

UCB algorithm

To state the algorithm formally, we will need some notation. For $i = 1, 2$, let $T_i(t)$ be the
number of times arm i is pulled up to time t

$$T_i(t) = \sum_{s \leq t} \mathbf{1}\{I_s = i\},$$

and let $X_{i,s}, s = 1, \ldots, n$, be i.i.d. samples from v_i. Assume that the reward at time t is

$$Z_t = \begin{cases} X_{1,T_1(t-1)+1} & \text{if } I_t = 1, \\ X_{2,T_2(t-1)+1} & \text{otherwise.} \end{cases}$$

In other words, $X_{i,s}$ is the sth observed reward from arm i. Let $\hat{\mu}_{i,s}$ be the sample mean of the
observed rewards after pulling s times on arm i

$$\hat{\mu}_{i,s} = \frac{1}{s} \sum_{r \leq s} X_{i,r}.$$

Since the $X_{i,s}$s are independent and $[0, 1]$-valued by assumption, by Hoeffding's inequality
(Theorem 2.4.10), for any $\beta > 0$,

$$\mathbb{P}[\hat{\mu}_{i,s} - \mu_i \geq \beta] \vee \mathbb{P}[\mu_i - \hat{\mu}_{i,s} \geq \beta] \leq \exp\left(-2s\beta^2\right).$$

The right-hand side can be made $\leq \delta$ provided

$$\beta \geq \sqrt{\frac{\log \delta^{-1}}{2s}} := \mathrm{H}(s, \delta).$$

We are now ready to state the α-UCB algorithm, where $\alpha > 1$ is the *exploration parameter*. At each time t, we pick

$$I_t \in \underset{i=1,2}{\arg\max} \left\{ \hat{\mu}_{i,T_i(t-1)} + \alpha\, \mathrm{H}(T_i(t-1), 1/t) \right\}.$$

The argument above implies that the true mean μ_i has probability less than $1/t^{\alpha^2}$ of being
higher than $\hat{\mu}_{i,T_i(t-1)} + \alpha\mathrm{H}(T_i(t-1), 1/t)$. The algorithm makes an "optimistic" decision: it
chooses the higher of the two values.

The following theorem shows that UCB achieves a pseudo-regret of the order of $O(\log n)$.
Define $\Delta_i = \mu^* - \mu_i$ and $\Delta_* = \Delta_1 \vee \Delta_2$.

Theorem 3.2.26 (Pseudo-regret of UCB). *In the two-arm stochastic bandit problem where
the rewards are in $[0, 1]$ with distinct means, α-UCB with $\alpha > 1$ achieves*

$$\overline{R}_n \leq \frac{2\alpha^2}{\Delta_*} \log n + \Delta_* C_\alpha,$$

for some constant $C_\alpha \in (0, +\infty)$ depending only on α.

This bound should not come entirely as a surprise. Indeed, a simple, alternative approach to UCB is to (1) first pull each arm $m_n = o(n)$ times and then (2) use the arm with largest estimated mean for the remainder. Assuming there is a known lower bound on Δ_*, then Hoeffding's inequality (Theorem 2.4.10) guarantees that m_n can be chosen of the order of $\frac{1}{\Delta_*^2} \log n$ to identify the largest mean with probability $1 - 1/n$. Because the rewards are bounded by 1, accounting for the contribution of the first phase and the probability of failure in the second phase, one gets a pseudo-regret of the order of $\Delta^* \frac{1}{\Delta_*^2} \log n + \frac{1}{n} \Delta_* n \approx \frac{1}{\Delta_*} \log n$. The UCB strategy, on the other hand, elegantly adapts to the gap Δ_* and the horizon n.

Analysis of the UCB algorithm

We break down the proof into a sequence of lemmas. We first rewrite the pseudo-regret as

$$
\begin{aligned}
\overline{R}_n &= n\mu^* - \mathbb{E}\left[\sum_{t=1}^n \mu_{I_t}\right] \\
&= \mathbb{E}\left[\sum_{t=1}^n (\mu^* - \mu_{I_t})\right] \\
&= \mathbb{E}\left[\sum_{t=1}^n \sum_{i=1,2} \mathbf{1}\{I_t = i\}\Delta_i\right] \\
&= \sum_{i=1,2} \Delta_i \mathbb{E}[T_i(n)].
\end{aligned}
\tag{3.2.17}
$$

Hence, the problem boils down to bounding $\mathbb{E}[T_i(n)]$, the expected number of times that arm i is pulled. Note that $T_i(n)$ is a complicated function of the observations. To analyze it, we will use the following sufficient condition. Let i^* be the optimal arm, that is, the one that achieves μ^*. Intuitively, if arm $i \neq i^*$ is pulled, it is because either our upper estimate of μ_{i^*} happens to be low or our lower estimate of μ_i happens to be high (i.e., our concentration inequality failed); or there is too much uncertainty in our estimate of μ_i (i.e., we haven't pulled arm i enough).

Lemma 3.2.27 *Under the α-UCB strategy, if arm $i \neq i^*$ is pulled at time t, then at least one of the following events hold:*

$$
\mathcal{E}_{t,1} = \{\hat{\mu}_{i^*,T_{i^*}(t-1)} + \alpha \, \mathrm{H}(T_{i^*}(t-1), 1/t) \leq \mu^*\},
\tag{3.2.18}
$$

$$
\mathcal{E}_{t,2} = \{\hat{\mu}_{i,T_i(t-1)} - \alpha \, \mathrm{H}(T_i(t-1), 1/t) > \mu_i\},
\tag{3.2.19}
$$

$$
\mathcal{E}_{t,3} = \left\{\alpha \, \mathrm{H}(T_i(t-1), 1/t) > \frac{\Delta_i}{2}\right\}.
\tag{3.2.20}
$$

Proof We argue by contradiction. Assume all the conditions above are false. Then,

$$
\begin{aligned}
\hat{\mu}_{i^*,T_{i^*}(t-1)} + \alpha \, \mathrm{H}(T_{i^*}(t-1), 1/t) &> \mu^* \\
&= \mu_i + \Delta_i \\
&\geq \hat{\mu}_{i,T_i(t-1)} + \alpha \, \mathrm{H}(T_i(t-1), 1/t).
\end{aligned}
$$

That implies that arm i would not be chosen. ∎

We first deal with $\mathcal{E}_{t,3}$. Let

$$u_n = \frac{2\alpha^2 \log n}{\Delta_*^2}.$$

Using the condition in Lemma 3.2.27, we get the following bound on $\mathbb{E}[T_i(n)]$.

Lemma 3.2.28 *Under the α-UCB strategy, for $i \neq i^*$,*

$$\mathbb{E}[T_i(n)] \leq u_n + \sum_{t=1}^{n} \mathbb{P}[\mathcal{E}_{t,1}] + \sum_{t=1}^{n} \mathbb{P}[\mathcal{E}_{t,2}].$$

Proof For $i \neq i^*$, by definition of $T_i(n)$,

$$\mathbb{E}[T_i(n)] = \mathbb{E}\left[\sum_{t=1}^{n} \mathbf{1}_{\{I_t=i\}}\right]$$

$$\leq \mathbb{E}\left[\sum_{t=1}^{n} \left[\mathbf{1}_{\{I_t=i\}\cap\mathcal{E}_{t,1}} + \mathbf{1}_{\{I_t=i\}\cap\mathcal{E}_{t,2}} + \mathbf{1}_{\{I_t=i\}\cap\mathcal{E}_{t,3}}\right]\right],$$

where we used that by Lemma 3.2.27,

$$\{I_t = i\} \subseteq \mathcal{E}_{t,1} \cup \mathcal{E}_{t,2} \cup \mathcal{E}_{t,3}.$$

The condition in $\mathcal{E}_{t,3}$ can be written equivalently as

$$\alpha\sqrt{\frac{\log t}{2T_i(t-1)}} > \frac{\Delta_i}{2} \iff T_i(t-1) < \frac{2\alpha^2 \log t}{\Delta_i^2}.$$

In particular, for all $t \leq n$, the event $\mathcal{E}_{t,3}$ implies that $T_i(t-1) < u_n$. As a result, since $T_i(t) = T_i(t-1)+1$ whenever $I_t = i$, the event $\{I_t = i\} \cap \mathcal{E}_{t,3}$ can occur at most u_n times and

$$\mathbb{E}[T_i(n)] \leq u_n + \mathbb{E}\left[\sum_{t=1}^{n} \left[\mathbf{1}_{\{I_t=i\}\cap\mathcal{E}_{t,1}} + \mathbf{1}_{\{I_t=i\}\cap\mathcal{E}_{t,2}}\right]\right]$$

$$\leq u_n + \sum_{t=1}^{n} \mathbb{P}[\mathcal{E}_{t,1}] + \sum_{t=1}^{n} \mathbb{P}[\mathcal{E}_{t,2}],$$

which proves the claim. ∎

It remains to bound $\mathbb{P}[\mathcal{E}_{t,1}]$ and $\mathbb{P}[\mathcal{E}_{t,2}]$ from above. This is not entirely straightforward because, while $\hat{\mu}_{i,T_i(t-1)}$ involves a sum of independent random variables, the number of terms $T_i(t-1)$ is itself a *random variable*. Moreover, $T_i(t-1)$ depends on the past rewards $Z_s, s \leq t-1$, in a complex way. So in order to apply a concentration inequality to $\hat{\mu}_{i,T_i(t-1)}$, we use a rather blunt approach: we bound the worst deviation over all possible (deterministic) values in the support of $T_i(t-1)$. That is,

$$\mathbb{P}[\mathcal{E}_{t,2}] = \mathbb{P}[\hat{\mu}_{i,T_i(t-1)} - \alpha\,\mathrm{H}(T_i(t-1),1/t) > \mu_i]$$

$$\leq \mathbb{P}\left[\bigcup_{s \leq t-1}\{\hat{\mu}_{i,s} - \alpha\,\mathrm{H}(s,1/t) > \mu_i\}\right]. \qquad (3.2.21)$$

We reformulate the previous bound as

$$\mathbb{P}\left[\bigcup_{s\leq t-1}\{\hat{\mu}_{i,s} - \alpha\,\mathrm{H}(s, 1/t) > \mu_i\}\right]$$

$$= \mathbb{P}\left[\sup_{s\leq t-1}\left(\hat{\mu}_{i,s} - \mu_i - \alpha\,\mathrm{H}(s, 1/t)\right) > 0\right]$$

$$= \mathbb{P}\left[\sup_{s\leq t-1}\left(\frac{1}{s}\sum_{r\leq s}X_{i,r} - \mu_i - \alpha\sqrt{\frac{\log t}{2s}}\right) > 0\right]$$

$$= \mathbb{P}\left[\sup_{s\leq t-1}\frac{1}{\sqrt{s}}\left(\frac{1}{\sqrt{s}}\sum_{r\leq s}(X_{i,r} - \mu_i) - \alpha\sqrt{\frac{\log t}{2}}\right) > 0\right]$$

$$= \mathbb{P}\left[\sup_{s\leq t-1}\frac{\sum_{r=1}^{s}(X_{i,r} - \mu_i)}{\sqrt{s}} > \alpha\sqrt{\frac{\log t}{2}}\right]. \tag{3.2.22}$$

Observe that the numerator on the left-hand side of the inequality on the last line is a martingale (see Example 3.1.29) with increments in $[-\mu_i, 1 - \mu_i]$. But the denominator depends on s.

We try two approaches:

- We could simply use $\sqrt{s} \geq 1$ on the denominator and apply the maximal Azuma–Hoeffding inequality (Theorem 3.2.1) to get

$$\sum_{t=1}^{n}\mathbb{P}[\mathcal{E}_{t,2}] \leq \sum_{t=1}^{n}\mathbb{P}\left[\sup_{s\leq t-1}\sum_{r=1}^{s}(X_{i,r} - \mu_i) > \alpha\sqrt{\frac{\log t}{2}}\right]$$

$$\leq \sum_{t=1}^{n}\exp\left(-\frac{2(\alpha\sqrt{(\log t)/2})^2}{t - 1}\right)$$

$$\leq \sum_{t=1}^{n}\exp\left(-\alpha^2\frac{\log t}{t - 1}\right). \tag{3.2.23}$$

That is of order $\Theta(n)$ for any α.

- On the other hand, we could use a union bound over s and apply the maximal Azuma–Hoeffding inequality to each term to get

$$\sum_{t=1}^{n}\mathbb{P}[\mathcal{E}_{t,2}] \leq \sum_{t=1}^{n}\sum_{s\leq t-1}\mathbb{P}\left[\sum_{r=1}^{s}(X_{i,r} - \mu_i) > \alpha\sqrt{\frac{s\log t}{2}}\right]$$

$$\leq \sum_{t=1}^{n}\sum_{s\leq t-1}\exp\left(-\frac{2(\alpha\sqrt{(s\log t)/2})^2}{s}\right)$$

$$= \sum_{t=1}^{n}(t - 1)\exp\left(-\alpha^2\log t\right)$$

$$\leq \sum_{t=1}^{n}\frac{1}{t^{\alpha^2-1}}. \tag{3.2.24}$$

The series converges for $\alpha > \sqrt{2}$. Therefore, in that case, this bound is $\Theta(1)$, which is much better than our previous attempt. For $1 < \alpha \leq \sqrt{2}$ however, we get a bound of order $\Theta(n^{\alpha^2})$, which is worse than before.

It turns out that doing something "in between" the two approaches above gives a bound that significantly improves over both of them in the $1 < \alpha \leq \sqrt{2}$ regime. This is known as the slicing (or peeling) method.

Slicing method

The *slicing method* is useful when bounding a weighted supremum. Its application is somewhat problem-specific so we will content ourselves with illustrating it in our case. Specifically, our goal is to control probabilities of the form

$$\mathbb{P}\left[\sup_{s \leq t-1} \frac{M_s}{w(s)} \geq \beta\right],$$

where $M_s := \sum_{r=1}^{s}(X_{i,r} - \mu_i)$, $w(s) := \sqrt{s}$, and $\beta := \alpha\sqrt{\frac{\log t}{2}}$. The idea is to divide up the supremum into *slices* $\gamma^{k-1} \leq s < \gamma^k$, $k \geq 1$, where the constant $\gamma > 1$ will be optimized below. That is, fixing $K_t = \lceil \frac{\log t}{\log \gamma} \rceil$ (which roughly solves $\gamma^{K_t} = t$), by a union bound over the slices

$$\mathbb{P}\left[\sup_{1 \leq s < t} \frac{M_s}{w(s)} \geq \beta\right] \leq \sum_{k=1}^{K_t} \mathbb{P}\left[\sup_{\gamma^{k-1} \leq s < \gamma^k} \frac{M_s}{w(s)} \geq \beta\right].$$

Because $w(s)$ is increasing, on each slice separately we can bound

$$\mathbb{P}\left[\sup_{\gamma^{k-1} \leq s < \gamma^k} \frac{M_s}{w(s)} \geq \beta\right] \leq \mathbb{P}\left[\sup_{\gamma^{k-1} \leq s < \gamma^k} \frac{M_s}{w(\gamma^{k-1})} \geq \beta\right]$$

$$= \mathbb{P}\left[\sup_{\gamma^{k-1} \leq s < \gamma^k} M_s \geq \beta w(\gamma^{k-1})\right]$$

$$\leq \mathbb{P}\left[\sup_{s \leq \gamma^k} M_s \geq \beta w(\gamma^{k-1})\right].$$

Now we apply the maximal Azuma–Hoeffding inequality (Theorem 3.2.1) to obtain

$$\mathbb{P}\left[\sup_{s \leq \gamma^k} M_s \geq \beta w(\gamma^{k-1})\right] \leq \exp\left(-\frac{2(\beta w(\gamma^{k-1}))^2}{\gamma^k}\right)$$

$$\leq \exp\left(-\frac{2\beta^2}{\gamma}\right)$$

$$= t^{-\alpha^2/\gamma},$$

where we used that $M_s - M_{s-1} = X_{i,s} - \mu_i \in [-\mu_i, 1 - \mu_i]$, an interval of length 1. Combining the last three displays we get

$$\mathbb{P}\left[\sup_{1 \leq s < t} \frac{M_s}{w(s)} \geq \beta\right] \leq \left\lceil \frac{\log t}{\log \gamma} \right\rceil t^{-\alpha^2/\gamma}. \tag{3.2.25}$$

Now we see the trade-off: increasing γ makes the slices larger and hence the tail inequality weaker, but it also makes the number of slices smaller, which helps with the union bound.

Combining (3.2.21), (3.2.22), and (3.2.25), we have proved:

Lemma 3.2.29 *For any $\gamma > 1$, it holds that*

$$\sum_{t=1}^{n} \mathbb{P}[\mathcal{E}_{t,2}] \leq \sum_{t=1}^{n} \left\lceil \frac{\log t}{\log \gamma} \right\rceil t^{-\alpha^2/\gamma},$$

and similarly for $\mathbb{P}[\mathcal{E}_{t,1}]$.

For $\alpha > 1$, we can choose $\gamma > 1$ such that $\alpha^2/\gamma > 1$. In that case, the series on the right-hand side is summable. This improves over both (3.2.23) and (3.2.24).

We are ready to prove the main result.

Proof of Theorem 3.2.26 By (3.2.17) and Lemmas 3.2.27, 3.2.28, and 3.2.29, we have

$$\overline{R}_n = \sum_{i=1,2} \Delta_i \mathbb{E}[T_i(n)] \leq \Delta_* \left(u_n + 2 \sum_{t=1}^{n} \left\lceil \frac{\log t}{\log \gamma} \right\rceil t^{-\alpha^2/\gamma} \right).$$

Recalling that $\alpha > 1$, choose $\gamma > 1$ such that $\alpha^2/\gamma > 1$. In that case, as noted above, the series on the right-hand side is summable and there is $C_\alpha \in (0, +\infty)$ such that

$$\overline{R}_n \leq \Delta_*(u_n + C_\alpha).$$

That proves the claim. ∎

Remark 3.2.30 *A slightly better – and provably optimal – multiplicative constant in the pseudo-regret bound has been obtained by [GC11] using a variant of UCB called KL-UCB. The matching lower bound is due to [LR85]. See also [BCB12, sections 2.3–2.4]. Further improvements can be obtained by using Bernstein's rather than Hoeffding's inequality [AMS09].*

3.2.6 Coda: Talagrand's Inequality

We end this section with a celebrated concentration inequality that applies under weaker conditions than McDiarmid's inequality (Theorem 3.2.9) – but is *not* proved using the martingale method. It is known as *Talagrand's inequality.*

Bounds on $\|D_i f\|_\infty$ are often expressed in terms of a Lipschitz condition under an appropriate metric. Let $0 < c_i < +\infty$, $i = 1, \ldots, n$ and $\mathbf{c} = (c_1, \ldots, c_n)$. The $\mathbf{c}$-*weighted Hamming distance* is defined as

WEIGHTED HAMMING DISTANCE

$$\rho_{\mathbf{c}}(\mathbf{x}, \mathbf{y}) := \sum_{i=1}^{n} c_i \mathbf{1}_{\{x_i \neq y_i\}}$$

for $\mathbf{x} = (x_1, \ldots, x_n), \mathbf{y} = (y_1, \ldots, y_n) \in \mathcal{X}_1 \times \cdots \times \mathcal{X}_n$. The proof of the following equivalence is left as an exercise (see Exercise 3.8).

Lemma 3.2.31 (Lipschitz condition). *A function* $f: \mathcal{X}_1 \times \cdots \times \mathcal{X}_n \to \mathbb{R}$ *satisfies the Lipschitz condition*

$$|f(\mathbf{x}) - f(\mathbf{y})| \leq \rho_{\mathbf{c}}(\mathbf{x}, \mathbf{y}) \quad \forall \mathbf{x}, \mathbf{y} \in \mathcal{X}_1 \times \cdots \times \mathcal{X}_n \tag{3.2.26}$$

if and only if

$$\|D_i f\|_\infty \leq c_i \quad \forall i.$$

Consider the following relaxed version of (3.2.26):

$$f(\mathbf{x}) - f(\mathbf{y}) \leq \sum_{i=1}^{n} c_i(\mathbf{x}) \mathbf{1}_{\{x_i \neq y_i\}} \quad \forall \mathbf{x}, \mathbf{y} \in \mathcal{X}_1 \times \cdots \times \mathcal{X}_n, \tag{3.2.27}$$

where now $c_i(\mathbf{x})$ is a finite, positive function over $\mathcal{X}_1 \times \cdots \times \mathcal{X}_n$. Notice the "one-sided" nature of this condition, in the sense that c_i depends on $\mathbf{x}$ but not on $\mathbf{y}$. A typical example where (3.2.27) is satisfied, but (3.2.26) is not, is given in Example 3.2.33.

We state Talagrand's inequality without proof.

Theorem 3.2.32 (Talagrand's inequality). *Let* $X_1, \ldots, X_n$ *be independent random variables where* X_i *is* $\mathcal{X}_i$-*valued for all* i, *and let* $X = (X_1, \ldots, X_n)$. *Assume* $f: \mathcal{X}_1 \times \cdots \times \mathcal{X}_n \to \mathbb{R}$ *is a measurable function such that* (3.2.27) *holds. Then* $f(X)$ *is sub-Gaussian with variance factor* $\| \sum_{i \leq n} c_i^2 \|_\infty$. *In fact, for all* $\beta > 0$ *the following upper and lower tail bounds hold*

$$\mathbb{P}[f(X) - \mathbb{E}f(X) \geq \beta] \leq \exp\left(-\frac{\beta^2}{2\| \sum_{i \leq n} c_i^2 \|_\infty}\right)$$

and

$$\mathbb{P}[f(X) - \mathbb{E}f(X) \leq -\beta] \leq \exp\left(-\frac{\beta^2}{2\,\mathbb{E}\left[\sum_{i \leq n} c_i(X)^2\right]}\right).$$

Compared to McDiarmid's inequality (Theorem 3.2.9), the upper tail in Theorem 3.2.32 has the sum over the coordinates *inside the supremum*, potentially a major improvement; the lower tail is even better, replacing the supremum with an *expectation*.

Example 3.2.33 (Spectral norm of a random matrix with bounded entries). Let A be an $n \times n$ random matrix. We assume that the entries $A_{i,j}, i, j = 1, \ldots, n$ are independent, centered random variables in $[-1, 1]$. In Theorem 2.4.28, we proved an upper tail bound on the spectral norm

$$\|A\|_2 = \sup_{\mathbf{x} \in \mathbb{R}^n \setminus \{0\}} \frac{\|A\mathbf{x}\|_2}{\|\mathbf{x}\|_2} = \sup_{\substack{\mathbf{x} \in \mathbb{S}^{n-1} \\ \mathbf{y} \in \mathbb{S}^{n-1}}} \langle A\mathbf{x}, \mathbf{y} \rangle$$

of such a matrix (in the more general sub-Gaussian case) using an ε-net argument. Theorem 2.4.28 also implies that $\mathbb{E}\|A\|_2 = O(\sqrt{n})$ by (B.5.1). (See Exercise 3.9 for a lower bound on the expectation.)

Here we use Talagrand's inequality (Theorem 3.2.32) directly to show concentration around the mean. For this, we need to check (3.2.27) where we think of the spectral norm as a function of n^2 independent random variables

$$\|A\|_2 = f(\{A_{i,j}\}_{i,j}).$$

Let $\mathbf{x}^*(A)$ and $\mathbf{y}^*(A)$ be unit vectors in $\mathbb{R}^n$ such that

$$\|A\|_2 = \langle A\mathbf{x}^*(A), \mathbf{y}^*(A) \rangle,$$

which exist by compactness.

Given two $n \times n$ matrices $A, \widetilde{A}$ with entries in $[-1, 1]$, we have

$$
\begin{aligned}
\|A\|_2 - \|\widetilde{A}\|_2 &= \langle A\mathbf{x}^*(A), \mathbf{y}^*(A) \rangle - \sup_{\substack{\mathbf{x} \in \mathbb{S}^{n-1} \\ \mathbf{y} \in \mathbb{S}^{n-1}}} \langle \widetilde{A}\mathbf{x}, \mathbf{y} \rangle \\
&\leq \langle A\mathbf{x}^*(A), \mathbf{y}^*(A) \rangle - \langle \widetilde{A}\mathbf{x}^*(A), \mathbf{y}^*(A) \rangle \\
&= \langle (A - \widetilde{A})\mathbf{x}^*(A), \mathbf{y}^*(A) \rangle \\
&\leq \sum_{i,j} |A_{ij} - \widetilde{A}_{ij}| |\mathbf{x}^*(A)_i| |\mathbf{y}^*(A)_j| \\
&\leq \sum_{i,j} \mathbf{1}_{A_{ij} \neq \widetilde{A}_{ij}} c_{ij}(A),
\end{aligned}
$$

where on the last line we set

$$c_{ij}(A) := 2|\mathbf{x}^*(A)_i| |\mathbf{y}^*(A)_j|,$$

and used the fact that $|A_{ij} - \widetilde{A}_{ij}| \leq 2$. Note that

$$\sum_{i,j} c_{ij}(A)^2 = 4 \sum_i \mathbf{x}^*(A)_i^2 \sum_j \mathbf{y}^*(A)_j^2 = 4.$$

Hence, Talagrand's inequality implies that $\|A\|_2$ is sub-Gaussian with variance factor 4. ◀

3.3 Potential Theory and Electrical Networks

In this section, we develop a classical link between random walks and electrical networks. The electrical interpretation is a useful physical analogy. The mathematical substance of the connection starts with the following observation.

Let (X_t) be a Markov chain with transition matrix P on a finite or countable state space V. Recall from Definition 3.1.6 that τ_B is the first visit time to $B \subseteq V$. For two disjoint subsets A, Z of V, the probability of hitting A before Z

$$h(x) = \mathbb{P}_x[\tau_A < \tau_Z], \tag{3.3.1}$$

seen as a function of the starting point $x \in V$, is harmonic (with respect to P) on $W :=$ HARMONIC
$(A \cup Z)^c := V \setminus (A \cup Z)$ in the sense that FUNCTION

$$h(x) = \sum_y P(x, y)h(y) \qquad \forall x \in W. \tag{3.3.2}$$

Indeed, note that $h = 1$ (respectively $= 0$) on A (respectively Z) and by the Markov property (Theorem 1.1.18), after the first step of the chain, for $x \in W$,

$$\mathbb{P}_x[\tau_A < \tau_Z] = \sum_{y \notin A \cup Z} P(x,y)\,\mathbb{P}_y[\tau_A < \tau_Z]$$

$$+ \sum_{y \in A} P(x,y) \cdot 1 + \sum_{y \in Z} P(x,y) \cdot 0$$

$$= \sum_y P(x,y)\,\mathbb{P}_y[\tau_A < \tau_Z]. \tag{3.3.3}$$

Quantities such as (3.3.1) arise naturally, for instance in the study of recurrence, and the connection to potential theory, the study of harmonic functions, proves fruitful in that context as we outline in this section. It turns out that harmonic functions and martingales are closely related. In Section 3.3.1, we elaborate on that connection.

But first we rewrite (3.3.2) to reveal the electrical interpretation. For this we switch to reversible chains. Recall that a reversible Markov chain is equivalent to a random walk on a network $\mathcal{N} = (G, c)$, where the edges of G correspond to transitions of positive probability. If the chain is reversible with respect to a stationary measure π, then the edge weights are $c(x,y) = \pi(x)P(x,y)$. In this notation (3.3.2) becomes

$$h(x) = \frac{1}{c(x)} \sum_{y \sim x} c(x,y)h(y) \qquad \forall x \in (A \cup Z)^c, \tag{3.3.4}$$

where $c(x) := \sum_{y \sim x} c(x,y) = \pi(x)$. In words, $h(x)$ is the weighted average of its neighboring values. Now comes the electrical analogy: if one interprets $c(x,y)$ as a conductance, a function satisfying (3.3.4) is known as a voltage. The voltages at A and Z are 1 and 0, respectively. We show in the next subsection by a martingale argument that, under appropriate conditions, such a voltage exists and is unique. We develop the electrical analogy and many of its applications in Section 3.3.2.

3.3.1 Martingales, the Dirichlet Problem and Lyapounov Functions

To see why martingales come in, let $\mathcal{F}_t = \sigma(X_0, \ldots, X_t)$ and let $\tau^* := \tau_{W^c}$. By a first-step calculation again, (3.3.2) implies

$$h(X_{t \wedge \tau^*}) = \mathbb{E}\left[h(X_{(t+1) \wedge \tau^*}) \,|\, \mathcal{F}_t\right] \qquad \forall t \geq 0, \tag{3.3.5}$$

that is, $(h(X_{t \wedge \tau^*}))_t$ is a martingale with respect to $(\mathcal{F}_t)$. Indeed, on $\{\tau^* \leq t\}$,

$$\mathbb{E}[h(X_{(t+1) \wedge \tau^*}) \,|\, \mathcal{F}_t] = h(X_{\tau^*}) = h(X_{t \wedge \tau^*}),$$

and on $\{\tau^* > t\}$,

$$\mathbb{E}[h(X_{(t+1) \wedge \tau^*}) \,|\, \mathcal{F}_t] = \sum_y P(X_t, y)h(y) = h(X_t) = h(X_{t \wedge \tau^*}).$$

Although the rest of Section 3.3 is concerned with reversible Markov chains, the current subsection applies to the non-reversible case as well. We give an overview of potential theory for general, countable-space, discrete-time Markov chains and its connections to martingales. As a major application, we introduce the concept of a Lyapounov function which is useful in bounding certain hitting times.

Existence and uniqueness of a harmonic extension

We begin with a special case, which will be generalized in Theorem 3.3.9.

Theorem 3.3.1 (Harmonic extension: existence and uniqueness). *Let P be an irreducible transition matrix on a finite or countably infinite spate space V. Let W be a finite, proper subset of V and let $h\colon W^c \to \mathbb{R}$ be a bounded function on W^c. Then there exists a unique extension of h to W that is harmonic on W, that is, which satisfies (3.3.2). The solution is given by*

$$h(x) = \mathbb{E}_x[h\left(X_{\tau_{W^c}}\right)].$$

Proof We first argue about uniqueness. Suppose h is defined over all of V and satisfies (3.3.2). Let $\tau^* := \tau_{W^c}$. Then the process $(h\left(X_{t\wedge\tau^*}\right))_t$ is a martingale by (3.3.5). Because W is finite and the chain is irreducible, we have $\tau^* < +\infty$ almost surely, as implied by Lemma 3.1.25. Moreover, the process is bounded because h is bounded on W^c and W is finite. Hence, by Doob's optional stopping theorem (Theorem 3.1.38 (ii)),

$$h(x) = \mathbb{E}_x[h(X_{\tau^*})] \qquad \forall x \in W,$$

which implies that h is unique, since the right-hand side depends only on the chain and the fixed values of h on W^c.

For the existence, simply define $h(x) := \mathbb{E}_x[h\left(X_{\tau^*}\right)]\forall x \in W$, and use a first-step argument similarly to (3.3.3). ∎

For some insights on what happens when the assumptions of Theorem 3.3.1 are not satisfied, see Exercise 3.11. For an alternative (arguably more intuitive) proof of uniqueness based on the maximum principle, see Exercise 3.12.

In the proof above it suffices to specify h on the outer boundary of W:

$$\partial_V W = \{z \in V\backslash W : \exists y \in W, P(y,z) > 0\}.$$

Introduce the *Laplacian* associated to P:

LAPLACIAN

$$\begin{aligned}
\Delta f(x) &= \left[\sum_y P(x,y)f(y)\right] - f(x) \\
&= \sum_y P(x,y)[f(y) - f(x)] \\
&= \mathbb{E}_x[f(X_1) - f(X_0)], \qquad (3.3.6)
\end{aligned}$$

provided the expectation exists. We have proved that, under the assumptions of Theorem 3.3.1, there exists a unique solution to

$$\begin{cases} \Delta f(x) = 0 & \forall x \in W, \\ f(x) = h(x) & \forall x \in \partial_V W, \end{cases} \qquad (3.3.7)$$

and that solution is given by $f(x) = \mathbb{E}_x[h\left(X_{\tau_{W^c}}\right)]$ for $x \in W \cup \partial_V W$. The system (3.3.7), in reference to its counterpart in the theory of partial differential equations, is referred to as a *Dirichlet problem*.

DIRICHLET PROBLEM

Example 3.3.2 (Simple random walk on $\mathbb{Z}^d$). The Laplacian above can be interpreted as a discretized version of the standard Laplacian. For instance, for simple random walk on $\mathbb{Z}$,

$$\Delta f(x) = \left[\sum_y P(x,y) f(y) \right] - f(x)$$

$$= \sum_y P(x,y)[f(y) - f(x)]$$

$$= \frac{1}{2}\{[f(x+1) - f(x)] - [f(x) - f(x-1)]\},$$

which is a discretized second derivative. More generally, for simple random walk on $\mathbb{Z}^d$, we get

$$\Delta f(x) = \left[\sum_y P(x,y) f(y) \right] - f(x)$$

$$= \sum_y P(x,y)[f(y) - f(x)]$$

$$= \frac{1}{2d} \sum_{i=1}^{d} \{[f(x + \mathbf{e}_i) - f(x)] - [f(x) - f(x - \mathbf{e}_i)]\},$$

where $\mathbf{e}_1, \ldots, \mathbf{e}_d$ is the standard basis in $\mathbb{R}^d$.   ◄

Theorem 3.3.1 has many applications. One of its consequences is that harmonic functions on a finite state space are constant.

Corollary 3.3.3 *Let P be an irreducible transition matrix on a finite state space V. If h is harmonic on* all *of V, then it is constant.*

Proof Fix the value of h at an arbitrary vertex z and set $W = V \backslash \{z\}$. Applying Theorem 3.3.1, for all $x \in W$, $h(x) = \mathbb{E}_x[h(X_{\tau_{W^c}})] = h(z)$.   ∎

As an example of application of this corollary, we prove the following surprising result: in a finite, irreducible Markov chain, the expected time to hit a target chosen at random according to the stationary distribution does not depend on the starting point.

Theorem 3.3.4 (Random target lemma). *Let (X_t) be an irreducible Markov chain on a finite state space V with transition matrix P and stationary distribution π. Then,*

$$h(x) := \sum_{y \in V} \pi(y) \, \mathbb{E}_x[\tau_y]$$

does not in fact depend on x.

Proof Because the chain is irreducible and has a finite state space, $\mathbb{E}_x[\tau_y] < +\infty$ for all x, y. By Corollary 3.3.3, it suffices to show that $h(x) := \sum_y \pi(y) \mathbb{E}_x[\tau_y]$ is harmonic on all of V. As before, it is natural to expand $\mathbb{E}_x[\tau_y]$ according to the first step of the chain,

$$\mathbb{E}_x[\tau_y] = \mathbf{1}_{\{x \neq y\}} \left(1 + \sum_z P(x,z) \, \mathbb{E}_z[\tau_y] \right).$$

Substituting into the definition of $h(x)$ gives

$$h(x) = (1 - \pi(x)) + \sum_z \sum_{y \neq x} \pi(y) P(x, z) \mathbb{E}_z[\tau_y]$$

$$= (1 - \pi(x)) + \sum_z P(x, z) (h(z) - \pi(x) \mathbb{E}_z[\tau_x]).$$

Rearranging, we get

$$\Delta h(x) = \left[\sum_z P(x, z) h(z) \right] - h(x)$$

$$= \pi(x) \left(1 + \sum_z P(x, z) \mathbb{E}_z[\tau_x] \right) - 1$$

$$= 0,$$

where we used $1/\pi(x) = \mathbb{E}_x[\tau_x^+] = 1 + \sum_z P(x, z) \mathbb{E}_z[\tau_x]$ by Theorem 3.1.19 and a first-step argument (recall that the first return time τ_x^+ was defined in Definition 3.1.6). ∎

Potential theory for Markov chains

More generally, many quantities of interest can be expressed in the following form. Consider again a subset $W \subset V$ and the stopping time

$$\tau_{W^c} = \inf\{t \geq 0 \colon X_t \in W^c\}.$$

Let also $h \colon W^c \to \mathbb{R}_+$ and $k \colon W \to \mathbb{R}_+$. Define the quantity

$$u(x) := \mathbb{E}_x \left[h(X_{\tau_{W^c}}) \mathbf{1}\{\tau_{W^c} < +\infty\} + \sum_{0 \leq t < \tau_{W^c}} k(X_t) \right]. \tag{3.3.8}$$

The first term on the right-hand side is a final cost incurred when we exit W (and it depends on where we do), while the second term is a unit time cost incurred along the sample path. Note that, in fact, it suffices to define h on $\partial_V W$, the outer boundary of W if we restrict ourselves to $x \in W$. Observe also that the function $u(x)$ may take the value $+\infty$; the expectation is well defined (in $\mathbb{R}_+ \cup \{+\infty\}$) by the non-negativity of the terms (see Appendix B).

Example 3.3.5 (Some special cases). Here are some important special cases:

- Revisiting (3.3.1), for two disjoint subsets A, Z of V, the probability

$$u(x) := \mathbb{P}_x[\tau_A < \tau_Z]$$

of hitting A before Z as a function of the starting point $x \in V$ is obtained by taking $W := (A \cup Z)^c$, $h = 1$ (respectively $= 0$) on A (respectively Z), and $k = 0$ on V. The further special case $Z = \emptyset$ leads to the *exit probability* from A:

EXIT
PROBABILITY

$$u(x) := \mathbb{P}_x[\tau_A < +\infty].$$

On the other hand, if A and Z form a disjoint partition of W^c (or $\partial_V W$ will suffice if $x \in W$), we get the *exit law* from W:

EXIT LAW

$$u(x) := \mathbb{P}_x[X_{\tau_{W^c}} \in A; \tau_{W^c} < +\infty].$$

• The *average occupation time* of $A \subseteq W$ before exiting W,

$$u(x) := \mathbb{E}_x \left[\sum_{0 \le t < \tau_{W^c}} \mathbf{1}_{\{X_t \in A\}} \right],$$

is obtained by taking $h = 0$ on V, and $k = 1$ (respectively $= 0$) on A (respectively on A^c). Revisiting (3.1.3), the Green function of the chain stopped at τ_{W^c}, that is,

$$u(x) := \mathscr{G}_{\tau_{W^c}}(x, y) = \mathbb{E}_x \left[\sum_{0 \le t < \tau_{W^c}} \mathbf{1}_{\{X_t = y\}} \right],$$

is obtained by taking $A = \{y\}$. Another special case is $A = W$, where we get the *mean exit time* from A

$$u(x) := \mathbb{E}_x [\tau_{A^c}].$$ ◄

The function u in (3.3.8) turns out to satisfy a generalized version of (3.3.7). The proof is usually called *first-step analysis* (of which we have already seen many instances).

Theorem 3.3.6 (First-step analysis). *Let P be a transition matrix on a finite or countable spate space V. Let W be a proper subset of V, and let $h\colon W^c \to \mathbb{R}_+$ and $k\colon W \to \mathbb{R}_+$ be bounded functions. Then the function $u \ge 0$, as defined in (3.3.8), satisfies the system of equations*

$$\begin{cases} u(x) = k(x) + \sum_y P(x, y)u(y) & \text{for } x \in W, \\ u(x) = h(x) & \text{for } x \in W^c. \end{cases} \tag{3.3.9}$$

Proof For $x \in W^c$, by definition $u(x) = h(x)$ since $\tau_{W^c} = 0$. Fix $x \in W$. By taking out what is known (Lemma B.6.13), the tower property (Lemma B.6.16) and the Markov property (Theorem 1.1.18),

$$u(x) = k(x) + \mathbb{E}_x \left[h(X_{\tau_{W^c}}) \mathbf{1}\{\tau_{W^c} < +\infty\} + \sum_{1 \le t < \tau_{W^c}} k(X_t) \right]$$

$$= k(x) + \mathbb{E}_x \left[\mathbb{E} \left[h(X_{\tau_{W^c}}) \mathbf{1}\{\tau_{W^c} < +\infty\} + \sum_{1 \le t < \tau_{W^c}} k(X_t) \,\middle|\, \mathcal{F}_1 \right] \right]$$

$$= k(x) + \mathbb{E}_x [u(X_1)],$$

which gives the claim.   ∎

If u is finite, the system (3.3.9) can be rewritten as the *Poisson equation* (once again as an analogue of its counterpart in the theory of partial differential equations)

$$\begin{cases} \Delta u = -k & \text{on } W, \\ u = h & \text{on } W^c. \end{cases} \tag{3.3.10}$$

This is well defined for instance if W is a finite subset and P is irreducible. Indeed, as we argued in the proof of Theorem 3.3.1, the stopping time τ_{W^c} then has a finite expectation. Because h is bounded, it follows that

$$
\begin{aligned}
u(x) := \mathbb{E}_x \left[h(X_{\tau_{W^c}}) \mathbf{1}\{\tau_{W^c} < +\infty\} + \sum_{0 \leq t < \tau_{W^c}} k(X_t) \right] \\
\leq \sup_{x \in W^c} h(x) + \sup_{x \in W} k(x) \sup_{x \in W} \mathbb{E}_x [\tau_{W^c}] \\
< +\infty,
\end{aligned}
$$

uniformly in x. Using (3.3.6) and rearranging (3.3.9) gives (3.3.10).

Remark 3.3.7 *A more general form of the statement which can be used to study certain moment-generating functions can be found, for example, in [Ebe, Theorem 1.3].*

In a generalization of Theorem 3.3.1, our next theorem allows one to establish uniqueness of the solution of the system (3.3.10) under some conditions (which we will not detail here, but see Exercise 3.13). Perhaps even more useful, it also gives an effective approach to bound the function u from above. This is based on the following supermartingale.

Lemma 3.3.8 (Locally superharmonic functions). *Let P be a transition matrix on a finite or countable spate space V. Let W be a proper subset of V, and let $h \colon W^c \to \mathbb{R}_+$ and $k \colon W \to \mathbb{R}_+$ be bounded functions. Suppose the non-negative function $\psi \colon V \to \mathbb{R}_+$ satisfies*

$$
\Delta \psi \leq -k \qquad on \ W.
$$

Then the process

$$
N_t := \psi(X_{t \wedge \tau_{W^c}}) + \sum_{0 \leq s < t \wedge \tau_{W^c}} k(X_s)
$$

is a non-negative supermartingale for any initial point $x \in V$.

Proof Observe that on $\{\tau_{W^c} \leq t\}$, we have $N_{t+1} = N_t$; while on $\{\tau_{W^c} > t\}$ we have $N_{t+1} - N_t = \psi(X_{t+1}) - \psi(X_t) + k(X_t)$ by cancellations in the sum. So, since $\{\tau_{W^c} > t\} \in \mathcal{F}_t$ by definition of a stopping time, it holds by taking out what is known that

$$
\begin{aligned}
\mathbb{E}[N_{t+1} - N_t \mid \mathcal{F}_t] &= \mathbb{E}[\mathbf{1}\{\tau_{W^c} > t\}(\psi(X_{t+1}) - \psi(X_t) + k(X_t)) \mid \mathcal{F}_t] \\
&= \mathbf{1}\{\tau_{W^c} > t\}(\mathbb{E}[\psi(X_{t+1}) - \psi(X_t) \mid \mathcal{F}_t] + k(X_t)) \\
&= \mathbf{1}\{\tau_{W^c} > t\}(\Delta \psi(X_t) + k(X_t)) \\
&\leq \mathbf{1}\{\tau_{W^c} > t\}(-k(X_t) + k(X_t)) \\
&= 0,
\end{aligned}
$$

where we used that, by (3.3.6) and the Markov property,

$$
\mathbb{E}[\psi(X_{t+1}) - \psi(X_t) \mid \mathcal{F}_t] = \Delta \psi(X_t), \tag{3.3.11}
$$

and that $X_t \in W$ on $\{\tau_{W^c} > t\}$. ∎

Theorem 3.3.9 (Poisson equation: bounding the solution). *Let P be a transition matrix on a finite or countable spate space V. Let W be a proper subset of V, and let $h \colon W^c \to \mathbb{R}_+$*

*and $k \colon W \to \mathbb{R}_+$ be bounded functions. Suppose the non-negative function $\psi \colon V \to \mathbb{R}_+$
satisfies the system of inequalities*

$$\begin{cases} \Delta \psi \leq -k & \text{on } W, \\ \psi \geq h & \text{on } W^c. \end{cases} \tag{3.3.12}$$

Then,

$$\psi \geq u, \quad \text{on } V, \tag{3.3.13}$$

where u is the function defined in (3.3.8).

Proof   The system (3.3.13) holds on W^c by Theorem 3.3.6 and (3.3.12) since in that case
$u(x) = h(x) \leq \psi(x)$.

Fix $x \in W$. Consider the non-negative supermartingale (N_t) in Lemma 3.3.8. By the convergence of non-negative supermartingales (Corollary 3.1.48), (N_t) converges almost surely
to a finite limit with expectation $\leq \mathbb{E}_x[N_0]$. In particular, the limit $N_{\tau_{W^c}}$ is well defined, non-negative, and finite, including on the event that $\{\tau_{W^c} = +\infty\}$. As a result,

$$\begin{aligned}
N_{\tau_{W^c}} &= \psi(X_{\tau_{W^c}})\mathbf{1}\{\tau_{W^c} < +\infty\} + \sum_{0 \leq s < \tau_{W^c}} k(X_s) \\
&\geq h(X_{\tau_{W^c}})\mathbf{1}\{\tau_{W^c} < +\infty\} + \sum_{0 \leq s < \tau_{W^c}} k(X_s),
\end{aligned}$$

where we used (3.3.12).

Hence, by definition of u,

$$\begin{aligned}
u(x) &= \mathbb{E}_x\left[h(X_{\tau_{W^c}})\mathbf{1}\{\tau_{W^c} < +\infty\} + \sum_{0 \leq t < \tau_{W^c}} k(X_t) \right] \\
&\leq \mathbb{E}_x\left[N_{\tau_{W^c}} \right] \\
&\leq \mathbb{E}_x[N_0] \\
&= \psi(x),
\end{aligned}$$

where, on the last line, we used that the initial state is $x \in W$. That proves the claim.   ■

Lyapounov functions

Here is an important application, bounding from above the hitting time τ_A to a set A in
expectation.

Theorem 3.3.10 (Controlling hitting times via Lyapounov functions).   *Let P be a transition
matrix on a finite or countably infinite state space V. Let A be a proper subset of V. Suppose
the non-negative function $\psi \colon V \to \mathbb{R}_+$ satisfies the system of inequalities*

$$\Delta \psi \leq -1 \quad \text{on } A^c. \tag{3.3.14}$$

Then,

$$\mathbb{E}_x[\tau_A] \leq \psi(x)$$

for all $x \in V$.

Proof Indeed, by (3.3.14) and non-negativity (in particular on A), the function ψ satisfies the assumptions of Theorem 3.3.9 with $W = A^c$, $h = 0$ on A, and $k = 1$ on A^c. Hence, by definition of u and the claim in Theorem 3.3.9,

$$\mathbb{E}_x[\tau_A] = \mathbb{E}_x\left[h(X_{\tau_A})\mathbf{1}\{\tau_A < +\infty\} + \sum_{0 \leq t < \tau_A} k(X_t)\right]$$

$$= u(x)$$

$$\leq \psi(x).$$

That establishes the claim. ∎

Recalling (3.3.11), condition (3.3.14) is equivalent to the following conditional expected decrease in ψ outside A:

$$\mathbb{E}[\psi(X_{t+1}) - \psi(X_t) \mid \mathcal{F}_t] \leq -1 \quad \text{on } \{X_t \in A^c\}. \tag{3.3.15}$$

A non-negative function satisfying an inequality of this type, also known as *drift condition*, is often referred to as a *Lyapounov function*. Intuitively, it tends to decrease along the sample path outside of A. Because it is non-negative, it cannot decrease forever and therefore the chain eventually enters A. We consider a simple example next.

LYAPOUNOV FUNCTION

Example 3.3.11 (A Markov chain on the non-negative integers). Let $(Z_t)_{t \geq 1}$ be i.i.d. integrable random variables taking values in $\mathbb{Z}$ such that $\mathbb{E}[Z_1] < 0$. Let $(X_t)_{t \geq 0}$ be the chain defined by $X_0 = x$ for some $x \in \mathbb{Z}_+$ and

$$X_{t+1} = (X_t + Z_{t+1})^+,$$

where recall that $z^+ = \max\{0, z\}$. In particular, $X_t \in \mathbb{Z}_+$ for all t. Let $(\mathcal{F}_t)$ be the corresponding filtration. When X_t is large, the "local drift" is close to $\mathbb{E}[Z_1] < 0$. By analogy to the biased case of the gambler's ruin (Example 3.1.43), we might expect that, from a large starting point x, it will take time roughly $x/|\mathbb{E}[Z_1]|$ in expectation to "return to a neighborhood of 0." We prove something along those lines here using a Lyapounov function.

Observe that, for any $y \in \mathbb{Z}_+$, we have on the event $\{X_t = y\}$ by the Markov property:

$$\mathbb{E}_x[X_{t+1} - X_t \mid \mathcal{F}_t] = \mathbb{E}[(y + Z_{t+1})^+ - y]$$

$$= \mathbb{E}[-y\mathbf{1}\{Z_{t+1} \leq -y\} + Z_{t+1}\mathbf{1}\{Z_{t+1} > -y\}]$$

$$\leq \mathbb{E}[Z_{t+1}\mathbf{1}\{Z_{t+1} > -y\}]$$

$$= \mathbb{E}[Z_1\mathbf{1}\{Z_1 > -y\}]. \tag{3.3.16}$$

For all y, the random variable $|Z_1\mathbf{1}\{Z_1 > -y\}|$ is bounded by $|Z_1|$, itself an integrable random variable. Moreover, $Z_1\mathbf{1}\{Z_1 > -y\} \to Z_1$ as $y \to +\infty$ almost surely. Hence, the dominated convergence theorem (Proposition B.4.14) implies that

$$\lim_{y \to +\infty} \mathbb{E}[Z_1\mathbf{1}\{Z_1 > -y\}] = \mathbb{E}[Z_1] < 0.$$

So for any $0 < \varepsilon < -\mathbb{E}[Z_1]$, there is $y_\varepsilon \in \mathbb{Z}_+$ large enough that $\mathbb{E}[Z_1\mathbf{1}\{Z_1 > -y\}] < -\varepsilon$ for all $y > y_\varepsilon$. Fix ε satisfying the previous constraint and define

$$A := \{0, 1, \ldots, y_\varepsilon\}.$$

We use Theorem 3.3.10 to bound τ_A in expectation. Define the Lyapounov function

$$\psi(x) = \frac{x}{\varepsilon} \quad \forall x \in \mathbb{Z}_+.$$

On the event $\{X_t = y\}$, we rewrite (3.3.16) as

$$\mathbb{E}[\psi(X_{t+1}) - \psi(X_t) \mid \mathcal{F}_t] \leq \frac{\mathbb{E}[Z_1 \mathbf{1}\{Z_1 > -y\}]}{\varepsilon}$$

$$\leq -1$$

for $y \in A^c$. This is the same as (3.3.15). Hence, we can apply Theorem 3.3.10 to get

$$\mathbb{E}_x[\tau_A] \leq \psi(x) = \frac{x}{\varepsilon}$$

for all $x \geq y_\varepsilon$.   ◄

A well-known, closely related result gives a criterion for positive recurrence. We state it without proof.

Theorem 3.3.12 (Foster's theorem). *Let P be an irreducible transition matrix on a countable state space V. Let A be a finite, proper subset of V. Suppose the non-negative function* $\psi : V \to \mathbb{R}_+$ *satisfies the system of inequalities*

$$\Delta\psi \leq -1 \quad \text{on } A^c,$$

as well as the condition

$$\sum_{y \in V} P(x,y)\psi(y) < +\infty \quad \forall x \in A.$$

Then, P is positive recurrent.

3.3.2 Basic Electrical Network Theory

We now develop the basic theory of electrical networks for the analysis of random walks. All results in this subsection (and the next one) concern *reversible* Markov chains, or random walks on networks (see Definition 1.2.7). We begin with a few definitions. Throughout, we will use the notation $h|_B$ for the function h restricted to the subset B. We also write $h \equiv c$ if h is identically equal to the constant c.

Definitions

Let $\mathcal{N} = (G, c)$ be a finite or countable network with $G = (V, E)$. Throughout this section we assume that $\mathcal{N}$ is connected and locally finite. In the context of electrical networks, edge CONDUCTANCE weights are called *conductances*. The reciprocal of the conductances are called *resistances* RESISTANCE and are denoted by $r(e) := 1/c(e)$ for all $e \in E$. For an edge $e = \{x, y\}$ we overload $c(x, y) := c(e)$ and $r(x, y) := r(e)$. Both c and r are symmetric as functions of x, y. Recall that the transition matrix of the random walk on $\mathcal{N}$ satisfies

$$P(x,y) = \frac{c(x,y)}{c(x)},$$

where

$$c(x) = \sum_{z:z \sim x} c(x,z).$$

Let A, Z be disjoint, non-empty subsets of V such that $W := (A \cup Z)^c$ is finite. For our purposes it will suffice to take A to be a singleton, that is, $A = \{a\}$ for some a. Then a is called the *source* and Z is called the *sink-set*, or *sink* for short. As an immediate corollary   SOURCE, of Theorem 3.3.1, we obtain the existence and uniqueness of a voltage function, defined   SINK formally in the next corollary. It will be useful to consider voltages taking an arbitrary value at a, but we always set the voltage on Z to 0.

Corollary 3.3.13 (Voltage) *Fix $v_0 > 0$. Let $\mathcal{N} = (G, c)$ be a finite or countable, connected network with $G = (V, E)$. Let $A := \{a\}$, Z be disjoint non-empty subsets of V such that $W = (A \cup Z)^c$ is non-empty and finite. Then there exists a unique* voltage *defined as follows:*   VOLTAGE *a function v on V such that v is harmonic on W, that is,*

$$v(x) = \frac{1}{c(x)} \sum_{y:y\sim x} c(x,y)v(y) \qquad \forall x \in W, \tag{3.3.17}$$

where

$$v(a) = v_0 \qquad and \qquad v|_Z \equiv 0. \tag{3.3.18}$$

Moreover,

$$\frac{v(x)}{v_0} = \mathbb{P}_x[\tau_a < \tau_Z] \tag{3.3.19}$$

for the corresponding random walk on $\mathcal{N}$.

Proof   Set $h(x) = v(x)$ on $A \cup Z$. Theorem 3.3.1 gives the result.   ∎

Note in the definition above that if v is a voltage with value v_0 at a, then $\tilde{v}(x) = v(x)/v_0$ is a voltage with value 1 at a.

Let v be a voltage function on $\mathcal{N}$ with source a and sink Z. The Laplacian-based formulation of harmonicity, (3.3.7), can be interpreted in terms of flows (see Definition 1.1.13). We define the *current* function   CURRENT

$$i(x, y) := c(x,y)[v(x) - v(y)], \tag{3.3.20}$$

or, equivalently, $v(x) - v(y) = r(x,y)\,i(x,y)$. The latter definition is usually referred to as *Ohm's "law."* Notice that the current function is defined on ordered pairs of vertices and is   OHM'S LAW anti-symmetric, that is, $i(x,y) = -i(y,x)$. In terms of the current function, the harmonicity of v is then expressed as

$$\sum_{y:y\sim x} i(x, y) = 0 \qquad \forall x \in W, \tag{3.3.21}$$

that is, i is a flow on W (without capacity constraints). This set of equations is known as *Kirchhoff's node law*. We also refer to these constraints as flow-conservation constraints.   KIRCHHOFF'S To be clear, the current function is not just any flow. It is a flow that can be written as a   NODE LAW potential difference according to Ohm's law. Such a current also satisfies *Kirchhoff's cycle*   KIRCHHOFF'S *law*: if $x_1 \sim x_2 \sim \cdots \sim x_k \sim x_{k+1} = x_1$ is a cycle, then   CYCLE LAW

$$\sum_{j=1}^{k} i(x_j, x_{j+1})\, r(x_j, x_{j+1}) = 0,$$

as can be seen by substituting Ohm's law.

STRENGTH

The *strength* of the current is defined as

$$\|i\| := \sum_{y:y\sim a} i(a,y).$$

Because $a \notin W$, it does not satisfy Kirchhoff's node law and the strength is not 0 in general. The definition of $i(x,y)$ ensures that the flow out of the source is non-negative as $\mathbb{P}_y[\tau_a < \tau_Z] \le 1 = \mathbb{P}_a[\tau_a < \tau_Z]$ for all $y \sim a$ so that

$$i(a,y) = c(a,y)[v(a) - v(y)] = c(a,y)\left[v_0\mathbb{P}_a[\tau_a < \tau_Z] - v_0\mathbb{P}_y[\tau_a < \tau_Z]\right] \ge 0.$$

Note that by multiplying the voltage by a constant we obtain a current which is similarly scaled. Up to that scaling, the current function is unique from the uniqueness of the voltage. We will often consider the *unit current* where we scale v and i so as to enforce that $\|i\| = 1$.

UNIT
CURRENT

Summing up the previous paragraphs, to determine the voltage it suffices to find functions v and i that simultaneously satisfy Ohm's law and Kirchhoff's node law. Here is an example.

Example 3.3.14 (Network reduction: birth-death chain). Let $\mathcal{N}$ be the line on $\{0, 1, \ldots, n\}$ with $j \sim k \iff |j - k| = 1$ and arbitrary (positive) conductances on the edges. Let (X_t) be the corresponding walk. We use the principle above to compute $\mathbb{P}_x[\tau_0 < \tau_n]$ for $1 \le x \le n - 1$. Consider the voltage function v when $v(0) = 1$ and $v(n) = 0$ with current i, which exists and is unique by Corollary 3.3.13. The desired quantity is $v(x)$.

Note that because i is a flow on $\mathcal{N}$, the flow into every vertex equals the flow out of that vertex, and we must have $i(y, y + 1) = i(0, 1) = \|i\|$ for all y. To compute $v(x)$, we note that it remains the same if we replace the path $0 \sim 1 \sim \cdots \sim x$ with a single edge of resistance $R_{0,x} = r(0, 1) + \cdots + r(x - 1, x)$. Indeed, leave the voltage unchanged on the remaining nodes (to the right of x) and define the current on the new edge as $\|i\|$. Kirchhoff's node law is automatically satisfied by the argument above. To check Ohm's law on the new "super-edge," note that on the original network $\mathcal{N}$ (with the original voltage function)

$$
\begin{aligned}
v(0) - v(x) &= (v(0) - v(1)) + \cdots + (v(x - 1) - v(x)) \\
&= r(x - 1, x)i(x - 1, x) + \cdots + r(0, 1)i(0, 1) \\
&= [r(0, 1) + \cdots + r(x - 1, x)]\|i\| \\
&= R_{0,x}\|i\|.
\end{aligned}
$$

Ohm's law is also satisfied on every other edge (to the right of x) because nothing has changed there. That proves the claim.

We do the same reduction on the other side of x by replacing $x \sim x + 1 \sim \cdots \sim n$ with a single edge of resistance $R_{x,n} = r(x, x + 1) + \cdots + r(n - 1, n)$. See Figure 3.4.

Because the voltage at x was not changed by this transformation, we can compute $v(x) = \mathbb{P}_x[\tau_0 < \tau_n]$ directly on the reduced network, where it is now a straightforward computation. Indeed, starting at x, the reduced walk jumps to 0 with probability proportional to the conductance on the new super-edge $0 \sim x$ (or the reciprocal of the resistance), that is,

$$
\begin{aligned}
\mathbb{P}_x[\tau_0 < \tau_n] &= \frac{R_{0,x}^{-1}}{R_{0,x}^{-1} + R_{x,n}^{-1}} \\
&= \frac{R_{x,n}}{R_{x,n} + R_{0,x}} \\
&= \frac{r(x, x + 1) + \cdots + r(n - 1, n)}{r(0, 1) + \cdots + r(n - 1, n)}.
\end{aligned}
$$

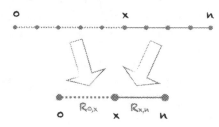

Figure 3.4 Reduced network.

Some special cases:

- *Simple random walk.* In the case of simple random walk, all resistances are equal and we get

$$\mathbb{P}_x[\tau_0 < \tau_n] = \frac{n-x}{n}.$$

- *Gambler's ruin.* The gambler's ruin example corresponds to taking $c(j, j+1) = (p/q)^j$ or $r(j, j+1) = (q/p)^j$ for some $0 < p < 1$. In this case we obtain

$$\mathbb{P}_x[\tau_0 < \tau_n] = \frac{\sum_{j=x}^{n-1}(q/p)^j}{\sum_{j=0}^{n-1}(q/p)^j} = \frac{(q/p)^x(1-(q/p)^{n-x})}{1-(q/p)^n} = \frac{(p/q)^{n-x}-1}{(p/q)^n-1},$$

when $p \neq q$ (otherwise we get back the simple random walk case). ◀

The Example 3.3.14 illustrates the *series law*: resistances in series add up. There is a similar *parallel law*: conductances in parallel add up. To formalize these laws, one needs to introduce multigraphs. This is straightforward, although to avoid complicating the notation further we will not do this here. (But see Example 3.3.22 for a simple case.) SERIES LAW, PARALLEL LAW

Another useful network reduction technique is illustrated in the next example.

Example 3.3.15 (Network reduction: binary tree). Let $\mathcal{N}$ be the rooted binary tree with n levels $\widehat{\mathbb{T}}_2^n$ and equal conductances on all edges. Let 0 be the root. Pick an arbitrary leaf and denote it by n. The remaining vertices on the path between 0 and n, which we refer to as the main path, will be denoted by $1, \ldots, n-1$ moving away from the root. We claim that, for all $0 < x < n$, it holds that

$$\mathbb{P}_x[\tau_0 < \tau_n] = (n-x)/n.$$

Indeed, let v be the voltage with values 1 and 0 at $a = 0$ and $Z = \{n\}$, respectively. Let i be the corresponding current. Notice that, for each $0 \leq y < n$, the current – as a flow – has "nowhere to go" on the subtree T_y hanging from y away from the main path. The leaves of the subtree are dead ends. Hence the current must be 0 on T_y and by Ohm's law the voltage must be constant on it, that is, every vertex in T_y has voltage $v(y)$.

Imagine collapsing all vertices in T_y, including y, into a single vertex (and removing the self-loops so created). Doing this for every vertex on the main path results in a new reduced network which is formed of a single path as in Example 3.3.14. Note that the voltage and the current can be taken to be the same as they were previously on the main path. Indeed, with this choice, Ohm's law is automatically satisfied. Moreover, because there is no current on the hanging subtrees in the original network, Kirchhoff's node law is also satisfied on the reduced network, as no current is "lost."

Hence, the answer can be obtained from Example 3.3.14. That proves the claim. (You should convince yourself that this result is obvious from a probabilistic point of view.) ◄

We gave a probabilistic interpretation of the voltage. What about the current? The following result says that, roughly speaking, $i(x,y)$ is the net traffic on the edge $\{x,y\}$ from x to y. We start with an important formula for the voltage at a. For the walk started at a, we use the shorthand

$$\mathbb{P}[a \to Z] := \mathbb{P}_a[\tau_Z < \tau_a^+]$$

for the *escape probability*. The next lemma can be interpreted as a sort of Ohm's law between a and Z, where $c(a)\,\mathbb{P}[a \to Z]$ is the "effective conductance." (We will be more formal in Definition 3.3.19.)

Lemma 3.3.16 (Effective Ohm's Law). *Let v be a voltage on $\mathcal{N}$ with source a and sink Z. Let i be the associated current. Then,*

$$\frac{v(a)}{\|i\|} = \frac{1}{c(a)\,\mathbb{P}[a \to Z]}. \tag{3.3.22}$$

Proof Using the usual first-step analysis,

$$\mathbb{P}[a \to Z] = \sum_{x:x\sim a} P(a,x)\mathbb{P}_x[\tau_Z < \tau_a]$$
$$= \sum_{x:x\sim a} \frac{c(a,x)}{c(a)}\left(1 - \frac{v(x)}{v(a)}\right)$$
$$= \frac{1}{c(a)v(a)} \sum_{x:x\sim a} c(a,x)[v(a) - v(x)]$$
$$= \frac{1}{c(a)v(a)} \sum_{x:x\sim a} i(a,x),$$

where we used Corollary 3.3.13 on the second line and Ohm's law on the last line. Rearranging gives the result. ■

Recall the Green function from (3.1.3).

Theorem 3.3.17 (Probabilistic interpretation of the current). *For $x \sim y$, let $N_{x\to y}^Z$ be the number of transitions from x to y up to the time of the first visit to the sink Z for the random walk on $\mathcal{N}$ started at a. Let v be the voltage corresponding to the unit current i. Then the following formulas hold:*

$$v(x) = \frac{\mathscr{G}_{\tau_Z}(a,x)}{c(x)} \qquad \forall x \tag{3.3.23}$$

and

$$i(x,y) = \mathbb{E}_a[N_{x\to y}^Z - N_{y\to x}^Z] \qquad \forall x \sim y.$$

Proof We prove the formula for the voltage by showing that $v(x)$ as defined in (3.3.23) is harmonic on $W = V \backslash (\{a\} \cup Z)$. Note first that, for all $z \in Z$, the expected number of visits to z before reaching Z (i.e., $\mathscr{G}_{\tau_Z}(a, z)$) is 0. Or, put differently, $0 = v(z) = \frac{\mathscr{G}_{\tau_Z}(a, z)}{c(z)}$. Moreover, to compute $\mathscr{G}_{\tau_Z}(a, a)$, note that the number of visits to a before the first visit to Z is geometric with success probability $\mathbb{P}[a \to Z]$ by the strong Markov property (Theorem 3.1.8) and hence

$$\mathscr{G}_{\tau_Z}(a, a) = \frac{1}{\mathbb{P}[a \to Z]},$$

and, by Lemma 3.3.16 and the fact that we are using the unit current, $v(a) = \frac{\mathscr{G}_{\tau_Z}(a, a)}{c(a)}$, as required.

To establish the formula for $x \in W$, we compute the quantity

$$\frac{1}{c(x)} \sum_{y : y \sim x} \mathbb{E}_a[N_{y \to x}^Z]$$

in two ways. First, because each visit to $x \in W$ must enter through one of x's neighbors (including itself in the presence of a self-loop), we get

$$\frac{1}{c(x)} \sum_{y : y \sim x} \mathbb{E}_a[N_{y \to x}^Z] = \frac{\mathscr{G}_{\tau_Z}(a, x)}{c(x)}. \tag{3.3.24}$$

On the other hand, by the Markov property (Theorem 1.1.18):

$$\mathbb{E}_a[N_{y \to x}^Z]$$

$$= \mathbb{E}_a \left[\sum_{0 \le t < \tau_Z} \mathbf{1}_{\{X_t = y, X_{t+1} = x\}} \right]$$

$$= \sum_{t \ge 0} \mathbb{P}_a[X_t = y, X_{t+1} = x, \tau_Z > t]$$

$$= \sum_{t \ge 0} \mathbb{P}_a[\tau_Z > t] \mathbb{P}_a[X_t = y \mid \tau_Z > t] \mathbb{P}_a[X_{t+1} = x \mid X_t = y, \tau_Z > t]$$

$$= \sum_{t \ge 0} \mathbb{P}_a[\tau_Z > t] \mathbb{P}_a[X_t = y \mid \tau_Z > t] P(y, x)$$

$$= \sum_{t \ge 0} \mathbb{P}_a[X_t = y, \tau_Z > t] P(y, x)$$

$$= P(y, x) \mathbb{E}_a \left[\sum_{0 \le t < \tau_Z} \mathbf{1}_{\{X_t = y\}} \right]$$

$$= P(y, x) \mathscr{G}_{\tau_Z}(a, y), \tag{3.3.25}$$

so that, summing over y, we obtain this time

$$\frac{1}{c(x)} \sum_{y : y \sim x} \mathbb{E}_a[N_{y \to x}^Z] = \frac{1}{c(x)} \sum_{y : y \sim x} P(y, x) \mathscr{G}_{\tau_Z}(a, y)$$

$$= \sum_{y : y \sim x} P(x, y) \frac{\mathscr{G}_{\tau_Z}(a, y)}{c(y)}, \tag{3.3.26}$$

where we used that $c(x,y) = c(x)P(x,y) = c(y)P(y,x)$ (see Definition 1.2.7). Equating (3.3.24) and (3.3.26) shows that $\frac{\mathscr{G}_{\tau_Z}(a,x)}{c(x)}$ is harmonic on W and hence must be equal to the voltage function by Corollary 3.3.13.

Finally, by (3.3.25),

$$
\begin{aligned}
\mathbb{E}_a[N^Z_{x \to y} - N^Z_{y \to x}] &= P(x,y)\mathscr{G}_{\tau_Z}(a,x) - P(y,x)\mathscr{G}_{\tau_Z}(a,y) \\
&= P(x,y)v(x)c(x) - P(y,x)v(y)c(y) \\
&= c(x,y)[v(x) - v(y)] \\
&= i(x,y).
\end{aligned}
$$

That concludes the proof. ∎

Example 3.3.18 (Network reduction: binary tree (continued)). Recall the setting of Example 3.3.15. We argued that the current on side edges, that is, edges of subtrees hanging from the main path, is 0. This is clear from the probabilistic interpretation of the current: in a walk from a to z, any traversal of a side edge must be undone at a later time. ◀

The network reduction techniques illustrated above are useful. But the power of the electrical network perspective is more apparent in what comes next: the definition of the effective resistance and, especially, its variational characterization.

Effective resistance

Before proceeding further, let us recall our original motivation. Let $\mathcal{N} = (G,c)$ be a countable, locally finite, connected network and let (X_t) be the corresponding walk. Recall that a vertex a in G is transient if $\mathbb{P}_a[\tau_a^+ < +\infty] < 1$.

Exhaustive sequence To relate this to our setting, consider an *exhaustive sequence* of induced subgraphs G_n of G which for our purposes is defined as: G_0 contains only a, $G_n \subseteq G_{n+1}$, $G = \bigcup_n G_n$, and every G_n is finite and connected. Such a sequence always exists by iteratively adding the neighbors of the previous vertices and using that G is locally finite and connected. Let Z_n be the set of vertices of G not in G_n. Then, by Lemma 3.1.25, $\mathbb{P}_a[\tau_{Z_n} \wedge \tau_a^+ = +\infty] = 0$ for all n by our assumptions on (G_n). Hence, the remaining possibilities are

$$
\begin{aligned}
1 &= \mathbb{P}_a[\exists n,\ \tau_a^+ < \tau_{Z_n}] + \mathbb{P}_a[\forall n,\ \tau_{Z_n} < \tau_a^+] \\
&= \mathbb{P}_a[\tau_a^+ < +\infty] + \lim_n \mathbb{P}[a \to Z_n].
\end{aligned}
$$

Therefore, a is transient if and only if $\lim_n \mathbb{P}[a \to Z_n] > 0$. Note that the limit exists because the sequence of events $\{\tau_{Z_n} < \tau_a^+\}$ is decreasing by construction. By a sandwiching argument the limit also does not depend on the exhaustive sequence. Hence, we define

$$
\mathbb{P}[a \to \infty] := \lim_n \mathbb{P}[a \to Z_n].
$$

We use Lemma 3.3.16 to characterize this limit using electrical network concepts.

But, first, here comes the key definition. In Lemma 3.3.16, $v(a)$ can be thought of as the potential difference between the source and the sink, and $\|i\|$ can be thought of as the total current flowing through the network from the source to the sink. Hence, viewing the network as a single "super-edge," (3.3.22) is the analogue of Ohm's law if we interpret $c(a)\mathbb{P}[a \to Z]$ as an "effective conductance."

Definition 3.3.19 (Effective resistance and conductance). *Let $\mathcal{N} = (G, c)$ be a finite or countable, locally finite, connected network. Let $A = \{a\}$ and Z be disjoint non-empty subsets of the vertex set V such that $W := V \backslash (A \cup Z)$ is finite. Let v be a voltage from source a to sink Z and let i be the corresponding current. The* effective resistance *between a and Z is defined as*

$$\mathcal{R}(a \leftrightarrow Z) := \frac{1}{c(a)\,\mathbb{P}[a \to Z]} = \frac{v(a)}{\|i\|},$$

EFFECTIVE
RESISTANCE

where the rightmost equality holds by Lemma 3.3.16. The reciprocal is called the effective conductance *and denoted by $\mathcal{C}(a \leftrightarrow Z) := 1/\mathcal{R}(a \leftrightarrow Z)$.*

EFFECTIVE
CONDUCTANCE

Going back to recurrence, for an exhaustive sequence (G_n) with (Z_n) as above, it is natural to define

$$\mathcal{R}(a \leftrightarrow \infty) := \lim_n \mathcal{R}(a \leftrightarrow Z_n),$$

where, once again, the limit does not depend on the choice of exhaustive sequence.

Theorem 3.3.20 (Recurrence and resistance). *Let $\mathcal{N} = (G, c)$ be a countable, locally finite, connected network. Vertex a (and hence all vertices) in $\mathcal{N}$ is transient if and only if $\mathcal{R}(a \leftrightarrow \infty) < +\infty$.*

Proof This follows immediately from the definition of the effective resistance. Recall that, on a connected network, all states have the same type (recurrent or transient). ∎

Note that the network reduction techniques we discussed previously leave both the voltage and the current strength unchanged on the reduced network. Hence, they also leave the effective resistance unchanged.

Example 3.3.21 (Gambler's ruin chain revisited). Extend the gambler's ruin chain of Example 3.3.14 to all of $\mathbb{Z}_+$. We determine when this chain is transient. Because it is irreducible, all states have the same type and it suffices to look at 0. Consider the exhaustive sequence obtained by letting G_n be the graph restricted to $\{0, 1, \ldots, n-1\}$ and letting $Z_n = \{n, n+1 \ldots\}$. To compute the effective resistance $\mathcal{R}(0 \leftrightarrow Z_n)$, we use the same reduction as in Example 3.3.14. The "super-edge" between 0 and n has resistance

$$\mathcal{R}(0 \leftrightarrow Z_n) = \sum_{j=0}^{n-1} r(j, j+1) = \sum_{j=0}^{n-1} (q/p)^j = \frac{(q/p)^n - 1}{(q/p) - 1},$$

when $p \neq q$, and similarly it has resistance n in the $p = q$ case. Hence, taking a limit as $n \to +\infty$,

$$\mathcal{R}(0 \leftrightarrow \infty) = \begin{cases} +\infty, & p \leq 1/2, \\ \frac{p}{2p-1}, & p > 1/2. \end{cases}$$

So 0 is transient if and only if $p > 1/2$. ◄

Example 3.3.22 (Biased walk on the b-ary tree). Fix $\lambda \in (0, +\infty)$. Consider the rooted, infinite b-ary tree with conductance λ^j on all edges between level $j - 1$ and j, for $j \geq 1$. We determine when this chain is transient. Because it is irreducible, all states have the same type and it suffices to look at the root. Denote the root by 0. For an exhaustive sequence, let G_n be the root together with the first $n - 1$ levels. Let Z_n be as before. To compute $\mathscr{R}(0 \leftrightarrow Z_n)$: (i) glue together all vertices of Z_n; (ii) glue together all vertices on the same level of G_n; (iii) replace parallel edges with a single edge whose conductance is the sum of the conductances; (iv) let the current on this edge be the sum of the currents; and (v) leave the voltages unchanged. It can be checked that Ohm's law and Kirchhoff's node law are still satisfied, and that hence we have not changed the effective resistance. (This is an application of the parallel law.)

The reduced network is now a line. Denote the new vertices $0, 1, \ldots, n$. The conductance on the edge between j and $j + 1$ is $b^{j+1}\lambda^j = b(b\lambda)^j$. So this is the chain from the previous example with $(p/q) = b\lambda$ where all conductances are scaled by a factor of b. Hence,

$$\mathscr{R}(0 \leftrightarrow \infty) = \begin{cases} +\infty, & b\lambda \leq 1, \\ \frac{1}{b(1-(b\lambda)^{-1})}, & b\lambda > 1. \end{cases}$$

So the root is transient if and only if $b\lambda > 1$.

A generalization is provided in Example 3.3.27.   ◄

3.3.3 Bounding the Effective Resistance via Variational Principles

The examples we analyzed so far were atypical in that it was possible to reduce the network down to a single edge using simple rules and read off the effective resistance. In general, we need more robust techniques to bound the effective resistance. The following two variational principles provide a powerful approach for this purpose. We derive them for finite networks, but will later on apply them to exhaustive sequences.

Variational principles

Recall from Definition 1.1.13 that a flow θ from source a to sink Z on a countable, locally finite, connected network $\mathcal{N} = (G, c)$ is a function on pairs of adjacent vertices such that θ is anti-symmetric, that is, $\theta(x, y) = -\theta(y, x)$ for all $x \sim y$; and it satisfies the flow-conservation constraint $\sum_{y:y\sim x} \theta(x, y) = 0$ on all vertices x except those in $\{a\} \cup Z$. The strength of the flow is $\|\theta\| = \sum_{y:y\sim a} \theta(a, y)$. The current is a special flow – one that can be written as a potential difference according to Ohm's law. As we show next, it can also be characterized

ENERGY

as a flow minimizing a certain energy. Specifically, the *energy* of a flow θ is defined as

$$\mathscr{E}(\theta) = \frac{1}{2} \sum_{x,y} r(x, y)\theta(x, y)^2.$$

The proof of the variational principle we present here employs a neat trick, convex duality. In particular, it reveals that the voltage and current are dual in the sense of convex analysis.

Theorem 3.3.23 (Thomson's principle). *Let* $\mathcal{N} = (G, c)$ *be a finite, connected network. The effective resistance between source a and sink Z is characterized by*

$$\mathscr{R}(a \leftrightarrow Z) = \inf \left\{ \mathscr{E}(\theta) : \theta \text{ is a unit flow between } a \text{ and } Z \right\}. \tag{3.3.27}$$

The unique minimizer is the unit current.

Proof It will be convenient to work in vector form. Let $1, \ldots, n$ be the vertices of G and order the edges arbitrarily as $e_1, \ldots, e_m$. Choose an arbitrary orientation of $\mathcal{N}$, that is, replace each edge $e_i = \{x, y\}$ with either $\vec{e}_i = (x, y)$ or (y, x). Let $\vec{G}$ be the corresponding directed graph. Think of the flow $\boldsymbol{\theta}$ as a vector with one coordinate for each oriented edge. Then the flow constraint can be written as a linear system $B\boldsymbol{\theta} = \boldsymbol{b}$. Here the matrix B has a column for each directed edge and a row for each vertex *except those in Z*. The entries of B are $B_{x,(x,y)} = 1$, $B_{y,(x,y)} = -1$, and 0 otherwise. We have already encountered this matrix: it is an oriented incidence matrix of G (see Definition 1.1.16) restricted to the rows in $V \setminus Z$. The vector $\boldsymbol{b}$ has 0s everywhere except for $b_a = 1$. Let $\boldsymbol{r}$ be the vector of resistances and let R be the diagonal matrix with diagonal $\boldsymbol{r}$. In vector form, $\mathscr{E}(\boldsymbol{\theta}) = \boldsymbol{\theta}^T R \boldsymbol{\theta}$ and the optimization problem (3.3.27) reads

$$\mathscr{E}^* = \inf\{\boldsymbol{\theta}^T R \boldsymbol{\theta} : B\boldsymbol{\theta} = \boldsymbol{b}\}.$$

We first characterize the optimal flow. We introduce the *Lagrangian*

$$\mathscr{L}(\boldsymbol{\theta}; \boldsymbol{h}) := \boldsymbol{\theta}^T R \boldsymbol{\theta} - 2\boldsymbol{h}^T (B\boldsymbol{\theta} - \boldsymbol{b}),$$

LAGRANGIAN

where $\boldsymbol{h}$ has an entry for all vertices *except those in Z*. For all $\boldsymbol{h}$,

$$\mathscr{E}^* \geq \inf_{\boldsymbol{\theta}} \mathscr{L}(\boldsymbol{\theta}; \boldsymbol{h}),$$

because those $\boldsymbol{\theta}$s with $B\boldsymbol{\theta} = \boldsymbol{b}$ make the second term vanish in $\mathscr{L}(\boldsymbol{\theta}; \boldsymbol{h})$. Since $\mathscr{L}(\boldsymbol{\theta}; \boldsymbol{h})$ is strictly convex as a function of $\boldsymbol{\theta}$, the solution to its minimization is characterized by the usual optimality conditions which in this case read $2R\boldsymbol{\theta} - 2B^T \boldsymbol{h} = 0$, or

$$\boldsymbol{\theta} = R^{-1} B^T \boldsymbol{h}. \tag{3.3.28}$$

Substituting into the Lagrangian and simplifying, we have proved that

$$\mathscr{E}(\boldsymbol{\theta}) \geq \mathscr{E}^* \geq -\boldsymbol{h}^T B R^{-1} B^T \boldsymbol{h} + 2\boldsymbol{h}^T \boldsymbol{b} =: \mathscr{L}^*(\boldsymbol{h}) \tag{3.3.29}$$

for all $\boldsymbol{h}$ and flow $\boldsymbol{\theta}$. This inequality is a statement of weak duality. To show that a flow $\boldsymbol{\theta}$ is optimal it suffices to find $\boldsymbol{h}$ such that $\mathscr{E}(\boldsymbol{\theta}) = \mathscr{L}^*(\boldsymbol{h})$.

Let $\boldsymbol{\theta} = \boldsymbol{i}$ be the unit current in vector form, which satisfies $B\boldsymbol{\theta} = \boldsymbol{b}$ by our choice of $\boldsymbol{b}$ and Kirchhoff's node law (i.e., (3.3.21)). The suitable dual turns out to be the corresponding voltage $\boldsymbol{h} = \boldsymbol{v}$ in vector form restricted to $V \setminus Z$. To see this, observe that $B^T \boldsymbol{h}$ is the vector of neighboring node differences

$$B^T \boldsymbol{h} = (h(x) - h(y))_{(x,y) \in \vec{G}}, \tag{3.3.30}$$

where implicitly $h|_Z \equiv 0$. Hence, the optimality condition (3.3.28) is nothing but Ohm's law (i.e., (3.3.20)) in vector form. Therefore, if $\boldsymbol{i}$ is the unit current and $\boldsymbol{v}$ is the associated voltage in vector form, it holds that

$$\mathscr{L}^*(\boldsymbol{v}) = \mathscr{L}(\boldsymbol{i}; \boldsymbol{v}) = \mathscr{E}(\boldsymbol{i}),$$

where the first equality follows from the fact that $\boldsymbol{i}$ minimizes $\mathscr{L}(\boldsymbol{i}; \boldsymbol{v})$ by (3.3.28) and the second equality follows from the fact that $B\boldsymbol{i} = \boldsymbol{b}$. So we must have $\mathscr{E}(\boldsymbol{i}) = \mathscr{E}^*$ by weak duality (i.e., (3.3.29)).

As for uniqueness, it can be checked that two minimizers $\boldsymbol{\theta}, \boldsymbol{\theta}'$ satisfy

$$\mathscr{E}^* = \frac{\mathscr{E}(\boldsymbol{\theta}) + \mathscr{E}(\boldsymbol{\theta}')}{2} = \mathscr{E}\left(\frac{\boldsymbol{\theta} + \boldsymbol{\theta}'}{2}\right) + \mathscr{E}\left(\frac{\boldsymbol{\theta} - \boldsymbol{\theta}'}{2}\right)$$

by definition of the energy. The first term in the rightmost expression is greater than or equal to $\mathscr{E}^*$ since the average of two unit flows is still a unit flow. The second term is non-negative by definition. Hence, the latter must be zero and the only way for this to happen is if $\boldsymbol{\theta} = \boldsymbol{\theta}'$.

To conclude the proof, it remains to compute the optimal value. The matrix $BR^{-1}B^T$ is related to the Laplacian associated to random walk on $\mathcal{N}$ (see Section 3.3.1) up to a row scaling. Multiplying by row $x \in V \setminus Z$ involves taking a conductance-weighted average of the neighboring values and subtracting the value at x, that is,

$$
\begin{aligned}
\left(BR^{-1}B^T v\right)_x &= \sum_{y:(x,y)\in \vec{G}} \left[c(x,y)(v(x) - v(y)) \right] \\
&\quad - \sum_{y:(y,x)\in \vec{G}} \left[c(y,x)(v(y) - v(x)) \right] \\
&= \sum_{y:y\sim x} \left[c(x,y)(v(x) - v(y)) \right],
\end{aligned}
$$

where we used (3.3.30) and the facts that $r(x,y)^{-1} = c(x,y)$ and $c(x,y) = c(y,x)$, and it is assumed implicitly that $v|_Z \equiv 0$. By Corollary 3.3.13, this is zero except for the row $x = a$, where it is

$$\sum_{y:y\sim a} c(a,y)[v(a) - v(y)] = \sum_{y:y\sim a} i(a,y) = 1,$$

where we used Ohm's law and the fact that the current has unit strength. We have finally

$$
\begin{aligned}
\mathscr{E}^* &= \mathscr{L}^*(v) \\
&= -v^T BR^{-1}B^T v + 2v^T b \\
&= -v(a) + 2v(a) \\
&= v(a) \\
&= \mathscr{R}(a \leftrightarrow Z),
\end{aligned}
$$

by (3.3.16). That concludes the proof.   ∎

Observe that the convex combination $\boldsymbol{\alpha}$ minimizing the sum of squares $\sum_j \alpha_j^2$ is constant. In a similar manner, Thomson's principle (Theorem 3.3.23) stipulates roughly speaking that the more the flow can be spread out over the network, the lower is the effective resistance (penalizing flow on edges with higher resistance). Pólya's theorem (Theorem 3.3.38) provides a vivid illustration. Here is a simple example suggesting that, in a sense, the current is indeed a well-distributed flow.

Example 3.3.24 (Random walk on the complete graph). Let $\mathcal{N}$ be the complete graph on $\{1, \ldots, n\}$ with unit resistances, and let $a = 1$ and $Z = \{n\}$. Assume $n > 2$. The effective resistance is straightforward to compute in this case. Indeed, the escape probability (with a slight abuse of notation) is

$$\mathbb{P}[1 \to n] = \frac{1}{n-1} + \frac{1}{2}\left(1 - \frac{1}{n-1}\right) = \frac{n}{2(n-1)},$$

as we either jump to n immediately or jump to one of the remaining nodes, in which case we reach n first with probability $1/2$ by symmetry. Hence, since $c(1) = n - 1$, we get

$$\mathscr{R}(1 \leftrightarrow n) = \frac{2}{n}$$

from the definition of the effective resistance (Definition 3.3.19).

We now look for the optimal flow in Thomson's principle. Pushing a flow of 1 through the edge $\{1, n\}$ gives an upper bound of 1, which is far from the optimal $\frac{2}{n}$. Spreading the flow a bit more by pushing $1/2$ through the edge $\{1, n\}$ and $1/2$ through the path $1 \sim 2 \sim n$ gives the slightly better bound $3 \cdot (1/2)^2 = 3/4$. Taking this further, pushing a flow of $\frac{1}{n-1}$ through $\{1, n\}$ as well as through each two-edge path to n via the remaining neighbors of 1 gives the yet improved bound

$$\left(\frac{1}{n-1}\right)^2 + 2(n-2)\left(\frac{1}{n-1}\right)^2 = \frac{2n-3}{(n-1)^2} = \frac{2}{n} \cdot \frac{2n^2 - 3n}{2n^2 - 4n + 2} > \frac{2}{n},$$

when $n > 2$. Because the direct path from 1 to n has a somewhat lower resistance, the optimal flow is obtained by increasing the flow on that edge slightly. Namely, for a flow α on $\{1, n\}$ (and the rest divided up evenly among the two-edge paths), we get an energy of $\alpha^2 + 2(n-2)[\frac{1-\alpha}{n-2}]^2$, which is minimized at $\alpha = \frac{2}{n}$, where it is indeed

$$\left(\frac{2}{n}\right)^2 + \frac{2}{n-2}\left(\frac{n-2}{n}\right)^2 = \frac{2}{n}\left(\frac{2}{n} + \frac{n-2}{n}\right) = \frac{2}{n}.$$

◀

As we noted above, the matrix $BR^{-1}B^T$ in the proof of Thomson's principle is related to the Laplacian. Because $B^T\boldsymbol{h}$ is the vector of neighboring node differences, we have

$$\boldsymbol{h}^T BR^{-1}B^T\boldsymbol{h} = \frac{1}{2}\sum_{x,y} c(x,y)[h(y) - h(x)]^2,$$

where we implicitly fix $h|_Z \equiv 0$, which is called the *Dirichlet energy*. Thinking of B^T as a "discrete gradient," the Dirichlet energy can be interpreted as the weighted norm of the gradient of $\boldsymbol{h}$. The following is a "dual" to Thomson's principle. Exercise 3.15 asks for a proof.

DIRICHLET
ENERGY

Theorem 3.3.25 (Dirichlet's principle). *Let $\mathcal{N} = (G, c)$ be a finite, connected network. The effective conductance between source a and sink Z is characterized by*

$$\mathscr{C}(a \leftrightarrow Z) = \inf\left\{\frac{1}{2}\sum_{x,y} c(x,y)[h(y) - h(x)]^2 : h(a) = 1, \, h|_Z \equiv 0\right\}.$$

The unique minimizer is the voltage v with $v(a) = 1$.

The following lower bound is a typical application of Thomson's principle. See Pólya's theorem (Theorem 3.3.38) for an example of its use. Recall from Section 1.1.1 that, on a finite graph, a cutset separating a from Z is a set of edges Π such that any path between

a and Z must include at least one edge in Π. Similarly, as defined in Section 2.3.3, on a countable, locally finite network, a cutset separating a from ∞ is a finite set of edges that must be crossed by any infinite (self-avoiding) path from a.

Corollary 3.3.26 (Nash–Williams inequality) *Let $\mathcal{N}$ be a finite, connected network and let $\{\Pi_j\}_{j=1}^n$ be a collection of disjoint cutsets separating source a from sink Z. Then,*

$$\mathscr{R}(a \leftrightarrow Z) \geq \sum_{j=1}^n \left(\sum_{e \in \Pi_j} c(e) \right)^{-1}.$$

Similarly, if $\mathcal{N}$ is a countable, locally finite, connected network, then for any collection $\{\Pi_j\}_j$ of finite, disjoint cutsets separating a from ∞,

$$\mathscr{R}(a \leftrightarrow \infty) \geq \sum_j \left(\sum_{e \in \Pi_j} c(e) \right)^{-1}.$$

Proof Consider the case where $\mathcal{N}$ is finite first. We will need the following claim, which follows immediately from Lemma 1.1.14: for any unit flow θ between a and Z and any cutset Π_j separating a from Z, it holds that

$$\sum_{e \in \Pi_j} |\theta(e)| \geq \|\theta\| = 1.$$

By Cauchy–Schwarz (Theorem B.4.8),

$$\sum_{e \in \Pi_j} c(e) \sum_{e' \in \Pi_j} r(e')\theta(e')^2 \geq \left(\sum_{e \in \Pi_j} \sqrt{c(e)r(e)} \, |\theta(e)| \right)^2$$

$$= \left(\sum_{e \in \Pi_j} |\theta(e)| \right)^2$$

$$\geq 1.$$

Rearranging, summing over j and using the disjointness of the cutsets,

$$\mathscr{E}(\theta) = \frac{1}{2} \sum_{x,y} r(x,y)\theta(x,y)^2 \geq \sum_{j=1}^n \sum_{e' \in \Pi_j} r(e')\theta(e')^2 \geq \sum_{j=1}^n \left(\sum_{e \in \Pi_j} c(e) \right)^{-1}.$$

Thomson's principle gives the result.

The infinite case follows from a similar argument using an exhaustive sequence. ∎

The following example is an application of Nash–Williams (Corollary 3.3.26) and Thomson's principle to recurrence.

Example 3.3.27 (Biased walk on general trees). Let $\mathcal{T}$ be a locally finite tree with root 0. Consider again the biased walk from Example 3.3.22, that is, conductance is λ^j on all edges between level $j - 1$ and j. Recall the branching number $\mathrm{br}(\mathcal{T})$ from Definition 2.3.10.

Assume $\lambda > \mathrm{br}(\mathscr{T})$. For any $\varepsilon > 0$, there is a cutset Π such that $\sum_{e \in \Pi} \lambda^{-|e|} \leq \varepsilon$. By Nash–Williams,

$$\mathscr{R}(0 \leftrightarrow \infty) \geq \left(\sum_{e \in \Pi} c(e) \right)^{-1} \geq \varepsilon^{-1}.$$

Since ε is arbitrary, the walk is recurrent by Theorem 3.3.20.

Suppose instead that $\lambda < \mathrm{br}(\mathscr{T})$ and let $\lambda < \lambda_* < \mathrm{br}(\mathscr{T})$. By the proof of Claim 2.3.11, for all $n \geq 1$, there exist $\varepsilon > 0$ and a unit flow ϕ_n from 0 to the n-level vertices ∂_n with capacity constraints $|\phi_n(x,y)| \leq \varepsilon^{-1} \lambda_*^{-|e|}$ for all edges $e = \{x,y\}$, where $|e|$ is the graph distance from the root to the endvertex of e furthest from it. Then, letting $F_m = \{e \colon |e| = m\}$, the energy of the flow is

$$
\begin{aligned}
\mathscr{E}(\phi_n) &= \frac{1}{2} \sum_{x,y} r(x,y)\phi_n(x,y)^2 \\
&\leq \sum_{m=1}^{n} \lambda^m \sum_{e=\{x,y\}\in F_m} |\phi_n(x,y)| \varepsilon^{-1} \lambda_*^{-|e|} \\
&= \varepsilon^{-1} \sum_{m=1}^{n} \left(\frac{\lambda}{\lambda_*} \right)^m \sum_{e=\{x,y\}\in F_m} |\phi_n(x,y)| \\
&\leq \varepsilon^{-1} \sum_{m=1}^{+\infty} \left(\frac{\lambda}{\lambda_*} \right)^m \\
&< +\infty,
\end{aligned}
$$

where, on the fourth line, we used Lemma 1.1.14 together with the fact that ϕ_n is a unit flow and F_m is a cutset separating 0 and ∂_n. Thomson's principle implies that $\mathscr{R}(0 \leftrightarrow \partial_n)$ is uniformly bounded in n. The walk is transient by Theorem 3.3.20. ◄

Another typical application of Thomson's principle is the following monotonicity property (which is not obvious from a probabilistic point of view).

Corollary 3.3.28 *Adding an edge to a finite, connected network cannot increase the effective resistance between a source a and a sink Z. In particular, if the added edge is not incident to a, then* $\mathbb{P}[a \to Z]$ *cannot decrease.*

Proof The additional edge enlarges the space of possible flows, so by Thomson's principle it can only lower the resistance or leave it as is. The second statement follows from the definition of the effective resistance. ∎

More generally:

Corollary 3.3.29 (Rayleigh's principle) *Let $\mathcal{N}$ and $\mathcal{N}'$ be two networks on the same finite, connected graph G such that, for each edge in G, the resistance in $\mathcal{N}'$ is greater than it is in $\mathcal{N}$. Then, for any source a and sink Z,*

$$\mathscr{R}_{\mathcal{N}}(a \leftrightarrow Z) \leq \mathscr{R}_{\mathcal{N}'}(a \leftrightarrow Z).$$

Proof Compare the energies of an arbitrary flow on $\mathcal{N}$ and $\mathcal{N}'$, and apply Thomson's principle.   ∎

Note that this corollary implies the previous one by thinking of an absent edge as one with infinite resistance.

Flows to infinity

Combining Theorem 3.3.20 and Thomson's principle, we derive a flow-based criterion for recurrence. To state the result, it is convenient to introduce the notion of a *unit flow θ from source a to ∞* on a countable, locally finite network: θ is anti-symmetric, it satisfies the flow-conservation constraint on all vertices but a, and $\|\theta\| := \sum_{y \sim a} \theta(a,y) = 1$. Note that the energy $\mathscr{E}(\theta)$ of such a flow is well defined in $[0, +\infty]$.

FLOW TO ∞

Theorem 3.3.30 (Recurrence and finite-energy flows). *Let $\mathcal{N} = (G, c)$ be a countable, locally finite, connected network. Vertex a (and hence all vertices) in $\mathcal{N}$ is transient if and only if there is a unit flow from a to ∞ of finite energy.*

Proof Suppose such a flow exists and has energy bounded by $B < +\infty$. Let (G_n) be an exhaustive sequence with associated sinks (Z_n). A unit flow from a to ∞ on $\mathcal{N}$ yields, by projection, a unit flow from a to Z_n. This projected flow also has energy bounded by B. Hence, Thomson's principle implies $\mathscr{R}(a \leftrightarrow Z_n) \leq B$ for all n and transience follows from Theorem 3.3.20.

Proving the other direction involves producing a flow to ∞. Suppose a is transient and let (G_n) be an exhaustive sequence as above. Then, Theorem 3.3.20 implies that $\mathscr{R}(a \leftrightarrow Z_n) \leq \mathscr{R}(a \leftrightarrow \infty) < B$ for some $B < +\infty$ and Thomson's principle guarantees in turn the existence of a flow θ_n from a to Z_n with energy bounded by B. In particular, there is a unit current i_n, and associated voltage v_n, of energy bounded by B. So it remains to use the sequence of current flows (i_n) to construct a flow to ∞ on the infinite network. The technical point is to show that the limit of (i_n) exists and is indeed a flow. For this, consider the random walk on $\mathcal{N}$ started at a. Let $Y_n(x)$ be the number of visits to x before hitting Z_n the first time. By the monotone convergence theorem (Proposition B.4.14), $\mathbb{E}_a Y_n(x) \to \mathbb{E}_a Y_\infty(x)$, where $Y_\infty(x)$ is the total number of visits to x. Moreover, $\mathbb{E}_a Y_\infty(x) < +\infty$ by transience and (3.1.2). By (3.3.23), $\mathbb{E}_a Y_n(x) = c(x) v_n(x)$. So we can now define

$$v_\infty(x) := \lim_n v_n(x) < +\infty,$$

and then

$$
\begin{aligned}
i_\infty(x, y) &:= c(x, y)[v_\infty(x) - v_\infty(y)] \\
&= \lim_n c(x, y)[v_n(x) - v_n(y)] \\
&= \lim_n i_n(x, y)
\end{aligned}
$$

by Ohm's law (when n is large enough that both x and y are in G_n). Because i_n is a flow for all n, by taking limits in the flow-conservation constraints we see that so is i_∞. Note that by construction of i_ℓ,

$$\frac{1}{2} \sum_{x,y \in G_n} c(x,y) i_\infty(x,y)^2 = \lim_{\ell \geq n} \frac{1}{2} \sum_{x,y \in G_n} c(x,y) i_\ell(x,y)^2$$

$$\leq \limsup_{\ell \geq n} \mathscr{E}(i_\ell)$$

$$< B$$

uniformly in n. Because the left-hand side converges to the energy of i_∞ as $n \to +\infty$, we are done. ∎

We give an application to Pólya's theorem in Section 3.3.4.

Finally, we derive a useful general result illustrating the robustness reaped from Thomson's principle. At a high level, a rough embedding from $\mathcal{N}$ to $\mathcal{N}'$ is a mapping of the edges of $\mathcal{N}$ to paths of $\mathcal{N}'$ of comparable overall resistance that do not overlap much. The formal definition follows. As we will see, the purpose of a rough embedding is to allow a flow on $\mathcal{N}$ to be morphed into a flow on $\mathcal{N}'$ of comparable energy.

Definition 3.3.31 (Rough embedding). *Let $\mathcal{N} = (G, c)$ and $\mathcal{N}' = (G', c')$ be networks with resistances r and r', respectively. We say that a map ϕ from the vertices of G to the vertices of G' is a* rough embedding *if there are constants $\alpha, \beta < +\infty$ and a map Φ defined on the edges of G such that* ROUGH EMBEDDING

1 for every edge $e = \{x, y\}$ in G, $\Phi(e)$ is a non-empty path of edges of G' between $\phi(x)$ and $\phi(y)$ such that

$$\sum_{e' \in \Phi(e)} r'(e') \leq \alpha \, r(e);$$

2 for every edge e' in G', there are no more than β edges in G whose image under Φ contains e'.

The map ϕ need not in general be a bijection.

We say that two networks are roughly equivalent *if there exist rough embeddings between them, one in each direction.* ROUGHLY EQUIVALENT

Example 3.3.32 (Independent-coordinate random walk). Let $\mathcal{N} = \mathbb{L}^d$ with unit resistances and let $\mathcal{N}'$ be the network corresponding to the *independent-coordinate random walk*

$$(Y_t^{(1)}, \ldots, Y_t^{(d)}),$$

where each coordinate $(Y_t^{(i)})$ is an independent simple random walk on $\mathbb{Z}$ started at 0. For example, the neighborhood of the origin in $\mathcal{N}'$ is $\{(x_1, \ldots, x_d) : x_i \in \{-1, 1\} \; \forall i\}$. Note that $\mathcal{N}'$ contains only those points of $\mathbb{Z}^d$ with coordinates of identical parities.

Despite encoding quite different random walks, we claim that the networks $\mathcal{N}$ and $\mathcal{N}'$ are roughly equivalent.

- $\mathcal{N}$ to $\mathcal{N}'$: Consider the map ϕ, which associates to each $x \in \mathcal{N}$ a closest point in $\mathcal{N}'$ chosen in some arbitrary manner. For Φ, associate to each edge $e = \{x, y\} \in \mathcal{N}$ a shortest path in $\mathcal{N}'$ between $\phi(x)$ and $\phi(y)$, again chosen arbitrarily. If $\phi(x) = \phi(y)$, choose an arbitrary, non-empty, shortest cycle through $\phi(x)$.

- $\mathcal{N}'$ *to* $\mathcal{N}$: Consider the map ϕ, which associates to each $x \in \mathcal{N}'$ the corresponding point x in $\mathcal{N}$. Construct Φ similarly to the previous case.

Exercise 3.19 asks for a rigorous proof of rough equivalence. See also Exercise 3.20 for an important generalization of this example. ◀

Our main result about roughly equivalent networks is that they have the same type.

Theorem 3.3.33 (Recurrence and rough equivalence). *Let $\mathcal{N}$ and $\mathcal{N}'$ be roughly equivalent, locally finite, connected networks. Then $\mathcal{N}$ is transient if and only if $\mathcal{N}'$ is transient.*

Proof Assume $\mathcal{N}$ is transient and let θ be a unit flow from some a to ∞ of finite energy. The existence of this flow is guaranteed by Theorem 3.3.30. Let ϕ, Φ be a rough embedding from $\mathcal{N}$ to $\mathcal{N}'$ with parameters α and β.

The basic idea of the proof is to map the flow θ onto $\mathcal{N}'$ using Φ. Because flows are directional, it will be convenient to think of edges as being directed. For $e = \{x, y\}$ in $\mathcal{N}$, let $\overrightarrow{\Phi}(x, y)$ be the path $\Phi(e)$ oriented from $\phi(x)$ to $\phi(y)$. So $(x', y') \in \overrightarrow{\Phi}(x, y)$ means that $\{x', y'\} \in \Phi(e)$ and that x' is visited before y' in the path $\Phi(e)$ from $\phi(x)$ to $\phi(y)$. (If $\phi(x) = \phi(y)$, choose an arbitrary orientation of the cycle $\Phi(e)$ for $\overrightarrow{\Phi}(x, y)$ and the reversed orientation for $\overrightarrow{\Phi}(y, x)$.) Then define, for x', y' with $\{x', y'\}$ in $\mathcal{N}'$,

$$\theta'(x', y') := \sum_{(x,y):(x',y') \in \overrightarrow{\Phi}(x,y)} \theta(x, y). \tag{3.3.31}$$

See Figure 3.5.

We claim that θ' is a flow to ∞ of finite energy on $\mathcal{N}'$. We first check that θ' is a flow.

1. *(Anti-symmetry)* By construction, $\theta'(y', x') = -\theta'(x', y')$, that is, θ' is antisymmetric, because θ itself is anti-symmetric. We used the fact that $\overrightarrow{\Phi}(y, x)$ is $\overrightarrow{\Phi}(x, y)$ oriented in the opposite direction.
2. *(Flow conservation)* Next we check the flow-conservation constraints. Fix z' in $\mathcal{N}'$. By Condition 2 in Definition 3.3.31, there are finitely many edges e in $\mathcal{N}$ such that $\Phi(e)$ visits z'. Let $e = \{x, y\}$ be such an edge. There are two cases:

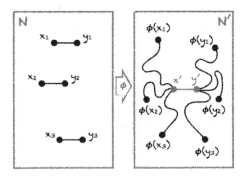

Figure 3.5 The flow on (x', y') is the sum of the flows on (x_1, y_1), (x_2, y_2), and (x_3, y_3).

- Assume first that $\phi(x), \phi(y) \neq z'$ and let $(u', z'), (z', w')$ be the directed edges incident with z' on $\overrightarrow{\Phi}(x, y)$. Observe that, in the definition of θ', (y, x) contributes $\theta(y, x) = -\theta(x, y)$ to $\theta'(z', u')$ and (x, y) contributes $\theta(x, y)$ to $\theta'(z', w')$. So these contributions cancel out in the flow-conservation constraint for z', that is, in the sum $\sum_{v': v' \sim z'} \theta'(z', v')$.
- If instead $e = \{x, y\}$ is such that $\phi(x) = z'$, let (z', w') be the first edge on the path $\overrightarrow{\Phi}(x, y)$. Edge (x, y) contributes $\theta(x, y)$ to $\theta'(z', w')$. A similar statement applies to $\phi(y) = z'$ by changing the role of x and y. This case also applies to $\phi(x) = \phi(y) = z'$.

From these two cases, summing over all paths visiting z' gives

$$\sum_{v': v' \sim z'} \theta'(z', v') = \sum_{z: \phi(z) = z'} \left(\sum_{v: v \sim z} \theta(z, v) \right).$$

Because θ is a flow, the sum in parentheses is 0 if $z \neq a$ and 1 otherwise. So the right-hand side is 0 unless $a \in \phi^{-1}(\{z'\})$, in which case it is 1.

We have shown that θ' is a unit flow from $\phi(a)$ to ∞. It remains to bound the energy of θ'. By (3.3.31), Cauchy–Schwarz, and Condition 2 in Definition 3.3.31,

$$\theta'(x', y')^2 = \left[\sum_{(x,y):(x',y') \in \overrightarrow{\Phi}(x,y)} \theta(x, y) \right]^2$$

$$\leq \left[\sum_{(x,y):(x',y') \in \overrightarrow{\Phi}(x,y)} 1 \right] \left[\sum_{(x,y):(x',y') \in \overrightarrow{\Phi}(x,y)} \theta(x, y)^2 \right]$$

$$\leq \beta \sum_{(x,y):(x',y') \in \overrightarrow{\Phi}(x,y)} \theta(x, y)^2.$$

Summing over all pairs and using Condition 1 in Definition 3.3.31 gives

$$\frac{1}{2} \sum_{x', y'} r'(x', y') \theta'(x', y')^2 \leq \beta \frac{1}{2} \sum_{x', y'} r'(x', y') \sum_{(x,y):(x',y') \in \overrightarrow{\Phi}(x,y)} \theta(x, y)^2$$

$$= \beta \frac{1}{2} \sum_{x, y} \theta(x, y)^2 \sum_{(x',y') \in \overrightarrow{\Phi}(x,y)} r'(x', y')$$

$$\leq \alpha \beta \frac{1}{2} \sum_{x, y} r(x, y) \theta(x, y)^2,$$

which is finite by assumption. That concludes the proof. ∎

As an application, we give a second proof of Pólya's theorem in Section 3.3.4.

Other applications

So far we have emphasized applications to recurrence. Here we show that electrical network theory can also be used to bound commute times. In Section 3.3.5, we give further applications beyond random walks on graphs.

An application of Lemma 3.1.24 gives another probabilistic interpretation of the effective resistance – and a useful formula.

Theorem 3.3.34 (Commute time identity). *Let $\mathcal{N} = (G, c)$ be a finite, connected network with vertex set V. For $x \neq y$, let the* commute time $\tau_{x,y}$ *be the time of the first return to x after the first visit to y. Then,*

$$\mathbb{E}_x[\tau_{x,y}] = \mathbb{E}_x[\tau_y] + \mathbb{E}_y[\tau_x] = c_{\mathcal{N}} \, \mathcal{R}(x \leftrightarrow y),$$

where $c_{\mathcal{N}} = 2 \sum_{e=\{x,y\} \in \mathcal{N}} c(e)$.

Proof This follows immediately from Lemma 3.1.24 and the definition of the effective resistance (Definition 3.3.19). Specifically,

$$\mathbb{E}_x[\tau_y] + \mathbb{E}_y[\tau_x] = \frac{1}{\pi_x \, \mathbb{P}_x[\tau_y < \tau_x^+]}$$

$$= \frac{1}{(2 \sum_{e=\{x,y\} \in \mathcal{N}} c(e))^{-1} c(x) \, \mathbb{P}_x[\tau_y < \tau_x^+]}$$

$$= c_{\mathcal{N}} \, \mathcal{R}(x \leftrightarrow y). \qquad \blacksquare$$

Example 3.3.35 (Random walk on the torus). Consider random walk on the d-dimensional torus $\mathbb{L}_n^d$ with unit resistances. We use the commute time identity to lower bound the mean hitting time $\mathbb{E}_x[\tau_y]$ for arbitrary vertices $x \neq y$ at graph distance k on $\mathbb{L}_n^d$. To use the commute time identity (Theorem 3.3.34), note that by symmetry $\mathbb{E}_x[\tau_y] = \mathbb{E}_y[\tau_x]$ so that

$$\mathbb{E}_x[\tau_y] = \frac{1}{2} c_{\mathcal{N}} \, \mathcal{R}(x \leftrightarrow y) = dn^d \, \mathcal{R}(x \leftrightarrow y), \tag{3.3.32}$$

where we used that the number of vertices is n^d and the graph is $2d$-regular.

To simplify, assume n is odd and identify the vertices of $\mathbb{L}_n^d$ with the box

$$B := \{-(n-1)/2, \ldots, (n-1)/2\}^d$$

in $\mathbb{L}^d$ centered at $x = 0$. Let $\partial B_j^\infty = \{z \in \mathbb{L}^d : \|z\|_\infty = j\}$ and let Π_j be the set of edges between ∂B_j^∞ and ∂B_{j+1}^∞. Note that on B the ℓ^1 norm of y is at most k (the graph distance between $x = 0$ and y). Since the ℓ^∞ norm is at least $1/d$ times the ℓ^1 norm on $\mathbb{L}^d$, there exists $J = O(k)$ such that all Π_js, $j \leq J$, are cutsets separating x from y. By the Nash–Williams inequality,

$$\mathcal{R}(x \leftrightarrow y) \geq \sum_{0 \leq j \leq J} |\Pi_j|^{-1} = \sum_{0 \leq j \leq J} \Omega\left(j^{-(d-1)}\right) = \begin{cases} \Omega(\log k), & d = 2, \\ \Omega(1), & d \geq 3. \end{cases}$$

From (3.3.32), we get:

Claim 3.3.36

$$\mathbb{E}_x[\tau_y] = \begin{cases} \Omega(n^d \log k), & d = 2, \\ \Omega(n^d), & d \geq 3. \end{cases} \qquad \blacktriangleleft$$

Remark 3.3.37 *The bounds in the previous example are tight up to constants. See [LPW06, Proposition 10.13]. Note that the case $d \geq 3$ does not in fact depend on the distance k.*

See Exercise 3.22 for an application of the commute time identity to cover times.

3.3.4 ▷ *Random Walks: Pólya's Theorem, Two Ways*

The following is a classical result.

Theorem 3.3.38 (Pólya's theorem). *Random walk on* $\mathbb{L}^d$ *is recurrent for* $d \leq 2$ *and transient for* $d \geq 3$.

We prove the theorem for $d = 2, 3$ using the tools developed in the previous subsection. The other cases follow by Rayleigh's principle (Corollary 3.3.29). There are elementary proofs of this result. But we showed above that the electrical network approach has the advantage of being robust to the details of the lattice. For a different argument, see Exercise 2.10.

The case $d = 2$ follows from the Nash–Williams inequality (Corollary 3.3.26) by letting Π_j be the set of edges connecting vertices of ℓ^∞ norm j and $j + 1$. Using the fact that all conductances are 1, that $|\Pi_j| = O(j)$, and that $\sum_j j^{-1}$ diverges, recurrence is established by Theorem 3.3.20.

First proof

Now consider the case $d = 3$ and let $a = 0$ be the origin. We start with a proof based on what is known as the random paths method.

We construct a finite-energy flow to ∞ using the *method of random paths*. Note that a simple way to produce a unit flow to ∞ is to push a flow of 1 through an infinite path (which, recall, are self-avoiding by definition). Taking this a step further, let μ be a probability measure on infinite paths and define the anti-symmetric function

$$\theta(x, y) := \mathbb{E}[\mathbf{1}_{(x,y)\in\Gamma} - \mathbf{1}_{(y,x)\in\Gamma}] = \mathbb{P}[(x, y) \in \Gamma] - \mathbb{P}[(y, x) \in \Gamma],$$

where Γ is a random path distributed according to μ, oriented away from 0. (We will give an explicit construction below where the appropriate formal probability space will be clear.) Observe that $\sum_{y \sim x}[\mathbf{1}_{(x,y)\in\Gamma} - \mathbf{1}_{(y,x)\in\Gamma}] = 0$ for any $x \neq 0$ because vertices visited by Γ are entered and exited exactly once. That same sum is 1 at $x = 0$. Hence, θ is a unit flow to ∞. For edge $e = \{x, y\}$, consider the following "edge marginal" of μ:

$$\mu(e) := \mathbb{P}[(x, y) \in \Gamma \text{ or } (y, x) \in \Gamma] = \mathbb{P}[(x, y) \in \Gamma] + \mathbb{P}[(y, x) \in \Gamma] \geq \theta(x, y),$$

where we used that a path Γ cannot visit both (x, y) and (y, x) by definition. Then we get the following bound.

Claim 3.3.39 (Method of random paths).

$$\mathscr{E}(\theta) \leq \sum_e \mu(e)^2. \tag{3.3.33}$$

For a measure μ concentrated on a single path, the sum in (3.3.33) is infinite. To obtain a useful bound, what we need is a large collection of spread out paths. On the lattice $\mathbb{L}^3$, we construct μ as follows. Let U be a uniformly random point on the unit sphere in $\mathbb{R}^3$ and let γ be the ray from 0 to ∞ going through U. Imagine centering a unit cube around each point in $\mathbb{Z}^3$ whose edges are aligned with the axes. Then, γ traverses an infinite number of such cubes. Let Γ be the corresponding path in the lattice $\mathbb{L}^3$. To see that this procedure indeed produces a path observe that γ, upon exiting a cube around a point $z \in \mathbb{Z}^3$, enters the cube of a neighboring point $z' \in \mathbb{Z}^3$ through a face corresponding to the edge between z and z' on the lattice $\mathbb{L}^3$ (unless it goes through a corner of the cube, but this has probability 0). To argue that μ distributes its mass among sufficiently spread out paths, we bound the probability that a vertex is visited by Γ. Let z be an arbitrary vertex in $\mathbb{Z}^3$. Because the sphere of radius $\|z\|_2$ around the origin in $\mathbb{R}^3$ has area $O(\|z\|_2^2)$ and its intersection with the unit cube centered around z has area $O(1)$, it follows that

$$\mathbb{P}[z \in \Gamma] = O\left(1/\|z\|_2^2\right).$$

That immediately implies a similar bound on the probability that an edge is visited by Γ. Moreover:

Lemma 3.3.40 *There are $O(j^2)$ edges with an endpoint at ℓ^2 distance within $[j, j+1]$ from the origin.*

Proof Consider a ball of ℓ^2 radius $1/2$ centered around each vertex of ℓ^2 norm within $[j, j+1]$. Those balls are non-intersecting and have total volume $\Theta(N_j)$, where N_j is the number of such vertices. On the other hand, the volume of the shell of ℓ^2 inner and outer radii $j - 1/2$ and $j + 3/2$ centered around the origin (where all those balls lie) is

$$\frac{4}{3}\pi(j + 3/2)^3 - \frac{4}{3}\pi(j - 1/2)^3 = O(j^2).$$

Hence, $N_j = O(j^2)$. Finally, note that each vertex has six incident edges. ∎

Plugging those bounds into (3.3.33), we get

$$\mathscr{E}(\theta) \le \sum_j O(j^2) \cdot \left[O(1/j^2)\right]^2 = O\left(\sum_j j^{-2}\right) < +\infty.$$

Transience follows from Theorem 3.3.30. (This argument clearly does not work on $\mathbb{L}$ where there are only two rays. You should convince yourself that it does not work on $\mathbb{L}^2$ either. But see Exercise 3.17.)

Second proof

We briefly describe a second proof based on the independent-coordinate random walk. Consider the networks $\mathcal{N}$ and $\mathcal{N}'$ in Example 3.3.32. Because they are roughly equivalent (Definition 3.3.31), they have the same type by Theorem 3.3.33. Recall that because the number of returns to 0 is geometric with success probability equal to the escape probability, random walk on $\mathcal{N}'$ is transient if and only if the expected number of visits to 0 is finite (see (3.1.2)). By independence of the coordinates, this expectation can be written as

$$\sum_{t \ge 0} \left(\mathbb{P}\left[Y_{2t}^{(1)} = 0\right]\right)^d = \sum_{t \ge 0} \left(\binom{2t}{t} 2^{-2t}\right)^d = \sum_{t \ge 0} \Theta(t^{-d/2}),$$

where we used Stirling's formula (see Appendix A). The rightmost sum is finite if and only if $d \ge 3$. That implies random walk on $\mathcal{N}'$ is transient under that condition. By rough equivalence, the same is true of $\mathcal{N}$.

3.3.5 ▷ *Randomized Algorithms: Wilson's Method for Generating Uniform Spanning Trees*

In this section, we describe an application of electrical network theory to spanning trees.
    With a slight abuse of notation, we use $e \in G$ to indicate that e is an edge of G.

Uniform spanning trees Let $G = (V, E)$ be a finite connected graph. Recall that a spanning tree is a subtree of G containing all its vertices. Such a tree has $|V| - 1$ edges. A *uniform spanning tree* is a spanning tree T chosen uniformly at random among all spanning trees of G.

We make some simple observations first. Because G is connected, it has at least one spanning tree by Corollary 1.1.6. Moreover, for any edge $e \in G$, there always exists at least one spanning tree including it. To see this, let T' be any spanning tree of G, which exists by the previous observation. If $e \notin T'$, then we obtain a new spanning tree by adding e to T' and removing one edge $\neq e$ in the cycle created. As a consequence, the probability of inclusion $\mathbb{P}[e \in T]$ in a uniform spanning tree T cannot be 0. It is, however, possible for $\mathbb{P}[e \in T]$ to equal to 1 if removing e disconnects the graph. Such an edge is called a *bridge*.

A fundamental property of uniform spanning trees is the following negative correlation between edges.

Claim 3.3.41 *For a uniform spanning tree T of a connected graph G,*

$$\mathbb{P}[e \in T \mid e' \in T] \leq \mathbb{P}[e \in T] \qquad \forall e \neq e' \in G.$$

This property is perhaps not surprising. For one, the number of edges in a spanning tree is fixed, so the inclusion of e' makes it seemingly less likely for other edges to be present. Yet proving Claim 3.3.41 is not trivial. The proof relies on the electrical network perspective. The key is a remarkable formula for the inclusion of an edge in a uniform spanning tree.

Theorem 3.3.42 (Kirchhoff's resistance formula). *Let $G = (V, E)$ be a finite, connected graph and let $\mathcal{N}$ be the network on G with unit resistances. If T is a uniform spanning tree on G, then for all $e = \{x, y\}$,*

$$\mathbb{P}[e \in T] = \mathscr{R}(x \leftrightarrow y).$$

Before explaining how this formula arises, we show that it implies Claim 3.3.41.

Proof of Claim 3.3.41. Recall that $\mathbb{P}[e' \in T] \neq 0$. By the law of total probability,

$$\mathbb{P}[e \in T] = \mathbb{P}[e \in T \mid e' \in T]\mathbb{P}[e' \in T] + \mathbb{P}[e \in T \mid e' \notin T]\mathbb{P}[e' \notin T],$$

so, since $\mathbb{P}[e' \in T] + \mathbb{P}[e' \notin T] = 1$, we can instead prove

$$\mathbb{P}[e \in T \mid e' \notin T] \geq \mathbb{P}[e \in T]. \tag{3.3.34}$$

Picking a uniform spanning tree on $\mathcal{N}$ conditioned on $\{e' \notin T\}$ is the same as picking a uniform spanning tree on the modified network $\mathcal{N}'$, where e' is removed. By Rayleigh's principle (in the form of Corollary 3.3.28),

$$\mathscr{R}_{\mathcal{N}'}(x \leftrightarrow y) \geq \mathscr{R}_{\mathcal{N}}(x \leftrightarrow y),$$

and Kirchhoff's resistance formula (Theorem 3.3.42) gives (3.3.34). ∎

Remark 3.3.43 *More generally, thinking of a uniform spanning tree T as a random subset of edges, the law of T has the property of negative associations, defined as follows. An event $\mathcal{A} \subseteq 2^E$ is said to be* increasing *if $\omega \cup \{e\} \in \mathcal{A}$ whenever $\omega \in \mathcal{A}$. The event $\mathcal{A}$ is said to depend only on $F \subseteq E$ if for all $\omega_1, \omega_2 \in 2^E$ that agree on F, either both are in $\mathcal{A}$ or neither is. The law $\mathbb{P}_T$ of T has* negative associations *in the sense that for any two increasing events*

$\mathcal{A}$ and $\mathcal{B}$ that depend only on disjoint sets of edges, we have $\mathbb{P}_T[\mathcal{A} \cap \mathcal{B}] \leq \mathbb{P}_T[\mathcal{A}]\mathbb{P}_T[\mathcal{B}]$. See [LP16, Exercise 4.6].

Let $e = \{x, y\}$. To get some insight into Kirchhoff's resistance formula, we first note that, if i is the unit current from x to y and v is the associated voltage, by definition of the effective resistance

$$\mathscr{R}(x \leftrightarrow y) = \frac{v(x)}{\|i\|} = c(e)(v(x) - v(y)) = i(x, y), \qquad (3.3.35)$$

where we used Ohm's law (i.e., (3.3.20)) as well as the fact that $c(e) = 1$, $v(y) = 0$, and $\|i\| = 1$. Note that $\|i\|$ and $i(x, y)$ are *not* the same quantity: although $\|i\| = 1$, $i(x, y)$ is only the current along the edge to y. Furthermore, by the probabilistic interpretation of the current (Theorem 3.3.17), with $Z = \{y\}$,

$$i(x, y) = \mathbb{E}_x[N^Z_{x \to y} - N^Z_{y \to x}] = \mathbb{P}_x\left[(x, y) \text{ is traversed before } \tau_y\right]. \qquad (3.3.36)$$

Indeed, started at x, $N^Z_{y \to x} = 0$ and $N^Z_{x \to y} \in \{0, 1\}$. Kirchhoff's resistance formula is then established by relating the random walk on $\mathcal{N}$ to the probability that e is present in a uniform spanning tree T. To do this we introduce a random-walk-based algorithm for generating uniform spanning trees. This rather miraculous procedure, known as *Wilson's method*, is of independent interest. (For a classical connection between random walks and spanning trees, see also Exercise 3.23.)

Wilson's method It will be somewhat easier to work in a more general context. Let $\mathcal{N} = (G, c)$ be a finite, connected network on G with arbitrary conductances and define the *weight* of a spanning tree T on $\mathcal{N}$ as

$$W(T) = \prod_{e \in T} c(e).$$

With a slight abuse, we continue to call a tree T picked at random among all spanning trees of G with probability proportional to $W(T)$ a "uniform" spanning tree on $\mathcal{N}$.

To state Wilson's method, we need the notion of *loop erasure*. Let $\mathcal{P} = x_0 \sim \cdots \sim x_k$ be a walk in $\mathcal{N}$. The loop erasure of $\mathcal{P}$ is obtained by removing cycles in the order they appear. That is, let j^* be the smallest j such that $x_j = x_\ell$ for some $\ell < j$. Remove the subwalk $x_{\ell+1} \sim \cdots \sim x_j$ from $\mathcal{P}$, and repeat. The result is self-avoiding, that is, a path, and is denoted by $\text{LE}(\mathcal{P})$.

Let v_0 be an arbitrary vertex of G, which we refer to as the root, and let T_0 be the subtree made up of v_0 alone. Starting with the root, order arbitrarily the vertices of G as $v_0, \ldots, v_{n-1}$. Wilson's method constructs an increasing sequence of subtrees as follows. See Figure 3.6. Let $T := T_0$.

1. Let v be the vertex of G not in T with lowest index. Perform random walk on $\mathcal{N}$ started at v until the first visit to a vertex of T. Let $\mathcal{P}$ be the resulting walk.
2. Add the loop erasure $\text{LE}(\mathcal{P})$ to T.
3. Repeat until all vertices of G are in T.

Let $T_0, \ldots, T_m$ be the sequence of subtrees produced by Wilson's method.

Claim 3.3.44 *Forgetting the root, T_m is a uniform spanning tree on $\mathcal{N}$.*

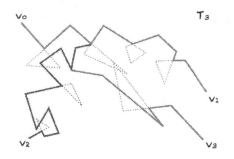

Figure 3.6 An illustration of Wilson's method. The dashed lines indicate erased loops.

This claim is far from obvious. Before proving it, we finish the proof of Kirchhoff's resistance formula.

Proof of Theorem 3.3.42   From (3.3.35) and (3.3.36), it suffices to prove that, for $e = \{x, y\}$,

$$\mathbb{P}_x\big[(x, y) \text{ is traversed before } \tau_y\big] = \mathbb{P}[e \in T],$$

where the probability on the left-hand side refers to random walk on $\mathcal{N}$ with unit resistances started at x and the probability on the right-hand side refers to a uniform spanning tree T on $\mathcal{N}$. Generate T using Wilson's method started at root $v_0 = y$ with the choice $v_1 = x$. If the walk from x to y during the first iteration of Wilson's method includes (x, y), then the loop erasure is simply $x \sim y$ and e is in T. On the other hand, if the walk from x to y does not include (x, y), then e cannot be used at a later stage because it would create a cycle. That immediately proves the theorem. ∎

It remains to prove the claim.

Proof of Claim 3.3.44   The idea of the proof is to cast Wilson's method in the more general framework of cycle popping algorithms. We begin by explaining how such algorithms work.

Let P be the transition matrix corresponding to random walk on $\mathcal{N} = (G, c)$ with $G = (V, E)$ and root v_0. To each vertex $x \neq v_0$ in V, we assign an independent stack of "colored directed edges"

$$\mathcal{S}_0^x := (\langle x, Y_1^x \rangle_1, \langle x, Y_2^x \rangle_2, \ldots),$$

where each Y_j^x is chosen independently at random from the distribution $P(x, \cdot)$. In particular, all Y_j^xs are neighbors of x in $\mathcal{N}$. The index j in $\langle x, Y_j^x \rangle_j$ is the *color* of the edge. It keeps track COLOR of the position of the edge in the original stack. (Picture $\mathcal{S}^x$ as a spring-loaded plate dispenser located on vertex x.)

We consider a process which involves popping edges off the stacks. We use the notation $\mathcal{S}^x$ to denote the *current* stack at x. The initial assignment of the stack is $\mathcal{S}^x := \mathcal{S}_0^x$ as above. Given the current stacks $(\mathcal{S}^x)_x$, we call *visible graph* the (colored) directed graph over V with VISIBLE edges Top$(\mathcal{S}^x)$ for all $x \neq v_0$, where Top$(\mathcal{S}^x)$ is the first edge in the current stack $\mathcal{S}^x$. The GRAPH latter are referred to as *visible edges*. We denote the current visible graph by $\overrightarrow{G}_\odot$. VISIBLE EDGE

Note that $\overrightarrow{G}_\odot$ has out-degree 1 for all $x \neq v_0$ and the root has out-degree 0. In particular, all undirected cycles in $\overrightarrow{G}_\odot$ are in fact directed cycles, and we refer to them simply as cycles. (Indeed, a set of edges forming an undirected cycle that is not directed must have a vertex of out-degree 2.) Also recall the following characterization from Corollary 1.1.8: an acyclic, undirected subgraph with $|V|$ vertices and $|V| - 1$ edges is a spanning tree of G. Hence, if there is no cycle in $\overrightarrow{G}_\odot$, then it must be a spanning tree (as an undirected graph) where, furthermore, all edges point toward the root. Such a tree is also known as a *spanning arborescence*. Once that happens, we are done.

SPANNING AR-
BORESCENCE

As the name suggests, a cycle popping algorithm proceeds by popping cycles in $\overrightarrow{G}_\odot$ off the tops of the stacks until a spanning arborescence is produced. That is, at every iteration, if $\overrightarrow{G}_\odot$ contains at least one cycle, then a cycle $\overrightarrow{C}$ is picked according to some rule, the top of each stack in $\overrightarrow{C}$ is popped, and a new visible graph $\overrightarrow{G}_\odot$ is revealed. See Figure 3.7 for an illustration.

With these definitions in place, the proof of the claim involves the following steps.

(i) *Wilson's method is a cycle popping algorithm.* Recasting Wilson's method, we can think of the initial stacks (S_0^x) as corresponding to picking – ahead of time – all potential transitions in the random walks. With this representation, the algorithm boils down to a recipe for choosing which cycle to pop next. Indeed, at each iteration, we start from a vertex v not in the current tree T. A key observation: following the visible edges from v traces a path *whose distribution is that of random walk on $\mathcal{N}$*. Loop erasure then corresponds to popping cycles as they are closed. We pop only those visible edges on the removed cycles, as they originate from vertices that will be visited again by the algorithm and for which a new transition will then be needed. Those visible edges in the resulting loop-erased path are not popped – note that they are part of the final arborescence.

(ii) *The popping order does not matter.* We just argued that Wilson's method is a cycle popping algorithm. In fact, we claim that any cycle popping algorithm, that is, no matter what popping choices are made along the way, produces the same final arborescence. To make this precise, we identify the popped cycles uniquely. This is where the colors come in. A *colored cycle* is a directed cycle over V made of colored edges from the stacks (not necessarily of the same color and not necessarily in the current visible graph). We say that a colored cycle $\overrightarrow{C}$ is *poppable* for a visible graph $\overrightarrow{G}_\odot$ if there exists a sequence of colored cycles $\overrightarrow{C}_1, \ldots,$ $\overrightarrow{C}_r = \overrightarrow{C}$ that can be popped in that order starting from $\overrightarrow{G}_\odot$. Note that, by this definition, $\overrightarrow{C}_1$ is a cycle in $\overrightarrow{G}_\odot$. Now we claim that if $\overrightarrow{C}_1'$ were popped first instead of $\overrightarrow{C}_1$, producing the new visible graph $\overrightarrow{G}_\odot'$, then $\overrightarrow{C}$ would still be poppable for $\overrightarrow{G}_\odot'$. This claim implies that, in any cycle popping algorithm, either an infinite number of cycles are popped or eventually all poppable cycles are popped – independently of the order – producing the same outcome. (Note that, while the same cycle may be popped more than once, the same *colored* cycle cannot.)

COLORED
CYCLE

POPPABLE
CYCLE

To prove the claim, note first that if $\overrightarrow{C}_1' = \overrightarrow{C}$ or if $\overrightarrow{C}_1'$ does not share a vertex with any of $\overrightarrow{C}_1, \ldots, \overrightarrow{C}_r$, there is nothing to prove. So let $\overrightarrow{C}_j$ be the first cycle

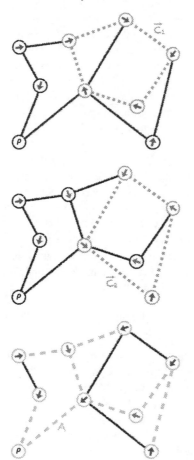

Figure 3.7 A realization of a cycle popping algorithm (from top to bottom). In all three figures, the underlying graph is G while the arrows depict the visible edges.

in the sequence sharing a vertex with $\vec{C}'_1$, say x. Let $\langle x, y \rangle_c$ and $\langle x, y' \rangle_{c'}$ be the colored edges emanating from x in $\vec{C}_j$ and $\vec{C}'_1$, respectively. By definition, x is not on any of $\vec{C}_1, \ldots, \vec{C}_{j-1}$ so the edge originating from x is *not popped* by that sequence and we must have $\langle x, y \rangle_c = \langle x, y' \rangle_{c'}$ as colored edges. In particular, the vertex y is also a shared vertex of $\vec{C}_j$ and $\vec{C}'_1$, and the same argument applies to it. Proceeding by induction leads to the conclusion that $\vec{C}'_1 = \vec{C}_j$ as colored cycles. But then $\vec{C}$ is clearly poppable for the visible graph resulting from popping $\vec{C}'_1$ first, because it can be popped with the rearranged sequence $\vec{C}'_1 = \vec{C}_j, \vec{C}_1, \ldots, \vec{C}_{j-1}, \vec{C}_{j+1}, \ldots, \vec{C}_r = \vec{C}$, where we used the fact that $\vec{C}'_1$ does not share a vertex with $\vec{C}_1, \ldots, \vec{C}_{j-1}$.

(iii) *Termination occurs in finite time almost surely.* We have shown so far that, in
any cycle popping algorithm, either an infinite number of cycles are popped or
eventually all poppable cycles are popped. But Wilson's method – a cycle popping
algorithm as we have shown – stops after a finite amount of time with probability
1. Indeed, because the network is finite and connected, the random walk started at
each iteration hits the current T in finite time almost surely (by Lemma 3.1.25). To
sum up, all cycle popping algorithms terminate and produce the same spanning
arborescence. It remains to compute the distribution of the outcome.

(iv) *The arborescence has the desired distribution.* Let $\mathcal{A}$ be the spanning arborescence produced by any cycle popping algorithm on the stacks $(\mathcal{S}_0^x)$. To compute
the distribution of $\mathcal{A}$, we first compute the distribution of a particular cycle popping realization leading to $\mathcal{A}$. Because the popping order does not matter, by "realization" we mean a collection $\mathcal{C}$ of colored cycles together with a final spanning
arborescence $\mathcal{A}$. Notice that what lies in the stacks "under" $\mathcal{A}$ is not relevant to
the realization, that is, the same outcome is produced no matter what is under $\mathcal{A}$.

So, from the distribution of the stacks, the probability of observing $(\mathcal{C}, \mathcal{A})$ is
simply the product of the transitions corresponding to the "popped edges" in $\mathcal{C}$
and the "final edges" in $\mathcal{A}$, that is,

$$\prod_{\vec{e} \in \mathcal{C} \cup \mathcal{A}} P(\vec{e}) = \Psi(\mathcal{A}) \prod_{\vec{\mathcal{C}} \in \mathcal{C}} \Psi\left(\vec{\mathcal{C}}\right),$$

where the function Ψ returns the product of the transition probabilities of a set of
directed edges. Thanks to the product form on the right-hand side, summing over
all possible $\mathcal{C}$s gives that the probability of producing $\mathcal{A}$ is proportional to $\Psi(\mathcal{A})$.

For this argument to work though, there are two small details to take care of.
First, note that we want the probability of the "uncolored" arborescence. But observe that, in fact, there is no need to keep track of the colors on the edges of $\mathcal{A}$
because these are determined by $\mathcal{C}$. Second, we need for the collection of possible
$\mathcal{C}$s *not to vary with* $\mathcal{A}$. But it is clear that any arborescence could lie under any $\mathcal{C}$.

To see that we are done, let T be the undirected spanning tree corresponding to the outcome, $\mathcal{A}$, of Wilson's method. Then, because $P(x, y) = \frac{c(x,y)}{c(x)}$, we get

$$\Psi(\mathcal{A}) = \frac{W(T)}{\prod_{x \neq v_0} c(x)},$$

where note that the denominator does not depend on T. So if we forget the orientation of $\mathcal{A}$,
which is determined by the root (i.e., sum over all choices of root), we get a spanning tree
whose distribution is proportional to $W(T)$, as required.   ■

Exercises

Exercise 3.1 (Reflection). Give a rigorous proof of Theorem 3.1.9 through a formal application of the strong Markov property (i.e., specify f_t and F_t in Theorem 3.1.8).

Exercise 3.2 (Time of kth return). Give a rigorous proof of (3.1.1) through a formal application of the strong Markov property (i.e., specify f_t and F_t in Theorem 3.1.8).

Exercise 3.3 (Tightness of Matthews' bounds). Show that the bounds (3.1.6) and (3.1.7) are tight up to smaller order terms for the coupon collector problem (Example 2.1.4). (Hint: State the problem in terms of the cover time of a random walk on the complete graph with self-loops.)

Exercise 3.4 (Pólya's urn: a suprisingly simple formula). Consider the setting of Example 3.1.49. Prove that

$$\mathbb{P}[G_t = m + 1] = \binom{t}{m} \frac{m!(t-m)!}{(t+1)!}.$$

(Hint: Consider the probability of one particular sequence of outcomes producing the desired event.)

Exercise 3.5 (Optional stopping theorem). Give a rigorous proof of the remaining cases of the optional stopping theorem (Theorem 3.1.38).

Exercise 3.6 (Supermartingale inequality). Let (M_t) be a non-negative, supermartingale. Show that, for any $b > 0$,

$$\mathbb{P}\left[\sup_{s \geq 0} M_s \geq b\right] \leq \frac{\mathbb{E}[M_0]}{b}.$$

(Hint: Mimic the proof of the submartingale case.)

Exercise 3.7 (Azuma–Hoeffding: a second proof). This exercise leads the reader through an alternative proof of the Azuma–Hoeffding inequality.

(i) Show that for all $x \in [-1, 1]$ and $a > 0$,

$$e^{ax} \leq \cosh a + x \sinh a.$$

(ii) Use a Taylor expansion to show that for all x,

$$\cosh x \leq e^{x^2/2}.$$

(iii) Let $X_1, \ldots, X_n$ be (not necessarily independent) random variables such that for all i, $|X_i| \leq c_i$ for some constant $c_i < +\infty$ and

$$\mathbb{E}\left[X_{i_1} \cdots X_{i_k}\right] = 0 \qquad \forall 1 \leq k \leq n, \ \forall 1 \leq i_1 < \cdots < i_k \leq n. \tag{3.3.37}$$

Show, using (i) and (ii), that for all $b > 0$,

$$\mathbb{P}\left[\sum_{i=1}^{n} X_i \geq b\right] \leq \exp\left(-\frac{b^2}{2 \sum_{i=1}^{n} c_i^2}\right).$$

(iv) Prove that (iii) implies the Azuma–Hoeffding inequality as stated in Theorem 3.2.1.
(v) Show that the random variables in Exercise 2.6 do not satisfy (3.3.37) (without using the claim in part (ii) of that exercise).

Exercise 3.8 (Lipschitz condition). Give a rigorous proof of Lemma 3.2.31.

Exercise 3.9 (Lower bound on expected spectral norm). Let A be an $n \times n$ random matrix. Assume that the entries $A_{i,j}$, $i,j = 1, \ldots, n$, are independent, centered random variables in $[-1, 1]$. Suppose further that there is $0 < \sigma^2 < +\infty$ such that $\mathrm{Var}[A_{ij}] \geq \sigma^2$ for all i,j. Show that there is $0 < c < +\infty$ such that

$$\mathbb{E}\|A\| \geq c\sqrt{n}$$

for n large enough. (Hint: Use the fact that $\|A\|^2 \geq \|A\mathbf{e}_1\|^2$ together with Chebyshev's inequality.)

Exercise 3.10 (Kirchhoff's laws). Consider a finite, connected network with a source and a sink. Show that an anti-symmetric function on the edges satisfying Kirchhoff's two laws is a current function (i.e., it corresponds to a voltage function through Ohm's law).

Exercise 3.11 (Dirichlet problem: non-uniqueness). Let (X_t) be the birth-and-death chain on $\mathbb{Z}_+$ with $P(x, x+1) = p$ and $P(x, x-1) = 1 - p$ for all $x \geq 1$, and $P(0, 1) = 1$, for some $0 < p < 1$. Fix $h(0) = 1$.

  (i) When $p > 1/2$, show that there is more than one bounded extension of h to $\mathbb{Z}_+\backslash\{0\}$ that is harmonic on $\mathbb{Z}_+\backslash\{0\}$. (Hint: Consider $\mathbb{P}_x[\tau_0 = +\infty]$.)
  (ii) When $p \leq 1/2$, show that there exists a unique bounded extension of h to $\mathbb{Z}_+\backslash\{0\}$ that is harmonic on $\mathbb{Z}_+\backslash\{0\}$.

Exercise 3.12 (Maximum principle). Let $\mathcal{N} = (G, c)$ be a finite or countable, connected network with $G = (V, E)$. Let W be a finite, connected, proper subset of V.

  (i) Let $h \colon V \to \mathbb{R}$ be a function on V. Prove the maximum principle: if h is harmonic on W, that is, it satisfies

$$h(x) = \frac{1}{c(x)} \sum_{y \sim x} c(x, y) h(y) \qquad \forall x \in W,$$

and if h achieves its supremum on W, then h is constant on $W \cup \partial_V W$, where

$$\partial_V W = \{z \in V \setminus W : \exists y \in W, y \sim z\}.$$

  (ii) Let $h \colon W^c \to \mathbb{R}$ be a bounded function on $W^c := V \setminus W$. Let h_1 and h_2 be extensions of h to W that are harmonic on W. Use part (i) to prove that $h_1 \equiv h_2$.

Exercise 3.13 (Poisson equation: uniqueness). Show that u is the unique solution of the system in Theorem 3.3.6 under the conditions of Theorem 3.3.1. (Hint: Use Theorem 3.3.9 and mimic the proof of Theorem 3.3.1.)

Exercise 3.14 (Effective resistance: metric). Show that effective resistances between pairs of vertices form a metric.

Exercise 3.15 (Dirichlet principle: proof). Prove Theorem 3.3.25.

Exercise 3.16 (Martingale problem). Let V be countable, let (X_t) be a stochastic process adapted to $(\mathcal{F}_t)$ and taking values in V, and let P be a transition probability on V with associated Laplacian operator Δ. Show that the following are equivalent:

  (i) The process (X_t) is a Markov chain with transition probability P.

(ii) For any bounded measurable function $f : V \to \mathbb{R}$, the process

$$M_t^f = f(X_t) - \sum_{s=0}^{t-1} \Delta f(X_s)$$

is a martingale with respect to $(\mathcal{F}_t)$.

Exercise 3.17 (Random walk on $\mathbb{L}^2$: effective resistance). Consider random walk on $\mathbb{L}^2$, which we showed is recurrent. Let (G_n) be the exhaustive sequence corresponding to vertices at distance at most n from the origin and let Z_n be the corresponding sink-set. Show that $\mathscr{R}(0 \leftrightarrow Z_n) = \Theta(\log n)$. (Hint: Use the Nash–Williams inequality and the method of random paths.)

Exercise 3.18 (Random walk on regular graphs: effective resistance). Let G be a d-regular graph with n vertices and $d > n/2$. Let $\mathcal{N}$ be the network (G, c) with unit conductances. Let a and z be arbitrary distinct vertices.

(i) Show that there are at least $2d - n$ vertices $x \neq a, z$ such that $a \sim x \sim z$ is a path.
(ii) Prove that

$$\mathscr{R}(a \leftrightarrow z) \leq \frac{2dn}{2d - n}.$$

Exercise 3.19 (Independent-coordinate random walk). Give a rigorous proof that the two networks in Example 3.3.32 are roughly equivalent.

Exercise 3.20 (Rough isometries). Graphs $G = (V, E)$ and $G' = (V', E')$ are *roughly iso-* ROUGH *metric* (or quasi-isometric) if there is a map $\phi : V \to V'$ and constants $0 < \alpha, \beta < +\infty$ ISOMETRY such that for all $x, y \in V$,

$$\alpha^{-1} d(x, y) - \beta \leq d'(\phi(x), \phi(y)) \leq \alpha d(x, y) + \beta,$$

where d and d' are the graph distances on G and G', respectively, and furthermore, all vertices in G' are within distance β of the image of V. Let $\mathcal{N} = (G, c)$ and $\mathcal{N}' = (G', c')$ be countable, connected networks with uniformly bounded conductances, resistances, and degrees. Prove that if G and G' are roughly isometric, then $\mathcal{N}$ and $\mathcal{N}'$ are roughly equivalent. (Hint: Start by proving that being roughly isometric is an equivalence relation.)

Exercise 3.21 (Random walk on the cycle: hitting time). Use the commute time identity (Theorem 3.3.34) to compute $\mathbb{E}_x[\tau_y]$ in Example 3.3.35 in the case $d = 1$. Give a second proof using a direct martingale argument.

Exercise 3.22 (Random walk on the binary tree: cover time). As in Example 3.3.15, let $\mathcal{N}$ be the rooted binary tree with n levels $\widehat{\mathbb{T}}_2^n$ and equal conductances on all edges.

(i) Show that the maximal hitting time $\mathbb{E}_a \tau_b$ is achieved for a and b such that their most recent common ancestor is the root 0. Furthermore, argue that in that case $\mathbb{E}_a[\tau_b] = \mathbb{E}_a[\tau_{a,0}]$, where recall that $\tau_{a,0}$ is the commute time between a and 0.
(ii) Use the commute time identity (Theorem 3.3.34) and Matthews' cover time bounds (Theorem 3.1.27) to give an upper bound on the mean cover time of the order of $O(n^2 2^n)$.

Exercise 3.23 (Markov chain tree theorem). Let P be the transition matrix of a finite, irreducible Markov chain with stationary distribution π. Let G be the directed graph corresponding to the positive transitions of P. For an arborescence $\mathcal{A}$ of G, define its weight as

$$\Psi(\mathcal{A}) = \prod_{\vec{e} \in \mathcal{A}} P(\vec{e}).$$

Consider the following process on spanning arborescences over G. Let ρ be the root of the current spanning arborescence $\mathcal{A}$. Pick an outgoing edge $\vec{e} = (\rho, x)$ of ρ according to $P(\rho, \cdot)$. Edge $\vec{e}$ is not in $\mathcal{A}$ by definition of an arborescence. Add $\vec{e}$ to $\mathcal{A}$. This creates a cycle. Remove the edge *of this cycle* originating from x, producing a new arborescence $\mathcal{A}'$ with root x. Repeat the process.

   (i) Show that this chain is irreducible.
   (ii) Show that Ψ is a stationary measure for this chain.
   (iii) Prove the *Markov chain tree theorem*: the stationary distribution π of P is proportional to

$$\pi_x = \sum_{\mathcal{A}\,:\,\mathrm{root}(\mathcal{A})=x} \Psi(\mathcal{A}).$$

Bibliographic Remarks

Section 3.1 Picking up where Appendix B leaves off, Sections 3.1.1 and 3.1.3 largely follow the textbooks [Wil91] and [Dur10], which contain excellent introductions to martingales. The latter also covers Markov chains, and includes the proofs we skipped here. Theorem 3.1.11 is proved in [Dur10, Theorem 4.3.2]. Many more results like Corollary 3.1.24 can be derived from the occupation measure identity; see, for example, [AF, chapter 2]. The upper bound in Theorem 3.1.27 was first proved by Matthews [Mat88].

Section 3.2 The Azuma–Hoeffding inequality is due to Hoeffding [Hoe63] and Azuma [Azu67]. The version of the inequality in Exercise 3.7 is from [Ste97]. The method of bounded differences has its origins in the works of Yurinskii [Yur76], Maurey [Mau79], Milman and Schechtman [MS86], Rhee and Talagrand [RT87], and Shamir and Spencer [SS87]. In its current form, it appears in [McD89]. Example 3.2.11 is taken from [MU05, section 12.5]. The presentation in Section 3.2.3 follows [AS11, section 7.3]. Claim 3.2.16 is due to Shamir and Spencer [SS87]. The 2-point concentration result alluded to in Section 3.2.3 is due to Alon and Krivelevich [AK97]. For the full story on the chromatic number of Erdős–Rényi graphs, see [JLR11, chapter 7]. Claim 3.2.21 is due to Bollobás, Riordan, Spencer, and Tusnády [BRST01]. It confirmed simulations of Barabási and Albert [BA99]. The expectation was analyzed by Dorogovtsev, Mendes, and Samukhin [DMS00]. For much more on preferential attachment models, see [Dur06], [CL06], or [vdH17]. Example 3.2.12 borrows from [BLM13, section 7.1] and [Pet, section 6.3]. General references on the concentration of measure phenomenon and concentration inequalities are [Led01] and [BLM13]. See [BCB12] or [LS20] for an introduction to bandit problems; or [AJKS22] for an introduction to the sample complexity of the more general reinforcement learning problem. The slicing argument in Section 3.2.5 is based on [Bub10]. A more general discussion of the slicing method, whose best known application is the proof of the law of the iterated logarithm (e.g., [Wil91, section 14.7]), can be found in [vH16]. Section 3.2.6 is based on [vH16, section 4.3]. In particular, a proof of Talagrand's inequality (Theorem 3.2.32) can be found there. See also [AS11, chapter 7] or [BLM13, chapter 7].

Section 3.3 Section 3.3.1 is based partly on [Nor98, sections 4.1–2], [Ebe, sections 0.3, 1.1-2, 3.1-2], and [Bre17, sections 7.3, 17.1]. The material in Sections 3.3.2–3.3.5 borrows from [LPW06, chapters 9, 10], [AF, chapters 2, 3], and, especially, [LP16, sections 2.1–2.6, 4.1–4.2, 5.5]. Foster's theorem (Theorem 3.3.12) is from [Fos53]. The classical reference on potential theory and its probabilistic counterpart is [Doo01]. For the discrete case and the electrical network point of view, the book of Doyle and Snell is excellent [DS84]. In particular, the series and parallel laws are defined and illustrated. See also [KSK76]. For an introduction to convex optimization and duality, see, for example, [BV04]. The Nash–Williams inequality is due to Nash–Williams [NW59]. The result in Example 3.3.27 is due to R. Lyons [Lyo90]. Theorem 3.3.33 is due to Kanai [Kan86]. The commute time identity was proved by Chandra, Raghavan, Ruzzo, Smolensky, and Tiwari [CRR+89]. An elementary proof of Pólya's theorem can be found in [Dur10, section 4.2]. The flow we used in the proof of Pólya's theorem is essentially due to T. Lyons [Lyo83]. Wilson's method is due to Wilson [Wil96]. A related method for generating uniform spanning trees was introduced by Aldous [Ald90] and Broder [Bro89]. A connection between loop-erased random walks and uniform spanning trees had previously been established by Pemantle [Pem91] using the Aldous–Broder method. For more on negative correlation in uniform spanning trees, see, for example, [LP16, section 4.2]. For a proof of the matrix tree theorem using Wilson's method, see [KRS]. For a discussion of the running time of Wilson's method and other spanning tree generation approaches, see [Wil96].

4

Coupling

In this chapter we move on to *coupling*, another probabilistic technique with a wide range of applications (far beyond discrete stochastic processes). The idea behind the coupling method is deceptively simple: to compare two probability measures μ and ν, it is sometimes useful to construct a *joint* probability space with marginals μ and ν. For instance, in the classical application of coupling to the convergence of Markov chains (Theorem 1.1.33), one simultaneously constructs *two* copies of a Markov chain – one of which is already at stationarity – and shows that they can be made to coincide after a random amount of time, called the coupling time. We begin in Section 4.1 by defining coupling formally and deriving its connection to the total variation distance through the *coupling inequality*. We illustrate the basic idea on a classical Poisson approximation result, which we apply to the degree sequence of an Erdős–Rényi graph. In Section 4.2, we introduce the concept of *stochastic domination* and some related *correlation inequalities*. We develop a key application in percolation theory. Coupling of Markov chains is the subject of Section 4.3, where it serves as a powerful tool to derive mixing time bounds. Finally, we end in Section 4.4 with the *Chen–Stein method* for Poisson approximations, a technique that applies in particular in some natural settings with dependent variables.

4.1 Background

We begin with some background on coupling. After defining the concept formally and giving a few simple examples, we derive the coupling inequality, which provides a fundamental approach to bounding the distance between two distributions. As an application, we analyze the degree distribution in the Erdős–Rényi graph model. Throughout this chapter, $(S, \mathcal{S})$ is a measurable space. Also we will denote by μ_Z the law of random variable Z.

4.1.1 Basic Definitions

A formal definition of coupling follows. Recall (see Appendix B) that for measurable spaces $(S_1, \mathcal{S}_1)$ $(S_2, \mathcal{S}_2)$, we can consider the product space $(S_1 \times S_2, \mathcal{S}_1 \times \mathcal{S}_2)$ where

$$S_1 \times S_2 := \{(s_1, s_2) \colon s_1 \in S_1, s_2 \in S_2\}$$

is the Cartesian product of S_1 and S_2, and $\mathcal{S}_1 \times \mathcal{S}_2$ is the smallest σ-algebra on $S_1 \times S_2$ containing the rectangles $A_1 \times A_2$ for all $A_1 \in \mathcal{S}_1$ and $A_2 \in \mathcal{S}_2$.

Definition 4.1.1 (Coupling). *Let μ and ν be probability measures on the same measurable space $(S, \mathcal{S})$. A coupling of μ and ν is a probability measure γ on the product space* COUPLING *$(S \times S, \mathcal{S} \times \mathcal{S})$ such that the* marginals *of γ coincide with μ and ν, that is,*

$$\gamma(A \times S) = \mu(A) \quad and \quad \gamma(S \times A) = \nu(A) \qquad \forall A \in \mathcal{S}.$$

For two random variables X and Y taking values in $(S, \mathcal{S})$, a coupling *of X and Y is a joint variable (X', Y') taking values in $(S \times S, \mathcal{S} \times \mathcal{S})$ whose law as a probability measure is a coupling of the laws of X and Y. Note that, under this definition, X and Y need not be defined on the same probability space (but X' and Y' do need to). We also say that (X', Y') is a coupling of μ and ν if the law of (X', Y') is a coupling of μ and ν.*

We give a few examples.

Example 4.1.2 (Coupling of Bernoulli variables). Let X and Y be Bernoulli random variables with parameters $0 \leq q < r \leq 1$, respectively. That is, $\mathbb{P}[X = 1] = q$ and $\mathbb{P}[Y = 1] = r$. Here $S = \{0, 1\}$ and $\mathcal{S} = 2^S$.

- *(Independent coupling)* One coupling of X and Y is (X', Y'), where $X' \overset{\mathrm{d}}{=} X$ and $Y' \overset{\mathrm{d}}{=} Y$ are *independent* of one another. Its law is

$$\left(\mathbb{P}[(X', Y') = (i, j)] \right)_{i, j \in \{0, 1\}} = \begin{pmatrix} (1-q)(1-r) & (1-q)r \\ q(1-r) & qr \end{pmatrix}.$$

- *(Monotone coupling)* Another possibility is to pick U uniformly at random in $[0, 1]$, and set $X'' = \mathbf{1}_{\{U \leq q\}}$ and $Y'' = \mathbf{1}_{\{U \leq r\}}$. Then, (X'', Y'') is a coupling of X and Y with law

$$\left(\mathbb{P}[(X'', Y'') = (i, j)] \right)_{i, j \in \{0, 1\}} = \begin{pmatrix} 1-r & r-q \\ 0 & q \end{pmatrix}. \qquad \blacktriangleleft$$

Example 4.1.3 (Bond percolation: monotonicity). Let $G = (V, E)$ be a countable graph. Denote by $\mathbb{P}_p$ the law of bond percolation (Definition 1.2.1) on G with density p. Let $x \in V$ and assume $0 \leq q < r \leq 1$. Using the monotone coupling in the previous example *on each edge independently* produces a coupling of $\mathbb{P}_q$ and $\mathbb{P}_r$. More precisely:

- Let $\{U_e\}_{e \in E}$ be independent uniforms on $[0, 1]$.
- For $p \in [0, 1]$, let W_p be the set of edges e such that $U_e \leq p$.

Thinking of W_p as specifying the open edges in the percolation process on G under $\mathbb{P}_p$, we see that (W_q, W_r) is a coupling of $\mathbb{P}_q$ and $\mathbb{P}_r$ with the property that $\mathbb{P}[W_q \subseteq W_r] = 1$. Let $\mathcal{C}_x^{(q)}$ and $\mathcal{C}_x^{(r)}$ be the open clusters of x under W_q and W_r, respectively. Because $\mathcal{C}_x^{(q)} \subseteq \mathcal{C}_x^{(r)}$,

$$\begin{aligned}
\theta(q) &:= \mathbb{P}_q[|\mathcal{C}_x| = +\infty] \\
&= \mathbb{P}[|\mathcal{C}_x^{(q)}| = +\infty] \\
&\leq \mathbb{P}[|\mathcal{C}_x^{(r)}| = +\infty] \\
&= \mathbb{P}_r[|\mathcal{C}_x| = +\infty] \\
&= \theta(r).
\end{aligned}$$

(We made this claim in Section 2.2.4.) $\qquad \blacktriangleleft$

Example 4.1.4 (Biased random walk on $\mathbb{Z}$). For $p \in [0,1]$, let $(S_t^{(p)})$ be nearest-neighbor random walk on $\mathbb{Z}$ started at 0 with probability p of jumping to the right and probability $1 - p$ of jumping to the left. (See the gambler's ruin problem in Example 3.1.43.) Assume $0 \leq q < r \leq 1$. Using again the monotone coupling of Bernoulli variables in Example 4.1.2 we produce a coupling of $S^{(q)}$ and $S^{(r)}$.

- Let $(X_i'', Y_i'')_i$ be an infinite sequence of i.i.d. monotone Bernoulli couplings with parameters q and r, respectively.
- Define $(Z_i^{(q)}, Z_i^{(r)}) := (2X_i'' - 1, 2Y_i'' - 1)$. Note that $\mathbb{P}[2X_1'' - 1 = 1] = \mathbb{P}[X_1'' = 1] = q$ and $\mathbb{P}[2X_1'' - 1 = -1] = \mathbb{P}[X_1'' = 0] = 1 - q$, and similarly for Y_i''.
- Let $\hat{S}_t^{(q)} = \sum_{i \leq t} Z_i^{(q)}$ and $\hat{S}_t^{(r)} = \sum_{i \leq t} Z_i^{(r)}$.

Then $(\hat{S}_t^{(q)}, \hat{S}_t^{(r)})$ is a coupling of $(S_t^{(q)}, S_t^{(r)})$ such that $\hat{S}_t^{(q)} \leq \hat{S}_t^{(r)}$ for all t almost surely. In particular, we deduce that for all y and all t,

$$\mathbb{P}[S_t^{(q)} \leq y] = \mathbb{P}[\hat{S}_t^{(q)} \leq y] \geq \mathbb{P}[\hat{S}_t^{(r)} \leq y] = \mathbb{P}[S_t^{(r)} \leq y]. \qquad \blacktriangleleft$$

4.1.2 ▷ *Random Walks: Harmonic Functions on Lattices and Infinite d-Regular Trees*

Let (X_t) be a Markov chain on a finite or countably infinite state space V with transition matrix P and let $\mathbb{P}_x$ be the law of (X_t) started at x. We say that a function $h \colon V \to \mathbb{R}$ is bounded if $\sup_{x \in V} |h(x)| < +\infty$. Recall from Section 3.3 that h is harmonic (with respect to P) on V if

$$h(x) = \sum_{y \in V} P(x,y) h(y) \qquad \forall x \in V.$$

We first give a coupling-based criterion for bounded harmonic functions to be constant. Recall that we treated the finite state-space case (where boundedness is automatic) in Corollary 3.3.3.

Lemma 4.1.5 (Coupling and bounded harmonic functions). *If, for all $y, z \in V$, there is a coupling $((Y_t)_t, (Z_t)_t)$ of $\mathbb{P}_y$ and $\mathbb{P}_z$ such that*

$$\lim_t \mathbb{P}[Y_t \neq Z_t] = 0,$$

then all bounded harmonic functions on V are constant.

Proof Let h be bounded and harmonic on V with $\sup_x |h(x)| = M < +\infty$. Let y, z be any points in V. Then, arguing as in Section 3.3.1, $(h(Y_t))$ and $(h(Z_t))$ are martingales and, in particular,

$$\mathbb{E}[h(Y_t)] = \mathbb{E}[h(Y_0)] = h(y) \quad \text{and} \quad \mathbb{E}[h(Z_t)] = \mathbb{E}[h(Z_0)] = h(z).$$

So by Jensen's inequality (Theorem B.4.15) and the boundedness assumption,

$$\begin{aligned}
|h(y) - h(z)| = |\mathbb{E}[h(Y_t)] - \mathbb{E}[h(Z_t)]| \\
\leq \mathbb{E}\,|h(Y_t) - h(Z_t)| \\
\leq 2M\,\mathbb{P}[Y_t \neq Z_t] \\
\to 0.
\end{aligned}$$

So $h(y) = h(z)$. ∎

Harmonic functions on $\mathbb{Z}^d$ Consider random walk on $\mathbb{L}^d$ for $d \geq 1$. In that case, we show that all bounded harmonic functions are constant.

Theorem 4.1.6 (Bounded harmonic functions on $\mathbb{Z}^d$). *All bounded harmonic functions on $\mathbb{L}^d$ are constant.*

Proof From (3.3.2), h is harmonic with respect to random walk on $\mathbb{L}^d$ if and only if it is harmonic with respect to *lazy* random walk (Definition 1.1.31), that is, the walk that stays put with probability $1/2$ at every step. Let $\mathbb{P}_y$ and $\mathbb{P}_z$ be the laws of lazy random walk on $\mathbb{L}^d$ started at y and z, respectively. We construct a coupling $((Y_t), (Z_t)) = ((Y_t^{(i)})_{i \in [d]}, (Z_t^{(i)})_{i \in [d]})$ of $\mathbb{P}_y$ and $\mathbb{P}_z$ as follows: at time t, pick a coordinate $I \in [d]$ uniformly at random, then

- if $Y_t^{(I)} = Z_t^{(I)}$ then do nothing with probability $1/2$, otherwise pick $W \in \{-1, +1\}$ uniformly at random, set $Y_{t+1}^{(I)} = Z_{t+1}^{(I)} := Z_t^{(I)} + W$, and leave the other coordinates unchanged;
- if instead $Y_t^{(I)} \neq Z_t^{(I)}$, pick $W \in \{-1, +1\}$ uniformly at random, and with probability $1/2$ set $Y_{t+1}^{(I)} := Y_t^{(I)} + W$ and leave Z_t and the other coordinates of Y_t unchanged, or otherwise set $Z_{t+1}^{(I)} := Z_t^{(I)} + W$ and leave Y_t and the other coordinates of Z_t unchanged.

It is straightforward to check that $((Y_t), (Z_t))$ is indeed a coupling of $\mathbb{P}_y$ and $\mathbb{P}_z$. To apply the previous lemma, it remains to bound $\mathbb{P}[Y_t \neq Z_t]$.

The key is to note that, for each coordinate i, the difference $(Y_t^{(i)} - Z_t^{(i)})$ is itself a nearest-neighbor random walk on $\mathbb{Z}$ started at $y^{(i)} - z^{(i)}$ with holding probability (i.e., probability of staying put) $1 - \frac{1}{d}$ – until it hits 0. Simple random walk on $\mathbb{Z}$ is irreducible and recurrent (Theorem 3.3.38). The holding probability does not affect the type of the walk. So $(Y_t^{(i)} - Z_t^{(i)})$ hits 0 in finite time with probability 1. Hence, letting $\tau^{(i)}$ be the first time $Y_t^{(i)} - Z_t^{(i)} = 0$, we have $\mathbb{P}[Y_t^{(i)} \neq Z_t^{(i)}] \leq \mathbb{P}[\tau^{(i)} > t] \to \mathbb{P}[\tau^{(i)} = +\infty] = 0$.

By a union bound,

$$\mathbb{P}[Y_t \neq Z_t] \leq \sum_{i \in [d]} \mathbb{P}[Y_t^{(i)} \neq Z_t^{(i)}] \to 0,$$

as desired. ∎

Exercise 4.1 asks for an example of a non-constant (necessarily unbounded) harmonic function on $\mathbb{Z}^d$.

Harmonic functions on $\mathbb{T}_d$ On trees, the situation is different. Let $\mathbb{T}_d$ be the infinite d-regular tree with root ρ. For $x \in \mathbb{T}_d$, we let T_x be the subtree, rooted at x, of descendants of x.

Theorem 4.1.7 (Bounded harmonic functions on $\mathbb{T}_d$). *For $d \geq 3$, let (X_t) be simple random walk on $\mathbb{T}_d$ and let P be the corresponding transition matrix. Let a be a neighbor of the root and consider the function*

$$h(x) := \mathbb{P}_x[X_t \in T_a \text{ for all but finitely many } t].$$

Then, h is a non-constant, bounded harmonic function on $\mathbb{T}_d$.

Proof The function h is bounded since it is defined as a probability, and by the usual first-step analysis,

$$h(x) = \sum_{y:y \sim x} \frac{1}{d} \mathbb{P}_y[X_t \in T_a \text{ for all but finitely many } t] = \sum_y P(x,y)h(y),$$

so h is harmonic on all of $\mathbb{T}_d$.

Let $b \neq a$ be a neighbor of the root. The key of the proof is the following lemma.

Lemma 4.1.8

$$q := \mathbb{P}_a[\tau_\rho = +\infty] = \mathbb{P}_b[\tau_\rho = +\infty] > 0.$$

Proof The equality of the two probabilities follows by symmetry. To see that $q > 0$, let (Z_t) be simple random walk on $\mathbb{T}_d$ started at a until the walk hits ρ and let L_t be the graph distance between Z_t and the root. Then (L_t) is a biased random walk on $\mathbb{Z}$ started at 1 jumping to the right with probability $1 - \frac{1}{d}$ and jumping to the left with probability $\frac{1}{d}$. The probability that (L_t) hits 0 in finite time is < 1 because $1 - \frac{1}{d} > \frac{1}{2}$ when $d \geq 3$ by the gambler's ruin (Example 3.1.43). ∎

Note that

$$h(\rho) \leq \left(1 - \frac{1}{d}\right)(1 - q) < 1.$$

Indeed, if on the first step the random walk started at ρ moves away from a, an event of probability $1 - \frac{1}{d}$, then it must come back to ρ in finite time to reach T_a. Similarly, by the strong Markov property (Theorem 3.1.8),

$$h(a) = q + (1 - q)h(\rho).$$

Since $h(\rho) \neq 1$ and $q > 0$, this shows that $h(a) > h(\rho)$. So h is not constant. ∎

4.1.3 Total Variation Distance and Coupling Inequality

In the examples of Section 4.1.1, we used coupling to prove monotonicity statements. Coupling is also useful to bound the distance between probability measures. For this, we need the coupling inequality.

Total variation distance Let μ and ν be probability measures on $(S, \mathcal{S})$. Recall the definition of the total variation distance

$$\|\mu - \nu\|_{\mathrm{TV}} := \sup_{A \in \mathcal{S}} |\mu(A) - \nu(A)|.$$

As the next lemma shows in the countable case, the total variation distance can be thought of as an ℓ^1 distance on probability measures as vectors (up to a constant factor).

Lemma 4.1.9 (Alternative definition of total variation distance). *If S is countable, then it holds that*

$$\|\mu - \nu\|_{\text{TV}} = \frac{1}{2} \sum_{x \in S} |\mu(x) - \nu(x)|.$$

Proof Let $E_* := \{x \,:\, \mu(x) \geq \nu(x)\}$. Then, for any $A \subseteq S$, by definition of E_*,

$$\mu(A) - \nu(A) \leq \mu(A \cap E_*) - \nu(A \cap E_*) \leq \mu(E_*) - \nu(E_*).$$

Similarly, we have

$$\begin{aligned}
\nu(A) - \mu(A) &\leq \nu(E_*^c) - \mu(E_*^c) \\
&= (1 - \nu(E_*)) - (1 - \mu(E_*)) \\
&= \mu(E_*) - \nu(E_*).
\end{aligned}$$

The two bounds above are equal so $|\mu(A) - \nu(A)| \leq \mu(E_*) - \nu(E_*)$. Equality is achieved when $A = E_*$. Also,

$$\begin{aligned}
\mu(E_*) - \nu(E_*) &= \frac{1}{2} \left[\mu(E_*) - \nu(E_*) + \nu(E_*^c) - \mu(E_*^c) \right] \\
&= \frac{1}{2} \sum_{x \in S} |\mu(x) - \nu(x)|.
\end{aligned}$$

That concludes the proof. ∎

Like the ℓ^1 distance, the total variation distance is a metric. In particular, it satisfies the triangle inequality.

Lemma 4.1.10 (Total variation distance: triangle inequality). *Let μ, ν, η be probability measures on $(S, \mathcal{S})$. Then,*

$$\|\mu - \nu\|_{\text{TV}} \leq \|\mu - \eta\|_{\text{TV}} + \|\eta - \nu\|_{\text{TV}}.$$

Proof From the definition,

$$\begin{aligned}
\sup_{A \in \mathcal{S}} |\mu(A) - \nu(A)| &\leq \sup_{A \in \mathcal{S}} \{|\mu(A) - \eta(A)| + |\eta(A) - \nu(A)|\} \\
&\leq \sup_{A \in \mathcal{S}} |\mu(A) - \eta(A)| + \sup_{A \in \mathcal{S}} |\eta(A) - \nu(A)|.
\end{aligned} \qquad ∎$$

Coupling inequality We come to an elementary, yet fundamental inequality.

Lemma 4.1.11 (Coupling inequality). *Let μ and ν be probability measures on $(S, \mathcal{S})$. For any coupling (X, Y) of μ and ν,*

$$\|\mu - \nu\|_{\text{TV}} \leq \mathbb{P}[X \neq Y].$$

Proof For any $A \in \mathcal{S}$,

$$\mu(A) - \nu(A) = \mathbb{P}[X \in A] - \mathbb{P}[Y \in A]$$
$$= \mathbb{P}[X \in A, \, X = Y] + \mathbb{P}[X \in A, \, X \neq Y]$$
$$\quad - \mathbb{P}[Y \in A, \, X = Y] - \mathbb{P}[Y \in A, \, X \neq Y]$$
$$= \mathbb{P}[X \in A, \, X \neq Y] - \mathbb{P}[Y \in A, \, X \neq Y]$$
$$\leq \mathbb{P}[X \neq Y],$$

and, similarly, $\nu(A) - \mu(A) \leq \mathbb{P}[X \neq Y]$. Hence,

$$|\mu(A) - \nu(A)| \leq \mathbb{P}[X \neq Y].$$

Taking a supremum over A gives the claim. $\blacksquare$

Here is a quick example.

Example 4.1.12 (A coupling of Poisson random variables). Let $X \sim \text{Poi}(\lambda)$ and $Y \sim \text{Poi}(\nu)$ with $\lambda > \nu$. Recall that a sum of independent Poisson is Poisson (see Exercise 6.7). This fact leads to a natural coupling: let $\hat{Y} \sim \text{Poi}(\nu)$, $\hat{Z} \sim \text{Poi}(\lambda - \nu)$ independently of $\hat{Y}$, and $\hat{X} = \hat{Y} + \hat{Z}$. Then, $(\hat{X}, \hat{Y})$ is a coupling of X and Y, and by the coupling inequality (Lemma 4.1.11),

$$\|\mu_X - \mu_Y\|_{\text{TV}} \leq \mathbb{P}[\hat{X} \neq \hat{Y}] = \mathbb{P}[\hat{Z} > 0] = 1 - e^{-(\lambda - \nu)} \leq \lambda - \nu,$$

where we used $1 - e^{-x} \leq x$ for all x (see Exercise 1.16). ◀

Remarkably, the inequality in Lemma 4.1.11 is tight. For simplicity, we prove this in the finite case only.

Lemma 4.1.13 (Maximal coupling). *Assume S is finite and let $\mathcal{S} = 2^S$. Let μ and ν be probability measures on $(S, \mathcal{S})$. Then,*

$$\|\mu - \nu\|_{\text{TV}} = \inf\{\mathbb{P}[X \neq Y]: \textit{coupling } (X, Y) \textit{ of } \mu \textit{ and } \nu\}.$$

Proof We construct a coupling which achieves equality in the coupling inequality. Such a coupling is called a *maximal coupling*.

MAXIMAL
COUPLING

Let $A = \{x \in S: \mu(x) > \nu(x)\}$, $B = \{x \in S: \mu(x) \leq \nu(x)\}$, and

$$p := \sum_{x \in S} \mu(x) \wedge \nu(x), \quad \alpha := \sum_{x \in A} [\mu(x) - \nu(x)], \quad \beta := \sum_{x \in B} [\nu(x) - \mu(x)].$$

Assume $p > 0$. First, two lemmas. See Figure 4.1 for a proof by picture.

Lemma 4.1.14

$$\sum_{x \in S} \mu(x) \wedge \nu(x) = 1 - \|\mu - \nu\|_{\text{TV}}.$$

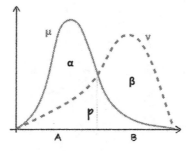

Figure 4.1 Proof by picture that: $1 - p = \alpha = \beta = \|\mu - \nu\|_{\mathrm{TV}}$.

Proof We have

$$
\begin{aligned}
2\|\mu - \nu\|_{\mathrm{TV}} &= \sum_{x \in S} |\mu(x) - \nu(x)| \\
&= \sum_{x \in A} [\mu(x) - \nu(x)] + \sum_{x \in B} [\nu(x) - \mu(x)] \\
&= \sum_{x \in A} \mu(x) + \sum_{x \in B} \nu(x) - \sum_{x \in S} \mu(x) \wedge \nu(x) \\
&= 2 - \sum_{x \in B} \mu(x) - \sum_{x \in A} \nu(x) - \sum_{x \in S} \mu(x) \wedge \nu(x) \\
&= 2 - 2 \sum_{x \in S} \mu(x) \wedge \nu(x),
\end{aligned}
$$

where we used that both μ and ν sum to 1. Rearranging gives the claim. ∎

Lemma 4.1.15

$$
\sum_{x \in A} [\mu(x) - \nu(x)] = \sum_{x \in B} [\nu(x) - \mu(x)] = \|\mu - \nu\|_{\mathrm{TV}} = 1 - p.
$$

Proof The first equality is immediate by the fact that μ and ν are probability measures. The second equality follows from the first one together with the second line in the proof of the previous lemma. The last equality is a restatement of the last lemma. ∎

The maximal coupling is defined as follows:

- With probability p, pick $X = Y$ from $\gamma_{\min}$, where

$$
\gamma_{\min}(x) := \frac{1}{p} \mu(x) \wedge \nu(x), \quad x \in S.
$$

- Otherwise, pick X from γ_A, where

$$
\gamma_A(x) := \frac{\mu(x) - \nu(x)}{1 - p}, \quad x \in A,
$$

and, independently, pick Y from

$$\gamma_B(x) := \frac{\nu(x) - \mu(x)}{1 - p}, \quad x \in B.$$

Note that $X \neq Y$ in that case because A and B are disjoint.

The marginal law of X is: for $x \in A$,

$$p\,\gamma_{\min}(x) + (1 - p)\,\gamma_A(x) = \nu(x) + \mu(x) - \nu(x) = \mu(x),$$

and for $x \in B$,

$$p\,\gamma_{\min}(x) + (1 - p)\,\gamma_A(x) = \mu(x).$$

A similar calculation holds for Y. Finally, $\mathbb{P}[X \neq Y] = 1 - p = \|\mu - \nu\|_{\mathrm{TV}}$. ∎

Remark 4.1.16 *A proof of this result for general Polish spaces can be found in [dH, section 2.5].*

We return to our coupling of Bernoulli variables.

Example 4.1.17 (Coupling of Bernoulli variables (continued)). Recall the setting of Example 4.1.2. To construct the maximal coupling as in Lemma 4.1.13, we note that

$$A := \{0\}, \qquad B := \{1\},$$

$$p := \sum_x \mu(x) \wedge \nu(x) = (1 - r) + q, \qquad 1 - p = \alpha = \beta := r - q,$$

$$(\gamma_{\min}(x))_{x=0,1} = \left(\frac{1 - r}{(1 - r) + q}, \frac{q}{(1 - r) + q} \right),$$

$$\gamma_A(0) := 1, \qquad \gamma_B(1) := 1.$$

The law of the maximal coupling (X''', Y''') is given by

$$\left(\mathbb{P}[(X''', Y''') = (i,j)] \right)_{i,j \in \{0,1\}} = \begin{pmatrix} p\,\gamma_{\min}(0) & (1 - p)\,\gamma_A(0)\gamma_B(1) \\ 0 & p\,\gamma_{\min}(1) \end{pmatrix}$$

$$= \begin{pmatrix} 1 - r & r - q \\ 0 & q \end{pmatrix}.$$

Notice that it happens to coincide with the monotone coupling. ◀

Poisson approximation Here is a classical application of coupling: the approximation of a sum of independent Bernoulli variables with a Poisson. It gives a quantitative bound in total variation distance. Let $X_1, \ldots, X_n$ be independent Bernoulli random variables with parameters $p_1, \ldots, p_n$, respectively. We are interested in the case where the p_is are "small." Let $S_n := \sum_{i \leq n} X_i$. We approximate S_n with a Poisson random variable Z_n as follows: let $W_1, \ldots, W_n$ be independent Poisson random variables with means $\lambda_1, \ldots, \lambda_n$, respectively, and define $Z_n := \sum_{i \leq n} W_i$. We choose $\lambda_i = -\log(1 - p_i)$ for reasons that will become clear below. Note that $Z_n \sim \mathrm{Poi}(\lambda)$, where $\lambda = \sum_{i \leq n} \lambda_i$.

Theorem 4.1.18 (Poisson approximation).

$$\|\mu_{S_n} - \text{Poi}(\lambda)\|_{\text{TV}} \leq \frac{1}{2}\sum_{i\leq n}\lambda_i^2.$$

Proof We couple the pairs (X_i, W_i) independently for $i \leq n$. Let

$$W_i' \sim \text{Poi}(\lambda_i) \quad \text{and} \quad X_i' = W_i' \wedge 1.$$

Because of our choice $\lambda_i = -\log(1 - p_i)$, which implies

$$1 - p_i = \mathbb{P}[X_i = 0] = \mathbb{P}[W_i = 0] = e^{-\lambda_i},$$

(X_i', W_i') is indeed a coupling of (X_i, W_i). Let $S_n' := \sum_{i\leq n} X_i'$ and $Z_n' := \sum_{i\leq n} W_i'$. Then (S_n', Z_n') is a coupling of (S_n, Z_n). By the coupling inequality

$$\|\mu_{S_n} - \mu_{Z_n}\|_{\text{TV}} \leq \mathbb{P}[S_n' \neq Z_n']$$

$$\leq \sum_{i\leq n} \mathbb{P}[X_i' \neq W_i']$$

$$= \sum_{i\leq n} \mathbb{P}[W_i' \geq 2]$$

$$= \sum_{i\leq n}\sum_{j\geq 2} e^{-\lambda_i}\frac{\lambda_i^j}{j!}$$

$$\leq \sum_{i\leq n}\frac{\lambda_i^2}{2}\sum_{\ell\geq 0} e^{-\lambda_i}\frac{\lambda_i^\ell}{\ell!}$$

$$= \sum_{i\leq n}\frac{\lambda_i^2}{2}. \qquad \blacksquare$$

Mappings reduce total variation distance The following lemma will be useful.

Lemma 4.1.19 (Mappings). *Let X and Y be random variables taking values in $(S, \mathcal{S})$, let h be a measurable map from $(S, \mathcal{S})$ to $(S', \mathcal{S}')$, and let $X' := h(X)$ and $Y' := h(Y)$. The following inequality holds*

$$\|\mu_{X'} - \mu_{Y'}\|_{\text{TV}} \leq \|\mu_X - \mu_Y\|_{\text{TV}}.$$

Proof From the definition of the total variation distance, we seek to bound

$$\sup_{A'\in\mathcal{S}'}\left|\mathbb{P}[X' \in A'] - \mathbb{P}[Y' \in A']\right|$$

$$= \sup_{A'\in\mathcal{S}'}\left|\mathbb{P}[h(X) \in A'] - \mathbb{P}[h(Y) \in A']\right|$$

$$= \sup_{A'\in\mathcal{S}'}\left|\mathbb{P}[X \in h^{-1}(A')] - \mathbb{P}[Y \in h^{-1}(A')]\right|.$$

Since $h^{-1}(A') \in \mathcal{S}$ by the measurability of h, this last expression is less than or equal to

$$\sup_{A\in\mathcal{S}}\left|\mathbb{P}[X \in A] - \mathbb{P}[Y \in A]\right|,$$

which proves the claim. $\qquad \blacksquare$

Coupling of Markov chains In the context of Markov chains, a natural way to couple is to do so step by step. We will refer to such couplings as Markovian. An important special case is a Markovian coupling of a chain with itself.

Definition 4.1.20 (Markovian coupling). *Let P and Q be transition matrices on the same state space V. A Markovian coupling of P and Q is a Markov chain $(X_t, Y_t)_t$ on $V \times V$ with transition matrix R satisfying: for all $x, y, x', y' \in V$,*

$$\sum_{z'} R((x, y), (x', z')) = P(x, x'),$$

$$\sum_{z'} R((x, y), (z', y')) = Q(y, y').$$

We will give many examples throughout this chapter. See also Example 4.2.14 for an example of a coupling of Markov chains that is *not* Markovian.

4.1.4 ▷ *Random Graphs: Degree Sequence in Erdős–Rényi Model*

Let $G_n \sim \mathbb{G}_{n,p_n}$ be an Erdős–Rényi graph with $p_n := \frac{\lambda}{n}$ and $\lambda > 0$ (see Definition 1.2.2). For $i \in [n]$, let $D_i(n)$ be the degree of vertex i and define

$$N_d(n) := \sum_{i=1}^{n} \mathbf{1}_{\{D_i(n)=d\}},$$

the number of vertices of degree d.

Theorem 4.1.21 (Erdős–Rényi graph: degree sequence).

$$\frac{1}{n}N_d(n) \to_{\mathrm{p}} f_d := e^{-\lambda}\frac{\lambda^d}{d!} \qquad \forall d \geq 0.$$

Proof We proceed in two steps:

1. We use the coupling inequality (Lemma 4.1.11) to show that the expectation of $\frac{1}{n}N_d(n)$ is close to f_d; and
2. we appeal to Chebyshev's inequality (Theorem 2.1.2) to show that $\frac{1}{n}N_d(n)$ is close to its expectation.

We justify each step as a lemma.

Lemma 4.1.22 (Convergence of the mean).

$$\lim_{n \to +\infty} \frac{1}{n}\mathbb{E}_{n,p_n}[N_d(n)] = f_d \qquad \forall d \geq 1.$$

Proof Note that the degrees $D_i(n)$, $i \in [n]$, are identically distributed (but not independent) so

$$\frac{1}{n}\mathbb{E}_{n,p_n}[N_d(n)] = \mathbb{P}_{n,p_n}[D_1(n) = d].$$

Moreover, by definition, $D_1(n) \sim \text{Bin}(n-1, p_n)$. Let $S_n \sim \text{Bin}(n, p_n)$ and $Z_n \sim \text{Poi}(\lambda)$. Using the Poisson approximation (Theorem 4.1.18) and a Taylor expansion,

$$
\begin{aligned}
\|\mu_{S_n} - \mu_{Z_n}\|_{\text{TV}} &\leq \frac{1}{2} \sum_{i \leq n} (-\log(1 - p_n))^2 \\
&= \frac{1}{2} \sum_{i \leq n} \left(\frac{\lambda}{n} + O(n^{-2}) \right)^2 \\
&= \frac{\lambda^2}{2n} + O(n^{-2}).
\end{aligned}
$$

We can further couple $D_1(n)$ and S_n as

$$
\left(\sum_{i \leq n-1} X_i, \sum_{i \leq n} X_i \right),
$$

where the X_is are i.i.d. $\text{Ber}(p_n)$, that is, Bernoulli with parameter p_n. By the coupling inequality (Theorem 4.1.11),

$$
\|\mu_{D_1(n)} - \mu_{S_n}\|_{\text{TV}} \leq \mathbb{P}\left[\sum_{i \leq n-1} X_i \neq \sum_{i \leq n} X_i \right] = \mathbb{P}[X_n = 1] = p_n = \frac{\lambda}{n}.
$$

By the triangle inequality for the total variation distance (Lemma 4.1.10) and the bounds above,

$$
\begin{aligned}
\frac{1}{2} \sum_{d \geq 0} |\mathbb{P}_{n,p_n}[D_1(n) = d] - f_d| &= \|\mu_{D_1(n)} - \mu_{Z_n}\|_{\text{TV}} \\
&\leq \|\mu_{D_1(n)} - \mu_{S_n}\|_{\text{TV}} + \|\mu_{S_n} - \mu_{Z_n}\|_{\text{TV}} \\
&\leq \frac{\lambda + \lambda^2/2}{n} + O(n^{-2}).
\end{aligned}
$$

Therefore, for all d,

$$
\left| \frac{1}{n} \mathbb{E}_{n,p_n}[N_d(n)] - f_d \right| \leq \frac{2\lambda + \lambda^2}{n} + O(n^{-2}) \to 0,
$$

as $n \to +\infty$. ∎

Lemma 4.1.23 (Concentration around the mean).

$$
\mathbb{P}_{n,p_n}\left[\left| \frac{1}{n} N_d(n) - \frac{1}{n} \mathbb{E}_{n,p_n}[N_d(n)] \right| \geq \varepsilon \right] \leq \frac{2\lambda + 1}{\varepsilon^2 n} \qquad \forall d \geq 1, \forall n.
$$

Proof By Chebyshev's inequality, for all $\varepsilon > 0$,

$$
\mathbb{P}_{n,p_n}\left[\left| \frac{1}{n} N_d(n) - \frac{1}{n} \mathbb{E}_{n,p_n}[N_d(n)] \right| \geq \varepsilon \right] \leq \frac{\text{Var}_{n,p_n}[\frac{1}{n} N_d(n)]}{\varepsilon^2}. \tag{4.1.1}
$$

To compute the variance, we note that

$$\mathrm{Var}_{n,p_n}\left[\frac{1}{n}N_d(n)\right]$$

$$= \frac{1}{n^2}\left\{\mathbb{E}_{n,p_n}\left[\left(\sum_{i\le n}\mathbf{1}_{\{D_i(n)=d\}}\right)^2\right] - (n\,\mathbb{P}_{n,p_n}[D_1(n)=d])^2\right\}$$

$$= \frac{1}{n^2}\Big\{n(n-1)\mathbb{P}_{n,p_n}[D_1(n)=d, D_2(n)=d]$$

$$\quad + n\,\mathbb{P}_{n,p_n}[D_1(n)=d] - n^2\mathbb{P}_{n,p_n}[D_1(n)=d]^2\Big\}$$

$$\le \frac{1}{n} + \Big\{\mathbb{P}_{n,p_n}[D_1(n)=d, D_2(n)=d] - \mathbb{P}_{n,p_n}[D_1(n)=d]^2\Big\}, \qquad (4.1.2)$$

where we used the crude bound $\mathbb{P}_{n,p_n}[D_1(n)=d] \le 1$. We bound the last line using a neat coupling argument. Let Y_1 and Y_2 be independent $\mathrm{Bin}(n-2,p_n)$, and let X_1 and X_2 be independent $\mathrm{Ber}(p_n)$. By separating the contribution of the edge between 1 and 2 from those of edges to other vertices, we see that the joint degrees $(D_1(n), D_2(n))$ have the same distribution as (X_1+Y_1, X_1+Y_2). So the term in curly bracket in (4.1.2) is equal to

$$\mathbb{P}[(X_1+Y_1, X_1+Y_2) = (d,d)] - \mathbb{P}[X_1+Y_1=d]^2$$

$$= \mathbb{P}[(X_1+Y_1, X_1+Y_2) = (d,d)] - \mathbb{P}[(X_1+Y_1, X_2+Y_2) = (d,d)]$$

$$\le \mathbb{P}[(X_1+Y_1, X_1+Y_2) = (d,d),\ (X_1+Y_1, X_2+Y_2) \ne (d,d)]$$

$$= \mathbb{P}[(X_1+Y_1, X_1+Y_2) = (d,d),\ X_2+Y_2 \ne d]$$

$$= \mathbb{P}[X_1 = 0,\ Y_1 = Y_2 = d,\ X_2 = 1]$$

$$\quad + \mathbb{P}[X_1 = 1,\ Y_1 = Y_2 = d-1,\ X_2 = 0]$$

$$\le \mathbb{P}[X_2 = 1] + \mathbb{P}[X_1 = 1]$$

$$= \frac{2\lambda}{n}.$$

Plugging back into (4.1.2) we get $\mathrm{Var}_{n,p_n}\left[\frac{1}{n}N_d(n)\right] \le \frac{2\lambda+1}{n}$, and (4.1.1) gives the claim. ∎

Combining the lemmas concludes the proof of Theorem 4.1.21. ∎

4.2 Stochastic Domination

In comparing two probability measures, a natural relationship is that of "domination." For instance, let $(X_i)_{i=1}^n$ be independent $\mathbb{Z}_+$-valued random variables with

$$\mathbb{P}[X_i \ge 1] \ge p,$$

and let $S = \sum_{i=1}^n X_i$ be their sum. Now consider a separate random variable

$$S_* \sim \mathrm{Bin}(n,p).$$

It is intuitively clear that one should be able to bound S from below by analyzing S_* instead – which may be considerably easier. Indeed, in some sense, S "dominates" S_*, that is, S should have a tendency to be bigger than S_*. One expects more specifically that

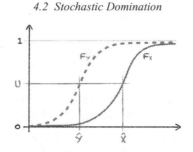

Figure 4.2 The law of X, represented here by its cumulative distribution function F_X in solid, stochastically dominates the law of Y, in dashed. The construction of a monotone coupling, $(\hat{X}, \hat{Y}) := (F_X^{-1}(U), F_Y^{-1}(U))$, where U is uniform in $[0, 1]$, is also depicted.

$$\mathbb{P}[S > x] \geq \mathbb{P}[S_* > x].$$

Coupling provides a formal characterization of this notion, as we detail in this section.

In particular, we study an important special case known as positive associations. Here a measure "dominates itself" in the following sense: conditioning on certain events makes other events more likely. That concept is formalized in Section 4.2.3.

4.2.1 Definitions

We start with the simpler case of real random variables and then consider partially ordered sets, a natural setting for this concept.

Ordering of real random variables Recall that, intuitively, stochastic domination captures the idea that one variable "tends to take larger values" than the other. For real random variables, it is defined in terms of tail probabilities, or equivalently in terms of cumulative distribution functions. See Figure 4.2 for an illustration.

Definition 4.2.1 (Stochastic domination). *Let μ and ν be probability measures on $\mathbb{R}$. The measure μ is said to* stochastically dominate ν, *denoted by $\mu \succeq \nu$, if for all $x \in \mathbb{R}$,*

$$\mu\big[(x, +\infty)\big] \geq \nu\big[(x, +\infty)\big].$$

A real random variable X stochastically dominates Y, denoted by $X \succeq Y$, if the law of X dominates the law of Y.

Example 4.2.2 (Bernoulli vs. Poisson). Let $X \sim \text{Poi}(\lambda)$ be Poisson with mean $\lambda > 0$ and let Y be a Bernoulli trial with success probability $p \in (0, 1)$. In order for X to stochastically dominate Y, we need to have

$$\mathbb{P}[X > \ell] \geq \mathbb{P}[Y > \ell] \qquad \forall \ell \geq 0.$$

This is always true for $\ell \geq 1$ since $\mathbb{P}[X > \ell] > 0$ but $\mathbb{P}[Y > \ell] = 0$. So it remains to consider the case $\ell = 0$. We have

$$1 - e^{-\lambda} = \mathbb{P}[X > 0] \geq \mathbb{P}[Y > 0] = p,$$

if and only if

$$\lambda \geq -\log(1-p).$$

◀

Note that stochastic domination does not require X and Y to be defined on the same probability space. However, the connection to coupling arises from the following characterization.

Theorem 4.2.3 (Coupling and stochastic domination). *The real random variable X stochastically dominates Y if and only if there is a coupling $(\hat{X}, \hat{Y})$ of X and Y such that*

$$\mathbb{P}[\hat{X} \geq \hat{Y}] = 1. \tag{4.2.1}$$

MONOTONE
COUPLING

We refer to $(\hat{X}, \hat{Y})$ as a monotone coupling *of X and Y.*

Proof Suppose there is such a coupling. Then, for all $x \in \mathbb{R}$,

$$\mathbb{P}[Y > x] = \mathbb{P}[\hat{Y} > x] = \mathbb{P}[\hat{X} \geq \hat{Y} > x] \leq \mathbb{P}[\hat{X} > x] = \mathbb{P}[X > x].$$

For the other direction, define the cumulative distribution functions $F_X(x) = \mathbb{P}[X \leq x]$ and $F_Y(x) = \mathbb{P}[Y \leq x]$. Assume $X \succeq Y$. The idea of the proof is to use the following standard way of generating a real random variable (see Theorem B.2.7):

$$X \stackrel{d}{=} F_X^{-1}(U), \tag{4.2.2}$$

where U is a $[0, 1]$-valued uniform random variable and

$$F_X^{-1}(u) := \inf\{x \in \mathbb{R} : F_X(x) \geq u\}$$

is a generalized inverse. It is natural to construct a coupling of X and Y by simply using the same uniform random variable U in this representation, that is, we define $\hat{X} = F_X^{-1}(U)$ and $\hat{Y} = F_Y^{-1}(U)$. See Figure 4.2. By (4.2.2), this is a coupling of X and Y. It remains to check (4.2.1). Because $F_X(x) \leq F_Y(x)$ for all x by definition of stochastic domination, by the definition of the generalized inverse,

$$\mathbb{P}[\hat{X} \geq \hat{Y}] = \mathbb{P}[F_X^{-1}(U) \geq F_Y^{-1}(U)] = 1,$$

as required. ∎

Example 4.2.4 Returning to the example in the first paragraph of Section 4.2, let $(X_i)_{i=1}^n$ be independent $\mathbb{Z}_+$-valued random variables with $\mathbb{P}[X_i \geq 1] \geq p$ and consider their sum $S := \sum_{i=1}^n X_i$. Furthermore, let $S_* \sim \text{Bin}(n, p)$. Write S_* as the sum $\sum_{i=1}^n Y_i$, where (Y_i) are independent Bernoullli variables with $\mathbb{P}[Y_i = 1] = p$. To couple S and S_*, first set $(\hat{Y}_i) := (Y_i)$ and $\hat{S}_* := \sum_{i=1}^n \hat{Y}_i$. Let $\hat{X}_i$ be 0 whenever $\hat{Y}_i = 0$. Otherwise (i.e., if $\hat{Y}_i = 1$), generate $\hat{X}_i$ according to the distribution of X_i conditioned on $\{X_i \geq 1\}$, independently of everything else. By construction, $\hat{X}_i \geq \hat{Y}_i$ almost surely for all i and as a result $\sum_{i=1}^n \hat{X}_i =: \hat{S} \geq \hat{S}_*$ almost surely, or $S \succeq S_*$ by Theorem 4.2.3. That implies for instance that $\mathbb{P}[S > x] \geq \mathbb{P}[S_* > x]$ as we claimed earlier. A slight modification of this argument gives the following useful fact about binomials:

$$n \geq m, \ q \geq p \implies \text{Bin}(n, q) \succeq \text{Bin}(m, p).$$

Exercise 4.2 asks for a formal proof. ◀

Example 4.2.5 (Poisson distribution). Let $X \sim \text{Poi}(\mu)$ and $Y \sim \text{Poi}(\nu)$ with $\mu > \nu$. Recall that a sum of independent Poisson is Poisson (see Exercise 6.7). This fact leads to a natural coupling: let $\hat{Y} \sim \text{Poi}(\nu)$, $\hat{Z} \sim \text{Poi}(\mu - \nu)$ independently of Y, and $\hat{X} = \hat{Y} + \hat{Z}$. Then $(\hat{X}, \hat{Y})$ is a coupling and $\hat{X} \geq \hat{Y}$ a.s. because $\hat{Z} \geq 0$. Hence, $X \succeq Y$. ◀

We record two useful consequences of Theorem 4.2.3.

Corollary 4.2.6 *Let X and Y be real random variables with $X \succeq Y$ and let $f : \mathbb{R} \to \mathbb{R}$ be a non-decreasing function. Then, $f(X) \succeq f(Y)$ and furthermore, provided $\mathbb{E}|f(X)|, \mathbb{E}|f(Y)| < +\infty$, we have that*

$$\mathbb{E}[f(X)] \geq \mathbb{E}[f(Y)].$$

Proof Let $(\hat{X}, \hat{Y})$ be the monotone coupling of X and Y whose existence is guaranteed by Theorem 4.2.3. Then, $f(\hat{X}) \geq f(\hat{Y})$ almost surely so that, provided the expectations exist,

$$\mathbb{E}[f(X)] = \mathbb{E}[f(\hat{X})] \geq \mathbb{E}[f(\hat{Y})] = \mathbb{E}[f(Y)],$$

and furthermore $(f(\hat{X}), f(\hat{Y}))$ is a monotone coupling of $f(X)$ and $f(Y)$. Hence, $f(X) \succeq f(Y)$. ∎

Corollary 4.2.7 *Let X_1, X_2 be independent random variables. Let Y_1, Y_2 be independent random variables such that $X_i \succeq Y_i$, $i = 1, 2$. Then,*

$$X_1 + X_2 \succeq Y_1 + Y_2.$$

Proof Let $(\hat{X}_1, \hat{Y}_1)$ and $(\hat{X}_2, \hat{Y}_2)$ be independent, monotone couplings of (X_1, Y_1) and (X_2, Y_2) on the same probability space. Then,

$$X_1 + X_2 \overset{d}{=} \hat{X}_1 + \hat{X}_2 \geq \hat{Y}_1 + \hat{Y}_2 \overset{d}{=} Y_1 + Y_2.$$

∎

Example 4.2.8 (Binomial vs. Poisson). A sum of n independent Poisson variables with mean λ is $\text{Poi}(n\lambda)$. A sum of n independent Bernoulli trials with success probability p is $\text{Bin}(n, p)$. Using Example 4.2.2 and Corollary 4.2.7, we get

$$\lambda \geq -\log(1 - p) \quad \Longrightarrow \quad \text{Poi}(n\lambda) \succeq \text{Bin}(n, p). \tag{4.2.3}$$

The following special case will be useful later. Let $0 < \Lambda < 1$ and let m be a positive integer. Then,

$$\frac{\Lambda}{m-1} \geq \frac{\Lambda}{m-\Lambda} = \frac{m}{m-\Lambda} - 1 \geq \log\left(\frac{m}{m-\Lambda}\right) = -\log\left(1 - \frac{\Lambda}{m}\right),$$

where we used that $\log x \leq x - 1$ for all $x \in \mathbb{R}_+$ (see Exercise 1.16). So, setting $\lambda := \frac{\Lambda}{m-1}$, $p := \frac{\Lambda}{m}$, and $n := m - 1$ in (4.2.3), we get

$$\Lambda \in (0, 1) \quad \Longrightarrow \quad \text{Poi}(\Lambda) \succeq \text{Bin}\left(m - 1, \frac{\Lambda}{m}\right). \tag{4.2.4}$$

◀

Ordering on partially ordered sets The definition of stochastic domination hinges on the totally ordered nature of $\mathbb{R}$. It also extends naturally to posets. Let $(\mathcal{X}, \leq)$ be a *poset*, that is,   POSET
for all $x, y, z \in \mathcal{X}$:

- *(Reflexivity)* $x \leq x$;
- *(Antisymmetry)* if $x \leq y$ and $y \leq x$, then $x = y$; and
- *(Transitivity)* if $x \leq y$ and $y \leq z$, then $x \leq z$.

For instance the set $\{0, 1\}^F$ is a poset when equipped with the relation $\boldsymbol{x} \leq \boldsymbol{y}$ if and only if $x_i \leq y_i$ for all $i \in F$, where $\boldsymbol{x} = (x_i)_{i \in F}$ and $\boldsymbol{y} = (y_i)_{i \in F}$. Equivalently, the subsets of F, denoted by 2^F, form a poset with the inclusion relation.

A *totally ordered* set satisfies in addition that, for any x, y, we have either $x \leq y$ or $y \leq x$. That is not satisfied in the previous example.

INCREASING   Let $\mathcal{F}$ be a σ-algebra over the poset $\mathcal{X}$. An event $A \in \mathcal{F}$ is *increasing* if $x \in A$ implies that any $y \geq x$ is also in A. A function $f : \mathcal{X} \to \mathbb{R}$ is *increasing* if $x \leq y$ implies $f(x) \leq f(y)$. Some properties of increasing events are derived in Exercise 4.4.

Definition 4.2.9 (Stochastic domination for posets). *Let $(\mathcal{X}, \leq)$ be a poset and let $\mathcal{F}$ be a σ-algebra on $\mathcal{X}$. Let μ and ν be probability measures on $(\mathcal{X}, \mathcal{F})$. The measure μ is said to stochastically dominate ν, denoted by $\mu \succeq \nu$, if for all increasing $A \in \mathcal{F}$,*

$$\mu(A) \geq \nu(A).$$

An $\mathcal{X}$-valued random variable X stochastically dominates Y, denoted by $X \succeq Y$, if the law of X dominates the law of Y.

As before, a *monotone coupling* $(\hat{X}, \hat{Y})$ of X and Y is one which satisfies $\hat{X} \geq \hat{Y}$ almost surely.

Example 4.2.10 (Monotonicity of the percolation function). We have already seen an example of stochastic domination in Section 2.2.4. We revisit this example now to illustrate our definitions. Consider bond percolation on the d-dimensional lattice $\mathbb{L}^d$ (Definition 1.2.1). Here the poset is the collection of all subsets of edges, specifying the open edges, with the inclusion relation. Recall that the percolation function is given by

$$\theta(p) := \mathbb{P}_p[|\mathcal{C}_0| = +\infty],$$

where $\mathcal{C}_0$ is the open cluster of the origin. We argued in Section 2.2.4 that $\theta(p)$ is non-decreasing by considering the following alternative representation of the percolation process under $\mathbb{P}_p$: to each edge e, assign a uniform $[0, 1]$-valued random variable U_e and declare the edge open if $U_e \leq p$. Using the same U_es for two different values of p, say $p_1 < p_2$, gives a monotone coupling of the processes for p_1 and p_2. It follows immediately that $\theta(p_1) \leq \theta(p_2)$, where we used that the event $\{|\mathcal{C}_0| = +\infty\}$ is increasing.   ◄

The existence of a monotone coupling is perhaps more surprising for posets. We prove the result in the finite case only, which will be enough for our purposes.

Theorem 4.2.11 (Strassen's theorem). *Let X and Y be random variables taking values in a finite poset $(\mathcal{X}, \leq)$ with the σ-algebra $\mathcal{F} = 2^{\mathcal{X}}$. Then, $X \succeq Y$ if and only if there exists a monotone coupling $(\hat{X}, \hat{Y})$ of X and Y.*

Proof Suppose there is such a coupling. Then for all increasing A,

$$\mathbb{P}[Y \in A] = \mathbb{P}[\hat{Y} \in A] = \mathbb{P}[\hat{X} \geq \hat{Y} \in A] \leq \mathbb{P}[\hat{X} \in A] = \mathbb{P}[X \in A].$$

The proof in the other direction relies on the max-flow min-cut theorem (Theorem 1.1.15). To see the connection with flows, let μ_X and μ_Y be the laws of X and Y, respectively, and denote by ν their joint distribution under the desired coupling. Noting that we want $\nu(x, y) > 0$ only if $x \geq y$, the marginal conditions on the coupling read

$$\sum_{y \leq x} \nu(x, y) = \mu_X(x) \qquad \forall x \in \mathcal{X} \qquad (4.2.5)$$

and

$$\sum_{x \geq y} \nu(x, y) = \mu_Y(y), \qquad \forall y \in \mathcal{X}. \qquad (4.2.6)$$

These equations can be interpreted as flow-conservation constraints. Consider the following directed graph. There are two vertices, $(w, 1)$ and $(w, 2)$, for each element w in $\mathcal{X}$ with edges connecting each $(x, 1)$ to those $(y, 2)$s with $x \geq y$. These edges have capacity $+\infty$. In addition, there is a source a and a sink z. The source has a directed edge of capacity $\mu_X(x)$ to $(x, 1)$ for each $x \in \mathcal{X}$ and, similarly, each $(y, 2)$ has a directed edge of capacity $\mu_Y(y)$ to the sink. The existence of a monotone coupling will follow once we show that there is a flow of strength 1 between a and z. Indeed, in that case, all edges from the source and all edges to the sink are at capacity. If we let $\nu(x, y)$ be the flow on edge $\langle (x, 1), (y, 2) \rangle$, the systems in (4.2.5) and (4.2.6) encode conservation of flow on the vertices $(\mathcal{X} \times \{1\}) \cup (\mathcal{X} \times \{2\})$. Hence, the flow between $\mathcal{X} \times \{1\}$ and $\mathcal{X} \times \{2\}$ yields the desired coupling. See Figure 4.3.

By the max-flow min-cut theorem (Theorem 1.1.15), it suffices to show that a minimum cut has capacity 1. Such a cut is of course obtained by choosing all edges out of the source. So it remains to show that no cut has capacity less than 1. This is where we use the fact that $\mu_X(A) \geq \mu_Y(A)$ for all increasing A. Because the edges between $\mathcal{X} \times \{1\}$ and $\mathcal{X} \times \{2\}$ have infinite capacity, they cannot be used in a minimum cut. So we can restrict our attention to those cuts containing edges from a to $A_* \times \{1\}$ and from $Z_* \times \{2\}$ to z for subsets $A_*, Z_* \subseteq \mathcal{X}$.

We must have

$$A_* \supseteq \{x \in \mathcal{X} : \exists y \in Z_*^c, \, x \geq y\}$$

to block all paths of the form $a \sim (x, 1) \sim (y, 2) \sim z$ with x and y as in the previous display. In fact, for a minimum cut, we further have

$$A_* = \{x \in \mathcal{X} : \exists y \in Z_*^c, \, x \geq y\},$$

as adding an x not satisfying this property is redundant. In particular, A_* is increasing: if $x_1 \in A_*$ and $x_2 \geq x_1$, then $\exists y \in Z_*^c$ such that $x_1 \geq y$ and, since $x_2 \geq x_1 \geq y$, we also have $x_2 \in A_*$.

Observe further that, because $y \geq y$, the set A_* also includes Z_*^c. If it were the case that $A_* \neq Z_*^c$, then we could construct a cut with lower or equal capacity by fixing A_* and setting $Z_* := A_*^c$: suppose $A_* \cap Z_*$ is non-empty; because A_* is increasing, any $y \in A_* \cap Z_*$ is such that paths of the form $a \sim (x, 1) \sim (y, 2) \sim z$ with $x \geq y$ are cut by $x \in A_*$; so we do not

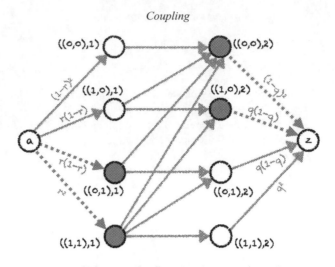

Figure 4.3 Construction of a monotone coupling through the max-flow representation for independent Bernoulli pairs with parameters r (on the left) and $q < r$ (on the right). Edge labels indicate capacity. Edges without labels have infinite capacity. The dotted edges depict a suboptimal cut. The dark vertices correspond to the sets A_* and Z_* for this cut. The capacity of the cut is $r^2 + r(1-r) + (1-q)^2 + (1-q)q = r + (1-q) > r + (1-r) = 1$.

need those ys in Z_*. Hence, for a minimum cut, we can assume that in fact $A_* = Z_*^c$. The capacity of the cut is

$$\mu_X(A_*) + \mu_Y(Z_*) = \mu_X(A_*) + 1 - \mu_Y(A_*) = 1 + (\mu_X(A_*) - \mu_Y(A_*)) \geq 1,$$

where the term in parenthesis is non-negative by assumption and the fact that A_* is increasing. That concludes the proof. ∎

Remark 4.2.12 *Strassen's theorem (Theorem 4.2.11) holds more generally on Polish spaces with a closed partial order. See, for example, [Lin02, section IV.1.2] for the details.*

The proof of Corollary 4.2.6 immediately extends to:

Corollary 4.2.13 *Let X and Y be $\mathcal{X}$-valued random variables with $X \succeq Y$ and let $f : \mathcal{X} \to \mathbb{R}$ be an increasing function. Then, $f(X) \succeq f(Y)$ and furthermore, provided $\mathbb{E}|f(X)|, \mathbb{E}|f(Y)| < +\infty$, we have that*

$$\mathbb{E}[f(X)] \geq \mathbb{E}[f(Y)].$$

Ordering of Markov chains Stochastic domination also arises in the context of Markov chains. We begin with an example. Recall the notion of a Markovian coupling from Definition 4.1.20. The following coupling of Markov chains is *not* Markovian.

Example 4.2.14 (Lazier chain). Consider a random walk (X_t) on the network $\mathcal{N} = ((V, E), c)$, where $V = \{0, 1, \ldots, n\}$ and $i \sim j$ if and only if $|i - j| \leq 1$ (including self-loops). Let $\mathcal{N}' = ((V, E), c')$ be a modified version of $\mathcal{N}$ on the same graph where, for all i, $c(i, i) \leq c'(i, i)$.

That is, if (X_t') is random walk on $\mathcal{N}'$, then (X_t') is "lazier" than (X_t) in that it is more likely to stay put. To simplify the calculations, assume $c(i,i) = 0$ for all i.

Assume that both (X_t) and (X_t') start at i_0 and define $M_s := \max_{t \le s} X_t$ and $M_s' := \max_{t \le s} X_t'$. Since (X_t') "travels less" than (X_t), the following claim is intuitively obvious.

Claim 4.2.15

$$M_s \succeq M_s'.$$

We prove this by producing a monotone coupling. First set $(\hat{X}_t)_{t \in \mathbb{Z}_+} := (X_t)_{t \in \mathbb{Z}_+}$. We then generate $(\hat{X}_t')_{t \in \mathbb{Z}_+}$ as a "sticky" version of $(\hat{X}_t)_{t \in \mathbb{Z}_+}$. That is, $(\hat{X}_t')$ follows exactly the same transitions as $(\hat{X}_t)$ (including the self-loops), but at each time it opts to stay where it currently is, say state j, for an extra time step with probability

$$\alpha_j := \frac{c'(j,j)}{\sum_{i:i \sim j} c'(i,j)},$$

which is in $[0,1]$ by assumption. Marginally, $(\hat{X}_t')$ is a random walk on $\mathcal{N}'$. Indeed, we have by construction of the coupling that the probability of staying put when in state j is

$$\alpha_j = \frac{c'(j,j)}{\sum_{i:i \sim j} c'(i,j)},$$

and, for $k \ne j$ with $k \sim j$, the probability of moving to state k when in state j is

$$
\begin{aligned}
(1 - \alpha_j) \frac{c(j,k)}{\sum_{i:i \sim j} c(i,j)} &= \left(\frac{[\sum_{i:i \sim j} c'(i,j)] - c'(j,j)}{\sum_{i:i \sim j} c'(i,j)} \right) \frac{c(j,k)}{\sum_{i:i \sim j} c(i,j)} \\
&= \left(\frac{\sum_{i:i \sim j} c(i,j)}{\sum_{i:i \sim j} c'(i,j)} \right) \frac{c'(j,k)}{\sum_{i:i \sim j} c(i,j)} \\
&= \frac{c'(j,k)}{\sum_{i:i \sim j} c'(i,j)},
\end{aligned}
$$

where, on the second line, we used that $c'(i,j) = c(i,j)$ for $i \ne j$ and $i \sim j$. This coupling satisfies almost surely

$$\widehat{M}_s := \max_{t \le s} \hat{X}_t \ge \max_{t \le s} \hat{X}_t' =: \widehat{M}_s'$$

because $(\hat{X}_t')_{t \le s}$ visits a subset of the states visited by $(\hat{X}_t)_{t \le s}$. In other words, $(\widehat{M}_s, \widehat{M}_s')$ is a monotone coupling of (M_s, M_s') and this proves the claim. ◄

As we indicated, the previous example involved an "asynchronous" coupling of the chains. Often a simpler step-by-step approach – that is, through the construction of a Markovian coupling – is possible. We specialize the notion of stochastic domination to that important case.

Definition 4.2.16 (Stochastic domination of Markov chains). *Let P and Q be transition matrices on a finite or countably infinite poset $(\mathcal{X}, \le)$. The transition matrix Q is said to stochastically dominate the transition matrix P if*

$$x \leq y \implies P(x, \cdot) \preceq Q(y, \cdot). \tag{4.2.7}$$

If the above condition is satisfied for $P = Q$, we say that P is stochastically monotone.

The analogue of Strassen's theorem in this case is the following theorem, which we prove in the finite case only again.

Theorem 4.2.17 *Let $(X_t)_{t \in \mathbb{Z}_+}$ and $(Y_t)_{t \in \mathbb{Z}_+}$ be Markov chains on a finite poset $(\mathcal{X}, \leq)$ with transition matrices P and Q, respectively. Assume that Q stochastically dominates P. Then, for all $x_0 \leq y_0$ there is a coupling $(\hat{X}_t, \hat{Y}_t)$ of (X_t) started at x_0 and (Y_t) started at y_0 such that almost surely*

$$\hat{X}_t \leq \hat{Y}_t \qquad \forall t.$$

Furthermore, if the chains are irreducible and have stationary distributions π and μ, respectively, then $\pi \preceq \mu$.

Observe that for a Markovian, monotone coupling to exist, it is not generally enough for the weaker condition $P(x, \cdot) \preceq Q(x, \cdot)$ to hold for all x, as should be clear from the proof. See also Exercise 4.3.

Proof of Theorem 4.2.17   Let

$$\mathcal{W} := \{(x, y) \in \mathcal{X} \times \mathcal{X} : x \leq y\}.$$

For all $(x, y) \in \mathcal{W}$, let $R((x, y), \cdot)$ be the joint law of a monotone coupling of $P(x, \cdot)$ and $Q(y, \cdot)$. Such a coupling exists by Strassen's theorem and Condition (4.2.7). Let $(\hat{X}_t, \hat{Y}_t)$ be a Markov chain on $\mathcal{W}$ with transition matrix R started at (x_0, y_0). By construction, $\hat{X}_t \leq \hat{Y}_t$ for all t almost surely. That proves the first half of the theorem.

For the second half, let A be increasing on $\mathcal{X}$. Note that the first half implies that for all $s \geq 1$,

$$P^s(x_0, A) = \mathbb{P}[\hat{X}_s \in A] \leq \mathbb{P}[\hat{Y}_s \in A] = Q^s(y_0, A),$$

because $\hat{X}_s \leq \hat{Y}_s$ and A is increasing. Then, by a standard convergence result for irreducible Markov chains (i.e., (1.1.5)),

$$\pi(A) = \lim_{t \to +\infty} \frac{1}{t} \sum_{s \leq t} P^s(x_0, A) \leq \lim_{t \to +\infty} \frac{1}{t} \sum_{s \leq t} Q^s(y_0, A) = \mu(A).$$

This proves the claim by definition of stochastic domination. ∎

An example of application of this theorem is given in the next subsection.

4.2.2 Ising Model: Boundary Conditions

Consider the d-dimensional lattice $\mathbb{L}^d$. Let Λ be a finite subset of vertices in $\mathbb{L}^d$ and define $\mathcal{X} := \{-1, +1\}^\Lambda$, which is a poset when equipped with the relation $\sigma \leq \sigma'$ if and only if $\sigma_i \leq \sigma'_i$ for all $i \in \Lambda$. Generalizing Example 1.2.5, for $\xi \in \{-1, +1\}^{\mathbb{L}^d}$, the (ferromagnetic) Ising model on Λ with *boundary conditions* ξ and inverse temperature β is the probability distribution over spin configurations $\sigma \in \mathcal{X}$ given by

BOUNDARY
CONDITIONS

$$\mu_{\beta,\Lambda}^{\xi}(\sigma) := \frac{1}{\mathcal{Z}_{\Lambda,\xi}(\beta)} e^{-\beta \mathcal{H}_{\Lambda,\xi}(\sigma)},$$

where

$$\mathcal{H}_{\Lambda,\xi}(\sigma) := -\sum_{\substack{i \sim j \\ i,j \in \Lambda}} \sigma_i \sigma_j - \sum_{\substack{i \sim j \\ i \in \Lambda, j \notin \Lambda}} \sigma_i \xi_j$$

is the Hamiltonian and

$$\mathcal{Z}_{\Lambda,\xi}(\beta) := \sum_{\sigma \in \mathcal{X}} e^{-\beta \mathcal{H}_{\Lambda,\xi}(\sigma)}$$

is the partition function. For shorthand, we occasionally write $+$ and $-$ instead of $+1$ and -1.

For the all-$(+1)$ and all-(-1) boundary conditions we denote the measure above by $\mu_{\beta,\Lambda}^{+}(\sigma)$ and $\mu_{\beta,\Lambda}^{-}(\sigma)$, respectively. In this section, we show that these two measures are "extreme" in the following sense.

Claim 4.2.18 *For all boundary conditions $\xi \in \{-1,+1\}^{\mathbb{L}^d}$,*

$$\mu_{\beta,\Lambda}^{+} \succeq \mu_{\beta,\Lambda}^{\xi} \succeq \mu_{\beta,\Lambda}^{-}.$$

Intuitively, because the ferromagnetic Ising model favors spin agreement, the all-$(+1)$ boundary condition tends to produce more $+1$s, which in turn makes increasing events more likely. And vice versa.

The idea of the proof is to use Theorem 4.2.17 with a suitable choice of Markov chain.

Stochastic domination In this context, vertices are often referred to as sites. Adapting Definition 1.2.8, we consider the single-site Glauber dynamics, which is the Markov chain on $\mathcal{X}$ which, at each time, selects a site $i \in \Lambda$ uniformly at random and updates the spin σ_i according to $\mu_{\beta,\Lambda}^{\xi}(\sigma)$ conditioned on agreeing with σ at all sites in $\Lambda \setminus \{i\}$. Specifically, for $\gamma \in \{-1,+1\}$, $i \in \Lambda$, and $\sigma \in \mathcal{X}$, let $\sigma^{i,\gamma}$ be the configuration σ with the state at i being set to γ. Then, letting $n = |\Lambda|$, the transition matrix of the Glauber dynamics is

$$Q_{\beta,\Lambda}^{\xi}(\sigma, \sigma^{i,\gamma}) := \frac{1}{n} \cdot \frac{e^{\gamma \beta S_i^{\xi}(\sigma)}}{e^{-\beta S_i^{\xi}(\sigma)} + e^{\beta S_i^{\xi}(\sigma)}},$$

where

$$S_i^{\xi}(\sigma) := \sum_{\substack{j:j \sim i \\ j \in \Lambda}} \sigma_j + \sum_{\substack{j:j \sim i \\ j \notin \Lambda}} \xi_j.$$

All other transitions have probability 0. It is straightforward to check that $Q_{\beta,\Lambda}^{\xi}$ is a stochastic matrix.

This chain is clearly irreducible. It is also reversible with respect to $\mu_{\beta,\Lambda}^{\xi}$. Indeed, for all $\sigma \in \mathcal{X}$ and $i \in \Lambda$, let

$$S_{\neq i}^{\xi}(\sigma) := \mathcal{H}_{\Lambda,\xi}(\sigma^{i,+}) + S_i^{\xi}(\sigma) = \mathcal{H}_{\Lambda,\xi}(\sigma^{i,-}) - S_i^{\xi}(\sigma).$$

Arguing as in Theorem 1.2.9, we have

$$
\begin{aligned}
\mu_{\beta,\Lambda}^{\xi}(\sigma^{i,-})\,Q_{\beta,\Lambda}^{\xi}(\sigma^{i,-},\sigma^{i,+}) &= \frac{e^{-\beta S_{\neq i}^{\xi}(\sigma)}e^{-\beta S_i^{\xi}(\sigma)}}{\mathcal{Z}_{\Lambda,\xi}(\beta)}\cdot\frac{e^{\beta S_i^{\xi}(\sigma)}}{n[e^{-\beta S_i^{\xi}(\sigma)}+e^{\beta S_i^{\xi}(\sigma)}]} \\
&= \frac{e^{-\beta S_{\neq i}^{\xi}(\sigma)}}{n\mathcal{Z}_{\Lambda,\xi}(\beta)[e^{-\beta S_i^{\xi}(\sigma)}+e^{\beta S_i^{\xi}(\sigma)}]} \\
&= \frac{e^{-\beta S_{\neq i}^{\xi}(\sigma)}e^{\beta S_i^{\xi}(\sigma)}}{\mathcal{Z}_{\Lambda,\xi}(\beta)}\cdot\frac{e^{-\beta S_i^{\xi}(\sigma)}}{n[e^{-\beta S_i^{\xi}(\sigma)}+e^{\beta S_i^{\xi}(\sigma)}]} \\
&= \mu_{\beta,\Lambda}^{\xi}(\sigma^{i,+})\,Q_{\beta,\Lambda}^{\xi}(\sigma^{i,+},\sigma^{i,-}).
\end{aligned}
$$

In particular, $\mu_{\beta,\Lambda}^{\xi}$ is the stationary distribution of $Q_{\beta,\Lambda}^{\xi}$.

Claim 4.2.19

$$
\xi' \geq \xi \;\Longrightarrow\; Q_{\beta,\Lambda}^{\xi'} \text{ stochastically dominates } Q_{\beta,\Lambda}^{\xi}. \tag{4.2.8}
$$

Proof   Because the Glauber dynamics updates a single site at a time, establishing stochastic domination reduces to checking simple one-site inequalities.

Lemma 4.2.20 *To establish* (4.2.8), *it suffices to show that for all i and all $\sigma \leq \tau$,*

$$
Q_{\beta,\Lambda}^{\xi}(\sigma,\sigma^{i,+}) \leq Q_{\beta,\Lambda}^{\xi'}(\tau,\tau^{i,+}). \tag{4.2.9}
$$

Proof   Assume (4.2.9) holds. Let A be increasing in $\mathcal{X}$ and let $\sigma \leq \tau$. Then, for the single-site Glauber dynamics, we have

$$
Q_{\beta,\Lambda}^{\xi}(\sigma,A) = Q_{\beta,\Lambda}^{\xi}(\sigma,A\cap B_{\sigma}), \tag{4.2.10}
$$

where

$$
B_{\sigma} := \{\sigma^{i,\gamma} : i \in \Lambda,\ \gamma \in \{-1,+1\}\},
$$

and similarly for τ, ξ'. Moreover, because A is increasing and $\tau \geq \sigma$,

$$
\sigma^{i,\gamma} \in A \;\Longrightarrow\; \tau^{i,\gamma} \in A \tag{4.2.11}
$$

and

$$
\sigma^{i,-} \in A \;\Longrightarrow\; \sigma^{i,+} \in A. \tag{4.2.12}
$$

Letting

$$
I_{\sigma,A}^{\pm} := \{i \in \Lambda : \sigma^{i,-} \in A\}, \qquad I_{\sigma,A}^{+} := \{i \in \Lambda : \sigma^{i,+} \in A\},
$$

and similarly for τ, we have by (4.2.9), (4.2.10), (4.2.11), and (4.2.12),

$$
\begin{aligned}
Q_{\beta,\Lambda}^{\xi}(\sigma, A) &= Q_{\beta,\Lambda}^{\xi}(\sigma, A \cap B_\sigma) \\
&= \sum_{i \in I_{\sigma,A}^+} Q_{\beta,\Lambda}^{\xi}(\sigma, \sigma^{i,+}) + \sum_{i \in I_{\sigma,A}^\pm} \left[Q_{\beta,\Lambda}^{\xi}(\sigma, \sigma^{i,-}) + Q_{\beta,\Lambda}^{\xi}(\sigma, \sigma^{i,+}) \right] \\
&\leq \sum_{i \in I_{\sigma,A}^+} Q_{\beta,\Lambda}^{\xi'}(\tau, \tau^{i,+}) + \sum_{i \in I_{\sigma,A}^\pm} \frac{1}{n} \\
&\leq \sum_{i \in I_{\tau,A}^+} Q_{\beta,\Lambda}^{\xi'}(\tau, \tau^{i,+}) + \sum_{i \in I_{\tau,A}^\pm} \left[Q_{\beta,\Lambda}^{\xi'}(\tau, \tau^{i,-}) + Q_{\beta,\Lambda}^{\xi'}(\tau, \tau^{i,+}) \right] \\
&= Q_{\beta,\Lambda}^{\xi'}(\tau, A),
\end{aligned}
$$

as claimed. ∎

Returning to the proof of Claim 4.2.19, observe that

$$
Q_{\beta,\Lambda}^{\xi}(\sigma, \sigma^{i,+}) = \frac{1}{n} \cdot \frac{e^{\beta S_i^{\xi}(\sigma)}}{e^{-\beta S_i^{\xi}(\sigma)} + e^{\beta S_i^{\xi}(\sigma)}} = \frac{1}{n} \cdot \frac{1}{e^{-2\beta S_i^{\xi}(\sigma)} + 1},
$$

which is increasing in $S_i^{\xi}(\sigma)$. Now $\sigma \leq \tau$ and $\xi \leq \xi'$ imply that $S_i^{\xi}(\sigma) \leq S_i^{\xi'}(\tau)$. That proves the claim by Lemma 4.2.20. ∎

Finally:

Proof of Claim 4.2.18 Combining Theorem 4.2.17 and Claim 4.2.19 gives the result. ∎

Remark 4.2.21 *One can make sense of the limit of $\mu_{\beta,\Lambda}^+$ and $\mu_{\beta,\Lambda}^-$ when $|\Lambda| \to +\infty$, which is known as an infinite-volume Gibbs measure. For more, see, for example, [RAS15, chapters 7–10].*

Observe that we have not used any special property of the d-dimensional lattice. Indeed, Claim 4.2.18 in fact holds for any countable, locally finite graph with positive coupling constants. We give another proof in Example 4.2.33.

4.2.3 Correlation Inequalities: FKG and Holley's Inequalities

A special case of stochastic domination is positive associations. In this section, we restrict ourselves to posets of the form $\{0, 1\}^F$ for F finite. We begin with an example.

Example 4.2.22 (Erdős–Rényi graph: positive associations). Consider an Erdős–Rényi graph $G \sim \mathbb{G}_{n,p}$. Let $\mathcal{E} = \{\{x, y\} : x, y \in [n], x \neq y\}$. Think of G as taking values in the poset $(\{0, 1\}^{\mathcal{E}}, \leq)$, where a 1 indicates that the corresponding edge is present. In fact, observe that the law of G, which we denote as usual by $\mathbb{P}_{n,p}$, is a product measure on $\{0, 1\}^{\mathcal{E}}$. The event $\mathcal{A}$ that G is connected is increasing because adding edges cannot disconnect an already connected graph. So is the event $\mathcal{B}$ of having a chromatic number larger than 4. Intuitively then, conditioning on $\mathcal{A}$ makes $\mathcal{B}$ more likely: the occurrence of $\mathcal{A}$ tends to be accompanied with a larger number of edges which in turn makes $\mathcal{B}$ more probable.

This is an example of a more general phenomenon. That is, for any non-empty increasing events $\mathcal{A}$ and $\mathcal{B}$, we have:

Claim 4.2.23

$$\mathbb{P}_{n,p}[\mathcal{B} \mid \mathcal{A}] \geq \mathbb{P}_{n,p}[\mathcal{B}]. \qquad (4.2.13)$$

Or, put differently, the conditional measure $\mathbb{P}_{n,p}[\cdot \mid \mathcal{A}]$ stochastically dominates the unconditional measure $\mathbb{P}_{n,p}[\cdot]$. This is a special case of what is known as Harris' inequality, proved below. Note that (4.2.13) is equivalent to $\mathbb{P}_{n,p}[\mathcal{A} \cap \mathcal{B}] \geq \mathbb{P}_{n,p}[\mathcal{A}]\,\mathbb{P}_{n,p}[\mathcal{B}]$, that is, to the fact that $\mathcal{A}$ and $\mathcal{B}$ are positively correlated.   ◄

More generally:

POSITIVE
ASSOCIATIONS

Definition 4.2.24 (Positive associations). *Let μ be a probability measure on $\{0, 1\}^F$ where F is finite. Then μ is said to have* positive associations, *or is positively associated, if for all increasing functions $f, g \colon \{0, 1\}^F \to \mathbb{R}$,*

$$\mu(fg) \geq \mu(f)\mu(g),$$

where

$$\mu(h) := \sum_{\omega \in \{0,1\}^F} \mu(\omega)h(\omega).$$

In particular, for any increasing events A and B it holds that

$$\mu(A \cap B) \geq \mu(A)\mu(B),$$

POSITIVELY
CORRELATED

that is, A and B are positively correlated. *Denoting by $\mu(A \mid B)$ the conditional probability of A given B, this is equivalent to*

$$\mu(A \mid B) \geq \mu(A).$$

Remark 4.2.25 *Note that positive associations is concerned only with* increasing *events. See Remark 4.2.45.*

Remark 4.2.26 *A notion of negative associations, which is a somewhat more delicate concept, was defined in Remark 3.3.43. See also [Pem00].*

DECREASING

Let μ be positively associated. Note that if A and B are *decreasing*, that is, their complements are increasing (see Exercise 4.4), then

$$
\begin{aligned}
\mu(A \cap B) &= 1 - \mu(A^c \cup B^c) \\
&= 1 - \mu(A^c) - \mu(B^c) + \mu(A^c \cap B^c) \\
&\geq 1 - \mu(A^c) - \mu(B^c) + \mu(A^c)\mu(B^c) \\
&= \mu(A)\mu(B),
\end{aligned}
$$

or $\mu(A \mid B) \geq \mu(A)$. Similarly, if A is increasing and B is decreasing, we have $\mu(A \cap B) \leq \mu(A)\mu(B)$, or

$$\mu(A \mid B) \leq \mu(A). \qquad (4.2.14)$$

Harris' inequality states that product measures on $\{0, 1\}^F$ have positive associations. We prove a more general result known as the *FKG inequality*. For two configurations ω, ω' in $\{0, 1\}^F$, we let $\omega \wedge \omega'$ and $\omega \vee \omega'$ be the coordinatewise minimum and maximum of ω and ω'.

Definition 4.2.27 (FKG condition). *Let* $\mathcal{X} = \{0, 1\}^F$, *where F is finite. A positive probability measure* μ *on* $\mathcal{X}$ *satisfies the* FKG *condition if*

FKG
CONDITION

$$\mu(\omega \vee \omega')\,\mu(\omega \wedge \omega') \geq \mu(\omega)\,\mu(\omega') \qquad \forall \omega, \omega' \in \mathcal{X}. \tag{4.2.15}$$

This property is also known as log-supermodularity. We call such a measure an FKG *measure.*

Theorem 4.2.28 (FKG inequality). *Let* $\mathcal{X} = \{0, 1\}^F$, *where F is finite. Suppose* μ *is a positive probability measure on* $\mathcal{X}$ *satisfying the FKG condition. Then* μ *has positive associations.*

FKG
INEQUALITY

Remark 4.2.29 *Strict positivity is not in fact needed [FKG71]. The FKG condition is equivalent to a strong form of positive associations. See Exercise 4.8.*

Note that product measures satisfy the FKG condition with equality. Indeed, if $\mu(\omega)$ is of the form $\prod_{f \in F} \mu_f(\omega_f)$, then

$$\mu(\omega \vee \omega')\,\mu(\omega \wedge \omega') = \prod_f \mu_f(\omega_f \vee \omega_f')\,\mu_f(\omega_f \wedge \omega_f')$$

$$= \prod_{f:\omega_f = \omega_f'} \mu_f(\omega_f)^2 \prod_{f:\omega_f \neq \omega_f'} \mu_f(\omega_f)\mu_f(\omega_f')$$

$$= \prod_{f:\omega_f = \omega_f'} \mu_f(\omega_f)\mu_f(\omega_f') \prod_{f:\omega_f \neq \omega_f'} \mu_f(\omega_f)\mu_f(\omega_f')$$

$$= \mu(\omega)\,\mu(\omega').$$

So the FKG inequality (Theorem 4.2.28) applies, for instance, to bond percolation and the Erdős–Rényi random graph model. The pointwise nature of the FKG condition also makes it relatively easy to check for measures which are defined explicitly up to a normalizing constant, such as the Ising model.

Example 4.2.30 (Ising model with boundary conditions: checking FKG). Consider again the setting of Section 4.2.2. We work on the space $\mathcal{X} := \{-1, +1\}^{\Lambda}$ rather than $\{0, 1\}^F$. Fix a finite $\Lambda \subseteq \mathbb{L}^d$, $\xi \in \{-1, +1\}^{\mathbb{L}^d}$ and $\beta > 0$.

Claim 4.2.31 *The measure* $\mu_{\beta,\Lambda}^{\xi}$ *satisfies the FKG condition and therefore has positive associations.*

Intuitively, taking the minimum (or maximum) of two spin configurations tends to increase agreement and therefore leads to a higher likelihood. For $\sigma, \sigma' \in \mathcal{X}$, let $\overline{\tau} = \sigma \vee \sigma'$ and $\underline{\tau} = \sigma \wedge \sigma'$. By taking logarithms in the FKG condition and rearranging, we arrive at

$$\mathcal{H}_{\Lambda,\xi}(\overline{\tau}) + \mathcal{H}_{\Lambda,\xi}(\underline{\tau}) \leq \mathcal{H}_{\Lambda,\xi}(\sigma) + \mathcal{H}_{\Lambda,\xi}(\sigma'), \tag{4.2.16}$$

and we see that proving the claim boils down to checking an inequality for each term in the Hamiltonian (which, confusingly, has a negative sign in it).

When $i \in \Lambda$ and $j \notin \Lambda$ such that $i \sim j$, we have

$$\overline{\tau}_i \xi_j + \underline{\tau}_i \xi_j = (\overline{\tau}_i + \underline{\tau}_i)\xi_j = (\sigma_i + \sigma_i')\xi_j = \sigma_i \xi_j + \sigma_i' \xi_j. \tag{4.2.17}$$

For $i, j \in \Lambda$ with $i \sim j$, note first that the case $\sigma_j = \sigma_j'$ reduces to the previous calculation (with $\sigma_j = \sigma_j'$ playing the role of ξ_j), so we assume $\sigma_i \neq \sigma_i'$ and $\sigma_j \neq \sigma_j'$. Then,

$$\overline{\tau}_i \overline{\tau}_j + \underline{\tau}_i \underline{\tau}_j = (+1)(+1) + (-1)(-1) = 2 \geq \sigma_i \sigma_j + \sigma_i' \sigma_j',$$

since 2 is the largest value the rightmost expression ever takes. We have established (4.2.16), which implies the claim.

Again, we have not used any special property of the lattice and the same result holds for countable, locally finite graphs with positive coupling constants. Note however that in the anti-ferromagnetic case, that is, if we multiply the Hamiltonian by -1, the above argument does not work. Indeed there is no reason to expect positive associations in that case.   ◄

The FKG inequality in turn follows from a more general result known as *Holley's inequality*.

Holley's inequality   **Theorem 4.2.32** (Holley's inequality). *Let $\mathcal{X} = \{0, 1\}^F$, where F is finite. Suppose μ_1 and μ_2 are positive probability measures on $\mathcal{X}$ satisfying*

$$\mu_2(\omega \vee \omega') \mu_1(\omega \wedge \omega') \geq \mu_2(\omega) \mu_1(\omega')   \qquad \forall \omega, \omega' \in \mathcal{X}. \tag{4.2.18}$$

Then, $\mu_1 \preceq \mu_2$.

Before proving Holley's inequality (Theorem 4.2.32), we check that it indeed implies the FKG inequality. See Exercise 4.5 for an elementary proof in the independent case, that is, of Harris' inequality.

Proof of Theorem 4.2.28   Assume that μ satisfies the FKG condition and let f, g be increasing functions. Because of our restriction to positive measures in Holley's inequality, we will work with positive functions. This is done without loss of generality. Indeed, letting **0** be the all-0 vector, note that f and g are increasing if and only if $f' := f - f(\mathbf{0}) + 1 > 0$ and $g' := g - g(\mathbf{0}) + 1 > 0$ are increasing and that, moreover,

$$\begin{aligned}
\mu(f'g') - \mu(f')\mu(g') &= \mu([f' - \mu(f')][g' - \mu(g')]) \\
&= \mu([f - \mu(f)][g - \mu(g)]) \\
&= \mu(fg) - \mu(f)\mu(g).
\end{aligned}$$

In Holley's inequality, we let $\mu_1 := \mu$ and define the positive probability measure

$$\mu_2(\omega) := \frac{g(\omega)\mu(\omega)}{\mu(g)}.$$

We check that μ_1 and μ_2 satisfy the conditions of Theorem 4.2.32. Note that $\omega' \leq \omega \vee \omega'$ for any ω so that, because g is increasing, we have $g(\omega') \leq g(\omega \vee \omega')$. Hence, for any ω, ω',

$$\begin{aligned}
\mu_1(\omega)\mu_2(\omega') &= \mu(\omega) \frac{g(\omega')\mu(\omega')}{\mu(g)} \\
&= \mu(\omega)\mu(\omega') \frac{g(\omega')}{\mu(g)} \\
&\leq \mu(\omega \wedge \omega')\mu(\omega \vee \omega') \frac{g(\omega \vee \omega')}{\mu(g)} \\
&= \mu_1(\omega \wedge \omega')\mu_2(\omega \vee \omega'),
\end{aligned}$$

where on the third line we used the FKG condition satisfied by μ.

So Holley's inequality implies that $\mu_2 \succeq \mu_1$. Hence, since f is increasing, by Corollary 4.2.13,

$$\mu(f) = \mu_1(f) \leq \mu_2(f) = \frac{\mu(fg)}{\mu(g)},$$

and the theorem is proved. ∎

Proof of Theorem 4.2.32 The idea of the proof is to use Theorem 4.2.17. This is similar to what was done in Section 4.2.2. Again we use a single-site dynamic. For $x \in \mathcal{X}$ and $\gamma \in \{0, 1\}$, we let $x^{i,\gamma}$ be x with coordinate i set to γ. We write $x \sim y$ if $\|x - y\|_1 = 1$. Let $n = |F|$. We use a scheme analogous to the Metropolis algorithm (see Example 1.1.30). A natural symmetric chain on $\mathcal{X}$ is to pick a coordinate uniformly at random, and flip its value. We modify it to guarantee reversibility with respect to the desired stationary distributions, namely, μ_1 and μ_2.

For $\alpha, \beta > 0$ small enough, the following transition matrix over $\mathcal{X}$ is irreducible and reversible with respect to its stationary distribution μ_2: for all $i \in F, y \in \mathcal{X}$,

$$Q(y^{i,0}, y^{i,1}) = \frac{1}{n}\alpha \{\beta\},$$

$$Q(y^{i,1}, y^{i,0}) = \frac{1}{n}\alpha \left\{ \beta \frac{\mu_2(y^{i,0})}{\mu_2(y^{i,1})} \right\},$$

$$Q(y, y) = 1 - \sum_{z:z\sim y} Q(y, z).$$

Let P be similarly defined with respect to μ_1 *with the same values of α and β*. For reasons that will be clear below, the value of $0 < \beta < 1$ is chosen small enough that the *sum* of the two expressions in brackets above is smaller than 1 *for all y, i* in both P and Q. The value of $\alpha > 0$ is then chosen small enough that $P(y, y), Q(y, y) \geq 0$ *for all y*. Reversibility follows immediately from the first two equations. We call the first transition above an *upward transition* and the second one a *downward transition*.

By Theorem 4.2.17, it remains to show that Q stochastically dominates P. That is, for any $x \preceq y$, we want to show that $P(x, \cdot) \preceq Q(y, \cdot)$. We produce a monotone coupling $(\hat{X}, \hat{Y})$ of these two distributions. Because $x \preceq y$, our goal is never to perform an upward transition in x *simultaneously* with a downward transition in y. Observe that

$$\frac{\mu_1(x^{i,0})}{\mu_1(x^{i,1})} \geq \frac{\mu_2(y^{i,0})}{\mu_2(y^{i,1})} \tag{4.2.19}$$

by taking $\omega = y^{i,0}$ and $\omega' = x^{i,1}$ in Condition (4.2.18).

The coupling works as follows. Fix $x \preceq y$. With probability $1 - \alpha$, set $(\hat{X}, \hat{Y}) := (x, y)$. Otherwise, pick a coordinate $i \in F$ uniformly at random. There are several cases to consider depending on the coordinates x_i, y_i (with $x_i \leq y_i$ by assumption):

- $(x_i, y_i) = (0, 0)$: With probability β, perform an upward transition in both coordinates, that is, set $\hat{X} := x^{i,1}$ and $\hat{Y} := y^{i,1}$. With probability $1 - \beta$, set $(\hat{X}, \hat{Y}) := (x, y)$ instead.

- $(x_i, y_i) = (1, 1)$: With probability $\beta \frac{\mu_2(y^{i,0})}{\mu_2(y^{i,1})}$, perform a downward transition in both coordinates, that is, set $\hat{X} := x^{i,0}$ and $\hat{Y} := y^{i,0}$. With probability

$$\beta \left(\frac{\mu_1(x^{i,0})}{\mu_1(x^{i,1})} - \frac{\mu_2(y^{i,0})}{\mu_2(y^{i,1})} \right),$$

perform a downward transition in x only, that is, set $\hat{X} := x^{i,0}$ and $\hat{Y} := y$. With the remaining probability, set $(\hat{X}, \hat{Y}) := (x, y)$ instead. Note that (4.2.19) and our choice of β guarantees that this step is well defined.

- $(x_i, y_i) = (0, 1)$: With probability β, perform an upward transition in x only, that is, set $\hat{X} := x^{i,1}$ and $\hat{Y} := y$. With probability $\beta \frac{\mu_2(y^{i,0})}{\mu_2(y^{i,1})}$, perform a downward transition in y only, that is, set $\hat{X} := x$ and $\hat{Y} := y^{i,0}$. With the remaining probability, set $(\hat{X}, \hat{Y}) := (x, y)$ instead. Again our choice of β guarantees that this step is well defined.

A little accounting shows that this is indeed a coupling of $P(x, \cdot)$ and $Q(y, \cdot)$. By construction, this coupling satisfies $\hat{X} \leq \hat{Y}$ almost surely. An application of Theorem 4.2.17 concludes the proof.   ∎

Example 4.2.33 (Ising model revisited). Holley's inequality implies Claim 4.2.18. To see this, just repeat the calculations of Example 4.2.30, where now (4.2.17) is replaced with an inequality. See Exercise 4.6.   ◀

4.2.4 ▷ *Random Graphs: Janson's Inequality and Application to the Clique Number in the Erdős–Rényi Model*

Let $G = (V, E) \sim \mathbb{G}_{n,p}$ be an Erdős–Rényi graph. By Claim 2.3.5, the property of being triangle-free has threshold n^{-1}. That is, the probability that G contains a triangle goes to 0 or 1 as $n \to +\infty$ according to whether $p \ll n^{-1}$ or $p \gg n^{-1}$, respectively. In this section, we investigate what happens *at* the threshold, by which we mean that we take $p = \lambda/n$ for some $\lambda > 0$ not depending on n.

For any subset S of three distinct vertices of G, let B_S be the event that S forms a triangle in G. So

$$\varepsilon := \mathbb{P}_{n,p}[B_S] = p^3 \to 0. \tag{4.2.20}$$

Denoting the unordered triples of distinct vertices by $\binom{V}{3}$, let $X_n = \sum_{S \in \binom{V}{3}} \mathbf{1}_{B_S}$ be the number of triangles in G. By the linearity of expectation, the mean number of triangles is

$$\mathbb{E}_{n,p} X_n = \binom{n}{3} p^3 = \frac{n(n-1)(n-2)}{6} \left(\frac{\lambda}{n} \right)^3 \to \frac{\lambda^3}{6},$$

as $n \to +\infty$. *If the events $\{B_S\}_S$ were mutually independent*, X_n would be binomially distributed and the event that G is triangle-free would have probability

$$\prod_{S \in \binom{V}{3}} \mathbb{P}_{n,p}[B_S^c] = (1 - p^3)^{\binom{n}{3}} \to e^{-\lambda^3/6}. \tag{4.2.21}$$

In fact, by the Poisson approximation to the binomial (e.g., Theorem 4.1.18), we would have that the number of triangles converges weakly to $\mathrm{Poi}(\lambda^3/6)$.

In reality, of course, the events $\{B_S\}$ are not *mutually* independent. Observe however that, for *most* pairs S, S', the events B_S and $B_{S'}$ are in fact *pairwise* independent. That is the case whenever $|S \cap S'| \leq 1$, that is, whenever the edges connecting S are disjoint from those connecting S'. Write $S \sim S'$ if $S \neq S'$ are not independent, that is, if $|S \cap S'| = 2$. The expected number of unordered pairs $S \sim S'$ both forming a triangle is

$$\Delta := \frac{1}{2} \sum_{\substack{S,S' \in \binom{V}{3} \\ S \sim S'}} \mathbb{P}_{n,p}[B_S \cap B_{S'}] = \frac{1}{2}\binom{n}{3}\binom{3}{2}(n-3)p^5 = \Theta(n^4 p^5) \to 0, \qquad (4.2.22)$$

where the $\binom{n}{3}$ comes from the number of ways of choosing S, the $\binom{3}{2}$ comes from the number of ways of choosing the vertices in common between S and S', and the $n - 3$ comes from the number of ways of choosing the third vertex of S'. Given that the events $\{B_S\}_{S \in \binom{V}{3}}$ are "mostly" independent, it is natural to expect that X_n behaves asymptotically as it would in the independent case. Indeed we prove:

Claim 4.2.34

$$\mathbb{P}_{n,p}[X_n = 0] \to e^{-\lambda^3/6}.$$

Remark 4.2.35 *In fact, $X_n \xrightarrow{d} \mathrm{Poi}(\lambda^3/6)$. See Exercises 2.18 and 4.9.*

The FKG inequality (Theorem 4.2.28) immediately gives one direction. Recall that $\mathbb{P}_{n,p}$, as a product measure over edge sets, satisfies the FKG condition and therefore has positive associations by the FKG inequality. Moreover, the events B_S^c are decreasing for all S. Hence, applying positive associations inductively,

$$\mathbb{P}_{n,p}\left[\bigcap_{S \in \binom{V}{3}} B_S^c\right] \geq \prod_{S \in \binom{V}{3}} \mathbb{P}_{n,p}[B_S^c] \to e^{-\lambda^3/6},$$

where the limit follows from (4.2.21). As it turns out, the FKG inequality also gives a bound in the other direction. This is known as *Janson's inequality*, which we state in a more general context.

Janson's inequality Let $\mathcal{X} := \{0, 1\}^F$, where F is finite. Let B_i, $i \in I$, be a finite collection of events of the form

$$B_i := \{\omega \in \mathcal{X} : \omega \geq \beta^{(i)}\}$$

for some $\beta^{(i)} \in \mathcal{X}$. Think of these as "bad events" corresponding to a certain subset of coordinates being set to 1. By definition, the B_is are increasing. Assume $\mathbb{P}$ is a positive product measure on $\mathcal{X}$. Write $i \sim j$ if $\beta_r^{(i)} = \beta_r^{(j)} = 1$ for at least one r and note that B_i is independent of B_j if $i \nsim j$. Set

$$\Delta := \sum_{\substack{\{i,j\} \\ i \sim j}} \mathbb{P}[B_i \cap B_j].$$

Theorem 4.2.36 (Janson's inequality). *Let $\mathcal{X} := \{0,1\}^F$, where F is finite and $\mathbb{P}$ be a positive product measure on $\mathcal{X}$. Let $\{B_i\}_{i \in I}$ and Δ be as above. Assume further that there is $\varepsilon > 0$ such that $\mathbb{P}[B_i] \leq \varepsilon$ for all $i \in I$. Then,*

$$\prod_{i \in I} \mathbb{P}[B_i^c] \leq \mathbb{P}[\cap_{i \in I} B_i^c] \leq e^{\frac{\Delta}{1-\varepsilon}} \prod_{i \in I} \mathbb{P}[B_i^c].$$

Before proving the theorem, we show that it implies Claim 4.2.34. We have already shown in (4.2.20) and (4.2.22) that $\varepsilon \to 0$ and $\Delta \to 0$. Janson's inequality (Theorem 4.2.36) immediately implies the claim by (4.2.21).

Proof of Theorem 4.2.36 The lower bound is the FKG inequality.

In the other direction, assume without loss of generality that $I = [m]$. The first step is to apply the chain rule to obtain

$$\mathbb{P}[\cap_{i \in I} B_i^c] = \prod_{i=1}^m \mathbb{P}[B_i^c \mid \cap_{j \in [i-1]} B_j^c].$$

The rest is clever manipulation. For $i \in [m]$, let $N(i) := \{\ell \in [m]: \ell \sim i\}$ and $N_<(i) := N(i) \cap [i-1]$. Note that B_i is independent of $\{B_\ell: \ell \in [i-1]\backslash N_<(i)\}$. Hence,

$$\begin{aligned}
\mathbb{P}[B_i \mid \cap_{j \in [i-1]} B_j^c] &= \frac{\mathbb{P}\left[B_i \cap \left(\cap_{j \in [i-1]} B_j^c\right)\right]}{\mathbb{P}[\cap_{j \in [i-1]} B_j^c]} \\
&\geq \frac{\mathbb{P}\left[B_i \cap \left(\cap_{j \in N_<(i)} B_j^c\right) \cap \left(\cap_{j \in [i-1]\backslash N_<(i)} B_j^c\right)\right]}{\mathbb{P}[\cap_{j \in [i-1]\backslash N_<(i)} B_j^c]} \\
&= \mathbb{P}\left[B_i \cap \left(\cap_{j \in N_<(i)} B_j^c\right) \mid \cap_{j \in [i-1]\backslash N_<(i)} B_j^c\right] \\
&= \mathbb{P}\left[B_i \mid \cap_{j \in [i-1]\backslash N_<(i)} B_j^c\right] \\
&\quad \times \mathbb{P}\left[\cap_{j \in N_<(i)} B_j^c \mid B_i \cap \left(\cap_{j \in [i-1]\backslash N_<(i)} B_j^c\right)\right] \\
&= \mathbb{P}[B_i] \, \mathbb{P}\left[\cap_{j \in N_<(i)} B_j^c \mid B_i \cap \left(\cap_{j \in [i-1]\backslash N_<(i)} B_j^c\right)\right],
\end{aligned}$$

where we used independence for the first term on the last line. By a union bound, the second term on the last line is

$$\begin{aligned}
\mathbb{P}\left[\cap_{j \in N_<(i)} B_j^c \mid B_i \cap \left(\cap_{j \in [i-1]\backslash N_<(i)} B_j^c\right)\right] \\
\geq 1 - \sum_{j \in N_<(i)} \mathbb{P}\left[B_j \mid B_i \cap \left(\cap_{j \in [i-1]\backslash N_<(i)} B_j^c\right)\right] \\
\geq 1 - \sum_{j \in N_<(i)} \mathbb{P}\left[B_j \mid B_i\right],
\end{aligned}$$

where the last line follows from the FKG inequality. This requires some explanations:

- On the event B_i, all coordinates ℓ with $\beta_\ell^{(i)} = 1$ are fixed to 1, and the other ones are free. So we can think of $\mathbb{P}[\cdot \mid B_i]$ as a positive product measure on $\{0,1\}^{F'}$ with $F' := \{\ell \in [m]: \beta_\ell^{(i)} = 0\}$.
- The event B_j is increasing, while the event $\cap_{j \in [i-1]\backslash N_<(i)} B_j^c$ is decreasing as the intersection of decreasing events (see Exercise 4.4).

- So we can apply the FKG inequality in the form (4.2.14) to $\mathbb{P}[\cdot \mid B_i]$.

Combining the last three displays and using $1 + x \leq e^x$ for all x (see Exercise 1.16), we get

$$
\begin{aligned}
\mathbb{P}[\cap_{i \in I} B_i^c] &\leq \prod_{i=1}^m \left[\mathbb{P}\left[B_i^c\right] + \sum_{j \in N_<(i)} \mathbb{P}\left[B_i \cap B_j\right] \right] \\
&\leq \prod_{i=1}^m \mathbb{P}\left[B_i^c\right] \left[1 + \frac{1}{1-\varepsilon} \sum_{j \in N_<(i)} \mathbb{P}\left[B_i \cap B_j\right] \right] \\
&\leq \prod_{i=1}^m \mathbb{P}\left[B_i^c\right] \exp\left(\frac{1}{1-\varepsilon} \sum_{j \in N_<(i)} \mathbb{P}\left[B_i \cap B_j\right] \right),
\end{aligned}
$$

where we used the assumption $\mathbb{P}[B_i] \leq \varepsilon$ on the second line. By the definition of Δ, we are done. ∎

4.2.5 ▷ *Percolation: RSW Theory and a Proof of Harris' Theorem*

Consider bond percolation (Definition 1.2.1) on the two-dimensional lattice $\mathbb{L}^2$. Recall that the percolation function is given by

$$
\theta(p) := \mathbb{P}_p[|\mathcal{C}_0| = +\infty],
$$

where $\mathcal{C}_0$ is the open cluster of the origin. We know from Example 4.2.10 that $\theta(p)$ is non-decreasing. Let

$$
p_c(\mathbb{L}^2) := \sup\{p \geq 0 : \theta(p) = 0\}
$$

be the critical value. We proved in Section 2.2.4 that there is a non-trivial transition, that is, $p_c(\mathbb{L}^2) \in (0, 1)$. See Exercise 2.3 for a proof that $p_c(\mathbb{L}^2) \in [1/3, 2/3]$.

Our goal in this section is to use the FKG inequality to improve this further to:

Theorem 4.2.37 (Harris' theorem).

$$
\theta(1/2) = 0.
$$

Or, put differently, $p_c(\mathbb{L}^2) \geq 1/2$.

Remark 4.2.38 *This bound is tight, that is, in fact $p_c(\mathbb{L}^2) = 1/2$. The other direction is known as* Kesten's theorem. *See, for example, [BR06a].*

Here we present a proof of Harris' theorem that uses an important tool in percolation theory, the *Russo–Seymour–Welsh (RSW) lemma*, an application of the FKG inequality.

Harris' theorem

To motivate the RSW lemma, we start with the proof of Harris' theorem.

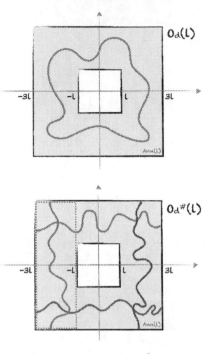

Figure 4.4 Top: the event $O_d(\ell)$. Bottom: the event $O_d^{\#}(\ell)$.

Proof of Theorem 4.2.37   Fix $p = 1/2$. We use the dual lattice $\widetilde{\mathbb{L}}^2$ as we did in Section 2.2.4. Consider the annulus

$$\text{Ann}(\ell) := [-3\ell, 3\ell]^2 \setminus [-\ell, \ell]^2.$$

The existence of a closed dual cycle inside $\text{Ann}(\ell)$, an event we denote by $O_d(\ell)$, prevents the possibility of an infinite open path from the origin in the primal lattice $\mathbb{L}^2$. See Figure 4.4. That is,

$$\mathbb{P}_{1/2}[|\mathcal{C}_0| = +\infty] \leq \prod_{k=0}^{K} \{1 - \mathbb{P}_{1/2}[O_d(3^k)]\} \tag{4.2.23}$$

for all K, where we took powers of 3 to make the annuli disjoint and therefore independent. To prove the theorem, it suffices to show that there is a constant $c^* > 0$ such that, for all ℓ, $\mathbb{P}_{1/2}[O_d(\ell)] \geq c^*$. Then the right-hand side of (4.2.23) tends to 0 as $K \to +\infty$.

To simplify further, thinking of $\text{Ann}(\ell)$ as a union of four rectangles $[-3\ell, -\ell) \times [-3\ell, 3\ell]$, $[-3\ell, 3\ell] \times (\ell, 3\ell]$, and so on, it suffices to consider the event $O_d^{\#}(\ell)$ that each one of these rectangles contains a closed dual path connecting its two shorter sides. To be more precise, for the first rectangle above for instance, the path connects $[-3\ell + 1/2, -\ell - 1/2] \times \{3\ell - 1/2\}$ to $[-3\ell + 1/2, -\ell - 1/2] \times \{-3\ell + 1/2\}$ and stays inside the rectangle. See Figure 4.4. By symmetry the probability that such a path exists is the same for all four rectangles. Denote that probability by ρ_ℓ. Moreover, the event that such a path exists is increasing so, although the four events are not independent, we can apply the FKG inequality (Theorem 4.2.28). Hence, since $O_d^\#(\ell) \subseteq O_d(\ell)$, we finally get the bound

$$\mathbb{P}_{1/2}[O_d(\ell)] \geq \rho_\ell^4.$$

The RSW lemma and some symmetry arguments, both of which are detailed below, imply:

Claim 4.2.39 *There is some $c > 0$ such that, for all ℓ,*

$$\rho_\ell \geq c.$$

That concludes the proof. ∎

It remains to prove Claim 4.2.39. We first state the RSW lemma.

RSW theory

We have reduced the proof of Harris' theorem to bounding the probability that certain closed paths exist in the dual lattice. To be consistent with the standard RSW notation, we switch to the primal lattice and consider open paths. We also let p take any value in $(0, 1)$.

Let $R_{n,\alpha}(p)$ be the probability that the rectangle

$$B(\alpha n, n) := [-n, (2\alpha - 1)n] \times [-n, n]$$

has an open path connecting its left and right sides with the path remaining inside the rectangle. Such a path is called an (open) *left-right crossing*. The event that a left-right crossing exists in a rectangle B is denoted by LR(B). We similarly define the event, TB(B), that a *top-bottom crossing* exists in B. In essence, the RSW lemma says this: if there is a significant probability that a left-right crossing exists in the square $B(n, n)$, then there is a significant probability that a left-right crossing exists in the rectangle $B(3n, n)$. More precisely, here is a version of the theorem that will be enough for our purposes. (See Exercise 4.10 for a generalization.)

Lemma 4.2.40 (RSW lemma). *For all $n \geq 2$ (divisible by 4) and $p \in (0, 1)$,*

$$R_{n,3}(p) \geq \frac{1}{256} R_{n,1}(p)^{11} R_{n/2,1}(p)^{12}. \tag{4.2.24}$$

The right-hand side of (4.2.24) depends only on the probability of crossing a square from left to right. By a duality argument, at $p = 1/2$, it turns out that this probability is at least $1/2$ *independently of* n. Before presenting a proof of the RSW lemma, we detail this argument and finish the proof of Harris' theorem.

Proof of Claim 4.2.39 The point of (4.2.24) is that if $R_{n,1}(1/2)$ is bounded away from 0 uniformly in n, then so is the left-hand side. By the argument in the proof of Harris' theorem, this then implies that a closed cycle exists in Ann(n) with a probability bounded away from 0 as well. Hence, to prove Claim 4.2.39 it suffices to give a lower bound on $R_{n,1}(1/2)$. It is crucial that this bound not depend on the "scale" n.

As it turns out, a simple duality-based symmetry argument does the trick. The following fact about $\mathbb{L}^2$ is a variant of the contour lemma (Lemma 2.2.14). Its proof is similar and Exercise 4.11 asks for the details (the "if" direction being the non-trivial implication).

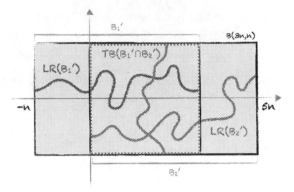

Figure 4.5 Illustration of the implication
$\mathrm{LR}(B'_1) \cap \mathrm{TB}(B'_1 \cap B'_2) \cap \mathrm{LR}(B'_2) \subseteq \mathrm{LR}(B(3n,n))$.

Lemma 4.2.41 *There is an open left-right crossing in the primal rectangle* $[0, n+1] \times [0,n]$ *if and only if there is* no *closed top-bottom crossing in the dual rectangle* $[1/2, n+1/2] \times [-1/2, n+1/2]$.

By symmetry, when $p = 1/2$, the two events in Lemma 4.2.41 have equal probability. So they must have probability $1/2$ because they form a partition of the space of outcomes. By monotonicity, that implies $R_{n,1}(1/2) \geq 1/2$ for all n. The RSW lemma then implies the required bound. ∎

The proof of the RSW lemma involves a clever choice of event that relates the existence of crossings in squares and rectangles. (Combining crossings of squares into crossings of rectangles is not as trivial as it might look. Try it before reading the proof.)

Proof of Lemma 4.2.40   There are several steps in the proof.

Step 1: It suffices to bound $R_{n,3/2}(p)$ We first reduce the proof to finding a bound on $R_{n,3/2}(p)$. Let $B'_1 := B(2n,n)$ and $B'_2 := [n,5n] \times [-n,n]$. Note that $B'_1 \cup B'_2 = B(3n,n)$ and $B'_1 \cap B'_2 = [n,3n] \times [-n,n]$. Then we have the implication

$$\mathrm{LR}(B'_1) \cap \mathrm{TB}(B'_1 \cap B'_2) \cap \mathrm{LR}(B'_2) \subseteq \mathrm{LR}(B(3n,n)).$$

See Figure 4.5. Each event on the left-hand side is increasing so the FKG inequality gives

$$R_{n,3}(p) \geq R_{n,2}(p)^2 R_{n,1}(p).$$

A similar argument over $B(2n,n)$ gives

$$R_{n,2}(p) \geq R_{n,3/2}(p)^2 R_{n,1}(p).$$

Combining the two, we have proved:

Lemma 4.2.42 (Proof of RSW: step 1).

$$R_{n,3}(p) \geq R_{n,3/2}(p)^4 R_{n,1}(p)^3. \tag{4.2.25}$$

Step 2: Bounding $R_{n,3/2}(p)$ The heart of the proof is to bound $R_{n,3/2}(p)$ using an event involving crossings of squares. Let

$$
\begin{aligned}
B_1 &:= B(n,n) = [-n,n] \times [-n,n], \\
B_2 &:= [0,2n] \times [-n,n], \\
B_{12} &:= B_1 \cap B_2 = [0,n] \times [-n,n], \\
S &:= [0,n] \times [0,n].
\end{aligned}
$$

Let Γ_1 be the event that there are paths P_1, P_2, where P_1 is a top-bottom crossing of S and P_2 is an open path connecting the left side of B_1 to P_1 and stays inside B_1. Similarly, let Γ'_2 be the event that there are paths P'_1, P'_2, where P'_1 is a top-bottom crossing of S and P'_2 is an open path connecting the right side of B_2 to P'_1 and stays inside B_2. Then we have the implication

$$
\Gamma_1 \cap \mathrm{LR}(S) \cap \Gamma'_2 \subseteq \mathrm{LR}(B(3n/2, n)).
$$

See Figure 4.6. By symmetry $\mathbb{P}_p[\Gamma_1] = \mathbb{P}_p[\Gamma'_2]$. Moreover, the events on the left-hand side are increasing so by the FKG inequality:

Lemma 4.2.43 (Proof of RSW: step 2).

$$
R_{n,3/2}(p) \geq \mathbb{P}_p[\Gamma_1]^2 R_{n/2,1}(p). \tag{4.2.26}
$$

Step 3: Bounding $\mathbb{P}_p[\Gamma_1]$ It remains to bound $\mathbb{P}_p[\Gamma_1]$. That requires several additional definitions. Let P_1 and P_2 be top-bottom crossings of S. There is a natural partial order over such crossings. The path P_1 divides S into two subgraphs: $[P_1\}$, which includes the left side of S (including edges on the left incident with P_1 but not those edges on P_1 itself) and $\{P_1]$, which includes the right side of S (and P_1 itself). Then, we write $P_1 \preceq P_2$ if $\{P_1] \subseteq \{P_2]$. Assuming $\mathrm{TB}(S)$ holds, one also gets the existence of a unique *rightmost crossing*. Roughly speaking, RIGHTMOST take the union of all top-bottom crossings of S as sets of edges; then the "right boundary" CROSSING of this set is a top-bottom crossing P_S^* such that $P_S^* \preceq P$ for all top-bottom crossings P of S. (We accept as a fact the existence of a unique rightmost crossing. See Exercise 4.11 for a related construction.)

Let I_S be the set of (not necessarily open) paths connecting the top and bottom of S and stay inside S. For $P \in I_S$, we let P' be the reflection of P in $B_{12} \backslash S$ through the x-axis and we let $\frac{P}{P'}$ be the union of P and P'. Define $[\frac{P}{P'}\}$ to be the subgraph of B_1 to the left of $\frac{P}{P'}$ (including edges on the left incident with $\frac{P}{P'}$ but not those edges on $\frac{P}{P'}$ itself). Let $\mathrm{LR}^+([\frac{P}{P'}\})$ be the event that there is a left-right crossing of $[\frac{P}{P'}\}$ ending on P, that is, that there is an open path connecting the left side of B_1 and P that stays within $[\frac{P}{P'}\}$. See Figure 4.6. Note that the existence of a left-right crossing of B_1 implies the existence of an open path connecting the left side of B_1 to $\frac{P}{P'}$. By symmetry we then get

$$
\mathbb{P}_p\left[\mathrm{LR}^+([\tfrac{P}{P'}\})\right] \geq \frac{1}{2}\mathbb{P}_p[\mathrm{LR}(B_1)] = \frac{1}{2}R_{n,1}(p). \tag{4.2.27}
$$

Now comes a subtle point. We turn to the rightmost crossing of S – for two reasons:

- First, by uniqueness of the rightmost crossing, $\{P_S^* = P\}_{P \in I_S}$ forms a *partition* of $\mathrm{TB}(S)$. Recall that we are looking to bound a probability *from below*, and therefore we have to be careful not to "double count."

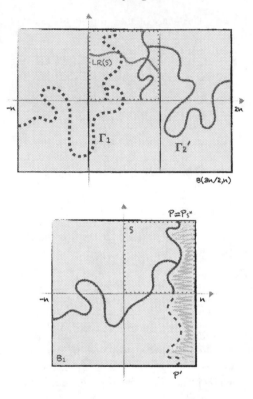

Figure 4.6 Top: illustration of the implication $\Gamma_1 \cap \mathrm{LR}(S) \cap \Gamma_2' \subseteq \mathrm{LR}(B(3n/2, n))$. Bottom: the event $\mathrm{LR}^+\big([\tfrac{P}{P'}]\big) \cap \{P = P_S^*\}$; the dashed path is the mirror image of the rightmost top-bottom crossing in S; the shaded region on the right is the complement in B_1 of the set $[\tfrac{P}{P'}]$. Note that, because in the bottom figure the left-right path must stay within $[\tfrac{P}{P'}]$ by definition of P_S^*, the configuration shown in the top figure where a left-right path (dotted) "travels behind" the top-bottom crossing of S cannot occur.

- Second, the rightmost crossing has a Markov-like property. Observe that, for $P \in I_S$, the event $\{P_S^* = P\}$ *depends only the bonds in* $\{P\}$. In particular, it is *independent of the bonds in* $[\tfrac{P}{P'}]$, for example, of the event $\mathrm{LR}^+([\tfrac{P}{P'}])$. Hence,

$$\mathbb{P}_p\big[\mathrm{LR}^+\big([\tfrac{P}{P'}]\big) \mid P_S^* = P\big] = \mathbb{P}_p\big[\mathrm{LR}^+\big([\tfrac{P}{P'}]\big)\big]. \tag{4.2.28}$$

Note that the event $\{P_S^* = P\}$ is *not increasing*, as adding more open bonds can shift the rightmost crossing rightward. Therefore, we cannot use the FKG inequality here.

Combining (4.2.27) and (4.2.28), we get

$$\mathbb{P}_p[\Gamma_1] \geq \sum_{P \in I_S} \mathbb{P}_p[P_S^* = P]\,\mathbb{P}_p\big[\mathrm{LR}^+\big([\tfrac{P}{P'}]\big) \mid P_S^* = P\big]$$

$$\geq \frac{1}{2}R_{n,1}(p)\sum_{P \in I_S}\mathbb{P}_p[P_S^* = P]$$

$$= \frac{1}{2} R_{n,1}(p)\, \mathbb{P}_p[\mathrm{TB}(S)]$$

$$= \frac{1}{2} R_{n,1}(p) R_{n/2,1}(p).$$

We have proved:

Lemma 4.2.44 (Proof of RSW: step 3).

$$\mathbb{P}_p[\Gamma_1] \geq \frac{1}{2} R_{n,1}(p) R_{n/2,1}(p). \tag{4.2.29}$$

Step 4: Putting everything together Combining (4.2.25), (4.2.26), and (4.2.29) gives

$$R_{n,3}(p) \geq R_{n,3/2}(p)^4 R_{n,1}(p)^3$$
$$\geq [\mathbb{P}_p[\Gamma_1]^2 R_{n/2,1}(p)]^4 R_{n,1}(p)^3$$
$$\geq \left[\left(\frac{1}{2} R_{n,1}(p) R_{n/2,1}(p) \right)^2 R_{n/2,1}(p) \right]^4 R_{n,1}(p)^3.$$

Collecting the terms concludes the proof of the RSW lemma. ∎

Remark 4.2.45 *This argument is quite subtle. It is instructive to read the remark after* [Gri97, Theorem 9.3].

4.3 Coupling of Markov Chains and Application to Mixing

As we have seen, coupling is useful to bound total variation distance. In this section we apply the technique to bound the mixing time of Markov chains.

4.3.1 Bounding the Mixing Time via Coupling

Let P be an irreducible, aperiodic Markov transition matrix on the finite state space V with stationary distribution π. Recall from Definition 1.1.35 that, for a fixed $0 < \varepsilon < 1/2$, the mixing time of P is

$$t_{\mathrm{mix}}(\varepsilon) := \min\{t \colon d(t) \leq \varepsilon\},$$

where

$$d(t) := \max_{x \in V} \| P^t(x, \cdot) - \pi \|_{\mathrm{TV}}.$$

It will be easier to work with

$$\bar{d}(t) := \max_{x,y \in V} \| P^t(x, \cdot) - P^t(y, \cdot) \|_{\mathrm{TV}}.$$

The quantities $d(t)$ and $\bar{d}(t)$ are related in the following way.

Lemma 4.3.1

$$d(t) \leq \bar{d}(t) \leq 2 d(t) \qquad \forall t.$$

Proof The second inequality follows from an application of the triangle inequality.

For the first inequality, note that by definition of the total variation distance and the stationarity of π,

$$\|P^t(x,\cdot) - \pi\|_{\mathrm{TV}} = \sup_{A \subseteq V} |P^t(x,A) - \pi(A)|$$

$$= \sup_{A \subseteq V} \left| \sum_{y \in V} \pi(y)[P^t(x,A) - P^t(y,A)] \right|$$

$$\leq \sup_{A \subseteq V} \sum_{y \in V} \pi(y)|P^t(x,A) - P^t(y,A)|$$

$$\leq \sum_{y \in V} \pi(y) \left\{ \sup_{A \subseteq V} |P^t(x,A) - P^t(y,A)| \right\}$$

$$\leq \sum_{y \in V} \pi(y)\|P^t(x,\cdot) - P^t(y,\cdot)\|_{\mathrm{TV}}$$

$$\leq \max_{x,y \in V} \|P^t(x,\cdot) - P^t(y,\cdot)\|_{\mathrm{TV}}.$$

$\blacksquare$

Coalescence Recall that a Markovian coupling of P with itself is a Markov chain $(X_t, Y_t)_t$ on $V \times V$ with transition matrix Q satisfying, for all $x, y, x', y' \in V$,

$$\sum_{z'} Q((x,y),(x',z')) = P(x,x'),$$

$$\sum_{z'} Q((x,y),(z',y')) = P(y,y').$$

COALESCING   We say that a Markovian coupling is *coalescing* if further, for all $z \in V$,

$$x' \neq y' \implies Q((z,z),(x',y')) = 0.$$

Let (X_t, Y_t) be a coalescing Markovian coupling of P. By the coalescing condition, if $X_s = Y_s$ then $X_t = Y_t$ for all $t \geq s$. That is, once (X_t) and (Y_t) meet, they remain equal. Let COALESCENCE τ_{coal} be the *coalescence time* (also called *coupling time*), that is, TIME

$$\tau_{\mathrm{coal}} := \inf\{t \geq 0 : X_t = Y_t\}.$$

The key to the coupling approach to mixing times is the following immediate consequence of the coupling inequality (Lemma 4.1.11). For any starting point (x,y),

$$\|P^t(x,\cdot) - P^t(y,\cdot)\|_{\mathrm{TV}} \leq \mathbb{P}_{(x,y)}[X_t \neq Y_t] = \mathbb{P}_{(x,y)}[\tau_{\mathrm{coal}} > t]. \tag{4.3.1}$$

Combining (4.3.1) and Lemma 4.3.1, we get the main tool of this section.

Theorem 4.3.2 (Bounding the mixing time: coupling method). *Let (X_t, Y_t) be a coalescing Markovian coupling of an irreducible transition matrix P on a finite state space V with stationary distribution π. Then,*

$$d(t) \leq \max_{x,y \in V} \mathbb{P}_{(x,y)}[\tau_{\mathrm{coal}} > t].$$

In particular,

$$t_{\text{mix}}(\varepsilon) \leq \inf\left\{t \geq 0 : \mathbb{P}_{(x,y)}[\tau_{\text{coal}} > t] \leq \varepsilon \; \forall x, y\right\}.$$

We give a few simple examples in the next subsection. First, we discuss a classical result.

Example 4.3.3 (Doeblin's condition). Let P be a transition matrix on a countable space V. One form of *Doeblin's condition* (also called a *minorization condition*) is: there is $s \in \mathbb{Z}_+$ and $\delta > 0$ such that

DOEBLIN'S CONDITION

$$\sup_{z \in V} \inf_{w \in V} P^s(w, z) > \delta.$$

In words, there is a state $z_0 \in V$ such that, starting from any state $w \in V$, the probability of reaching z_0 in exactly s steps is at least δ (which does not depend on w). Assume such a z_0 exists.

We construct a coalescing Markovian coupling (X_t, Y_t) of P. Assume first that $s = 1$ and let

$$\tilde{P}(w, z) = \frac{1}{1 - \delta}\left[P(w, z) - \delta \mathbf{1}\{z = z_0\}\right].$$

It can be checked that $\tilde{P}$ is a stochastic matrix on V provided z_0 satisfies the condition above (see Exercise 4.13). We use a technique known as *splitting*. While $X_t \neq Y_t$, at the next time step: (i) with probability δ we set $X_{t+1} = Y_{t+1} = z_0$, (ii) otherwise we pick $X_{t+1} \sim \tilde{P}(X_t, \cdot)$ and $Y_{t+1} \sim \tilde{P}(Y_t, \cdot)$ independently. On the other hand, if $X_t = Y_t$, we maintain the equality and pick the next state according to P. Put differently, the coupling Q is defined as: if $x \neq y$,

SPLITTING

$$Q((x, y), (x', y')) = \delta \mathbf{1}\{x' = y' = z_0\} + (1 - \delta)\tilde{P}(x, x')\tilde{P}(y, y'),$$

while if $x = y$,

$$Q((x, x), (x', x')) = P(x, x').$$

Observe that, in case (i) above, coalescence occurs at time $t + 1$. In case (ii), coalescence may or may not occur at time $t + 1$. In other words, while $X_t \neq Y_t$, coalescence occurs at the next step with probability *at least* δ. So τ_{coal} is stochastically dominated by a geometric random variable with success probability δ, or

$$\max_{x, y \in V} \mathbb{P}_{(x,y)}[\tau_{\text{coal}} > t] \leq (1 - \delta)^t.$$

By Theorem 4.3.2,

$$\max_{x \in V} \|P^t(x, \cdot) - \pi\|_{\text{TV}} \leq (1 - \delta)^t.$$

Exponential decay of the worst-case total variation distance to the stationary distribution is referred to as *uniform geometric ergodicity*.

UNIFORM GEOMETRIC ERGODICITY

Suppose now that $s > 1$. We apply the argument above to the chain P^s this time. We get

$$\max_{x, y \in V} \mathbb{P}_{(x,y)}[\tau_{\text{coal}} > ts] \leq (1 - \delta)^t,$$

so that, after a change of variable,

$$\max_{x \in V} \|P^t(x, \cdot) - \pi\|_{\text{TV}} \leq (1 - \delta)^{\lfloor t/s \rfloor}.$$

So, we have shown that uniform geometric ergodicity is implied by Doeblin's condition.

We note however that the rate of decay derived from this technique can be very slow. For instance, the condition always holds when P is finite, irreducible, and aperiodic (as follows from Lemma 1.1.32), but a straight application of the technique may lead to a bound depending badly on the size of the state space V (see Exercise 4.14).   ◄

4.3.2 ▷ *Random Walks: Mixing on Cycles, Hypercubes, and Trees*

In this section, we consider lazy simple random walk on various graphs. By this we mean that the walk stays put with probability $1/2$ and otherwise picks an adjacent vertex uniformly at random. In each case, we construct a coupling to bound the mixing time. As a reference, we compare our upper bounds to the diameter-based lower bound we will derive in Section 5.2.3. Specifically, by Claim 5.2.25, for a finite, reversible Markov chain with stationary distribution π and diameter Δ we have the lower bound

$$t_{\mathrm{mix}}(\varepsilon) = \Omega\left(\frac{\Delta^2}{\log(n \vee \pi_{\mathrm{min}}^{-1})}\right),$$

where π_{min} is the smallest value taken by π.

Cycle

Let (Z_t) be lazy simple random walk on the cycle of size n, $\mathbb{Z}_n := \{0, 1, \ldots, n-1\}$, where $i \sim j$ if $|j - i| = 1 \pmod n$. For any starting points x, y, we construct a Markovian coupling (X_t, Y_t) of this chain. Set $(X_0, Y_0) := (x, y)$. At each time, flip a fair coin. On heads, Y_t stays put and X_t moves one step, the direction of which is uniform at random. On tails, proceed similarly with the roles of X_t and Y_t reversed. Let D_t be the clockwise distance between X_t and Y_t. Observe that, by construction, (D_t) is simple random walk on $\{0, \ldots, n\}$ and $\tau_{\mathrm{coal}} = \tau_{\{0,n\}}^D$, the first time (D_t) hits $\{0, n\}$.

We use Markov's inequality (Theorem 2.1.1) to bound $\mathbb{P}_{(x,y)}[\tau_{\{0,n\}}^D > t]$. Denote by $D_0 = d_{x,y}$ the starting distance. By Wald's second identity (Theorem 3.1.40),

$$\mathbb{E}_{(x,y)}\left[\tau_{\{0,n\}}^D\right] = d_{x,y}(n - d_{x,y}).$$

Applying Theorem 4.3.2 and Markov's inequality, we get

$$
\begin{aligned}
d(t) &\leq \max_{x,y \in V} \mathbb{P}_{(x,y)}[\tau_{\mathrm{coal}} > t] \\
&\leq \max_{x,y \in V} \frac{\mathbb{E}_{(x,y)}\left[\tau_{\{0,n\}}^D\right]}{t} \\
&= \max_{x,y \in V} \frac{d_{x,y}(n - d_{x,y})}{t} \\
&\leq \frac{n^2}{4t},
\end{aligned}
$$

or:

Claim 4.3.4

$$t_{\mathrm{mix}}(\varepsilon) \leq \frac{n^2}{4\varepsilon}.$$

By the diameter-based lower bound on mixing in Section 5.2.3, this bound gives the correct order of magnitude in n up to logarithmic factors. Indeed, the diameter is $\Delta = n/2$ and $\pi_{\min} = 1/n$ so that Claim 5.2.25 gives

$$t_{\mathrm{mix}}(\varepsilon) \geq \frac{n^2}{64 \log n}$$

for n large enough. Exercise 4.15 sketches a tighter lower bound.

Hypercube

Let $(Z_t)_{t \in \mathbb{Z}_+}$ be lazy simple random walk on the n-dimensional hypercube $\mathbb{Z}_2^n := \{0, 1\}^n$, where $i \sim j$ if $\|i - j\|_1 = 1$. We denote the coordinates of Z_t by $(Z_t^{(1)}, \ldots, Z_t^{(n)})$. This is equivalent to performing the Glauber dynamics chain on an empty graph (see Definition 1.2.8): at each step, we first pick a coordinate uniformly at random, then refresh its value. Because of the way the updating is done, the chain stays put with probability $1/2$ at each time as required.

Inspired by this observation, the coupling (X_t, Y_t) started at (x, y) is the following. At each time t, pick a coordinate i uniformly at random in $[n]$, pick a bit value b in $\{0, 1\}$ uniformly at random independent of the coordinate choice. Set *both* i coordinates to b, that is, $X_t^{(i)} = Y_t^{(i)} = b$. By design we reach coalescence when all coordinates have been updated at least once.

The following standard bound from the coupon collector's problem (see Example 2.1.4) is what is needed to conclude.

Lemma 4.3.5 *Let τ_{coll} be the time it takes to update each coordinate at least once. Then, for any $c > 0$,*

$$\mathbb{P}[\tau_{\mathrm{coll}} > \lceil n \log n + cn \rceil] \leq e^{-c}.$$

Proof Let B_i be the event that the ith coordinate has not been updated by time $\lceil n \log n + cn \rceil$. Then, using that $1 - x \leq e^{-x}$ for all x (see Exercise 1.16),

$$\mathbb{P}[\tau_{\mathrm{coll}} > \lceil n \log n + cn \rceil] \leq \sum_i \mathbb{P}[B_i]$$

$$= \sum_i \left(1 - \frac{1}{n}\right)^{\lceil n \log n + cn \rceil}$$

$$\leq n \exp\left(-\frac{n \log n + cn}{n}\right)$$

$$= e^{-c}.$$

$\blacksquare$

Applying Theorem 4.3.2, we get

$$d(\lceil n \log n + cn \rceil) \leq \max_{x,y \in V} \mathbb{P}_{(x,y)}[\tau_{\mathrm{coal}} > \lceil n \log n + cn \rceil]$$

$$\leq \mathbb{P}[\tau_{\mathrm{coll}} > \lceil n \log n + cn \rceil]$$

$$\leq e^{-c}.$$

Hence, for $c_\varepsilon > 0$ large enough:

Claim 4.3.6

$$t_{\mathrm{mix}}(\varepsilon) \leq \lceil n \log n + c_\varepsilon n \rceil.$$

Again we get a quick lower bound using the diameter-based result from Section 5.2.3. Here $\Delta = n$ and $\pi_{\min} = 1/2^n$ so that Claim 5.2.25 gives

$$t_{\mathrm{mix}}(\varepsilon) \geq \frac{n^2}{12 \log n + (4 \log 2)n} = \Omega(n)$$

for n large enough. So the upper bound we derived above is off at most by a logarithmic factor in n. In fact:

Claim 4.3.7

$$t_{\mathrm{mix}}(\varepsilon) \geq \frac{1}{2} n \log n - O(n).$$

HAMMING
WEIGHT

Proof For simplicity, we assume that n is odd. Let W_t be the number of 1s, or *Hamming weight*, at time t. Let A be the event that the Hamming weight is $\leq n/2$. To bound the mixing time, we use the fact that for any z_0,

$$d(t) \geq \| P^t(z_0, \cdot) - \pi \|_{\mathrm{TV}} \geq | P^t(z_0, A) - \pi(A) |. \tag{4.3.2}$$

Under the stationary distribution, the Hamming weight is equal in distribution to a $\mathrm{Bin}(n, 1/2)$. In particular, the probability that a majority of coordinates are 0 is $1/2$. That is, $\pi(A) = 1/2$.

On the other hand, let (Z_t) start at z_0, the all-1 vector. Let U_t be the number of updated coordinates up to time t in the Glauber dynamics representation of the chain discussed before the statement of Lemma 4.3.5. By the definition of A,

$$| P^t(z_0, A) - \pi(A) | = | \mathbb{P}[W_t \leq n/2] - 1/2 |. \tag{4.3.3}$$

We use Chebyshev's inequality (Theorem 2.1.2) to bound the probability on the right-hand side. So we need to compute the expectation and variance of W_t.

Observe that, conditioned on U_t, the Hamming weight W_t is equal in distribution to $\mathrm{Bin}(U_t, 1/2) + (n - U_t)$ as the updated coordinates are uniform and the other ones are 1. Thus we have

$$\begin{aligned}
\mathbb{E}[W_t] &= \mathbb{E}[\mathbb{E}[W_t \mid U_t]] \\
&= \mathbb{E}\left[\frac{1}{2} U_t + (n - U_t) \right] \\
&= \mathbb{E}\left[n - \frac{1}{2} U_t \right] \\
&= n - \frac{1}{2} n \left[1 - \left(1 - \frac{1}{n} \right)^t \right] \\
&= \frac{n}{2} \left[1 + \left(1 - \frac{1}{n} \right)^t \right], \tag{4.3.4}
\end{aligned}$$

where on the fourth line we used the fact that $\mathbb{E}[U_t] = n \left[1 - \left(1 - \frac{1}{n} \right)^t \right]$ by summing over the coordinates and using linearity of expectation.

As to the variance, using again the observation above about the distribution of W_t given U_t,

$$\text{Var}[W_t] = \mathbb{E}[\text{Var}[W_t \mid U_t]] + \text{Var}[\mathbb{E}[W_t \mid U_t]]$$
$$= \frac{1}{4}\mathbb{E}[U_t] + \frac{1}{4}\text{Var}[U_t]. \tag{4.3.5}$$

It remains to compute $\text{Var}[U_t]$. Let $I_t^{(i)}$ be 1 if coordinate i has *not* been updated up to time t and 0 otherwise. Note that for $i \neq j$,

$$\text{Cov}[I_t^{(i)}, I_t^{(j)}] = \mathbb{E}[I_t^{(i)}I_t^{(j)}] - \mathbb{E}[I_t^{(i)}]\mathbb{E}[I_t^{(j)}]$$
$$= \left(1 - \frac{2}{n}\right)^t - \left(1 - \frac{1}{n}\right)^{2t}$$
$$= \left(1 - \frac{2}{n}\right)^t - \left(1 - \frac{2}{n} + \frac{1}{n^2}\right)^t$$
$$\leq 0,$$

that is, $I_t^{(i)}$ and $I_t^{(j)}$ are negatively correlated, while

$$\text{Var}[I_t^{(i)}] = \mathbb{E}[(I_t^{(i)})^2] - (\mathbb{E}[I_t^{(i)}])^2 \leq \mathbb{E}[I_t^{(i)}] = \left(1 - \frac{1}{n}\right)^t.$$

Then, writing $n - U_t$ as the sum of these indicators, we have

$$\text{Var}[U_t] = \text{Var}[n - U_t]$$
$$= \sum_{i=1}^n \text{Var}[I_t^{(i)}] + 2\sum_{i<j} \text{Cov}[I_t^{(i)}, I_t^{(j)}]$$
$$\leq n\left(1 - \frac{1}{n}\right)^t.$$

Plugging this back into (4.3.5), we get

$$\text{Var}[W_t] \leq \frac{n}{4}\left[1 - \left(1 - \frac{1}{n}\right)^t\right] + \frac{n}{4}\left(1 - \frac{1}{n}\right)^t = \frac{n}{4}.$$

For $t_\alpha = \frac{1}{2}n\log n - n\log\frac{\alpha}{2}$ with $\alpha > 0$, by (4.3.4),

$$\mathbb{E}[W_{t_\alpha}] = \frac{n}{2} + e^{t_\alpha(-1/n + \Theta(1/n^2))} = \frac{n}{2} + \frac{\alpha}{2}\sqrt{n} + o(1),$$

where we used that by a Taylor expansion, for $|z| \leq 1/2$, $\log(1 - z) = -z + \Theta(z^2)$. Fix $0 < \varepsilon < 1/2$. By Chebyshev's inequality, for $t_\alpha = \frac{1}{2}n\log n - n\log\frac{\alpha}{2}$ and n large enough,

$$\mathbb{P}[W_{t_\alpha} \leq n/2] \leq \mathbb{P}[|W_{t_\alpha} - \mathbb{E}[W_{t_\alpha}]| \geq (\alpha/2)\sqrt{n}] \leq \frac{n/4}{(\alpha/2)^2 n} \leq \frac{1}{2} - \varepsilon$$

for α large enough. By (4.3.2) and (4.3.3), that implies $d(t_\alpha) \geq \varepsilon$ and we are done. ∎

The previous proof relies on a "distinguishing statistic." Recall from Lemma 4.1.19 that for any random variables X, Y and mapping h it holds that

$$\|\mu_{h(X)} - \mu_{h(y)}\|_{\mathrm{TV}} \le \|\mu_X - \mu_Y\|_{\mathrm{TV}},$$

where μ_Z is the law of Z. The mapping used in the proof of the claim is the Hamming weight. In essence, we gave a lower bound on the total variation distance between the laws of the Hamming weight at stationarity and under $P^t(z_0, \cdot)$. See Exercise 4.16 for a more general treatment of the distinguishing statistic approach.

CUTOFF

Remark 4.3.8 *The upper bound in Claim 4.3.6 is indeed off by a factor of 2. See [LPW06, Theorem 18.3] for an improved upper bound and a discussion of the so-called* cutoff *phenomenon. The latter refers to the fact that for all $0 < \varepsilon < 1/2$ it can be shown in this case that*

$$\lim_{n \to +\infty} \frac{\mathrm{t}^{(n)}_{\mathrm{mix}}(\varepsilon)}{\mathrm{t}^{(n)}_{\mathrm{mix}}(1-\varepsilon)} = 1,$$

where $\mathrm{t}^{(n)}_{\mathrm{mix}}(\varepsilon)$ is the mixing time on the n-dimensional hypercube. In words, for large n, the total variation distance drops from 1 to 0 in a short time window. See Exercise 5.10 for a necessary condition for cutoff.

b-ary tree

Let $(Z_t)_{t \in \mathbb{Z}_+}$ be lazy simple random walk on the ℓ-level rooted b-ary tree, $\widehat{\mathbb{T}}^\ell_b$, with $\ell \ge 2$. The root, 0, is on level 0 and the leaves, L, are on level ℓ. All vertices have degree $b + 1$, except for the root which has degree b and the leaves which have degree 1. By Example 1.1.29 (noting that laziness makes no difference), the stationary distribution is

$$\pi(x) := \frac{\delta(x)}{2(n-1)},$$

where n is the number of vertices and $\delta(x)$ is the degree of x. We used that a tree on n vertices has $n - 1$ edges (Corollary 1.1.7). We construct a coupling (X_t, Y_t) of this chain started at (x, y). Assume without loss of generality that x is no further from the root than y, which we denote by $x \preccurlyeq y$ (which, here, does *not* mean that y is a descendant of x). The coupling has two stages:

- In the first stage, at each time, flip a fair coin. On heads, Y_t stays put and X_t moves one step chosen uniformly at random among its neighbors. Similarly, on tails, reverse the roles of X_t and Y_t. Do this until X_t and Y_t are on the same level.
- In the second stage, that is, once the two chains are on the same level, at each time first let X_t move as a lazy simple random walk on $\widehat{\mathbb{T}}^\ell_b$. Then let Y_t move in the same direction as X_t, that is, if X_t moves closer to the root, so does Y_t, and so on.

By construction, $X_t \preccurlyeq Y_t$ for all t. The key observation is the following. Let τ^* be the first time (X_t) visits the root *after visiting the leaves*. By time τ^*, the two chains have necessarily met: because $X_t \preccurlyeq Y_t$, when X_t reaches the leaves, so does Y_t; after that time, the coupling is in the second stage so X_t and Y_t remain on the same level; in particular, when X_t reaches the root (after visiting the leaves), so does Y_t. Hence $\tau_{\mathrm{coal}} \le \tau^*$. Intuitively, the mixing time is indeed dominated by the time it takes to reach the root from the worst starting point, a leaf. See Figure 4.7 and the corresponding lower bound argument.

To estimate $\mathbb{P}_{(x,y)}[\tau^* > t]$, we use Markov's inequality (Theorem 2.1.1), for which we need a bound on $\mathbb{E}_{(x,y)}[\tau^*]$. We note that $\mathbb{E}_{(x,y)}[\tau^*]$ is less than the mean time for the walk to go from the root to the leaves and back. Let L_t be the level of X_t and let $\mathcal{N}$ be the corresponding network (where the conductances are equal to the number of edges on each level of the tree). In terms of L_t, the quantity we seek to bound is the mean of $\tau_{0,\ell}$, the commute time of the chain (L_t) between the states 0 and ℓ. By the commute time identity (Theorem 3.3.34),

$$\mathbb{E}[\tau_{0,\ell}] = c_{\mathcal{N}} \, \mathcal{R}(0 \leftrightarrow \ell), \tag{4.3.6}$$

where

$$c_{\mathcal{N}} = 2 \sum_{e=\{x,y\}\in\mathcal{N}} c(e) = 4(n-1),$$

where we simply counted the number of edges in $\widehat{\mathbb{T}}_b^\ell$ and the extra factor of 2 accounts for self-loops. Using network reduction techniques, we computed the effective resistance $\mathcal{R}(0 \leftrightarrow \ell)$ in Examples 3.3.21 and 3.3.22 – without self-loops. Of course, adding self-loops does not affect the effective resistance as we can use the same voltage and current. So, ignoring them, we get

$$\mathcal{R}(0 \leftrightarrow \ell) = \sum_{j=0}^{\ell-1} r(j,j+1) = \sum_{j=0}^{\ell-1} b^{-(j+1)} = \frac{1}{b} \cdot \frac{1-b^{-\ell}}{1-b^{-1}}, \tag{4.3.7}$$

which implies

$$\frac{1}{b} \leq \mathcal{R}(0 \leftrightarrow \ell) \leq \frac{1}{b-1} \leq 1.$$

Finally, applying Theorem 4.3.2 and Markov's inequality and using (4.3.6), we get

$$
\begin{aligned}
d(t) &\leq \max_{x,y\in V} \mathbb{P}_{(x,y)}[\tau^* > t] \\
&\leq \max_{x,y\in V} \frac{\mathbb{E}_{(x,y)}[\tau^*]}{t} \\
&\leq \frac{\mathbb{E}[\tau_{0,\ell}]}{t} \\
&\leq \frac{4n}{t},
\end{aligned}
$$

or:

Claim 4.3.9

$$t_{\mathrm{mix}}(\varepsilon) \leq \frac{4n}{\varepsilon}.$$

This time the diameter-based bound is far off. We have $\Delta = 2\ell = \Theta(\log n)$ and $\pi_{\min} = 1/2(n-1)$ so that Claim 5.2.25 gives

$$t_{\mathrm{mix}}(\varepsilon) \geq \frac{(2\ell)^2}{12\log n + 4\log(2(n-1))} = \Omega(\log n)$$

for n large enough.

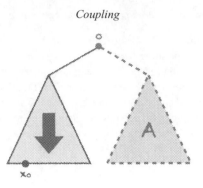

Figure 4.7 Setup for the lower bound on the mixing time on a *b*-ary tree. (Here $b = 2$.)

Here is a better lower bound. We take $b = 2$ to simplify. Intuitively, the mixing time is significantly greater than the squared diameter because the chain tends to be pushed away from the root. Consider the time it takes to go from the leaves on one side of the root to the leaves on the other, both of which have substantial weight under the stationary distribution. That typically takes time exponential in the diameter – that is, linear in n. Indeed, one first has to reach the root, which by the gambler's ruin problem (Example 3.1.43), takes an exponential in ℓ number of "excursions" (see Claim 3.1.44 (ii)).

Formally, let x_0 be a leaf of $\widehat{\mathbb{T}}_b^\ell$ and let A be the set of vertices "on the other side of root (inclusively)," that is, vertices whose graph distance from x_0 is at least ℓ. See Figure 4.7. Then, $\pi(A) \geq 1/2$ by symmetry. We use the fact that

$$\|P^t(x_0, \cdot) - \pi\|_{\mathrm{TV}} \geq |P^t(x_0, A) - \pi(A)|$$

to bound the mixing time from below. We claim that, started at x_0, the walk takes time linear in n to reach A with non-trivial probability.

Consider again the level L_t of X_t. Using definition of the effective resistance (Definition 3.3.19) as well as the expression for it in (4.3.7), we have

$$\mathbb{P}_\ell[\tau_0 < \tau_\ell^+] = \frac{1}{c(\ell)\mathscr{R}(0 \leftrightarrow \ell)} = \frac{1}{b^\ell} \cdot \frac{b-1}{1-b^{-\ell}} = \frac{b-1}{b^\ell - 1} = O\left(\frac{1}{n}\right).$$

Hence, started from the leaves, the number of excursions back to the leaves needed to reach the root for the first time is geometric with success probability $O(n^{-1})$. Each such excursion takes time at least 2 (which corresponds to going right back to the leaves after the first step). So $P^t(x_0, A)$ is bounded above by the probability that at least one such excursion was successful among the first $t/2$ attempts. That is,

$$P^t(x_0, A) \leq 1 - \left(1 - O\left(n^{-1}\right)\right)^{t/2} < \frac{1}{2} - \varepsilon,$$

for all $t \leq \alpha_\varepsilon n$ with $\alpha_\varepsilon > 0$ small enough and

$$\|P^{\alpha_\varepsilon n}(x_0, \cdot) - \pi\|_{\mathrm{TV}} \geq |P^{\alpha_\varepsilon n}(x_0, A) - \pi(A)| > \varepsilon.$$

We have proved that $t_{\mathrm{mix}}(\varepsilon) \geq \alpha_\varepsilon n$.

4.3.3 Path Coupling

Path coupling is a method for constructing Markovian couplings from "simpler" couplings. The building blocks are one-step couplings starting from pairs of initial states that are close in some "dissimilarity graph."

Let (X_t) be an irreducible Markov chain on a finite state space V with transition matrix P and stationary distribution π. Assume that we have a *dissimilarity graph* $H_0 = (V_0, E_0)$ on $V_0 := V$ with edge weights $w_0 \colon E_0 \to \mathbb{R}_+$. This graph need not have the same edges as the transition graph of (X_t). We extend w_0 to the *path metric*

DISSIMILARITY GRAPH, PATH METRIC

$$w_0(x,y) := \inf \left\{ \sum_{i=0}^{m-1} w_0(x_i, x_{i+1}) \colon x = x_0, x_1, \ldots, x_m = y \text{ is a path in } H_0 \right\},$$

where the infimum is over all paths connecting x and y in H_0. We call a path achieving the infimum a *minimum-weight path*. It is straightforward to check that w_0 satisfies the triangle inequality. Let

$$\Delta_0 := \max_{x,y} w_0(x,y)$$

be the *weighted diameter* of H_0.

Theorem 4.3.10 (Path coupling method). *Assume that*

$$w_0(u,v) \geq 1$$

for all $\{u,v\} \in E_0$. Assume further that there exists $\kappa \in (0,1)$ such that:

- (Local couplings) *For all x, y with $\{x,y\} \in E_0$, there is a coupling (X^*, Y^*) of $P(x, \cdot)$ and $P(y, \cdot)$ satisfying the* contraction *property*

$$\mathbb{E}[w_0(X^*, Y^*)] \leq \kappa\, w_0(x,y). \tag{4.3.8}$$

Then,

$$d(t) \leq \Delta_0\, \kappa^t,$$

or

$$\mathsf{t}_{\mathrm{mix}}(\varepsilon) \leq \left\lceil \frac{\log \Delta_0 + \log \varepsilon^{-1}}{\log \kappa^{-1}} \right\rceil.$$

Proof The crux of the proof is to extend (4.3.8) to arbitrary pairs of vertices.

Claim 4.3.11 (Global coupling). *For all $x, y \in V$ there is a coupling (X^*, Y^*) of $P(x, \cdot)$ and $P(y, \cdot)$ such that (4.3.8) holds.*

Iterating the coupling in this last claim immediately implies the existence of a coalescing Markovian coupling (X_t, Y_t) of P such that

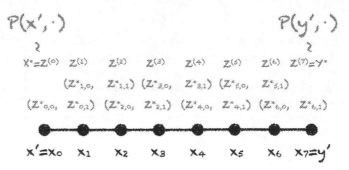

Figure 4.8 Coupling of $P(x', \cdot)$ and $P(y', \cdot)$ constructed from a sequence of local couplings $(Z^*_{0,0}, Z^*_{0,1}), \ldots, (Z^*_{0,m-1}, Z^*_{0,m-1})$.

$$\begin{aligned}
\mathbb{E}_{(x,y)}[w_0(X_t, Y_t)] &= \mathbb{E}_{(x,y)} \left[\mathbb{E}[w_0(X_t, Y_t) \mid X_{t-1}, Y_{t-1}] \right] \\
&\leq \mathbb{E}_{(x,y)} \left[\kappa \, w_0(X_{t-1}, Y_{t-1}) \right] \\
&\leq \cdots \\
&\leq \kappa^t \, \mathbb{E}_{(x,y)}[w_0(X_0, Y_0)] \\
&= \kappa^t \, w_0(x, y) \\
&\leq \kappa^t \, \Delta_0.
\end{aligned}$$

By assumption, $\mathbf{1}_{\{x \neq y\}} \leq w_0(x, y)$ so that by the coupling inequality and Lemma 4.3.1, we have

$$d(t) \leq \bar{d}(t) \leq \max_{x,y} \mathbb{P}_{(x,y)}[X_t \neq Y_t] \leq \max_{x,y} \mathbb{E}_{(x,y)}[w_0(X_t, Y_t)] \leq \kappa^t \, \Delta_0,$$

which implies the theorem.

Remark 4.3.12 *In essence, w_0 satisfies a form of Lyapounov condition (i.e., (3.3.15)) with a "geometric drift." See, for example, [MT09, chapter 15].*

It remains to prove Claim 4.3.11.

Proof of Claim 4.3.11 Fix $x', y' \in V$ such that $\{x', y'\}$ is *not* an edge in the dissimilarity graph H_0. The idea is to combine the local couplings on a minimum-weight path between x' and y' in H_0. Let $x' = x_0 \sim \cdots \sim x_m = y'$ be such a path. For all $i = 0, \ldots, m-1$, let $(Z^*_{i,0}, Z^*_{i,1})$ be a coupling of $P(x_i, \cdot)$ and $P(x_{i+1}, \cdot)$ satisfying the contraction property (4.3.8).

Then we proceed as follows. Set $Z^{(0)} := Z^*_{0,0}$ and $Z^{(1)} := Z^*_{0,1}$. Then iteratively pick $Z^{(i+1)}$ according to the law $\mathbb{P}[Z^*_{i,1} \in \cdot \mid Z^*_{i,0} = Z^{(i)}]$. By induction on i, $(X^*, Y^*) := (Z^{(0)}, Z^{(m)})$ is then a coupling of $P(x', \cdot)$ and $P(y', \cdot)$. See Figure 4.8.

To be more formal, define the transition matrix

$$R_i(z^{(i)}, z^{(i+1)}) := \mathbb{P}[Z^*_{i,1} = z^{(i+1)} \mid Z^*_{i,0} = z^{(i)}].$$

Observe that

$$\sum_{z^{(i+1)}} R_i(z^{(i)}, z^{(i+1)}) = 1 \tag{4.3.9}$$

and

$$\sum_{z^{(i)}} P(x_i, z^{(i)}) R_i(z^{(i)}, z^{(i+1)}) = P(x_{i+1}, z^{(i+1)}), \tag{4.3.10}$$

by construction of the coupling $(Z^*_{i,0}, Z^*_{i,1})$ and the definition of R_i. The law of the full coupling

$$(Z^{(0)}, \ldots, Z^{(m)})$$

is

$$\mathbb{P}[(Z^{(0)}, \ldots, Z^{(m)}) = (z^{(0)}, \ldots, z^{(m)})]$$
$$= P(x_0, z^{(0)}) R_0(z^{(0)}, z^{(1)}) \cdots R_{m-1}(z^{(m-1)}, z^{(m)}).$$

Using (4.3.9) and (4.3.10) inductively gives, respectively,

$$\mathbb{P}[X^* = z^{(0)}] = \mathbb{P}[Z^{(0)} = z^{(0)}] = P(x_0, z^{(0)})$$

and

$$\mathbb{P}[Y^* = z^{(m)}] = \mathbb{P}[Z^{(m)} = z^{(m)}] = P(x_m, z^{(m)}),$$

as required.

By the triangle inequality for w_0, the coupling (X^*, Y^*) satisfies

$$\mathbb{E}[w_0(X^*, Y^*)] = \mathbb{E}\left[w_0(Z^{(0)}, Z^{(m)})\right]$$
$$\leq \sum_{i=0}^{m-1} \mathbb{E}\left[w_0(Z^{(i)}, Z^{(i+1)})\right]$$
$$\leq \sum_{i=0}^{m-1} \kappa \, w_0(x_i, x_{i+1})$$
$$= \kappa \, w_0(x', y'),$$

where, on the third line, we used (4.3.8) for adjacent pairs and the last line follows from the fact that we chose a minimum-weight path. ∎

That concludes the proof of the theorem. ∎

We illustrate the path coupling method in the next subsection. See Exercise 4.17 for an optimal transport perspective on the path coupling method.

4.3.4 ▷ *Ising Model: Glauber Dynamics at High Temperature*

Let $G = (V, E)$ be a finite, connected graph with maximal degree $\bar{\delta}$. Define $\mathcal{X} := \{-1, +1\}^V$. Recall from Example 1.2.5 that the (ferromagnetic) Ising model on V with inverse temperature β is the probability distribution over spin configurations $\sigma \in \mathcal{X}$ given by

$$\mu_\beta(\sigma) := \frac{1}{\mathcal{Z}(\beta)} e^{-\beta \mathcal{H}(\sigma)},$$

where

$$\mathcal{H}(\sigma) := -\sum_{i \sim j} \sigma_i \sigma_j.$$

is the Hamiltonian and

$$\mathcal{Z}(\beta) := \sum_{\sigma \in \mathcal{X}} e^{-\beta \mathcal{H}(\sigma)}$$

is the partition function. In this context, recall that vertices are often referred to as sites. The single-site Glauber dynamics (Definition 1.2.8) of the Ising model is the Markov chain on $\mathcal{X}$, which, at each time, selects a site $i \in V$ uniformly at random and updates the spin σ_i according to $\mu_\beta(\sigma)$ conditioned on agreeing with σ at all sites in $V \setminus \{i\}$. Specifically, for $\gamma \in \{-1, +1\}$, $i \in V$, and $\sigma \in \mathcal{X}$, let $\sigma^{i,\gamma}$ be the configuration σ with the state at i being set to γ. Then, letting $n = |V|$, the transition matrix of the Glauber dynamics is

$$Q_\beta(\sigma, \sigma^{i,\gamma}) := \frac{1}{n} \cdot \frac{e^{\gamma \beta S_i(\sigma)}}{e^{-\beta S_i(\sigma)} + e^{\beta S_i(\sigma)}}$$

$$= \frac{1}{n} \left\{ \frac{1}{2} + \frac{1}{2} \tanh(\gamma \beta S_i(\sigma)) \right\}, \qquad (4.3.11)$$

where

$$S_i(\sigma) := \sum_{j \sim i} \sigma_j.$$

All other transitions have probability 0. Recall that this chain is irreducible and reversible with respect to μ_β. In particular, μ_β is the stationary distribution of Q_β.

In this section, we give an upper bound on the mixing time, $t_{mix}(\varepsilon)$, of Q_β using path coupling. We say that the Glauber dynamics is *fast mixing* if $t_{mix}(\varepsilon) = O(n \log n)$. We first make a simple observation:

Claim 4.3.13 (Glauber dynamics: lower bound on mixing).

$$t_{mix}(\varepsilon) = \Omega(n) \qquad \forall \beta > 0.$$

Proof Similarly to what we did in Section 4.3.2 in the context of random walk on the hypercube (but for a lower bound this time), we use a coupon collecting argument (see Example 2.1.4). Let $\bar{\sigma}$ be the all-(-1) configuration and let A be the set of configurations where at least half of the sites are $+1$. Then, by symmetry, $\mu_\beta(A) = \mu_\beta(A^c) = 1/2$, where we assumed for simplicity that n is odd. By definition of the total variation distance,

$$d(t) \geq \|Q_\beta^t(\bar{\sigma}, \cdot) - \mu_\beta(\cdot)\|_{TV}$$

$$\geq |Q_\beta^t(\bar{\sigma}, A) - \mu_\beta(A)|$$

$$= |Q_\beta^t(\bar{\sigma}, A) - 1/2|. \qquad (4.3.12)$$

So it remains to show that by time $c n$, for $c > 0$ small, the chain is unlikely to have reached A. That happens if, say, fewer than a third of the sites have been updated. Using the notation of Example 2.1.4, we are seeking a bound on $T_{n,n/3}$, that is, the time to collect $n/3$ coupons out of n.

We can write this random variable as a sum of $n/3$ independent geometric variables $T_{n,n/3} = \sum_{i=1}^{n/3} \tau_{n,i}$, where $\mathbb{E}[\tau_{n,i}] = \left(1 - \frac{i-1}{n}\right)^{-1}$ and $\text{Var}[\tau_{n,i}] \leq \left(1 - \frac{i-1}{n}\right)^{-2}$. Hence, approximating the Riemann sums in the next two displays by integrals, we get

$$\mathbb{E}[T_{n,n/3}] = \sum_{i=1}^{n/3} \left(1 - \frac{i-1}{n}\right)^{-1} = n \sum_{j=2n/3+1}^{n} j^{-1} = \Theta(n) \qquad (4.3.13)$$

and

$$\text{Var}[T_{n,n/3}] \leq \sum_{i=1}^{n/3} \left(1 - \frac{i-1}{n}\right)^{-2} = n^2 \sum_{j=2n/3+1}^{n} j^{-2} = \Theta(n). \qquad (4.3.14)$$

So by Chebyshev's inequality (Theorem 2.1.2),

$$\mathbb{P}[|T_{n,n/3} - \mathbb{E}[T_{n,n/3}]| \geq \varepsilon \, n] \leq \frac{\text{Var}[T_{n,n/3}]}{(\varepsilon \, n)^2} \to 0,$$

by (4.3.14). In view of (4.3.13), taking $\varepsilon > 0$ small enough and n large enough, we have shown that for $t \leq c_\varepsilon n$ for some $c_\varepsilon > 0$,

$$Q_\beta^t(\bar{\sigma}, A) \leq 1/3,$$

which proves the claim by (4.3.12) and the definition of the mixing time (Definition 1.1.35).

∎

Remark 4.3.14 *In fact, Ding and Peres proved that* $\text{t}_{\text{mix}}(\varepsilon) = \Omega(n \log n)$ *for any graph on* n *vertices [DP11]. In Claim 4.3.7, we treated the special case of the empty graph, which is equivalent to lazy random walk on the hypercube. See also Section 5.3.4 for a much stronger lower bound at low temperature for certain graphs with good "expansion properties."*

In our main result of this section, we show that the Glauber dynamics of the Ising model is fast mixing when the inverse temperature β is small enough as a function of the maximum degree.

Claim 4.3.15 (Glauber dynamics: fast mixing at high temperature).

$$\beta < \bar{\delta}^{-1} \implies \text{t}_{\text{mix}}(\varepsilon) = O(n \log n).$$

Proof We use path coupling. Let $H_0 = (V_0, E_0)$, where $V_0 := \mathcal{X}$ and $\{\sigma, \omega\} \in E_0$ if $\frac{1}{2}\|\sigma - \omega\|_1 = 1$ (i.e., they differ in exactly one coordinate) with unit weight on all edges. To avoid confusion, we reserve the notation $\sim$ for adjacency in G.

Let $\{\sigma, \omega\} \in E_0$ differ at coordinate i. We construct a coupling (X^*, Y^*) of $Q_\beta(\sigma, \cdot)$ and $Q_\beta(\omega, \cdot)$. We first pick the same coordinate i_* to update. If i_* is such that all its neighbors in G have the same state in σ and ω, that is, if $\sigma_j = \omega_j$ for all $j \sim i_*$, we update X^* from σ according to the Glauber rule and set $Y^* := X^*$. Note that this includes the case $i_* = i$. Otherwise, that is, if $i_* \sim i$, we proceed as follows. From the state σ, the probability of updating site i_* to state $\gamma \in \{-1, +1\}$ is given by the expression in brackets in (4.3.11), and similarly for ω. Unlike the previous case, we cannot guarantee that the update is identical in both chains. In order to minimize the chance of increasing the distance between the two chains, we use a monotone coupling, which recall from Example 4.1.17 is maximal in the two-state case. Specifically, we pick a uniform random variable U in $[-1, 1]$ and set

$$X_{i_*}^* := \begin{cases} +1 & \text{if } U \leq \tanh(\beta S_{i_*}(\sigma)), \\ -1 & \text{otherwise}, \end{cases}$$

and

$$Y_{i_*}^* := \begin{cases} +1 & \text{if } U \le \tanh(\beta S_{i_*}(\omega)), \\ -1 & \text{otherwise.} \end{cases}$$

We set $X_j^* := \sigma_j$ and $Y_j^* := \omega_j$ for all $j \ne i^*$. The expected distance between X^* and Y^* is then

$$\mathbb{E}[w_0(X^*, Y^*)]$$
$$= 1 - \underbrace{\frac{1}{n}}_{(a)} + \underbrace{\frac{1}{n} \sum_{j:j\sim i} \frac{1}{2} \left| \tanh(\beta S_j(\sigma)) - \tanh(\beta S_j(\omega)) \right|}_{(b)}, \qquad (4.3.15)$$

where (a) in Equation (4.3.14) corresponds to $i_* = i$, in which case $w_0(X^*, Y^*) = 0$; and (b) in Equation (4.3.14) corresponds to $i_* \sim i$, in which case $w_0(X^*, Y^*) = 2$ with probability

$$\frac{1}{2} \left| \tanh(\beta S_{i_*}(\sigma)) - \tanh(\beta S_{i_*}(\omega)) \right|$$

by our coupling; and otherwise $w_0(X^*, Y^*) = w_0(\sigma, \omega) = 1$. To bound (b) in Equation (4.3.14), we note that for any $j \sim i$,

$$\left| \tanh(\beta S_j(\sigma)) - \tanh(\beta S_j(\omega)) \right| = \tanh(\beta(s+2)) - \tanh(\beta s), \qquad (4.3.16)$$

where

$$s := S_j(\sigma) \wedge S_j(\omega).$$

The derivative of tanh is maximized at 0, where it is equal to 1. So the right-hand side of (4.3.16) is $\le \beta(s+2) - \beta s = 2\beta$. Plugging this back into (4.3.15) and using $1 - x \le e^{-x}$ for all x (see Exercise 1.16), we get

$$\mathbb{E}[w_0(X^*, Y^*)] \le 1 - \frac{1 - \bar{\delta}\beta}{n} \le \exp\left(-\frac{1 - \bar{\delta}\beta}{n}\right) = \kappa \, w_0(\sigma, \omega),$$

where

$$\kappa := \exp\left(-\frac{1 - \bar{\delta}\beta}{n}\right) < 1$$

by our assumption on β. The diameter of H_0 is $\Delta_0 = n$. By Theorem 4.3.10,

$$t_{\text{mix}}(\varepsilon) \le \left\lceil \frac{\log \Delta_0 + \log \varepsilon^{-1}}{\log \kappa^{-1}} \right\rceil = \left\lceil \frac{n(\log n + \log \varepsilon^{-1})}{1 - \bar{\delta}\beta} \right\rceil,$$

which implies the claim. ∎

Remark 4.3.16 *A slighlty more careful analysis shows that the condition $\bar{\delta} \tanh(\beta) < 1$ is enough for the claim to hold. See [LPW06, Theorem 15.1].*

4.4 Chen–Stein Method

The Chen–Stein method serves to establish Poisson approximation results with quantitative bounds in certain settings with dependent variables that are common, for instance, in random graphs and string statistics.

Setting The basic setup is a sum of Bernoulli (i.e., $\{0, 1\}$-valued) random variables $\{X_i\}_{i=1}^n$

$$W = \sum_{i=1}^n X_i, \tag{4.4.1}$$

where the X_is are *not* assumed independent or identically distributed. Define

$$p_i = \mathbb{P}[X_i = 1] \tag{4.4.2}$$

and

$$\mathbb{E}[W] = \lambda := \sum_{i=1}^n p_i. \tag{4.4.3}$$

Letting μ denote the law of W and π be the Poisson distribution with mean λ, our goal is to bound $\|\mu - \pi\|_{\mathrm{TV}}$.

We first state the main bounds and give some examples of its use. We then motivate and prove the result, and return to further applications. Throughout the next two subsections, we use the notation in (4.4.1), (4.4.2), and (4.4.3).

4.4.1 Main Bounds and Examples

We begin with an elementary observation.

Theorem 4.4.1 (Stein equation for the Poisson distribution). *Let $\lambda > 0$. A non-negative integer-valued random variable Z is $\mathrm{Poi}(\lambda)$ if and only if for all g bounded,*

$$\mathbb{E}[\lambda g(Z + 1) - Z g(Z)] = 0. \tag{4.4.4}$$

The "only if" follows a direct calculation. The "if" follows from taking $g(z) := \mathbf{1}_{\{z=k\}}$ for all $k \geq 1$ and deriving a recursion. Exercise 4.18 asks for the details. One might expect that if the left-hand side of (4.4.4) is "small for many gs," then Z is approximately Poisson.

The following key result in some sense helps to formalize this intuition. We prove it by constructing a Markov chain that "interpolates" between μ and π, where (4.4.4) will arise naturally (see Section 4.4.2).

Theorem 4.4.2 (Chen–Stein method). *Let $W \sim \mu$ and $\pi \sim \mathrm{Poi}(\lambda)$. Then there exists a function $h \colon \{0, 1, \ldots, n + 1\} \to \mathbb{R}$ such that*

$$\|\mu - \pi\|_{\mathrm{TV}} = \mathbb{E}\left[-\lambda h(W + 1) + W h(W)\right]. \tag{4.4.5}$$

Moreover, h satisfies the following Lipschitz condition: for all $y, y' \in \{0, 1, \ldots, n + 1\}$,

$$|h(y') - h(y)| \leq (1 \wedge \lambda^{-1})|y' - y|. \tag{4.4.6}$$

By bounding the right-hand side of (4.4.5) for any function satisfying (4.4.6), we get a Poisson approximation result for μ.

One way to do this is to construct a certain type of coupling. We begin with a definition, which will be justified in Corollary 4.4.4. We write $X \sim Y|A$ to mean that X is distributed as Y conditioned on the event A.

Definition 4.4.3 (Stein coupling). *A Stein coupling is a pair (U_i, V_i), for each $i = 1, \ldots, n$, such that*

$$U_i \sim W, \qquad V_i \sim W - 1 | X_i = 1.$$

Each pair (U_i, V_i) is defined on a joint probability space, but different pairs do not need to.

How such a coupling is constructed will become clearer in the examples below.

Corollary 4.4.4 *Let (U_i, V_i), $i = 1, \ldots, n$, be a Stein coupling. Then,*

$$\|\mu - \pi\|_{\text{TV}} \le (1 \wedge \lambda^{-1}) \sum_{i=1}^{n} p_i \, \mathbb{E} |U_i - V_i|. \tag{4.4.7}$$

Proof By (4.4.5), using the facts that $\lambda = \sum_{i=1}^{n} p_i$ and $W = \sum_{i=1}^{n} X_i$, we get

$$\begin{aligned}
&\|\mu - \pi\|_{\text{TV}} \\
&= \mathbb{E}\left[-\lambda h(W + 1) + W h(W)\right] \\
&= \mathbb{E}\left[-\left(\sum_{i=1}^{n} p_i\right) h(W + 1) + \left(\sum_{i=1}^{n} X_i\right) h(W)\right] \\
&= \sum_{i=1}^{n} \left(-p_i \mathbb{E}\left[h(W + 1)\right] + \mathbb{E}\left[X_i h(W)\right]\right) \\
&= \sum_{i=1}^{n} \left(-p_i \mathbb{E}\left[h(W + 1)\right] + \mathbb{E}\left[h(W) \,|\, X_i = 1\right] \mathbb{P}[X_i = 1]\right) \\
&= \sum_{i=1}^{n} p_i \left(-\mathbb{E}\left[h(W + 1)\right] + \mathbb{E}\left[h(W) \,|\, X_i = 1\right]\right).
\end{aligned}$$

Let (U_i, V_i), $i = 1, \ldots, n$, be a Stein coupling (Definition 4.4.3). Then, we can rewrite this last expression as

$$\begin{aligned}
&= \sum_{i=1}^{n} p_i \left(-\mathbb{E}\left[h(U_i + 1)\right] + \mathbb{E}\left[h(V_i + 1)\right]\right) \\
&\le \sum_{i=1}^{n} p_i \mathbb{E}\left[|h(U_i + 1) - h(V_i + 1)|\right].
\end{aligned}$$

By (4.4.6), we finally get

$$\|\mu - \pi\|_{\text{TV}} \le (1 \wedge \lambda^{-1}) \sum_{i=1}^{n} p_i \, \mathbb{E} |U_i - V_i|,$$

which concludes the proof.   ∎

As a first example, we derive a Poisson approximation result in the independent case. Compare to Theorem 4.1.18.

Example 4.4.5 (Independent X_is). Assume the X_is are independent. We prove the following:

Claim 4.4.6

$$\|\mu - \pi\|_{\mathrm{TV}} \le (1 \wedge \lambda^{-1}) \sum_{i=1}^{n} p_i^2.$$

We use the following Stein coupling. For each $i = 1, \dots, n$, we let

$$U_i = W$$

and

$$V_i = \sum_{j:j\neq i} X_j.$$

By independence,

$$V_i = W - X_i \sim W - 1 | X_i = 1,$$

as desired. Plugging into (4.4.7), we obtain the bound

$$\|\mu - \pi\|_{\mathrm{TV}} \le (1 \wedge \lambda^{-1}) \sum_{i=1}^{n} p_i \, \mathbb{E} |U_i - V_i|$$

$$\le (1 \wedge \lambda^{-1}) \sum_{i=1}^{n} p_i \, \mathbb{E} \left| W - \sum_{j\neq i} X_j \right|$$

$$\le (1 \wedge \lambda^{-1}) \sum_{i=1}^{n} p_i \, \mathbb{E} \, |X_i|$$

$$\le (1 \wedge \lambda^{-1}) \sum_{i=1}^{n} p_i^2.$$

◀

Here is a less straightforward example.

Example 4.4.7 (Balls in boxes). Suppose we throw k balls uniformly at random in n boxes independently. Let

$$X_i = \mathbf{1}\{\text{box } i \text{ is empty}\},$$

and let $W = \sum_{i=1}^{n} X_i$ be the number of empty boxes. Note that the X_is are *not* independent. In particular, we cannot use Theorem 4.1.18. Note that

$$p_i = \left(1 - \frac{1}{n}\right)^k$$

for all i and, hence,

$$\lambda = n \left(1 - \frac{1}{n}\right)^k.$$

For each $i = 1, \dots, n$, we generate the coupling (U_i, V_i) in the following way. We let $U_i = W$. If box i is empty, then $V_i = W - 1$. Otherwise, we redistribute all balls in box i among the remaining boxes and let V_i count the number of empty boxes $\neq i$. By construction, both conditions of the Stein coupling are satisfied. Moreover, we have almost surely $V_i \leq U_i$ so that

$$\sum_{i=1}^{n} p_i \mathbb{E}|U_i - V_i| = \sum_{i=1}^{n} p_i \mathbb{E}[U_i - V_i] = \lambda^2 - \sum_{i=1}^{n} p_i \mathbb{E}[V_i].$$

By the fact that $V_i \sim U_i - 1 | X_i = 1$ and Bayes' rule,

$$\sum_{i=1}^{n} p_i \mathbb{E}[V_i] = \sum_{i=1}^{n} \mathbb{P}[X_i = 1] \sum_{k=1}^{n} (k-1) \mathbb{P}[V_i = k-1]$$

$$= \sum_{i=1}^{n} \sum_{k=1}^{n} (k-1) \mathbb{P}[U_i = k \mid X_i = 1] \, \mathbb{P}[X_i = 1]$$

$$= \sum_{i=1}^{n} \sum_{k=1}^{n} (k-1) \mathbb{P}[X_i = 1 \mid U_i = k] \, \mathbb{P}[U_i = k].$$

Now we use the fact that $\mathbb{P}[X_i = 1 \mid U_i = k] = \mathbb{E}[X_i \mid U_i = k]$ because X_i is an indicator variable. So the last line above is

$$= \sum_{i=1}^{n} \sum_{k=1}^{n} (k-1) \mathbb{E}[X_i \mid W = k] \, \mathbb{P}[W = k]$$

$$= \sum_{k=1}^{n} (k-1) \mathbb{E}\left[\sum_{i=1}^{n} X_i \,\middle|\, W = k \right] \mathbb{P}[W = k]$$

$$= \sum_{k=1}^{n} (k-1) k \, \mathbb{P}[W = k]$$

$$= \mathbb{E}[W^2] - \mathbb{E}[W].$$

It remains to compute $\mathbb{E}[W^2]$. We have by symmetry

$$\mathbb{E}[W^2] = n \, \mathbb{E}[X_1^2] + n(n-1) \mathbb{E}[X_1 X_2]$$

$$= \lambda + n(n-1) \left(1 - \frac{2}{n} \right)^k,$$

so by Corollary 4.4.4,

$$\|\mu - \pi\|_{\mathrm{TV}} \leq (1 \wedge \lambda^{-1}) \left\{ n^2 \left(1 - \frac{1}{n} \right)^{2k} - n(n-1) \left(1 - \frac{2}{n} \right)^k \right\}.$$

When $k = n \log n + Cn$ for instance, it can be checked that $\|\mu - \pi\|_{\mathrm{TV}} = O(\log n / n)$. ◀

This last example is generalized in Exercise 4.22.

In special settings, one can give useful general bounds by constructing an appropriate Stein coupling. We give an important example next. Recall that $[n] = \{1, \ldots, n\}$.

Theorem 4.4.8 (Chen–Stein method: dissociated case). *Suppose that for each i there is a neighborhood $\mathcal{N}_i \subseteq [n] \setminus \{i\}$ such that*

$$X_i \text{ is independent of } \{X_j \colon j \notin \mathcal{N}_i \cup \{i\}\}.$$

Then,

$$\|\mu - \pi\|_{\text{TV}} \le (1 \wedge \lambda^{-1}) \sum_{i=1}^{n} \left\{ p_i^2 + \sum_{j \in \mathcal{N}_i} (p_i p_j + \mathbb{E}[X_i X_j]) \right\}.$$

Proof   We use the following Stein coupling. Let

$$U_i = W.$$

Then generate

$$(Y_j^{(i)})_{j \in \mathcal{N}_i} \sim (X_j)_{j \in \mathcal{N}_i} | \{X_k : k \notin \mathcal{N}_i \cup \{i\}\}, X_i = 1,$$

and set

$$V_i = \sum_{k \notin \mathcal{N}_i \cup \{i\}} X_k + \sum_{j \in \mathcal{N}_i} Y_j^{(i)}.$$

Because the law of $\{X_k : k \notin \mathcal{N}_i \cup \{i\}\}$ (and therefore of the first term in V_i) is independent of the event $\{X_i = 1\}$, the above scheme satisfies the conditions of the Stein coupling.

Hence we can apply Corollary 4.4.4. The construction of (U_i, V_i) guarantees that $U_i - V_i$ depends *only* on "i and its neighborhood." Specifically, we get

$$\|\mu - \pi\|_{\text{TV}} \le (1 \wedge \lambda^{-1}) \sum_{i=1}^{n} p_i \, \mathbb{E}|U_i - V_i|$$

$$= (1 \wedge \lambda^{-1}) \sum_{i=1}^{n} p_i \, \mathbb{E} \left| \sum_{j=1}^{n} X_j - \sum_{k \notin \mathcal{N}_i \cup \{i\}} X_k - \sum_{j \in \mathcal{N}_i} Y_j^{(i)} \right|$$

$$= (1 \wedge \lambda^{-1}) \sum_{i=1}^{n} p_i \, \mathbb{E} \left| X_i + \sum_{j \in \mathcal{N}_i} (X_j - Y_j^{(i)}) \right|$$

$$\le (1 \wedge \lambda^{-1}) \sum_{i=1}^{n} p_i \left(\mathbb{E}|X_i| + \sum_{j \in \mathcal{N}_i} (\mathbb{E}|X_j| + \mathbb{E}|Y_j^{(i)}|) \right),$$

where we used the triangle inequality. Recalling that $p_i = \mathbb{P}[X_i = 1] = \mathbb{E}[X_i] = \mathbb{E}|X_i|$ and the definition of $Y_j^{(i)}$, the last expression above is

$$= (1 \wedge \lambda^{-1}) \sum_{i=1}^{n} p_i \left(p_i + \sum_{j \in \mathcal{N}_i} [p_j + \mathbb{E}[|X_j||X_i = 1]] \right)$$

$$= (1 \wedge \lambda^{-1}) \sum_{i=1}^{n} \left\{ p_i^2 + \sum_{j \in \mathcal{N}_i} (p_i p_j + p_i \mathbb{E}[X_j | X_i = 1]) \right\}$$

$$= (1 \wedge \lambda^{-1}) \sum_{i=1}^{n} \left\{ p_i^2 + \sum_{j \in \mathcal{N}_i} (p_i p_j + \mathbb{E}[X_i X_j]) \right\}.$$

That concludes the proof.   ∎

Next we give an example of the previous theorem.

Example 4.4.9 (Longest head run). Let $0 < q < 1$ and let $Z_1, Z_2, \ldots$ be i.i.d. Bernoulli random variables with success probability $q = \mathbb{P}[Z_i = 1]$. We are interested in the distribution of R, the length of the longest run of 1s starting in the first n tosses. For any positive integer t, let $X_1^{(t)} := Z_1 \cdots Z_t$ and

$$X_i^{(t)} := (1 - Z_{i-1})Z_i \cdots Z_{i+t-1}, \qquad i \geq 2.$$

The event $\{X_i^{(t)} = 1\}$ indicates that a head run of length at least t *starts* at the ith toss. Now define

$$W^{(t)} := \sum_{i=1}^{n} X_i^{(t)}.$$

The key observation is that

$$\{R < t\} = \{W^{(t)} = 0\}. \tag{4.4.8}$$

Notice that, for fixed t, the $X_i^{(t)}$s are neither independent nor identically distributed. However, they exhibit a natural neighborhood structure as in Theorem 4.4.8. Indeed, let

$$\mathcal{N}_i^{(t)} := \{\alpha \in [n]: |\alpha - i| \leq t\} \setminus \{i\}.$$

Then, $X_i^{(t)}$ is independent of $\{X_j^{(t)}: j \notin \mathcal{N}_i \cup \{i\}\}$. For example,

$$X_i^{(t)} = (1 - Z_{i-1})Z_i \cdots Z_{i+t-1}$$

and

$$X_{i+t+1}^{(t)} = (1 - Z_{i+t})Z_{i+t+1} \cdots Z_{i+2t}$$

do not depend on any common Z_j, while $X_i^{(t)}$ and

$$X_{i+t}^{(t)} = (1 - Z_{i+t-1})Z_{i+t} \cdots Z_{i+2t-1}$$

both depend on Z_{i+t-1}.

We compute the quantities needed to apply Theorem 4.4.8. We have

$$p_1^{(t)} := \mathbb{E}[Z_1 \cdots Z_t] = \prod_{j=1}^{t} \mathbb{E}[Z_j] = q^t,$$

and, for $i \geq 2$,

$$\begin{aligned} p_i^{(t)} &:= \mathbb{E}[(1 - Z_{i-1})Z_i \cdots Z_{i+t-1}] \\ &= \mathbb{E}[1 - Z_{i-1}] \prod_{j=i}^{i+t-1} \mathbb{E}[Z_j] \\ &= (1 - q)q^t \\ &\leq q^t. \end{aligned}$$

For $i \geq 1$ and $j \in \mathcal{N}_i^{(t)}$, observe that a head run of length at least t cannot start simultaneously at i and j. So $\mathbb{E}[X_i^{(t)} X_j^{(t)}] = 0$ in that case. We also have

$$\lambda^{(t)} := \mathbb{E}[W^{(t)}] = q^t + (n-1)(1-q)q^t \in [n(1-q)q^t, nq^t]$$

and

$$\left| \mathcal{N}_i^{(t)} \right| \leq 2t.$$

We are ready to apply Theorem 4.4.8. We get

$$\|\mu - \pi\|_{\mathrm{TV}} \leq (1 \wedge (\lambda^{(t)})^{-1}) \sum_{i=1}^{n} \left\{ (p_i^{(t)})^2 + \sum_{j \in \mathcal{N}_i^{(t)}} \left(p_i^{(t)} p_j^{(t)} + \mathbb{E}[X_i^{(t)} X_j^{(t)}] \right) \right\}$$

$$\leq (1 \wedge (n(1-q)q^t)^{-1}) \left[nq^{2t} + 2tnq^{2t} \right]$$

$$\leq \frac{1}{(1-q)n} (1 \wedge (nq^t)^{-1})[2t+1](nq^t)^2.$$

This bound is non-asymptotic – it holds for any q, n, t. One special regime of note is $t = \log_{1/q} n + C$ with large n. In that case, we have $nq^t \to C'$ as $n \to +\infty$ for some $0 < C' < +\infty$ and the total variation above is of the order of $O(\log n / n)$.

Going back to (4.4.8), we finally obtain when $t = \log_{1/q} n + C$ that

$$\left| \mathbb{P}[R < t] - e^{-\lambda^{(t)}} \right| = O\left(\frac{\log n}{n} \right),$$

where recall that R and $\lambda^{(t)}$ implicitly depend on n. ◀

4.4.2 Some Motivation and Proof

The idea behind the Chen–Stein method is to interpolate between μ and π in Theorem 4.4.2 by constructing a Markov chain with initial distribution μ and stationary distribution π. Here we use a discrete-time, finite Markov chain.

Proof of Theorem 4.4.2 We seek a bound on

$$\|\mu - \pi\|_{\mathrm{TV}} = \sup_{A \subseteq \mathbb{Z}_+} |\mu(A) - \pi(A)|$$

$$= \mu(A^*) - \pi(A^*)$$

$$= \sum_{z \in A^*} (\mu(z) - \pi(z)), \tag{4.4.9}$$

where $A^* = \{z \in \mathbb{Z}_+ : \mu(z) > \pi(z)\}$, by Lemma 4.1.15. Since $W \leq n$ almost surely, $\mu(z) = 0$ for all $z > n$, which implies that $A^* \subseteq \{0, 1, \ldots, n\}$. In particular, it will suffice to bound $\mu(z) - \pi(z)$ for $0 \leq z \leq n$. We also assume $\lambda < n$ (the case $\lambda = n$ being uninteresting).

Constructing the Markov chain It will be convenient to truncate π at n, that is, we define

$$\bar{\pi}(z) = \begin{cases} \pi(z), & 0 \leq z \leq n, \\ 1 - \Pi(n), & z = n+1, \\ 0 & \text{otherwise,} \end{cases}$$

where $\Pi(z) = \sum_{w \leq z} \pi(w)$ is the cumulative distribution function of the Poisson distribution with mean λ. We construct a Markov chain with stationary distribution $\bar{\pi}$. We will also need the chain to be aperiodic and irreducible over $\{0, 1, \ldots, n+1\}$.

We choose the transition matrix $(P(x, y))_{0 \leq x, y \leq n+1}$ to be that of a birth–death chain reversible with respect to $\bar{\pi}$, that is, we require $P(x, y) = 0$ unless $|x - y| \leq 1$ and

$$\frac{P(x, x+1)}{P(x+1, x)} = \frac{\bar{\pi}(x+1)}{\bar{\pi}(x)} \qquad \forall x \in [n]. \tag{4.4.10}$$

For $x < n$,

$$\frac{\bar{\pi}(x+1)}{\bar{\pi}(x)} = \frac{\pi(x+1)}{\pi(x)} = \frac{e^{-\lambda} \lambda^{x+1}/(x+1)!}{e^{-\lambda} \lambda^x/x!} = \frac{\lambda}{x+1}.$$

In view of this, we want $P(x, x+1) \propto \lambda$ and $P(x, x-1) \propto x$. We choose the proportionality constant to ensure that all transition probabilities are in $[0, 1]$. Specifically, for $x \neq y$, the non-zero transition probabilities take values

$$P(x, y) = \begin{cases} \frac{1}{2n}\lambda & \text{if } 0 \leq x \leq n, y = x+1, \\ \frac{1}{2n}x & \text{if } 1 \leq x \leq n, y = x-1, \\ \frac{1}{2n}\lambda \frac{\pi(n)}{1-\Pi(n)} & \text{if } x = n+1, y = n. \end{cases} \tag{4.4.11}$$

The probability of staying put is $1 - \frac{1}{2n}\lambda$ if $x = 0$, $1 - \frac{1}{2n}x - \frac{1}{2n}\lambda$ if $1 \leq x \leq n$, and $1 - \frac{1}{2n}\lambda \frac{\pi(n)}{1-\Pi(n)}$ if $x = n+1$. Those are all strictly positive when $\lambda < n$. Hence, by construction P is aperiodic and irreducible, and it satisfies the detailed balance conditions (4.4.10).

Recalling (3.3.6), the Laplacian is

$$\Delta f(x) = \sum_y P(x, y)[f(y) - f(x)]$$

$$= P(x, x+1)[f(x+1) - f(x)] - P(x, x-1)[f(x) - f(x-1)]$$

$$= \lambda g(x+1) - x g(x)$$

for $0 \leq x \leq n$, where we defined

$$g(x) := \frac{f(x) - f(x-1)}{2n}, \qquad x \in \{1, \ldots, n+1\} \tag{4.4.12}$$

and $g(0)$ is arbitrary. At $x = n+1$,

$$\Delta f(n+1) = -\lambda n \frac{\pi(n)}{1 - \Pi(n)} g(n+1).$$

It is a standard fact (see Exercise 4.19) that the expectation of the Laplacian under the stationary distribution is 0. Inverting the relationship (4.4.12), for any $g: \{0, \ldots, n+1\} \to \mathbb{R}$, there is a corresponding f (unique up to an additive constant). So we have shown that if $Z \sim \bar{\pi}$ then

$$\mathbb{E}\left[(\lambda g(Z+1) - Zg(Z))\mathbf{1}_{\{Z \leq n\}} - \lambda n \frac{\pi(n)}{1 - \Pi(n)} g(Z)\mathbf{1}_{\{Z = n+1\}}\right] = 0,$$

that is,

$$\mathbb{E}\left[(\lambda g(Z+1) - Zg(Z))\mathbf{1}_{\{Z \leq n\}}\right] = \lambda n \pi(n) g(n+1).$$

Notice that, if g is extended to a bounded function on $\mathbb{Z}_+$, λ is fixed and $Z \sim \text{Poi}(\lambda)$, then taking $n \to +\infty$ recovers Theorem 4.4.1 by dominated convergence (Proposition B.4.14).[1]

Markov chains calculations By the convergence theorem for Markov chains (Theorem 1.1.33),

$$P^t(y, z) \to \bar{\pi}(z)$$

for all $0 \leq y \leq n+1$ and $0 \leq z \leq n+1$ as $t \to +\infty$. Hence, letting $\delta_z(x) = \mathbf{1}_{\{x=z\}}$, by telescoping

$$\delta_z(y) - \bar{\pi}(z) = \lim_{t \to +\infty} \mathbb{E}_y[\delta_z(X_0) - \delta_z(X_t)]$$

$$= \lim_{t \to +\infty} \sum_{s=0}^{t-1} \mathbb{E}_y[\delta_z(X_s) - \delta_z(X_{s+1})], \qquad (4.4.13)$$

where the subscript of $\mathbb{E}$ indicates the initial state. We will later take expectations over μ to interpolate between μ and π.

First, we use standard Markov chains facts to compute (4.4.13). Define for $y \in \{1, \ldots, n+1\}$,

$$g_z^t(y) := \frac{1}{2n} \sum_{s=0}^{t-1} (\mathbb{E}_y[\delta_z(X_s)] - \mathbb{E}_{y-1}[\delta_z(X_s)]) \qquad (4.4.14)$$

and $g_z^t(0) := 0$. The function $g_z^t(y)$ is, up to a factor (whose purpose will be clearer below), the difference between the expected number of visits to z up to time $t-1$ when started at y and $y-1$, respectively. It depends on μ only through λ and n. By Chapman–Kolmogorov (Theorem 1.1.20) applied to the first step of the chain,

$$\mathbb{E}_y[\delta_z(X_{s+1})] = P(y, y+1)\,\mathbb{E}_{y+1}[\delta_z(X_s)]$$
$$+ P(y, y)\,\mathbb{E}_y[\delta_z(X_s)] + P(y, y-1)\,\mathbb{E}_{y-1}[\delta_z(X_s)].$$

Using that $P(y, y+1) + P(y, y) + P(y, y-1) = 1$ and rearranging we get for $0 \leq y \leq n$ and $0 \leq z \leq n+1$,

---

[1] The above argument is more natural in the setting of continuous-time Markov chains, but we will not introduce them here.

$$\sum_{s=0}^{t-1} \mathbb{E}_y[\delta_z(X_s) - \delta_z(X_{s+1})]$$

$$= \sum_{s=0}^{t-1} \left\{ - P(y, y+1)(\mathbb{E}_{y+1}[\delta_z(X_s)] - \mathbb{E}_y[\delta_z(X_s)]) \right.$$

$$\left. + P(y, y-1)(\mathbb{E}_y[\delta_z(X_s)] - \mathbb{E}_{y-1}[\delta_z(X_s)]) \right\}$$

$$= -2nP(y, y+1)g_z^t(y+1) + 2nP(y, y-1)g_z^t(y)$$

$$= -\lambda g_z^t(y+1) + y g_z^t(y), \tag{4.4.15}$$

where we used (4.4.11) on the last line.

We establish after the proof of the theorem that $g_z^t(y)$ has a well-defined limit. That fact is not immediately obvious as the limit is the "difference of two infinities." But a simple coupling argument does the trick.

Lemma 4.4.10 Let $g_z^t \colon \{0, 1, \ldots, n+1\} \to \mathbb{R}$ be defined in (4.4.14). Then there exists a bounded function $g_z^\infty \colon \{0, 1, \ldots, n+1\} \to \mathbb{R}$ such that for all $0 \le z \le n+1$ and $0 \le y \le n+1$,

$$g_z^\infty(y) = \lim_{t \to +\infty} g_z^t(y).$$

In fact, an explicit expression for g_z^∞ can be derived via the following recursion. That expression will be helpful to establish the Lipschitz condition in Theorem 4.4.2.

Lemma 4.4.11 For all $0 \le y \le n$ and $0 \le z \le n+1$,

$$\delta_z(y) - \bar{\pi}(z) = -\lambda g_z^\infty(y+1) + y g_z^\infty(y).$$

Proof Combine (4.4.13), (4.4.15), and Lemma 4.4.10. ∎

Lemma 4.4.11 leads to the following formula for g_z^∞, which we establish after the proof of the theorem.

Lemma 4.4.12 For $1 \le y \le n+1$ and $0 \le z \le n+1$,

$$g_z^\infty(y) = \begin{cases} \frac{\Pi(y-1)}{y\pi(y)} \bar{\pi}(z) & \text{if } z \ge y, \\ -\frac{1-\Pi(y-1)}{y\pi(y)} \bar{\pi}(z) & \text{if } z < y. \end{cases} \tag{4.4.16}$$

and $g_z^\infty(0) = 0$.

Interpolating between μ and π For $A \subseteq \{0, 1, \ldots, n\}$, define

$$g_A^\infty(y) := \sum_{z \in A} g_z^\infty(y).$$

We obtain the following key bound.

Lemma 4.4.13 (Chen's equation). *Let $W \sim \mu$ and $\pi \stackrel{\mathrm{d}}{=} \mathrm{Poi}(\lambda)$. Then,*

$$\|\mu - \pi\|_{\mathrm{TV}} = \mathbb{E}\left[-\lambda g_{A^*}^\infty(W+1) + W g_{A^*}^\infty(W)\right], \tag{4.4.17}$$

where $A^* = \{z \in \mathbb{Z}_+ : \mu(z) > \pi(z)\}$.

Proof Fix $z \in \{0, 1, \ldots, n\}$. Multiplying both sides in Lemma 4.4.11 by $\mu(y)$ and summing over y in $\{0, 1, \ldots, n\}$ gives

$$\mu(z) - \pi(z) = \mathbb{E}\left[-\lambda g_z^\infty(W + 1) + W g_z^\infty(W)\right].$$

Now summing over z in $A^* \subseteq \{0, 1, \ldots, n\}$ and using (4.4.9) gives the claim. ∎

Lemma 4.4.12 can be used to derive a Lipschitz constant for g_A^∞. That lemma is also established after the proof of the theorem.

Lemma 4.4.14 *For $A \subseteq \{0, 1, \ldots, n\}$ and $y, y' \in \{0, 1, \ldots, n + 1\}$,*

$$|g_A^\infty(y') - g_A^\infty(y)| \le (1 \wedge \lambda^{-1})|y' - y|.$$

Lemmas 4.4.13 and 4.4.14 imply the theorem with $h := g_{A^*}^\infty$. ∎

Proofs of technical lemmas It remains to prove Lemmas 4.4.10, 4.4.12, and 4.4.14.

Proof of Lemma 4.4.10 We use a coupling argument. Let $(Y_s, \tilde{Y}_s)_{s=0}^{+\infty}$ be an independent Markovian coupling of (Y_s), the chain started at $y-1$, and $(\tilde{Y}_s)$, the chain started at y. Let τ be the first time s that $Y_s = \tilde{Y}_s$. Because Y_s and $\tilde{Y}_s$ are independent and P is a birth–death chain with strictly positive nearest-neighbor and staying-put transition probabilities, the coupled chain $(Y_s, \tilde{Y}_s)_{s=0}^{+\infty}$ is aperiodic and irreducible over $\{0, 1, \ldots, n + 1\}^2$. By the exponential tail of hitting times, Lemma 3.1.25, it holds that $\mathbb{E}[\tau] < +\infty$.

Modify the coupling $(Y_s, \tilde{Y}_s)$ to enforce $\tilde{Y}_s = Y_s$ for all $s \ge \tau$ (while not changing (Y_s)), that is, to make it coalescing. By the Strong Markov property (Theorem 3.1.8), the resulting chain $(Y_s^*, \tilde{Y}_s^*)$ is also a Markovian coupling of the chain started at $y - 1$ and y, respectively. Using this coupling, we rewrite

$$g_z^t(y) = \frac{1}{2n} \sum_{s=0}^{t-1} (\mathbb{E}_y[\delta_z(X_s)] - \mathbb{E}_{y-1}[\delta_z(X_s)])$$

$$= \frac{1}{2n} \sum_{s=0}^{t-1} \mathbb{E}[\delta_z(\tilde{Y}_s^*) - \delta_z(Y_s^*)]$$

$$= \frac{1}{2n} \mathbb{E}\left[\sum_{s=0}^{t-1} (\delta_z(\tilde{Y}_s^*) - \delta_z(Y_s^*))\right].$$

The random variable inside the expectation is bounded in absolute value by

$$\left|\sum_{s=0}^{t-1} (\delta_z(\tilde{Y}_s^*) - \delta_z(Y_s^*))\right| \le \tau$$

uniformly in t. Indeed, after $s = \tau$, the terms in the sum are 0, while before $s = \tau$ the terms are bounded by 1 in absolute value. By the integrability of τ, the dominated convergence theorem (Proposition B.4.14) allows to take the limit, leading to

$$g_z^\infty(y) = \lim_{t \to +\infty} \frac{1}{2n} \mathbb{E}\left[\sum_{s=0}^{t-1}(\delta_z(\tilde{Y}_s^*) - \delta_z(Y_s^*))\right]$$

$$= \frac{1}{2n} \mathbb{E}\left[\sum_{s=0}^{+\infty}(\delta_z(\tilde{Y}_s^*) - \delta_z(Y_s^*))\right]$$

$$< +\infty.$$

That concludes the proof.   ∎

Proof of Lemma 4.4.12   Our starting point is Lemma 4.4.11, from which we deduce the recursive formula

$$g_z^\infty(y+1) = \frac{1}{\lambda}\left\{yg_z^\infty(y) + \pi(z) - \delta_z(y)\right\} \tag{4.4.18}$$

for $0 \le y \le n$ and $0 \le z \le n$.

We guess a general formula and then check it. By (4.4.18),

$$g_z^\infty(1) = \frac{1}{\lambda}\left\{\pi(z) - \delta_z(0)\right\}, \tag{4.4.19}$$

$$g_z^\infty(2) = \frac{1}{\lambda}\left\{g_z^\infty(1) + \pi(z) - \delta_z(1)\right\}$$

$$= \frac{1}{\lambda}\left\{\frac{1}{\lambda}\left\{\pi(z) - \delta_z(0)\right\} + \pi(z) - \delta_z(1)\right\}$$

$$= \frac{1}{\lambda^2}\left\{\pi(z) - \delta_z(0)\right\} + \frac{1}{\lambda}\{\pi(z) - \delta_z(1)\},$$

$$g_z^\infty(3) = \frac{1}{\lambda}\left\{2g_z^\infty(2) + \pi(z) - \delta_z(2)\right\}$$

$$= \frac{1}{\lambda}\left\{2\frac{1}{\lambda^2}\left\{\pi(z) - \delta_z(0)\right\} + 2\frac{1}{\lambda}\{\pi(z) - \delta_z(1)\} + \pi(z) - \delta_z(2)\right\}$$

$$= \frac{2}{\lambda^3}\left\{\pi(z) - \delta_z(0)\right\} + \frac{2}{\lambda^2}\{\pi(z) - \delta_z(1)\} + \frac{1}{\lambda}\{\pi(z) - \delta_z(2)\},$$

and so forth. We posit the general formula

$$g_z^\infty(y) = \frac{(y-1)!}{\lambda^y}\sum_{k=0}^{y-1}\frac{\lambda^k}{k!}\{\pi(z) - \delta_z(k)\} \tag{4.4.20}$$

for $1 \le y \le n+1$ and $0 \le z \le n$.

The formula is straightforward to confirm by induction. Indeed, it holds for $y = 1$ as can be seen in (4.4.19) (and recalling that $0! = 1$ by convention) and, assuming it holds for y, we have by (4.4.18),

$$g_z^\infty(y+1) = \frac{1}{\lambda}\left\{yg_z^\infty(y) + \pi(z) - \delta_z(y)\right\}$$

$$= \frac{1}{\lambda}\left\{y\frac{(y-1)!}{\lambda^y}\sum_{k=0}^{y-1}\frac{\lambda^k}{k!}\{\pi(z) - \delta_z(k)\} + \pi(z) - \delta_z(y)\right\}$$

$$= \frac{y!}{\lambda^{y+1}}\sum_{k=0}^{y-1}\frac{\lambda^k}{k!}\{\pi(z) - \delta_z(k)\} + \frac{1}{\lambda}\{\pi(z) - \delta_z(y)\}$$

$$= \frac{y!}{\lambda^{y+1}}\sum_{k=0}^{y}\frac{\lambda^k}{k!}\{\pi(z) - \delta_z(k)\},$$

as desired.

We rewrite (4.4.20) according to whether the term $\delta_z(y) = \mathbf{1}\{z = y\}$ plays a role in the equation. For $z \geq y > 0$, the equation simplifies to

$$g_z^\infty(y) = \frac{(y-1)!}{\lambda^y}\sum_{k=0}^{y-1}\frac{\lambda^k}{k!}\pi(z)$$

$$= \frac{1}{y}\frac{y!}{e^{-\lambda}\lambda^y}\sum_{k=0}^{y-1}\frac{e^{-\lambda}\lambda^k}{k!}\pi(z)$$

$$= \frac{\Pi(y-1)}{y\pi(y)}\pi(z).$$

For $0 \leq z < y$, we get instead

$$g_z^\infty(y) = \frac{(y-1)!}{\lambda^y}\left\{\left(\sum_{k=0}^{y-1}\frac{\lambda^k}{k!}\pi(z)\right) - \frac{\lambda^z}{z!}\right\}$$

$$= \frac{1}{y}\frac{y!}{e^{-\lambda}\lambda^y}\left\{\left(\sum_{k=0}^{y-1}\frac{e^{-\lambda}\lambda^k}{k!}\pi(z)\right) - \pi(z)\right\}$$

$$= \frac{\Pi(y-1) - 1}{y\pi(y)}\pi(z).$$

The cases $z = n + 1$ are analogous. ∎

Proof of Lemma 4.4.14 It suffices to prove that, for $A \subseteq \{0, 1, \ldots, n\}$ and $y \in \{0, 1, \ldots, n\}$,

$$|g_A^\infty(y+1) - g_A^\infty(y)| \leq (1 \wedge \lambda^{-1}), \tag{4.4.21}$$

and then use the triangle inequality.

We start with the cases $y \geq 1$. We use the expression derived in Lemma 4.4.12. For $1 \leq y < z$,

$$g_z^\infty(y+1) - g_z^\infty(y) = \frac{\Pi(y)}{(y+1)\pi(y+1)}\bar{\pi}(z) - \frac{\Pi(y-1)}{y\pi(y)}\bar{\pi}(z)$$

$$= \bar{\pi}(z)\frac{1}{y\pi(y)}\left\{\frac{y}{\lambda}\Pi(y) - \Pi(y-1)\right\},$$

where we used that $\pi(y + 1)/\pi(y) = \lambda/(y + 1)$. We show that the expression in curly brackets is non-negative. Indeed, taking out the term $k' = 0$ in the first sum below and changing variables, we get

$$\frac{y}{\lambda} \sum_{k'=0}^{y} \frac{e^{-\lambda}\lambda^{k'}}{(k')!} - \sum_{k=0}^{y-1} \frac{e^{-\lambda}\lambda^{k}}{k!}$$

$$= \frac{y}{\lambda} e^{-\lambda} + \sum_{k=0}^{y-1} \frac{e^{-\lambda}\lambda^{(k+1)-1}}{(k+1)!/y} - \sum_{k=0}^{y-1} \frac{e^{-\lambda}\lambda^{k}}{k!}$$

$$\geq \frac{y}{\lambda} e^{-\lambda} + \sum_{k=0}^{y-1} \frac{e^{-\lambda}\lambda^{k}}{k!} - \sum_{k=0}^{y-1} \frac{e^{-\lambda}\lambda^{k}}{k!}$$

$$\geq 0.$$

So $g_z^{\infty}(y+1) - g_z^{\infty}(y) \geq 0$ for $1 \leq y < z$. A similar calculation, which we omit, shows that the same inequality holds for $z < y \leq n$. The cases $y = 0$, which are analogous, are detailed below.

For notational convenience, it will be helpful to define $g_z^{\infty}(n + 2)$ for all z. Then, for $y = n + 1$ and $z \leq n$, we get

$$g_z^{\infty}(n + 2) - g_z^{\infty}(n + 1) = 0 + \frac{1 - \Pi(n)}{(n + 1)\pi(n + 1)}\pi(z) \geq 0.$$

Moreover, by telescoping,

$$0 = g_z^{\infty}(n + 2) - g_z^{\infty}(0) = \sum_{y=0}^{n+1} \{g_z^{\infty}(y + 1) - g_z^{\infty}(y)\}.$$

We have argued that all the terms in this last sum are non-negative – with the sole exception of the term $y = z$. Hence, for a fixed $0 \leq z \leq n$, it must be that the maximum of $|g_z^{\infty}(y + 1) - g_z^{\infty}(y)|$ is achieved at $z = y$. The case $z = n + 1$ is analogous. By definition of g_z^{∞}, for $0 \leq y \leq n$, the previous display holds with a sum over z rather y and it must be that the maximum of $|g_A^{\infty}(y + 1) - g_A^{\infty}(y)|$ over $A \subseteq \{0, 1, \ldots, n\}$ is achieved at $A = \{y\}$. It remains to bound that last case.

We have, using $\pi(y + 1)/\pi(y) = \lambda/(y + 1)$ again, that

$$|g_y^{\infty}(y + 1) - g_y^{\infty}(y)|$$

$$= \left| -\frac{1 - \Pi(y)}{(y + 1)\pi(y + 1)}\pi(y) - \frac{\Pi(y - 1)}{y\pi(y)}\pi(y) \right|$$

$$= \frac{1}{\lambda} \sum_{k \geq y+1} e^{-\lambda}\frac{\lambda^{k}}{k!} + \frac{1}{y} \sum_{k=0}^{y-1} e^{-\lambda}\frac{\lambda^{k}}{k!}$$

$$= \frac{e^{-\lambda}}{\lambda} \left\{ \sum_{k'=1}^{y} \frac{\lambda^{k'}}{(k')!}\frac{k'}{y} + \sum_{k \geq y+1} \frac{\lambda^{k}}{k!} \right\}$$

$$\leq \frac{e^{-\lambda}}{\lambda} \left\{ \sum_{k \geq 1} \frac{\lambda^{k}}{k!} \right\}$$

$$= \frac{e^{-\lambda}}{\lambda} \{e^{\lambda} - 1\}$$

$$= \frac{1 - e^{-\lambda}}{\lambda}.$$

For $\lambda \geq 1$, we have $\frac{1-e^{-\lambda}}{\lambda} \leq \frac{1}{\lambda} = (1 \wedge \lambda^{-1})$, while for $0 < \lambda < 1$ we have $\frac{1-e^{-\lambda}}{\lambda} \leq \frac{\lambda}{\lambda} = 1 = (1 \wedge \lambda^{-1})$ by Exercise 1.16. The case $y = 0$ is analogous, as detailed next.

It remains to consider the cases $y = 0$. Recall that $g_z^\infty(0) = 0$. By Lemma 4.4.12, for $z \geq 1$,

$$
\begin{aligned}
g_z^\infty(1) - g_z^\infty(0) &= g_z^\infty(1) \\
&= \frac{\Pi(0)}{\pi(1)} \pi(z) \\
&= e^{-\lambda} \frac{e^{-\lambda} \lambda^z / z!}{e^{-\lambda} \lambda} \\
&= \frac{1}{\lambda} e^{-\lambda} \frac{\lambda^z}{z!}.
\end{aligned}
$$

And

$$
g_0^\infty(1) - g_0^\infty(0) = g_0^\infty(1) = -\frac{1 - \Pi(0)}{\pi(1)} \pi(0) = -\frac{1 - e^{-\lambda}}{\lambda}.
$$

So we have established (4.4.21) and that concludes the proof. ∎

4.4.3 ▷ *Random Graphs: Clique Number at the Threshold in the Erdős–Rényi Model*

We revisit the subgraph containment problem of Section 2.3.2 (and Section 4.2.4). Let $G_n \sim \mathbb{G}_{n,p_n}$ be an Erdős–Rényi graph with n vertices and density p_n. Let $\omega(G)$ be the *clique number* of a graph G, that is, the size of its largest clique. We showed previously that the property $\omega(G) \geq 4$ has threshold function $n^{-2/3}$. Here we consider what happens when

$$
p_n = C n^{-2/3}
$$

for some constant $C > 0$. We use the Chen–Stein method in the form of Theorem 4.4.8.

For an enumeration $S_1, \ldots, S_m$ of the 4-tuples of vertices in G_n, let $A_1, \ldots, A_m$ be the events that the corresponding 4-cliques are present and define $Z_i = \mathbf{1}_{A_i}$. Then, $W = \sum_{i=1}^m Z_i$ is the number of 4-cliques in G_n. We argued previously (see Claim 2.3.4) that

$$
q_i := \mathbb{E}[Z_i] = p_n^6
$$

and

$$
\lambda := \mathbb{E}[W] = \binom{n}{4} p_n^6.
$$

In our regime of interest, λ is of constant order.

Observe that the Z_is are not independent because the 4-tuples may share potential edges. However, they admit a neighborhood structure as in Theorem 4.4.8. Specifically, for $i = 1, \ldots, m$, define

$$
\mathcal{N}_i = \{j : S_i \text{ and } S_j \text{ share at least two vertices}\} \setminus \{i\}.
$$

Then, the conditions of Theorem 4.4.8 are satisfied, that is, X_i is independent of $\{Z_j : j \notin \mathcal{N}_i \cup \{i\}\}$. We argued previously (again see Claim 2.3.4) that

$$|\mathcal{N}_i| = \binom{4}{3}(n-4) + \binom{4}{2}\binom{n-4}{2} = \Theta(n^2),$$

where the first term counts the number of S_js sharing exactly three vertices with S_i, in which case $\mathbb{E}[Z_i Z_j] = p_n^9$, and the second term counts those sharing two, in which case $\mathbb{E}[Z_i Z_j] = p_n^{11}$.

We are ready to apply the bound in Theorem 4.4.8. Let π be the Poisson distribution with mean λ. Using the formulas above, we get when $p_n = Cn^{-2/3}$,

$$\|\mu - \pi\|_{\mathrm{TV}}$$

$$\leq (1 \wedge \lambda^{-1}) \sum_{i=1}^{n} \left\{ q_i^2 + \sum_{j \in \mathcal{N}_i} (q_i q_j + \mathbb{E}[Z_i Z_j]) \right\}$$

$$\leq (1 \wedge \lambda^{-1}) \binom{n}{4}$$

$$\times \left[p_n^{12} + \left\{ \binom{4}{3}(n-4)(p_n^{12} + p_n^9) + \binom{4}{2}\binom{n-4}{2}(p_n^{12} + p_n^{11}) \right\} \right]$$

$$= (1 \wedge \lambda^{-1}) \Theta(n^4 p_n^{12} + n^5 p_n^9 + n^6 p_n^{11})$$

$$= (1 \wedge \lambda^{-1}) \Theta(n^4 n^{-8} + n^5 n^{-6} + n^6 n^{-22/3})$$

$$= (1 \wedge \lambda^{-1}) \Theta(n^{-1}),$$

which goes to 0 as $n \to +\infty$.

See Exercise 4.21 for an improved bound.

Exercises

Exercise 4.1 (Harmonic function on $\mathbb{Z}^d$: unbounded). Give an example of an unbounded harmonic function on $\mathbb{Z}$. Give one on $\mathbb{Z}^d$ for general d. (Hint: What is the simplest function after the constant one?)

Exercise 4.2 (Binomial vs. Binomial). Use coupling to show that

$$n \geq m, \ q \geq p \implies \mathrm{Bin}(n, q) \succeq \mathrm{Bin}(m, p).$$

Exercise 4.3 (A chain that is not stochastically monotone). Consider random walk on a network $\mathcal{N} = ((V, E), c)$, where $V = \{0, 1, \dots, n\}$ and $i \sim j$ if and only if $|i - j| = 1$ (in particular, not including self-loops). Show that the transition matrix is, in general, *not* stochastically monotone (see Definition 4.2.16).

Exercise 4.4 (Increasing events: properties). Let $\mathcal{F}$ be a σ-algebra over the poset $\mathcal{X}$. Recall that an event $A \in \mathcal{F}$ is increasing if $x \in A$ implies that any $y \geq x$ is also in A and that a function $f : \mathcal{X} \to \mathbb{R}$ is increasing if $x \leq y$ implies $f(x) \leq f(y)$.

    (i) Show that an event $A \in \mathcal{F}$ is increasing if and only if the indicator function $\mathbf{1}_A$ is increasing.

(ii) Let $A, B \in \mathcal{F}$ be increasing. Show that $A \cap B$ and $A \cup B$ are increasing.

(iii) An event A is decreasing if $x \in A$ implies that any $y \leq x$ is also in A. Show that A is decreasing if and only if A^c is increasing.

(iv) Let $A, B \in \mathcal{F}$ be decreasing. Show that $A \cap B$ and $A \cup B$ are decreasing.

Exercise 4.5 (Harris' inequality: alternative proof). We say that $f \colon \mathbb{R}^n \to \mathbb{R}$ is *coordinate-wise non-decreasing* if it is non-decreasing in each variable while keeping the other variables fixed.

(i) *(Chebyshev's association inequality)* Let $f \colon \mathbb{R} \to \mathbb{R}$ and $g \colon \mathbb{R} \to \mathbb{R}$ be coordinate-wise non-decreasing and let X be a real random variable. Show that
$$\mathbb{E}[f(X)g(X)] \geq \mathbb{E}[f(X)]\mathbb{E}[g(X)].$$

(Hint: Consider the quantity $(f(X)-f(X'))(g(X)-g(X'))$, where X' is an independent copy of X.)

(ii) *(Harris' inequality)* Let $f \colon \mathbb{R}^n \to \mathbb{R}$ and $g \colon \mathbb{R}^n \to \mathbb{R}$ be coordinatewise non-decreasing and let $X = (X_1, \ldots, X_n)$ be independent real random variables. Show by induction on n that
$$\mathbb{E}[f(X)g(X)] \geq \mathbb{E}[f(X)]\mathbb{E}[g(X)].$$

Exercise 4.6 Provide the details for Example 4.2.33.

Exercise 4.7 (FKG: sufficient conditions). Let $\mathcal{X} := \{0, 1\}^F$ where F, is finite and let μ be a positive probability measure on $\mathcal{X}$. We use the notation introduced in the proof of Holley's inequality (Theorem 4.2.32).

(i) To check the FKG condition, show that it suffices to check that, for all $x \leq y \in \mathcal{X}$ and $i \in F$,
$$\frac{\mu(y^{i,1})}{\mu(y^{i,0})} \geq \frac{\mu(x^{i,1})}{\mu(x^{i,0})}.$$

(Hint: Write $\mu(\omega \vee \omega')/\mu(\omega)$ as a telescoping product.)

(ii) To check the FKG condition, show that it suffices to check (4.2.15) only for those $\omega, \omega' \in \mathcal{X}$ such that $\|\omega - \omega'\|_1 = 2$ and neither $\omega \leq \omega'$ nor $\omega' \leq \omega$. (Hint: Use (i).)

Exercise 4.8 (FKG and strong positive associations). Let $\mathcal{X} := \{0, 1\}^F$, where F is finite and let μ be a positive probability measure on $\mathcal{X}$. For $\Lambda \subseteq F$ and $\xi \in \mathcal{X}$, let
$$\mathcal{X}_\Lambda^\xi := \{\omega_\Lambda \times \xi_{\Lambda^c} : \omega_\Lambda \in \{0, 1\}^\Lambda\},$$
where $\omega_\Lambda \times \xi_{\Lambda^c}$ agrees with ω on coordinates in Λ and with ξ on coordinates in $F \backslash \Lambda$. Define the measure μ_Λ^ξ over $\{0, 1\}^\Lambda$ as
$$\mu_\Lambda^\xi(\omega_\Lambda) := \frac{\mu(\omega_\Lambda \times \xi_{\Lambda^c})}{\mu(\mathcal{X}_\Lambda^\xi)}.$$

That is, μ_Λ^ξ is μ conditioned on agreeing with ξ on $F \backslash \Lambda$. The measure μ is said to be *strongly positively associated* if $\mu_\Lambda^\xi(\omega_\Lambda)$ is positively associated for all Λ and ξ. Prove that the FKG condition is equivalent to strong positive associations. (Hint: Use Exercise 4.7 as well as the FKG inequality.)

Exercise 4.9 (Triangle-freeness: a second proof). Consider again the setting of Section 4.2.4.

  (i) Let e_t be the minimum number of edges in a t-vertex union of k not mutually vertex-disjoint triangles. Show that, for any $k \geq 2$ and $k \leq t < 3k$, it holds that $e_t > t$.
  (ii) Use Exercise 2.18 to give a second proof of the fact that $\mathbb{P}[X_n = 0] \to e^{-\lambda^3/6}$.

Exercise 4.10 (RSW lemma: general α). Let $R_{n,\alpha}(p)$ be as defined in Section 4.2.5. Show that for all $n \geq 2$ (divisible by 4) and $p \in (0, 1)$,

$$R_{n,\alpha}(p) \geq \left(\frac{1}{2}\right)^{2\alpha-2} R_{n,1}(p)^{6\alpha-7} R_{n/2,1}(p)^{6\alpha-6}.$$

Exercise 4.11 (Primal and dual crossings). Modify the proof of Lemma 2.2.14 to prove Lemma 4.2.41.

Exercise 4.12 (Square-root trick). Let μ be an FKG measure on $\{0, 1\}^F$, where F is finite. Let A_1 and A_2 be increasing events with $\mu(A_1) = \mu(A_2)$. Show that

$$\mu(A_1) \geq 1 - \sqrt{1 - \mu(A_1 \cup A_2)}.$$

Exercise 4.13 (Splitting: details). Show that $\tilde{P}$, as defined in Example 4.3.3, is a transition matrix on V provided z_0 satisfies the condition there.

Exercise 4.14 (Doeblin's condition in finite case). Let P be a transition matrix on a finite state space.

  (i) Show that Doeblin's condition (see Example 4.3.3) holds when P is finite, irreducible, and aperiodic.
  (ii) Show that Doeblin's condition holds for lazy random walk on the hypercube with $s = n$. Use it to derive a bound on the mixing time.

Exercise 4.15 (Mixing on cycles: lower bound). Let (Z_t) be lazy, simple random walk on the cycle of size n, $\mathbb{Z}_n := \{0, 1, \ldots, n - 1\}$, where $i \sim j$ if $|j - i| = 1 \pmod n$.

  (i) Let $A = \{n/2, \ldots, n - 1\}$. By coupling (Z_t) with lazy, simple random walk on $\mathbb{Z}$, show that

$$P^{\alpha n^2}(n/4, A) < \frac{1}{2} - \varepsilon$$

for $\alpha \leq \alpha_\varepsilon$ for some $\alpha_\varepsilon > 0$. (Hint: Use Kolmogorov's maximal inequality (Corollary 3.1.46).)
  (ii) Deduce that

$$t_{\mathrm{mix}}(\varepsilon) \geq \alpha_\varepsilon n^2.$$

Exercise 4.16 (Lower bound on mixing: distinguishing statistic). Let X and Y be random variables on a finite state space S. Let $h\colon S \to \mathbb{R}$ be a measurable real-valued map. Assume that

$$\mathbb{E}[h(Y)] - \mathbb{E}[h(X)] \geq r\sigma,$$

where $r > 0$ and $\sigma^2 := \max\{\mathrm{Var}[h(X)], \mathrm{Var}[h(Y)]\}$. Show that

$$\|\mu_X - \mu_Y\|_{\mathrm{TV}} \geq 1 - \frac{8}{r^2}.$$

(Hint: Consider the interval on one side of the midpoint between $\mathbb{E}[h(X)]$ and $\mathbb{E}[h(Y)]$.)

Exercise 4.17 (Path coupling and optimal transport). Let V be a finite state space and let P be an irreducible transition matrix on V with stationary distribution π. Let w_0 be a metric on V. For probability measures μ, ν on V, let

$$W_0(\mu, \nu) := \inf\{\mathbb{E}[w_0(X, Y)] : (X, Y) \text{ is a coupling of } \mu \text{ and } \nu\}$$

be the so-called *Wasserstein distance* (or *transportation metric*) between μ and ν.

WASSERSTEIN
DISTANCE

  (i) Show that W_0 is a metric. (Hint: See the proof of Claim 4.3.11.)

  (ii) Assume that the conditions of Theorem 4.3.10 hold. Show that for any probability measures μ, ν,

$$W_0(\mu P, \nu P) \le \kappa\, W_0(\mu, \nu).$$

(iii) Use (i) and (ii) to prove Theorem 4.3.10.

Exercise 4.18 (Stein equation for the Poisson distribution). Let $\lambda > 0$. Show that a non-negative integer-valued random variable Z is $\mathrm{Poi}(\lambda)$ if and only if for all g bounded,

$$\mathbb{E}[\lambda g(Z + 1) - Zg(Z)] = 0.$$

Exercise 4.19 (Laplacian and stationarity). Let P be an irreducible transition matrix on a finite or countably infinite spate space V. Recall the Laplacian operator is

$$\Delta f(x) = \left[\sum_y P(x, y)f(y)\right] - f(x)$$

provided the sum is finite. Show that a probability distribution μ over V is stationary for P if and only if for all bounded measurable functions,

$$\sum_{x \in V} \mu(x)\Delta f(x) = 0.$$

Exercise 4.20 (Chen–Stein method for positively related variables). Using the notation in (4.4.1), (4.4.2), and (4.4.3), suppose that for each i we can construct a coupling $\{(X_j^{(i)} : j = 1, \ldots, n), (Y_j^{(i)} : j \ne i)\}$ with $(X_j^{(i)})_j \sim (X_j)_j$ such that

$$(Y_j^{(i)}, j \ne i) \sim (X_j^{(i)}, j \ne i)|X_i^{(i)} = 1 \qquad \text{and} \qquad Y_j^{(i)} \ge X_j^{(i)}, \ \forall j \ne i.$$

Show that

$$\|\mu - \pi\|_{\mathrm{TV}} \le (1 \wedge \lambda^{-1})\left\{\mathrm{Var}(W) - \lambda + 2\sum_{i=1}^n p_i^2\right\}.$$

Exercise 4.21 (Chen–Stein and 4-cliques). Use Exercise 4.20 to give an improved asymptotic bound in the setting of Section 4.4.3.

Exercise 4.22 (Chen–Stein for negatively related variables). Using the notation in (4.4.1), (4.4.2), and (4.4.3), suppose that for each i we can construct a coupling $\{(X_j^{(i)} : j = 1, \ldots, n),$ $(Y_j^{(i)} : j \neq i)\}$ with $(X_j^{(i)})_j \sim (X_j)_j$ such that

$$(Y_j^{(i)}, j \neq i) \sim (X_j^{(i)}, j \neq i)|X_i^{(i)} = 1 \qquad \text{and} \qquad Y_j^{(i)} \leq X_j^{(i)}, \forall j \neq i.$$

Show that

$$\|\mu - \pi\|_{\text{TV}} \leq (1 \wedge \lambda^{-1}) \{\lambda - \text{Var}(W)\}.$$

Bibliographic Remarks

Section 4.1 The coupling method is generally attributed to Doeblin [Doe38]. The standard reference on coupling is [Lin02]. See that reference for a history of coupling and a facsimile of Doeblin's paper. See also [dH]. Section 4.1.2 is based on [Per, section 6] and Section 4.1.4 is based on [vdH17, section 5.3].

Section 4.2 Strassen's theorem is due to Strassen [Str65]. Harris' inequality is due to Harris [Har60]. The FKG inequality is due to Fortuin, Kasteleyn, and Ginibre [FKG71]. A "four-function" version of Holley's inequality, which also extends to distributive lattices, was proved by Ahlswede and Daykin [AD78]. See, for example, [AS11, section 6.1]. An exposition of submodularity and its connections to convexity can be found in [Lov83]. For more on Markov random fields, see, for example, [RAS15]. Section 4.2.4 follows [AS11, sections 8.1, 8.2, 10.1]. Janson's inequality is due to Janson [Jan90]. Boppana and Spencer [BS89] gave the proof presented here. For more on Janson's inequality, see [JLR11, section 2.2]. The presentation in Section 4.2.5 follows closely [BR06b, sections 3 and 4]. See also [BR06a, chapter 3]. Broadbent and Hammersley [BH57, Ham57] initiated the study of the critical value of percolation. Harris' theorem proved by Harris [Har60] and Kesten's theorem was proved two decades later by Kesten [Kes80], confirming non-rigorous work of Sykes and Essam [SE64]. The RSW lemma was obtained independently by Russo [Rus78] and Seymour and Welsh [SW78]. The proof we gave here is due to Bollobás and Riordan [BR06b]. Another short proof of a version of the RSW lemma for critical site percolation on a triangular lattice was given by Smirnov; see, for example, [Ste]. The type of "scale invariance" seen in the RSW lemma plays a key role in the contemporary theory of critical two-dimensional percolation and of two-dimensional lattice models more generally. See, for example, [Law05, Gri10a].

Section 4.3 The material in Section 4.3 borrows heavily from [LPW06, chapters 5, 14, 15] and [AF, chapter 12]. Aldous [Ald83] was the first author to make explicit use of coupling to bound total variation distance to stationarity of finite Markov chains. The link between couplings of Markov chains and total variation distance was also used by Griffeath [Gri75] and Pitman [Pit76]. Example 4.3.3 is based on [Str14] and [JH01]. For a more general treatment, see [MT09, chapter 16]. The proof of Claim 4.3.7 is partly based on [LPW06, Proposition 7.13]. See also [DGG+00] and [HS07] for alternative proofs. Path coupling is due to Bubley and Dyer [BD97]. The optimal transport perspective on the path coupling method in Exercise 4.17 is from [LPW06, chapter 14]. For more on optimal transport, see, for example, [Vil09]. The main result in Section 4.3.4 is taken from [LPW06, Theorem 15.1].

For more background on the so-called critical slowdown of the Glauber dynamics of Ising and Potts models on various graphs, see [CDL$^+$12, LS12].

Section 4.4 The Chen–Stein method was introduced by Chen in [Che75] as an adaptation of the Stein method [Ste72] to the Poisson distribution. The presentation in Section 4.4 is inspired heavily by [Dey] and [vH16]. Example 4.4.9 is taken from [AGG89]. Further applications of the Chen–Stein and Stein methods to random graphs can be found in [JLR11, chapter 6].

5

Spectral Methods

In this chapter, we develop spectral techniques. We highlight some applications to Markov chain mixing and network analysis. The main tools are the *spectral theorem* and the *variational characterization of eigenvalues*, which we review in Section 5.1 together with some related results. We also give a brief introduction to *spectral graph theory* and detail an application to community recovery. In Section 5.2, we apply the spectral theorem to reversible Markov chains. In particular, we define the *spectral gap* and establish its close relationship to the mixing time. Roughly speaking, we show through an eigendecomposition of the transition matrix that the gap between the eigenvalue 1 (which is the largest in absolute value) and the rest of the spectrum drives how fast P^t converges to the stationary distribution. We give several examples. We then show in Section 5.3 that the spectral gap can be bounded using certain isoperimetric properties of the underlying network. We prove *Cheeger's inequality*, which quantifies this relationship, and introduce expander graphs, an important family of graphs with good "expansion." Applications to mixing times are also discussed. One specific technique is the "canonical paths method," which bounds the spectral graph by formalizing a notion of congestion in the network.

5.1 Background

We first review some important concepts from linear algebra. In particular, we recall the spectral theorem as well as the variational characterization of eigenvalues. We also derive a few perturbation results. We end this section with an application to community recovery in network analysis.

5.1.1 Eigenvalues and Their Variational Characterization

SYMMETRIC MATRIX

ORTHOGONAL MATRIX

When a $d \times d$ matrix A is *symmetric*, that is, $a_{ij} = a_{ji}$ for all i, j, a remarkable result is that A is similar to a diagonal matrix by an orthogonal transformation. Put differently, there exists an orthonormal basis of $\mathbb{R}^d$ made of eigenvectors of A. Recall that a matrix $Q \in \mathbb{R}^{d \times d}$ is *orthogonal* if $QQ^T = I_{d \times d}$ and $Q^T Q = I_{d \times d}$, where $I_{d \times d}$ is the $d \times d$ identity matrix. In words, its columns form an orthonormal basis of $\mathbb{R}^d$. For a vector $\mathbf{z} = (z_1, \ldots, z_d)$, we let $\text{diag}(\mathbf{z}) = \text{diag}(z_1, \ldots, z_d)$ be the diagonal matrix with diagonal entries $z_1, \ldots, z_d$. Unless specified otherwise, a vector is by default a "column vector" and its transpose is a "row vector."

256

Theorem 5.1.1 (Spectral theorem). *Let $A \in \mathbb{R}^{d \times d}$ be a symmetric matrix, that is, $A^T =$* SPECTRAL
A. Then A has d orthonormal eigenvectors $\mathbf{q}_1, \ldots, \mathbf{q}_d$ with corresponding (not necessarily THEOREM
*distinct) real eigenvalues $\lambda_1 \geq \lambda_2 \geq \cdots \geq \lambda_d$. In matrix form, this is written as the matrix
factorization*

$$A = Q \Lambda Q^T = \sum_{i=1}^{d} \lambda_i \mathbf{q}_i \mathbf{q}_i^T,$$

*where Q has columns $\mathbf{q}_1, \ldots, \mathbf{q}_d$ and $\Lambda = \mathrm{diag}(\lambda_1, \ldots, \lambda_d)$. We refer to this factorization as
a spectral decomposition of A.*

The proof uses a greedy sequence maximizing the quadratic form $\langle \mathbf{v}, A\mathbf{v} \rangle$. To see why that
might come about, note that for a unit eigenvector $\mathbf{v}$ with eigenvalue λ we have $\langle \mathbf{v}, A\mathbf{v} \rangle =
\langle \mathbf{v}, \lambda \mathbf{v} \rangle = \lambda$.

We will need the following formula. Consider the block matrices

$$\begin{pmatrix} \mathbf{y} \\ \mathbf{z} \end{pmatrix} \quad \text{and} \quad \begin{pmatrix} A & B \\ C & D \end{pmatrix},$$

where $\mathbf{y} \in \mathbb{R}^{d_1}$, $\mathbf{z} \in \mathbb{R}^{d_2}$, $A \in \mathbb{R}^{d_1 \times d_1}$, $B \in \mathbb{R}^{d_1 \times d_2}$, $C \in \mathbb{R}^{d_2 \times d_1}$, and $D \in \mathbb{R}^{d_2 \times d_2}$. Then it
follows by direct calculation that

$$\begin{pmatrix} \mathbf{y} \\ \mathbf{z} \end{pmatrix}^T \begin{pmatrix} A & B \\ C & D \end{pmatrix} \begin{pmatrix} \mathbf{y} \\ \mathbf{z} \end{pmatrix} = \mathbf{y}^T A \mathbf{y} + \mathbf{y}^T B \mathbf{z} + \mathbf{z}^T C \mathbf{y} + \mathbf{z}^T D \mathbf{z}. \tag{5.1.1}$$

We will also need the following linear algebra fact. Let $\mathbf{v}_1, \ldots, \mathbf{v}_j$ be orthonormal vectors
in $\mathbb{R}^d$, with $j < d$. Then they can be completed into an orthonormal basis $\mathbf{v}_1, \ldots, \mathbf{v}_d$ of $\mathbb{R}^d$.

Proof of Theorem 5.1.1   We proceed by induction.

A first eigenvector   Let $A_1 = A$. Maximizing over the objective function $\langle \mathbf{v}, A_1 \mathbf{v} \rangle$, we let

$$\mathbf{v}_1 \in \arg\max\{\langle \mathbf{v}, A_1 \mathbf{v} \rangle \colon \|\mathbf{v}\|_2 = 1\}$$

and

$$\lambda_1 = \max\{\langle \mathbf{v}, A_1 \mathbf{v} \rangle \colon \|\mathbf{v}\|_2 = 1\}.$$

Complete $\mathbf{v}_1$ into an orthonormal basis of $\mathbb{R}^d$, $\mathbf{v}_1, \hat{\mathbf{v}}_2, \ldots, \hat{\mathbf{v}}_d$, and form the block matrix

$$\hat{W}_1 := \begin{pmatrix} \mathbf{v}_1 & \hat{V}_1 \end{pmatrix},$$

where the columns of $\hat{V}_1$ are $\hat{\mathbf{v}}_2, \ldots, \hat{\mathbf{v}}_d$. Note that $\hat{W}_1$ is orthogonal by construction.

Getting one step closer to diagonalization   We show next that $\hat{W}_1$ gets us one step closer
to a diagonal matrix by similarity transformation. Note first that

$$\hat{W}_1^T A_1 \hat{W}_1 = \begin{pmatrix} \lambda_1 & \mathbf{w}_1^T \\ \mathbf{w}_1 & A_2 \end{pmatrix},$$

where $\mathbf{w}_1 := \hat{V}_1^T A_1 \mathbf{v}_1$ and $A_2 := \hat{V}_1^T A_1 \hat{V}_1$. The key claim is that $\mathbf{w}_1 = \mathbf{0}$. This follows from
an argument by contradiction.

Suppose $\mathbf{w}_1 \neq \mathbf{0}$ and consider the unit vector

$$\mathbf{z} := \hat{W}_1 \times \frac{1}{\sqrt{1 + \delta^2 \|\mathbf{w}_1\|_2^2}} \begin{pmatrix} 1 \\ \delta \mathbf{w}_1 \end{pmatrix},$$

which achieves objective value

$$\mathbf{z}^T A_1 \mathbf{z} = \frac{1}{1 + \delta^2 \|\mathbf{w}_1\|_2^2} \begin{pmatrix} 1 \\ \delta \mathbf{w}_1 \end{pmatrix}^T \begin{pmatrix} \lambda_1 & \mathbf{w}_1^T \\ \mathbf{w}_1 & A_2 \end{pmatrix} \begin{pmatrix} 1 \\ \delta \mathbf{w}_1 \end{pmatrix}$$

$$= \frac{1}{1 + \delta^2 \|\mathbf{w}_1\|_2^2} \left(\lambda_1 + 2\delta \|\mathbf{w}_1\|_2^2 + \delta^2 \mathbf{w}_1^T A_2 \mathbf{w}_1 \right),$$

where we used (5.1.1). By the Taylor expansion,

$$\frac{1}{1 + \epsilon^2} = 1 - \epsilon^2 + O(\epsilon^4),$$

for δ small enough,

$$\mathbf{z}^T A_1 \mathbf{z} = (\lambda_1 + 2\delta \|\mathbf{w}_1\|_2^2 + \delta^2 \mathbf{w}_1^T A_2 \mathbf{w}_1)(1 - \delta^2 \|\mathbf{w}_1\|_2^2 + O(\delta^4))$$

$$= \lambda_1 + 2\delta \|\mathbf{w}_1\|_2^2 + O(\delta^2)$$

$$> \lambda_1.$$

That gives the desired contradiction.

So, letting $W_1 := \hat{W}_1$, we get

$$W_1^T A_1 W_1 = \begin{pmatrix} \lambda_1 & \mathbf{0} \\ \mathbf{0} & A_2 \end{pmatrix}.$$

Finally, note that $A_2 = \hat{V}_1^T A_1 \hat{V}_1$ is symmetric since

$$A_2^T = (\hat{V}_1^T A_1 \hat{V}_1)^T = \hat{V}_1^T A_1^T \hat{V}_1 = \hat{V}_1^T A_1 \hat{V}_1 = A_2,$$

by the symmetry of A_1 itself.

Next step of the induction Apply the same argument to the symmetric submatrix $A_2 \in \mathbb{R}^{(d-1) \times (d-1)}$, let $\hat{W}_2 \in \mathbb{R}^{(d-1) \times (d-1)}$ be the corresponding orthogonal matrix, and define λ_2 and A_3 through the equation

$$\hat{W}_2^T A_2 \hat{W}_2 = \begin{pmatrix} \lambda_2 & \mathbf{0} \\ \mathbf{0} & A_3 \end{pmatrix}.$$

Define the block matrix

$$W_2 = \begin{pmatrix} 1 & \mathbf{0} \\ \mathbf{0} & \hat{W}_2 \end{pmatrix}$$

and observe that

$$W_2^T W_1^T A_1 W_1 W_2 = W_2^T \begin{pmatrix} \lambda_1 & \mathbf{0} \\ \mathbf{0} & A_2 \end{pmatrix} W_2$$

$$= \begin{pmatrix} \lambda_1 & \mathbf{0} \\ \mathbf{0} & \hat{W}_2^T A_2 \hat{W}_2 \end{pmatrix}$$

$$= \begin{pmatrix} \lambda_1 & 0 & \mathbf{0} \\ 0 & \lambda_2 & \mathbf{0} \\ \mathbf{0} & \mathbf{0} & A_3 \end{pmatrix}.$$

Proceeding similarly by induction gives the claim, with the final Q being the product of the W_is (which is orthogonal as the product of orthogonal matrices). ∎

We derive an important variational characterization inspired by the proof of the spectral theorem. We will need the following quantity.

Definition 5.1.2 (Rayleigh quotient). *Let $A \in \mathbb{R}^{d \times d}$ be a symmetric matrix. The* Rayleigh quotient *is defined as*

RAYLEIGH QUOTIENT

$$\mathcal{R}_A(\mathbf{u}) = \frac{\langle \mathbf{u}, A\mathbf{u} \rangle}{\langle \mathbf{u}, \mathbf{u} \rangle},$$

which is defined for any $\mathbf{u} \neq \mathbf{0}$ in $\mathbb{R}^d$.

We let the *span* of a collection of vectors be defined as

SPAN

$$\text{span}(\boldsymbol{u}_1, \dots, \boldsymbol{u}_n) := \left\{ \sum_{i=1}^n \alpha_i \boldsymbol{u}_i : \alpha_1, \dots, \alpha_n \in \mathbb{R} \right\}.$$

COURANT–FISCHER

Theorem 5.1.3 (Courant–Fischer theorem). *Let $A \in \mathbb{R}^{d \times d}$ be a symmetric matrix with spectral decomposition $A = \sum_{i=1}^d \lambda_i \mathbf{v}_i \mathbf{v}_i^T$, where $\lambda_1 \geq \cdots \geq \lambda_d$. For each $k = 1, \dots, d$, define the subspace*

$$\mathcal{V}_k = \text{span}(\mathbf{v}_1, \dots, \mathbf{v}_k) \quad \text{and} \quad \mathcal{W}_{d-k+1} = \text{span}(\mathbf{v}_k, \dots, \mathbf{v}_d).$$

Then, for all $k = 1, \dots, d$,

$$\lambda_k = \min_{\mathbf{u} \in \mathcal{V}_k} \mathcal{R}_A(\mathbf{u}) = \max_{\mathbf{u} \in \mathcal{W}_{d-k+1}} \mathcal{R}_A(\mathbf{u}).$$

Furthermore, we have the following min-max formulas, which do not depend on the choice of spectral decomposition, for all $k = 1, \dots, d$:

$$\lambda_k = \max_{\dim(\mathcal{V})=k} \min_{\mathbf{u} \in \mathcal{V}} \mathcal{R}_A(\mathbf{u}) = \min_{\dim(\mathcal{W})=d-k+1} \max_{\mathbf{u} \in \mathcal{W}} \mathcal{R}_A(\mathbf{u}).$$

Note that, in all these formulas, the vector $\mathbf{u} = \mathbf{v}_k$ is optimal. To derive the "local" formula, the first ones above, we expand a vector in $\mathcal{V}_k$ into the basis $\mathbf{v}_1, \dots, \mathbf{v}_k$ and use the fact that $\mathcal{R}_A(\mathbf{v}_i) = \lambda_i$ and that eigenvalues are in non-increasing order. The "global" formulas then follow from a dimension argument.

We will need the following dimension-based fact. Let $\mathcal{U}, \mathcal{V} \subseteq \mathbb{R}^d$ be linear subspaces such that $\dim(\mathcal{U}) + \dim(\mathcal{V}) > d$, where $\dim(\mathcal{U})$ denotes the dimension of $\mathcal{U}$. Then there exists a non-zero vector in the intersection $\mathcal{U} \cap \mathcal{V}$. That is,

$$\dim(\mathcal{U}) + \dim(\mathcal{V}) > d \implies (\mathcal{U} \cap \mathcal{V}) \setminus \{\mathbf{0}\} \neq \emptyset. \tag{5.1.2}$$

Proof of Theorem 5.1.3   We first prove the local formulas, that is, the ones involving a *specific* decomposition.

Local formulas   Since $\mathbf{v}_1, \ldots, \mathbf{v}_k$ form an orthonormal basis of $\mathcal{V}_k$, any non-zero vector $\mathbf{u} \in \mathcal{V}_k$ can be written as $\mathbf{u} = \sum_{i=1}^{k} \langle \mathbf{u}, \mathbf{v}_i \rangle \mathbf{v}_i$ and it follows that

$$\langle \mathbf{u}, \mathbf{u} \rangle = \sum_{i=1}^{k} \langle \mathbf{u}, \mathbf{v}_i \rangle^2$$

and

$$\langle \mathbf{u}, A\mathbf{u} \rangle = \left\langle \mathbf{u}, \sum_{i=1}^{k} \langle \mathbf{u}, \mathbf{v}_i \rangle \lambda_i \mathbf{v}_i \right\rangle = \sum_{i=1}^{k} \lambda_i \langle \mathbf{u}, \mathbf{v}_i \rangle^2.$$

Thus,

$$\mathcal{R}_A(\mathbf{u}) = \frac{\langle \mathbf{u}, A\mathbf{u} \rangle}{\langle \mathbf{u}, \mathbf{u} \rangle} = \frac{\sum_{i=1}^{k} \lambda_i \langle \mathbf{u}, \mathbf{v}_i \rangle^2}{\sum_{i=1}^{k} \langle \mathbf{u}, \mathbf{v}_i \rangle^2} \geq \lambda_k \frac{\sum_{i=1}^{k} \langle \mathbf{u}, \mathbf{v}_i \rangle^2}{\sum_{i=1}^{k} \langle \mathbf{u}, \mathbf{v}_i \rangle^2} = \lambda_k,$$

where we used $\lambda_1 \geq \cdots \geq \lambda_k$ and the fact that $\langle \mathbf{u}, \mathbf{v}_i \rangle^2 \geq 0$. Moreover, $\mathcal{R}_A(\mathbf{v}_k) = \lambda_k$. So we have established

$$\lambda_k = \min_{\mathbf{u} \in \mathcal{V}_k} \mathcal{R}_A(\mathbf{u}).$$

The expression in terms of $\mathcal{W}_{d-k+1}$ is proved similarly.

Global formulas   Since $\mathcal{V}_k$ has dimension k, it follows from the local formula that

$$\lambda_k = \min_{\mathbf{u} \in \mathcal{V}_k} \mathcal{R}_A(\mathbf{u}) \leq \max_{\dim(\mathcal{V})=k} \min_{\mathbf{u} \in \mathcal{V}} \mathcal{R}_A(\mathbf{u}).$$

Let $\mathcal{V}$ be any subspace with dimension k. Because $\mathcal{W}_{d-k+1}$ has dimension $d - k + 1$, we have that $\dim(\mathcal{V}) + \dim(\mathcal{W}_{d-k+1}) > d$ and there must be non-zero vector $\mathbf{u}_0$ in the intersection $\mathcal{V} \cap \mathcal{W}_{d-k+1}$ by the dimension-based fact above. We then have by the other local formula that

$$\lambda_k = \max_{\mathbf{u} \in \mathcal{W}_{d-k+1}} \mathcal{R}_A(\mathbf{u}) \geq \mathcal{R}_A(\mathbf{u}_0) \geq \min_{\mathbf{u} \in \mathcal{V}} \mathcal{R}_A(\mathbf{u}).$$

Since this inequality holds for any subspace of dimension k, we have

$$\lambda_k \geq \max_{\dim(\mathcal{V})=k} \min_{\mathbf{u} \in \mathcal{V}} \mathcal{R}_A(\mathbf{u}).$$

Combining with the inequality in the other direction above gives the claim. The other global formula is proved similarly.   ∎

5.1.2 Elements of Spectral Graph Theory

We apply the variational characterization of eigenvalues to matrices arising in graph theory. In this section, graphs have no self-loop.

Unweighted graphs As we have previously seen, a convenient way of specifying a graph is through a matrix representation. Assume the undirected graph $G = (V, E)$ has $n = |V|$ vertices. Recall that the adjacency matrix A of G is the $n \times n$ symmetric matrix defined as

$$A_{xy} = \begin{cases} 1 & \text{if } \{x, y\} \in E, \\ 0 & \text{otherwise.} \end{cases}$$

Another matrix of interest is the Laplacian matrix. It is related to the Laplace operator we encountered previously. We will show in particular that it contains useful information about the connectedness of the graph. Recall that, given a graph $G = (V, E)$, the quantity $\delta(v)$ denotes the degree of $v \in V$.

Definition 5.1.4 (Graph Laplacian). *Let $G = (V, E)$ be a graph with vertices $V = \{1, \ldots, n\}$ and adjacency matrix $A \in \mathbb{R}^{n \times n}$. Let $D = \operatorname{diag}(\delta(1), \ldots, \delta(n))$ be the degree matrix. The graph Laplacian (or Laplacian matrix, or Laplacian for short) associated to G is defined as $L = D - A$. Its entries are* GRAPH LAPLACIAN

$$l_{ij} = \begin{cases} \delta(i) & \text{if } i = j, \\ -1 & \text{if } \{i, j\} \in E, \\ 0 & \text{otherwise.} \end{cases}$$

Observe that the Laplacian L of a graph G is a symmetric matrix:

$$L^T = (D - A)^T = D^T - A^T = D - A,$$

where we used that both D and A are themselves symmetric. The associated quadratic form is particularly simple and will play an important role.

Lemma 5.1.5 (Laplacian quadratic form). *Let $G = (V, E)$ be a graph with $n = |V|$ vertices. Its Laplacian L is a positive semi-definite matrix and furthermore we have the following formula for the Laplacian quadratic form (or Dirichlet energy)* LAPLACIAN QUADRATIC FORM

$$\mathbf{x}^T L \mathbf{x} = \sum_{e = \{i, j\} \in E} (x_i - x_j)^2$$

for any $\mathbf{x} = (x_1, \ldots, x_n) \in \mathbb{R}^n$.

Proof of Lemma 5.1.5 Let B be an oriented incidence matrix of G (see Definition 1.1.16). We claim that $BB^T = L$. Indeed, for $i \neq j$, entry (i, j) of BB^T is a sum over all edges containing i and j as endvertices, of which there is at most one. When $e = \{i, j\} \in E$, that entry is -1, since one of i or j has a 1 in the column of B corresponding to e and the other one has a -1. For $i = j$, letting b_{xy} be entry (x, y) of B,

$$(BB^T)_{ii} = \sum_{e = \{x, y\} \in E : i \in e} b_{xy}^2 = \delta(i).$$

That shows that $BB^T = L$ entry-by-entry.

For any $\mathbf{x}$, we have $(B^T \mathbf{x})_k = x_v - x_u$ if the edge $e_k = \{u, v\}$ is oriented as (u, v) under B. That implies

$$\mathbf{x}^T L \mathbf{x} = \mathbf{x}^T B B^T \mathbf{x} = \|B^T \mathbf{x}\|_2^2 = \sum_{e=\{i,j\}\in E} (x_i - x_j)^2.$$

Since the latter quantity is always non-negative, it also implies that L is positive semi-definite. ∎

As a convention, we denote the eigenvalues of a Laplacian matrix L by

$$0 \le \mu_1 \le \mu_2 \le \cdots \le \mu_n,$$

LAPLACIAN
EIGENVALUES
and we will refer to them as *Laplacian eigenvalues*. Here is a simple observation. For any $G = (V, E)$, the constant unit vector

$$\mathbf{y}_1 = \frac{1}{\sqrt{n}}(1, \ldots, 1)$$

is an eigenvector of the Laplacian with eigenvalue 0. Indeed, let B be an oriented incidence matrix of G and recall from the proof of Lemma 5.1.5 that $L = BB^T$. By construction, $B^T \mathbf{y}_1 = \mathbf{0}$ since each column of B has exactly one 1 and one -1. So $L\mathbf{y}_1 = BB^T \mathbf{y}_1 = \mathbf{0}$ as claimed. In general, the constant vector may not be the only eigenvector with eigenvalue one.

We are now ready to derive connectivity consequences. Recall that, for any graph G, the Laplacian eigenvalue $\mu_1 = 0$.

Lemma 5.1.6 (Laplacian and connectivity). *If G is connected, then the Laplacian eigenvalue $\mu_2 > 0$.*

Proof Let $G = (V, E)$ with $n = |V|$ and let $L = \sum_{i=1}^{n} \mu_i \mathbf{y}_i \mathbf{y}_i^T$ be a spectral decomposition of its Laplacian L with $0 = \mu_1 \le \cdots \le \mu_n$. Suppose by way of contradiction that $\mu_2 = 0$. Any eigenvector $\mathbf{y} = (y_1, \ldots, y_n)$ with 0 eigenvalue satisfies $L\mathbf{y} = \mathbf{0}$ by definition. By Lemma 5.1.5 then,

$$0 = \mathbf{y}^T L \mathbf{y} = \sum_{e=\{i,j\}\in E} (y_i - y_j)^2.$$

In order for this to hold, it must be that any two adjacent vertices i and j have $y_i = y_j$. That is, $\{i, j\} \in E$ implies $y_i = y_j$. Furthermore, because G is connected, between any two of its vertices u and v (adjacent or not) there is a path $u = w_0 \sim \cdots \sim w_k = v$ along which the y_ws must be the same. Thus, $\mathbf{y}$ is a constant vector.

But that is a contradiction since the eigenvectors $\mathbf{y}_1, \ldots, \mathbf{y}_n$ are in fact linearly independent, so that $\mathbf{y}_1$ and $\mathbf{y}_2$ cannot both be a constant vector. ∎

FIEDLER
VECTOR
The quantity μ_2 is sometimes referred to as the *algebraic connectivity* of the graph. The corresponding eigenvector, $\mathbf{y}_2$, is known as the *Fiedler vector*.

We will be interested in more quantitative results of this type. Before proceeding, we start with a simple observation. By our proof of Theorem 5.1.1, the largest eigenvalue μ_n of the Laplacian L is the solution to the optimization problem

$$\mu_n = \max\{\langle \mathbf{x}, L\mathbf{x}\rangle : \|\mathbf{x}\|_2 = 1\}.$$

Such extremal characterization is useful in order to bound the eigenvalue μ_n, since any choice of $\mathbf{x}$ with $\|\mathbf{x}\|_2 = 1$ gives a lower bound through the quantity $\langle \mathbf{x}, L\mathbf{x} \rangle$. We give a simple consequence.

Lemma 5.1.7 (Laplacian and degree). *Let $G = (V, E)$ be a graph with maximum degree $\bar{\delta}$. Let μ_n be the largest Laplacian eigenvalue. Then,*

$$\mu_n \geq \bar{\delta} + 1.$$

Proof Let $u \in V$ be a vertex with degree $\bar{\delta}$. Let $\mathbf{z}$ be the vector with entries

$$z_i = \begin{cases} \bar{\delta} & \text{if } i = u, \\ -1 & \text{if } \{i, u\} \in E, \\ 0 & \text{otherwise,} \end{cases}$$

and let $\mathbf{x}$ be the unit vector $\mathbf{z}/\|\mathbf{z}\|_2$. By definition of the degree of u, $\|\mathbf{z}\|_2^2 = \bar{\delta}^2 + \bar{\delta}(-1)^2 = \bar{\delta}(\bar{\delta} + 1)$.

Using the Lemma 5.1.5,

$$\begin{aligned}
\langle \mathbf{z}, L\mathbf{z} \rangle &= \sum_{e=\{i,j\}\in E} (z_i - z_j)^2 \\
&\geq \sum_{i:\{i,u\}\in E} (z_i - z_u)^2 \\
&= \sum_{i:\{i,u\}\in E} (-1 - \bar{\delta})^2 \\
&= \bar{\delta}(\bar{\delta} + 1)^2,
\end{aligned}$$

where we restricted the sum to those edges incident with u and used the fact that all terms in the sum are non-negative. Finally,

$$\langle \mathbf{x}, L\mathbf{x} \rangle = \left\langle \frac{\mathbf{z}}{\|\mathbf{z}\|_2}, L \frac{\mathbf{z}}{\|\mathbf{z}\|_2} \right\rangle = \frac{1}{\|\mathbf{z}\|_2^2} \langle \mathbf{z}, L\mathbf{z} \rangle = \frac{\bar{\delta}(\bar{\delta} + 1)^2}{\bar{\delta}(\bar{\delta} + 1)} = \bar{\delta} + 1,$$

so that

$$\mu_n = \max\{\langle \mathbf{x}', L\mathbf{x}' \rangle : \|\mathbf{x}'\|_2 = 1\} \geq \langle \mathbf{x}, L\mathbf{x} \rangle = \bar{\delta} + 1,$$

as claimed. ∎

A special case of Courant–Fischer (Theorem 5.1.3) for the Laplacian matrix is the following.

Corollary 5.1.8 (Variational characterization of μ_2) *Let $G = (V, E)$ be a graph with $n = |V|$ vertices. Assume the Laplacian L of G has spectral decomposition $L = \sum_{i=1}^n \mu_i \mathbf{y}_i \mathbf{y}_i^T$ with $0 = \mu_1 \leq \mu_2 \leq \cdots \leq \mu_n$ and $\mathbf{y}_1 = \frac{1}{\sqrt{n}}(1, \ldots, 1)$. Then,*

$$\mu_2 = \min \left\{ \frac{\sum_{\{u,v\}\in E}(x_u - x_v)^2}{\sum_{u=1}^n x_u^2} : \mathbf{x} = (x_1, \ldots, x_n) \neq \mathbf{0}, \sum_{u=1}^n x_u = 0 \right\}.$$

Proof   By Theorem 5.1.3,

$$\mu_2 = \min_{\mathbf{x} \in \mathcal{W}_{d-1}} \mathcal{R}_L(\mathbf{x}).$$

Since $\mathbf{y}_1$ is constant and $\mathcal{W}_{d-1}$ is the subspace orthogonal to it, this is equivalent to restricting the minimization to those non-zero $\mathbf{x}$s such that

$$0 = \langle \mathbf{x}, \mathbf{y}_1 \rangle = \frac{1}{\sqrt{n}} \sum_{u=1}^{m} x_u.$$

Moreover, by Lemma 5.1.5,

$$\langle \mathbf{x}, L\mathbf{x} \rangle = \sum_{\{u,v\} \in E} (x_u - x_v)^2$$

so the Rayleigh quotient is

$$\mathcal{R}_L(\mathbf{x}) = \frac{\langle \mathbf{x}, L\mathbf{x} \rangle}{\langle \mathbf{x}, \mathbf{x} \rangle} = \frac{\sum_{\{u,v\} \in E}(x_u - x_v)^2}{\sum_{u=1}^{n} x_u^2}.$$

That proves the claim.   ∎

One application of this extremal characterization is a graph drawing heuristic. Consider the entries of the second Laplacian eigenvector $\mathbf{y}_2$ normalized to have unit norm. The entries are centered around 0 by the condition $\sum_{u=1}^{n} x_u = 0$. Because it minimizes the quantity

$$\frac{\sum_{\{u,v\} \in E}(x_u - x_v)^2}{\sum_{u=1}^{n} x_u^2},$$

over all centered unit vectors, $\mathbf{y}_2$ tends to assign similar coordinates to adjacent vertices. A similar reasoning applies to the third Laplacian eigenvector, which in addition is orthogonal to the second one. See Figure 5.1 for an illustration.

Example 5.1.9 (Two-component graph).   Let $G = (V, E)$ be a graph with two connected components $\emptyset \neq V_1, V_2 \subseteq V$. By the properties of connected components, we have $V_1 \cap V_2 = \emptyset$ and $V_1 \cup V_2 = V$. Assume the Laplacian L of G has spectral decomposition $L = \sum_{i=1}^{n} \mu_i \mathbf{y}_i \mathbf{y}_i^T$ with $0 = \mu_1 \leq \mu_2 \leq \cdots \leq \mu_n$ and $\mathbf{y}_1 = \frac{1}{\sqrt{n}}(1, \ldots, 1)$. We claimed earlier that for such a graph $\mu_2 = 0$. We prove this here using Corollary 5.1.8:

$$\mu_2 = \min \left\{ \sum_{\{u,v\} \in E} (x_u - x_v)^2 : \right.$$

$$\left. \mathbf{x} = (x_1, \ldots, x_n) \in \mathbb{R}^n, \ \sum_{u=1}^{n} x_u = 0, \ \sum_{u=1}^{n} x_u^2 = 1 \right\}.$$

Based on this characterization, it suffices to find a vector $\mathbf{x}$ satisfying $\sum_{u=1}^{n} x_u = 0$ and $\sum_{u=1}^{n} x_u^2 = 1$ such that $\sum_{\{u,v\} \in E}(x_u - x_v)^2 = 0$. Indeed, since $\mu_2 \geq 0$ and any such $\mathbf{x}$ gives an upper bound on μ_2, we then necessarily have that $\mu_2 = 0$.

For $\sum_{\{u,v\} \in E}(x_u - x_v)^2$ to be 0, one might be tempted to take a constant vector $\mathbf{x}$. But then we could not satisfy $\sum_{u=1}^{n} x_u = 0$ and $\sum_{u=1}^{n} x_u^2 = 1$. Instead, we modify this guess slightly. Because the graph has two connected components, there is no edge between V_1 and V_2.

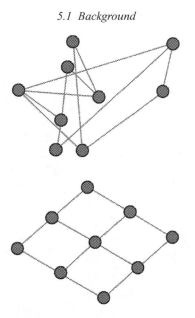

Figure 5.1 Top: A 3-by-3 grid graph with vertices located at independent uniformly random points in a square. Bottom: The same 3-by-3 grid graph with vertices located at the coordinates corresponding to the second and third eigenvectors of the Laplacian matrix. That is, vertex i is located at position $(y_{2,i}, y_{3,i})$.

Hence, we can assign a different value to each component and still get $\sum_{\{u,v\} \in E}(x_u - x_v)^2 = 0$. So we look for a vector $\mathbf{x} = (x_1, \ldots, x_n)$ of the form

$$x_u = \begin{cases} \alpha & \text{if } u \in V_1, \\ \beta & \text{if } u \in V_2. \end{cases}$$

To satisfy the constraints on $\mathbf{x}$, we require

$$\sum_{u=1}^{n} x_u = \sum_{u \in V_1} \alpha + \sum_{u \in V_2} \beta = |V_1|\alpha + |V_2|\beta = 0$$

and

$$\sum_{u=1}^{n} x_u^2 = \sum_{u \in V_1} \alpha^2 + \sum_{u \in V_2} \beta^2 = |V_1|\alpha^2 + |V_2|\beta^2 = 1.$$

Replacing the first equation in the second one, we get

$$|V_1| \left(\frac{-|V_2|\beta}{|V_1|} \right)^2 + |V_2|\beta^2 = \frac{|V_2|^2 \beta^2}{|V_1|} + |V_2|\beta^2 = 1$$

or

$$\beta^2 = \frac{|V_1|}{|V_2|(|V_2| + |V_1|)} = \frac{|V_1|}{n|V_2|}.$$

Take

$$\beta = -\sqrt{\frac{|V_1|}{n|V_2|}}, \qquad \alpha = \frac{-|V_2|\beta}{|V_1|} = \sqrt{\frac{|V_2|}{n|V_1|}}.$$

The vector $\mathbf{x}$ we constructed is in fact an eigenvector of L. Indeed, let B be an oriented incidence matrix of G. Then, for $e_k = \{u, v\}$, $(B^T\mathbf{x})_k$ is either $x_u - x_v$ or $x_v - x_u$. In both cases, that is 0. So $L\mathbf{x} = BB^T\mathbf{x} = \mathbf{0}$, that is, $\mathbf{x}$ is an eigenvector of L with eigenvalue 0.

We have shown that $\mu_2 = 0$ when G has two connected components. A slight modification of this argument shows that $\mu_2 = 0$ whenever G is not connected.   ◀

Networks In the case of a network (i.e., edge-weighted graph) $G = (V, E, w)$, the Laplacian can be defined as follows. As usual, we assume that $w\colon E \to \mathbb{R}_+$ is a function that assigns positive real weights to the edges. We write $w_e = w_{ij}$ for the weight of edge $e = \{i, j\}$. Recall that the degree of a vertex i is

$$\delta(i) = \sum_{j:\{i,j\}\in E} w_{ij}.$$

The adjacency matrix A of G is the $n \times n$ symmetric matrix defined as

$$A_{ij} = \begin{cases} w_{ij} & \text{if } \{i, j\} \in E, \\ 0 & \text{otherwise.} \end{cases}$$

Definition 5.1.10 (Network Laplacian). *Let $G = (V, E, w)$ be a network with $n = |V|$ vertices and adjacency matrix A. Let $D = \mathrm{diag}(\delta(1), \ldots, \delta(n))$ be the degree matrix. The network Laplacian (or Laplacian matrix, or Laplacian for short) associated to G is defined as $L = D - A$.*

It can be shown (see Exercise 5.2) that the Laplacian quadratic form satisfies in the edge-weighted case

$$\langle \mathbf{x}, L\mathbf{x} \rangle = \sum_{\{i,j\}\in E} w_{ij}(x_i - x_j)^2 \tag{5.1.3}$$

for $\mathbf{x} = (x_1, \ldots, x_n) \in \mathbb{R}^n$. (The keen observer will have noticed that we already encountered this quantity as the "Dirichlet energy" in Section 3.3.3; more on this in Section 5.3.) As a positive semi-definite matrix (see again Exercise 5.2), the network Laplacian has an orthonormal basis of eigenvectors with non-negative eigenvalues that satisfy the variational characterization we derived above. In particular, if we denote the eigenvalues $0 = \mu_1 \leq \mu_2 \leq \cdots \leq \mu_n$, it follows from Courant–Fischer (Theorem 5.1.3) that

$$\mu_2 = \min\left\{ \sum_{\{u,v\}\in E} w_{uv}(x_u - x_v)^2 : \right.$$
$$\left. \mathbf{x} = (x_1, \ldots, x_n)^T \in \mathbb{R}^n, \sum_{u=1}^{n} x_u = 0, \sum_{u=1}^{n} x_u^2 = 1 \right\}.$$

Other variants of the Laplacian are useful. We introduce the normalized Laplacian next.

Definition 5.1.11 (Normalized Laplacian). *The* normalized Laplacian *of* $G = (V, E, w)$ *with*
adjacency matrix A *and degree matrix* D *is defined as*

$$\mathcal{L} = I - D^{-1/2} A D^{-1/2}.$$

The entries of $\mathcal{L}$ are

$$\mathcal{L}_{i,j} = \begin{cases} 1 & \text{if } i = j, \\ -\frac{w_{ij}}{\sqrt{\delta(i)\delta(j)}} & \text{otherwise.} \end{cases}$$

We also note the following relation to the (unnormalized) Laplacian:

$$\mathcal{L} = D^{-1/2} L D^{-1/2}. \tag{5.1.4}$$

We check that the normalized Laplacian is symmetric:

$$\begin{aligned} \mathcal{L}^T &= I^T - (D^{-1/2} A D^{-1/2})^T \\ &= I - (D^{-1/2})^T A^T (D^{-1/2})^T \\ &= I - D^{-1/2} A D^{-1/2} \\ &= \mathcal{L}. \end{aligned}$$

It is also positive semi-definite. Indeed,

$$\mathbf{x}^T \mathcal{L} \mathbf{x} = \mathbf{x}^T D^{-1/2} L D^{-1/2} \mathbf{x} = (D^{-1/2}\mathbf{x})^T L (D^{-1/2}\mathbf{x}) \geq 0,$$

by the properties of the Laplacian. Hence, by the spectral theorem (Theorem 5.1.1), we can write

$$\mathcal{L} = \sum_{i=1}^{n} \eta_i \mathbf{z}_i \mathbf{z}_i^T,$$

where the $\mathbf{z}_i$s are orthonormal eigenvectors of $\mathcal{L}$ and the eigenvalues satisfy

$$0 \leq \eta_1 \leq \eta_2 \leq \cdots \leq \eta_n.$$

One more observation: because the constant vector is an eigenvector of L with eigenvalue 0, we get from (5.1.4) that $D^{1/2}\mathbf{1}$ is an eigenvector of $\mathcal{L}$ with eigenvalue 0. So $\eta_1 = 0$ and we set

$$(\mathbf{z}_1)_i = \left(\frac{D^{1/2}\mathbf{1}}{\|D^{1/2}\mathbf{1}\|_2} \right)_i = \sqrt{\frac{\delta(i)}{\sum_{i \in V} \delta(i)}} \quad \forall i \in [n],$$

which makes $\mathbf{z}_1$ into a unit norm vector. The relationship to the Laplacian implies (see Exercise 5.3) that

$$\mathbf{x}^T \mathcal{L} \mathbf{x} = \sum_{\{i,j\} \in E} w_{ij} \left(\frac{x_i}{\sqrt{\delta(i)}} - \frac{x_j}{\sqrt{\delta(j)}} \right)^2$$

for $\mathbf{x} = (x_1, \ldots, x_n) \in \mathbb{R}^n$. Through the change of variables,

$$y_i = \frac{x_i}{\sqrt{\delta(i)}},$$

Courant–Fischer (Theorem 5.1.3) gives this time

$$\eta_2 = \min \left\{ \sum_{\{u,v\} \in E} w_{uv}(y_u - y_v)^2 : \right.$$

$$\left. \mathbf{y} = (y_1, \ldots, y_n) \in \mathbb{R}^n, \ \sum_{u=1}^{n} \delta(u)y_u = 0, \ \sum_{u=1}^{n} \delta(u)y_u^2 = 1 \right\}. \tag{5.1.5}$$

5.1.3 Perturbation Results

We will need some perturbation results for eigenvalues and eigenvectors. Recall the following definition. Define $\mathbb{S}^{m-1} = \{\mathbf{x} \in \mathbb{R}^m : \|\mathbf{x}\|_2 = 1\}$. The spectral norm (or induced 2-norm or 2-norm) of a matrix $A \in \mathbb{R}^{n \times m}$ is

$$\|A\|_2 := \max_{\mathbf{0} \neq \mathbf{x} \in \mathbb{R}^m} \frac{\|A\mathbf{x}\|}{\|\mathbf{x}\|} = \max_{\mathbf{x} \in \mathbb{S}^{m-1}} \|A\mathbf{x}\|.$$

The induced 2-norm of a matrix has many other useful properties.

Lemma 5.1.12 (Properties of the induced norm). *Let $A, B \in \mathbb{R}^{n \times m}$ and $\alpha \in \mathbb{R}$. The following hold:*

  (i) $\|A\mathbf{x}\|_2 \leq \|A\|_2 \|\mathbf{x}\|_2 \ \forall \mathbf{0} \neq \mathbf{x} \in \mathbb{R}^m$
 (ii) $\|A\|_2 \geq 0$
 (iii) $\|A\|_2 = 0$ *if and only if $A = 0$*
 (iv) $\|\alpha A\|_2 = |\alpha| \|A\|_2$
  (v) $\|A + B\|_2 \leq \|A\|_2 + \|B\|_2$
 (vi) $\|AB\|_2 \leq \|A\|_2 \|B\|_2$.

Proof These properties all follow from the definition of the induced norm and the corresponding properties for the vector norm:

- Claims (i) and (ii) are immediate from the definition.
- For (ii) note that $\|A\|_2 = 0$ implies $\|A\mathbf{x}\|_2 = 0, \forall \mathbf{x} \in \mathbb{S}^{m-1}$, so that $A\mathbf{x} = \mathbf{0}, \forall \mathbf{x} \in \mathbb{S}^{m-1}$. In particular, $A_{ij} = \mathbf{e}_i^T A \mathbf{e}_j = 0, \forall i, j$.
- For (iv), (v), and (vi) observe that for all $\mathbf{x} \in \mathbb{S}^{m-1}$,

$$\|\alpha A\mathbf{x}\|_2 = |\alpha| \|A\mathbf{x}\|_2,$$

$$\|(A + B)\mathbf{x}\|_2 = \|A\mathbf{x} + B\mathbf{x}\|_2 \leq \|A\mathbf{x}\|_2 + \|B\mathbf{x}\|_2 \leq \|A\|_2 + \|B\|_2,$$

$$\|(AB)\mathbf{x}\|_2 = \|A(B\mathbf{x})\|_2 \leq \|A\|_2 \|B\mathbf{x}\|_2 \leq \|A\|_2 \|B\|_2.$$

∎

Perturbations of eigenvalues For a symmetric matrix $C \in \mathbb{R}^{d \times d}$, we let $\lambda_j(C), j = 1, \ldots, d$, be the eigenvalues of C in non-increasing order with corresponding orthonormal eigenvectors $\mathbf{v}_j(C), j = 1, \ldots, d$. As in the Courant–Fischer theorem (Theorem 5.1.3), define the subspaces

$$\mathcal{V}_k(C) = \mathrm{span}(\mathbf{v}_1(C), \ldots, \mathbf{v}_k(C))$$

and

$$\mathcal{W}_{d-k+1}(C) = \mathrm{span}(\mathbf{v}_k(C), \ldots, \mathbf{v}_d(C)).$$

The following lemma is one version of what is known as *Weyl's inequality*.

WEYL'S
INEQUALITY

Lemma 5.1.13 (Weyl's inequality). *Let $A \in \mathbb{R}^{d \times d}$ and $B \in \mathbb{R}^{d \times d}$ be symmetric matrices. Then, for all $j = 1, \ldots, d$,*

$$\max_{j \in [d]} \left| \lambda_j(B) - \lambda_j(A) \right| \le \|B - A\|_2.$$

Proof Let $H = B - A$. We prove only one upper bound. The other one follows from interchanging the roles of A and B. Because

$$\dim(\mathcal{V}_j(B)) + \dim(\mathcal{W}_{d-j+1}(A)) = j + (d - j + 1) = d + 1 > d,$$

it follows from (5.1.2) that $\mathcal{V}_j(B) \cap \mathcal{W}_{d-j+1}(A)$ contains a non-zero vector. Let $\mathbf{v}$ be a unit vector in that intersection.

By Theorem 5.1.3,

$$\lambda_j(B) \le \langle \mathbf{v}, (A + H)\mathbf{v} \rangle = \langle \mathbf{v}, A\mathbf{v} \rangle + \langle \mathbf{v}, H\mathbf{v} \rangle \le \lambda_j(A) + \langle \mathbf{v}, H\mathbf{v} \rangle.$$

Moreover, by Cauchy–Schwarz (Theorem B.4.8), since $\|\mathbf{v}\|_2 = 1$,

$$\langle \mathbf{v}, H\mathbf{v} \rangle \le \|\mathbf{v}\|_2 \|H\mathbf{v}\|_2 \le \|H\|_2,$$

which proves the claim after rearranging. ∎

Perturbations of eigenvectors While Weyl's inequality (Lemma 5.1.13) indicates that the eigenvalues of A and B are close when $\|A - B\|_2$ is small, it says nothing about the eigenvectors. The following theorem remediates that. It is traditionally stated in terms of the angle between the eigenvectors (whereby the name). Here we give a version that is more suited to the applications we will encounter. We do not optimize the constants. We use the same notation as in the previous paragraph. Recall Parseval's identity: if $u_1, \ldots, u_d$ is an orthonormal basis of $\mathbb{R}^d$, then $\|x\|^2 = \sum_{i=1}^d \langle x, u_i \rangle^2$.

Theorem 5.1.14 (Davis–Kahan $\sin \theta$ theorem). *Let $A \in \mathbb{R}^{d \times d}$ and $B \in \mathbb{R}^{d \times d}$ be symmetric matrices. For an $i \in \{1, \ldots, d\}$, assume that*

$$\delta := \min_{j \ne i} |\lambda_i(A) - \lambda_j(A)| > 0.$$

Then

$$\min_{s \in \{+1, -1\}} \|\mathbf{v}_i(A) - s\mathbf{v}_i(B)\|_2^2 \le \frac{8\|A - B\|_2^2}{\delta^2}.$$

Proof Expand $\mathbf{v}_i(B)$ in the basis formed by the eigenvectors of A, that is,

$$\mathbf{v}_i(B) = \sum_{j=1}^{d} \langle \mathbf{v}_i(B), \mathbf{v}_j(A) \rangle \, \mathbf{v}_j(A),$$

where we used the orthonormality of the $\mathbf{v}_j(A)$s. On the one hand,

$$\|(A - \lambda_i(A)I_{d \times d}) \, \mathbf{v}_i(B)\|_2^2$$

$$= \left\| \sum_{j=1}^{d} \langle \mathbf{v}_i(B), \mathbf{v}_j(A) \rangle (A - \lambda_i(A)I_{d \times d}) \, \mathbf{v}_j(A) \right\|_2^2$$

$$= \left\| \sum_{j=1, j \neq i}^{d} \langle \mathbf{v}_i(B), \mathbf{v}_j(A) \rangle (\lambda_j(A) - \lambda_i(A)) \, \mathbf{v}_j(A) \right\|_2^2$$

$$= \sum_{j=1, j \neq i}^{d} \langle \mathbf{v}_i(B), \mathbf{v}_j(A) \rangle^2 (\lambda_j(A) - \lambda_i(A))^2$$

$$\geq \delta^2 (1 - \langle \mathbf{v}_i(B), \mathbf{v}_i(A) \rangle^2),$$

where, on the last two lines, we used the orthonormality of the $\mathbf{v}_j(A)$s and $\mathbf{v}_j(B)$s through Parseval's identity, as well as the definition of δ.

On the other hand, letting $E = A - B$, by the triangle inequality,

$$\|(A - \lambda_i(A)I) \, \mathbf{v}_i(B)\|_2 = \|(B + E - \lambda_i(A)I) \, \mathbf{v}_i(B)\|_2$$

$$\leq \|(B - \lambda_i(A)I) \, \mathbf{v}_i(B)\|_2 + \|E \, \mathbf{v}_i(B)\|_2$$

$$\leq |\lambda_i(B) - \lambda_i(A)| \|\mathbf{v}_i(B)\|_2 + \|E\|_2 \|\mathbf{v}_i(B)\|_2$$

$$= 2\|E\|_2,$$

where we used Lemma 5.1.12 and Weyl's inequality.

Combining the last two inequalities gives

$$(1 - \langle \mathbf{v}_i(B), \mathbf{v}_i(A) \rangle^2) \leq \frac{4\|E\|_2^2}{\delta^2}.$$

The result follows by noting that since $|\langle \mathbf{v}_i(B), \mathbf{v}_i(A) \rangle| \leq 1$ by Cauchy–Schwarz (Theorem B.4.8), we have

$$\min_{s \in \{+1, -1\}} \|\mathbf{v}_i(A) - s\mathbf{v}_i(B)\|^2 = 2 - 2|\langle \mathbf{v}_i(B), \mathbf{v}_i(A) \rangle|$$

$$\leq 2(1 - \langle \mathbf{v}_i(B), \mathbf{v}_i(A) \rangle^2)$$

$$\leq \frac{8\|E\|_2^2}{\delta^2}. \qquad \blacksquare$$

5.1.4 ▷ *Data Science: Community Recovery*

A common task in network analysis is to recover hidden community structure. Informally, we seek groups of vertices with more edges within the groups than to the rest of the graph. More rigorously, providing statistical guarantees on the output of a community recovery algorithm requires some underlying random graph model. The standard model for this purpose is the *stochastic blockmodel*, a generalization of the Erdős–Renyí graph model with a "planted partition."

Stochastic blockmodel and recovery requirement We restrict ourselves to the simple case of two strictly balanced communities. Consider a random graph on n (even) nodes where there are two communities, labeled $+1$ and -1, consisting of $n/2$ nodes. Each vertex $i \in V$ is assigned a community label $X_i \in \{1, -1\}$ as follows: a subset of $n/2$ vertices is chosen uniformly at random among all such subsets to form community $+1$, and the rest of the vertices form community -1. For two nodes i, j, the edge $\{i, j\}$ is present with probability p if they belong to the same community, and with probability q otherwise. All edges are independent. The following 2×2 matrix describes the edge density within and across the two communities:

$$W = \begin{array}{c} {} \\ +1 \\ -1 \end{array} \begin{array}{c} +1 \quad\; -1 \\ \begin{bmatrix} p & q \\ q & p \end{bmatrix} \end{array}.$$

We assume that $p \geq q$, encoding the fact that vertices belonging to the same community are more likely to share an edge. To summarize, we say that $(X, G) \sim \mathrm{SBM}_{n,p,q}$ if:

1. *(Communities)* The assignment $X = (X_1, \ldots, X_n)$ is uniformly random over

$$\Pi_2^n := \{\mathbf{x} \in \{+1, -1\}^n : \mathbf{x}^T \mathbf{1} = 0\},$$

   where $\mathbf{1} = (1, \ldots, 1)$ is the all-one vector.
2. *(Graph)* Conditioned on X, the graph $G = ([n], E)$ has independent edges where $\{i, j\}$ is present with probability W_{X_i, X_j} for $\forall i < j$.

We denote the corresponding measure by $\mathbb{P}_{n,p,q}$. We allow p and q to depend on n (although we do not make that dependence explicit).

Roughly speaking, the *community recovery problem* is the following: given G, output X. There are different notions of recovery.

Definition 5.1.15 (Agreement). *The agreement between two community assignment vectors* $\mathbf{x}, \mathbf{y} \in \{+1, -1\}^n$ *is the largest fraction of common assignments between* $\mathbf{x}$ *and* $\pm\mathbf{y}$, *that is,*

$$\alpha(\mathbf{x}, \mathbf{y}) = \max_{s \in \{+1, -1\}} \frac{1}{n} \sum_{i=1}^n \mathbf{1}\{x_i = s y_i\}.$$

The role of s in this formula is to account for the fact that the community names are not meaningful.

Now consider the following recovery requirements. These are asymptotic notions, as $n \to +\infty$.

Definition 5.1.16 (Recovery requirement). *Let* $(X, G) \sim \mathrm{SBM}_{n,p,q}$. *For any estimator* $\hat{X} :=$ $\hat{X}(G) \in \Pi_2^n$, *we say that it achieves:*

- *exact recovery* if $\mathbb{P}_{n,p,q}[\alpha(X,\hat{X}) = 1] = 1 - o(1)$; or
- *almost exact recovery* if $\mathbb{P}_{n,p,q}[\alpha(X,\hat{X}) = 1 - o(1)] = 1 - o(1)$.

Next we establish sufficient conditions for almost exact recovery. First we describe a natural estimator $\hat{X}$.

MAP estimator and spectral clustering A natural starting point is the maximum a posteriori (MAP) estimator. Let $\Omega(X)$ be the balanced partition of $[n]$ corresponding to X and $\hat{\Omega}(G)$ be the one corresponding to $\hat{X}(G)$. The probability of error, that is, the probability of not recovering the true partition, is given by

$$\mathbb{P}[\Omega(X) \neq \hat{\Omega}(G)] = \sum_g \mathbb{P}[\hat{\Omega}(g) \neq \Omega(X) \mid G = g]\,\mathbb{P}[G = g], \qquad (5.1.6)$$

where the sum is over all graphs on n vertices (i.e., all possible subsets of edges present) and we dropped the subscript n, p, q to simplify the notation. The MAP estimator $\hat{\Omega}^{\mathrm{MAP}}(G)$ is obtained by minimizing each term $\mathbb{P}[\hat{\Omega}(g) \neq \Omega(X) \mid G = g]$ individually (note that $\mathbb{P}[G = g] > 0$ for all g by definition of the $\mathrm{SBM}_{n,p,q}$, a probability which does not depend on the estimator). Equivalently, we choose for each g a partition γ that maximizes the posterior probability

$$\mathbb{P}[\Omega(X) = \gamma \mid G = g] = \frac{\mathbb{P}[G = g \mid \Omega(X) = \gamma]\,\mathbb{P}[\Omega(X) = \gamma]}{\mathbb{P}[G = g]}$$

$$= \mathbb{P}[G = g \mid \Omega(X) = \gamma] \cdot \frac{1}{|\Pi_2^n|\,\mathbb{P}[G = g]}, \qquad (5.1.7)$$

where we applied Bayes' rule on the first line and the uniformity of the partition X on the second line.

Based on (5.1.7), we seek a partition that maximizes $\mathbb{P}[G = g \mid \Omega(X) = \gamma]$. We compute this last probability explicitly. For fixed g, let $M := M(g)$ be the number of edges in g. For any γ, denote by $M_{\mathrm{in}} := M_{\mathrm{in}}(g, \gamma)$ and $M_{\mathrm{out}} := M_{\mathrm{out}}(g, \gamma)$ the number of edges within and across communities, respectively, and note that $M_{\mathrm{in}} = M - M_{\mathrm{out}}$. By definition of the $\mathrm{SBM}_{n,p,q}$ model, the probability of a graph g given a partition γ is expressed simply as

$$\mathbb{P}[G = g \mid \Omega(X) = \gamma]$$

$$= q^{M_{\mathrm{out}}}(1 - q)^{\left(\frac{n}{2}\right)^2 - M_{\mathrm{out}}} p^{M_{\mathrm{in}}}(1 - p)^{\left\{\binom{n}{2} - \left(\frac{n}{2}\right)^2\right\} - M_{\mathrm{in}}}$$

$$= q^{M_{\mathrm{out}}}(1 - q)^{\left(\frac{n}{2}\right)^2 - M_{\mathrm{out}}} p^{M - M_{\mathrm{out}}}(1 - p)^{\left\{\binom{n}{2} - \left(\frac{n}{2}\right)^2\right\} - \{M - M_{\mathrm{out}}\}}$$

$$= \left[\frac{q}{1 - q} \cdot \frac{1 - p}{p}\right]^{M_{\mathrm{out}}} \left\{(1 - q)^{\left(\frac{n}{2}\right)^2} p^M (1 - p)^{\left\{\binom{n}{2} - \left(\frac{n}{2}\right)^2\right\} - M}\right\}.$$

The expression in curly brackets does not depend on the partition γ. Moreover, since we assume that $p \geq q$, we have that $\left[\frac{q}{1-q} \cdot \frac{1-p}{p}\right] \leq 1$ (which can be checked directly by rearranging and canceling). Therefore, to maximize $\mathbb{P}[G = g \mid \Omega(X) = \gamma]$ over γ for a fixed g, we need to choose a partition that results in the smallest possible value of M_{out}, the number of edges across the two communities. This problem is well known in combinatorial optimization, where it is referred to as the *minimum bisection problem*. It is unfortunately NP-hard and we consider a relaxation that admits a polynomial-time algorithmic solution.

MINIMUM BISECTION PROBLEM

To see how this comes about, observe that the minimum bisection problem can be refor-mulated as

$$\max_{\mathbf{x} \in \{+1, -1\}^n, \, \mathbf{x}^T \mathbf{1} = 0} \mathbf{x}^T A \mathbf{x},$$

where A is the $n \times n$ adjacency matrix. Replacing the combinatorial constraint $\mathbf{x} \in \{+1, -1\}^n$ by $\mathbf{x} \in \mathbb{R}^n$ with $\|\mathbf{x}\|_2 = n$ leads to the relaxation

$$\max_{\mathbf{z} \in \mathbb{R}^n, \, \mathbf{z}^T \mathbf{1} = 0, \, \|\mathbf{z}\|_2 = n} \mathbf{z}^T A \mathbf{z}$$

$$= \max_{0 \neq \mathbf{z} \in \mathbb{R}^n, \, \mathbf{z}^T \mathbf{1} = 0} \left(n \frac{\mathbf{z}}{\|\mathbf{z}\|_2} \right)^T A \left(n \frac{\mathbf{z}}{\|\mathbf{z}\|_2} \right)$$

$$= n^2 \max_{0 \neq \mathbf{z} \in \mathbb{R}^n, \, \mathbf{z}^T \mathbf{1} = 0} \frac{\mathbf{z}^T A \mathbf{z}}{\mathbf{z}^T \mathbf{z}},$$

where we changed the notation from $\mathbf{x}$ to $\mathbf{z}$ to emphasize that the solution no longer encodes a partition. We recognize the Rayleigh quotient of A as the objective function in the final for-mulation. At this point, it is tempting to use Courant–Fischer (Theorem 5.1.3) and conclude that the maximum above is achieved at the second eigenvalue of A. Note, however, that the vector $\mathbf{1}$ (appearing in the orthogonality constraint $\mathbf{z}^T \mathbf{1} = 0$) is not in general an eigenvector of A (unless the graph happens to be regular). To leverage the variational characterization of eigenvalues in a statistically justified way, we instead turn to the expected adjacency matrix and then establish concentration.

Lemma 5.1.17 (Expected adjacency). *Let* $(X, G) \sim \mathrm{SBM}_{n,p,q}$, *let* A *be the adjacency matrix of* G, *and let* $\mathcal{A}_X = \mathbb{E}_{n,p,q}[A \mid X]$. *Then,*

$$\mathcal{A}_X = n \frac{p + q}{2} \mathbf{u}_1 \mathbf{u}_1^T + n \frac{p - q}{2} \mathbf{u}_2 \mathbf{u}_2^T - p I,$$

where

$$\mathbf{u}_1 = \frac{1}{\sqrt{n}} \mathbf{1}, \quad \mathbf{u}_2 = \frac{1}{\sqrt{n}} X.$$

Proof For any distinct pair i, j, the term

$$\left(n \frac{p + q}{2} \mathbf{u}_1 \mathbf{u}_1^T \right)_{i,j} = n \frac{p + q}{2} \left(\frac{1}{\sqrt{n}} \right)^2 = \frac{p + q}{2},$$

while the term

$$\left(n \frac{p - q}{2} \mathbf{u}_2 \mathbf{u}_2^T \right)_{i,j} = n \frac{p - q}{2} \left(\frac{1}{\sqrt{n}} \right)^2 X_i X_j = \frac{p - q}{2} X_i X_j.$$

The product $X_i X_j$ is 1 when i and j belong to the same community and is -1 otherwise. In the former case, summing the two terms indeed gives p, while in the latter case it gives q. Finally, the term $-pI$ accounts for the fact that A has zeros on the diagonal. ∎

Now condition on X and observe that $\mathbf{u}_1$ and $\mathbf{u}_2$ in Lemma 5.1.17 are orthogonal by our assumption that X corresponds to a balanced partition (i.e., with two communities of equal size). Hence we deduce that an eigenvector decomposition of $\mathcal{A}_X$ is formed of $\mathbf{u}_1, \mathbf{u}_2$, and any orthonormal basis of the orthogonal complement of the span of $\mathbf{u}_1$ and $\mathbf{u}_2$, with respective eigenvalues

$$n\frac{p+q}{2} - p, \qquad n\frac{p-q}{2} - p, \qquad -p.$$

So the second largest eigenvalue of $\mathcal{A}_X$ is $\lambda_2(\mathcal{A}_X) = n\frac{p-q}{2} - p$ (independently of X), and Courant–Fischer implies

$$\max_{0 \neq \mathbf{z} \in \mathbb{R}^n, \mathbf{z}^T \mathbf{1} = 0} \frac{\mathbf{z}^T \mathcal{A}_X \mathbf{z}}{\mathbf{z}^T \mathbf{z}} = \lambda_2(\mathcal{A}_X).$$

The corresponding eigenvector, up to scaling and sign, is precisely what we are trying to recover, namely, the community assignment X.

SPECTRAL CLUSTERING

These observations motivate the following *spectral clustering* approach.

1. *Input:* graph G with adjacency matrix A.
2. Compute an eigenvector decomposition of A.
3. Let $\hat{\mathbf{u}}_2$ be the eigenvector corresponding to the second largest eigenvalue.
4. *Output:* $\hat{X}(G) = \operatorname{sgn}(\hat{\mathbf{u}}_2)$.

Here we used the notation

$$(\operatorname{sgn}(\mathbf{z}))_i = \begin{cases} +1 & \text{if } z_i \geq 0, \\ -1 & \text{otherwise.} \end{cases}$$

Because we used A rather than $\mathcal{A}_X$ (which we do not know), it is not immediate that this approach will work. Below, we use Davis–Kahan (Theorem 5.1.14) to show that, under some conditions, the second eigenvector of A is concentrated around that of $\mathcal{A}_X$ – and therefore almost exact recovery holds.

Before getting to the analysis, we make a final algorithmic remark. The "clustering" above, specifically taking the sign of the second eigenvector, works in this toy model but is perhaps somewhat naive. More generally, in a spectral clustering method, one uses the top eigenvectors (deciding how many is a bit of an art) of the adjacency matrix (or of another matrix associated to the graph such as the Laplacian or normalized Laplacian) to obtain a low-dimensional representation of the input. Then in a second step, one uses a clustering algorithm, for example, k-means clustering, to extract communities in the low-dimensional space.

Almost exact recovery We prove the following. We restrict ourselves to the case where p and q are constants not depending on n.

Theorem 5.1.18 *Let* $(X, G) \sim \mathrm{SBM}_{n,p,q}$ *and let A be the adjacency matrix of G. Let $\mu :=$ $\min\left\{q, \frac{p-q}{2}\right\} > 0$. Clustering according to the sign of the second eigenvector of A identifies the two communities of G with probability at least $1 - e^{-n}$, except for C/μ^2 misclassified nodes for some constant $C > 0$.*

There are two key ingredients to the proof: concentration of the adjacency matrix and perturbation arguments.

We start with the former.

Lemma 5.1.19 (Norm of the adjacency). *Let $(X, G) \sim \mathrm{SBM}_{n,p,q}$, let A be the adjacency matrix of G and let $\mathcal{A}_X = \mathbb{E}_{n,p,q}[A \mid X]$. There is a constant $C' > 0$ such that, conditioned on X,*

$$\|A - \mathcal{A}_X\|_2 \leq C'\sqrt{n},$$

with probability at least $1 - e^{-n}$.

Proof Condition on X. We use Theorem 2.4.28 on the random matrix $R := A - \mathcal{A}_X$. The entries of R are centered and independent (conditionally on X). Moreover, they are bounded. Indeed, for $i \neq j$, $A_{ij} \in \{0, 1\}$ while $(\mathcal{A}_X)_{ij} \in \{q, p\}$. So $R_{ij} \leq [-p, 1 - q]$. On the diagonal, $R_{ii} = 0$. Hence, by Hoeffding's lemma (Lemma 2.4.12), the entries are sub-Gaussian with variance factor

$$\frac{1}{4}(1 - q - (-p))^2 \leq 1.$$

Taking $t = \sqrt{n}$ in Theorem 2.4.28, there is a constant $C > 0$ such that with probability $1 - e^{-n}$,

$$\|A - \mathcal{A}_X\|_2 \leq C\sqrt{1}(\sqrt{n} + \sqrt{n} + \sqrt{n}).$$

Adjusting the constant gives the claim.   ∎

We are ready to prove the theorem.

Proof of Theorem 5.1.18 Condition on X. To apply the Davis–Kahan theorem (Theorem 5.1.14), we need to bound the smallest gap δ between the second largest eigenvalue of $\mathcal{A}_X$ and its other eigenvalues. Recall that the eigenvalues are

$$n\frac{p+q}{2} - p, \qquad n\frac{p-q}{2} - p, \qquad -p,$$

so

$$\delta = \min\left\{n\frac{p-q}{2}, nq\right\} = n\mu > 0.$$

By Davis–Kahan and Lemma 5.1.19, with probability at least $1 - e^{-n}$, there is $\theta \in \{+1, -1\}$ such that

$$\|\mathbf{u}_2 - \theta\,\hat{\mathbf{u}}_2\|_2^2 \leq \frac{8\|A - \mathcal{A}_X\|_2^2}{\delta^2} \leq \frac{8(C'\sqrt{n})^2}{(n\mu)^2} = \frac{C}{n\,\mu^2},$$

by adjusting the constant. Note that this bound holds for *any* X.

Rearranging and expanding the norm, we get

$$\sum_i \left|\sqrt{n}\,(\mathbf{u}_2)_i - \sqrt{n}\,\theta\,(\hat{\mathbf{u}}_2)_i\right|^2 \leq \frac{C}{\mu^2}.$$

If the signs of $(\mathbf{u}_2)_i$ and $\theta\,(\hat{\mathbf{u}}_2)_i$ disagree, then the ith term in the sum above is ≥ 1. So there can be at most C/μ^2 such disagreements. That establishes the desired bound on the number of misclassified nodes.   ∎

Remark 5.1.20 *It was shown in [YP14, MNS15a, AS15] that almost exact recovery in the balanced two-community model* SBM_{n,p_n,q_n} *with* $p_n = a_n/n$ *and* $q_n = b_n/n$ *is achievable and computationally efficiently so if and only if*

$$\frac{(a_n - b_n)^2}{(a_n + b_n)} = \omega(1).$$

On the other hand, it was shown in [ABH16, MNS15a] that exact recovery in the SBM_{n,p_n,q_n} *with* $p_n = \alpha \log n/n$ *and* $q_n = \beta \log n/n$ *is achievable and computationally efficiently so if* $\sqrt{\alpha} - \sqrt{\beta} > 2$ *and not achievable if* $\sqrt{\alpha} - \sqrt{\beta} < 2$.

5.2 Spectral Techniques for Reversible Markov Chains

In this section, we apply the spectral theorem to reversible Markov chains. Throughout (X_t) is a Markov chain on a state space V with transition matrix P reversible with respect to a positive stationary measure $\pi > 0$. Recall that this means that $\pi(x)P(x,y) = \pi(y)P(y,x)$ for all $x,y \in V$. We also assume that P is irreducible.

A Hilbert space It will be convenient to introduce a Hilbert space of functions over V. Let $\ell^2(V,\pi)$ be the space of functions $f : V \to \mathbb{R}$ such that $\sum_{x\in V} \pi(x)f(x)^2 < +\infty$. Equipped with the following inner product, it forms a Hilbert space (i.e., a real inner product space that is also a complete metric space (see Theorem B.4.10) with respect to the induced metric; we will work mostly in finite dimension where it is merely a slight generalization of Euclidean space). For $f,g \in \ell^2(V,\pi)$, define

$$\langle f,g \rangle_\pi := \sum_{x\in V} \pi(x)f(x)g(x)$$

and

$$\|f\|_\pi^2 := \langle f,f \rangle_\pi.$$

The inner product is well defined since the series is summable by Hölder's inequality (Theorem B.4.8), which implies the Cauchy–Schwarz inequality

$$\langle f,g \rangle_\pi \leq \|f\|_\pi \|g\|_\pi.$$

Minkowski's inequality (Theorem B.4.9) implies the triangle inequality

$$\|f + g\|_\pi \leq \|f\|_\pi + \|g\|_\pi.$$

The integral with respect to π (see Appendix B) reduces in this case to a sum

$$\pi(f) := \sum_{x\in V} \pi(x)f(x),$$

provided $\pi(|f|) < +\infty$ or $f \geq 0$. Here $|f|$ is defined as $|f|(x) := |f(x)|$ for all $x \in V$. We also write $\pi f = \pi(f)$ to simplify the notation.

We recall some standard Hilbert space facts. The countable collection of functions $\{f_i\}_{i=1}^\infty$ in $\ell^2(V,\pi)$ is an orthonormal basis if: (i) $\langle f_i,f_j \rangle_\pi = 0$ if $i \neq j$ and $= 1$ if $i = j$; and (ii) any

$f \in \ell^2(V, \pi)$ can be written as $\lim_{n \to +\infty} \sum_{i=1}^{n} \langle f_i, f \rangle_\pi f_i = f$, where the limit is in the norm. We then have *Parseval's identity*: for any $g \in \ell^2(V, \pi)$,

$$\|g\|_\pi^2 = \sum_{j=1}^{\infty} \langle g, f_j \rangle_\pi^2. \tag{5.2.1}$$

Think of P as an operator on $\ell^2(V, \pi)$. That is, let $Pf : V \to \mathbb{R}$ be defined as

$$(Pf)(x) := \sum_{y \in V} P(x, y) f(y),$$

for $x \in V$. For any $f \in \ell^2(V, \pi)$, Pf is well defined and further we have

$$\|Pf\|_\pi \le \|f\|_\pi. \tag{5.2.2}$$

Indeed by Cauchy–Schwarz, stochasticity, Fubini and stationarity,

$$
\begin{aligned}
\|P|f|\|_\pi^2 &= \sum_x \pi(x) \left[\sum_y P(x, y) |f(y)| \right]^2 \\
&\le \sum_x \pi(x) \left[\sum_y P(x, y) |f(y)|^2 \sum_z P(x, z) \right] \\
&= \sum_y \sum_x \pi(x) P(x, y) f(y)^2 \\
&= \sum_y \pi(y) f(y)^2 \\
&= \|f\|_\pi^2 < +\infty.
\end{aligned}
\tag{5.2.3}
$$

This shows that Pf is well defined since $\pi > 0$ and hence the series in square brackets on the first line is finite for all x. Applying the same argument to $\|Pf\|_\pi^2$ gives the inequality above.

Everything above holds whether or not P is reversible, so long as π is a stationary measure. Now we use reversibility. We claim that, when P reversible, then it is *self-adjoint*, that is,

$$\langle f, Pg \rangle_\pi = \langle Pf, g \rangle_\pi \qquad \forall f, g \in \ell^2(V, \pi). \tag{5.2.4}$$

This follows immediately by reversibility

$$
\begin{aligned}
\langle f, Pg \rangle_\pi &= \sum_{x \in V} \pi(x) f(x) \sum_{y \in V} P(x, y) g(y) \\
&= \sum_{x \in V} \sum_{y \in V} \pi(y) P(y, x) f(x) g(y) \\
&= \sum_{y \in V} \pi(y) g(y) \sum_{x \in V} P(y, x) f(x) \\
&= \langle Pf, g \rangle_\pi,
\end{aligned}
$$

where we argue as in (5.2.3) to justify using Fubini.

Throughout this section, we denote by 0 and 1 the all-zero and all-one functions, respectively.

5.2.1 Spectral Gap

In this subsection, we restrict ourselves to a finite state space V. Our goal is to bound the mixing time of (X_t) in terms of the eigenvalues of the transition matrix P. We assume that π is now the stationary distribution, that is, $\sum_{x \in V} \pi(x) = 1$ (which is unique by Theorem 1.1.24 and irreducibility). We also let $n := |V| < +\infty$.

Spectral decomposition Self-adjointness generalizes the notion of a symmetric matrix, with one consequence being that a version of the spectral theorem applies to P (at least in this finite-dimensional case; see Section 5.2.5 for more discussion on this). For completeness, we derive it from Theorem 5.1.1. It will be convenient to assume without loss of generality that $V = [n]$ and identify functions in $\ell^2(V, \pi)$ with vectors in $\mathbb{R}^n$.

Theorem 5.2.1 (Reversibility: spectral theorem). *There is an orthonormal basis of $\ell^2(V, \pi)$ formed of real eigenfunctions $\{f_j\}_{j=1}^n$ of P with real eigenvalues $\{\lambda_j\}_{j=1}^n$.*

Proof   Let D_π be the diagonal matrix with π on the diagonal. By reversibility

$$M(x,y) := (D_\pi^{1/2} P D_\pi^{-1/2})_{x,y}$$

$$= \sqrt{\frac{\pi(x)}{\pi(y)}} P(x,y)$$

$$= \sqrt{\frac{\pi(y)}{\pi(x)}} P(y,x)$$

$$= (D_\pi^{1/2} P D_\pi^{-1/2})_{y,x}$$

$$= M(y,x).$$

So $M = (M(x,y))_{x,y} = D_\pi^{1/2} P D_\pi^{-1/2}$ is a symmetric matrix. By the spectral theorem (Theorem 5.1.1), it has real eigenvectors $\{\phi_j\}_{j=1}^n$ forming an orthonormal basis of $\mathbb{R}^n$ with corresponding real eigenvalues $\{\lambda_j\}_{j=1}^n$. Define $f_j := D_\pi^{-1/2} \phi_j$. Then,

$$Pf_j = P D_\pi^{-1/2} \phi_j$$

$$= D_\pi^{-1/2} D_\pi^{1/2} P D_\pi^{-1/2} \phi_j$$

$$= D_\pi^{-1/2} M \phi_j$$

$$= \lambda_j D_\pi^{-1/2} \phi_j$$

$$= \lambda_j f_j$$

and

$$\langle f_i, f_j \rangle_\pi = \langle D_\pi^{-1/2} \phi_i, D_\pi^{-1/2} \phi_j \rangle_\pi$$

$$= \sum_{x \in V} \pi(x) [\pi(x)^{-1/2} \phi_i(x)][\pi(x)^{-1/2} \phi_j(x)]$$

$$= \langle \phi_i, \phi_j \rangle.$$

Because $\{\phi_j\}_{j=1}^n$ is an orthonormal basis of $\mathbb{R}^n$, we have that $\{f_j\}_{j=1}^n$ is an orthonormal basis of $(\mathbb{R}^n, \langle \cdot, \cdot \rangle_\pi)$. ∎

We collect a few more facts about the eigenbasis. Recall that

$$\|f\|_\infty = \max_{x \in V} |f(x)|.$$

Lemma 5.2.2 *Any eigenvalue λ of P satisfies $|\lambda| \le 1$.*

Proof It holds that

$$Pf = \lambda f \implies |\lambda| \|f\|_\infty = \|Pf\|_\infty = \max_x \left| \sum_y P(x,y)f(y) \right| \le \|f\|_\infty.$$

Rearranging gives the claim. ∎

We order the eigenvalues $1 \ge \lambda_1 \ge \cdots \ge \lambda_n \ge -1$. The second eigenvalue will play an important role below.

Lemma 5.2.3 *We have $\lambda_1 = 1$ and $\lambda_2 < 1$. Also we can take $f_1 = 1$.*

Proof Because P is stochastic, the all-one vector is a right eigenvector with eigenvalue 1. Any eigenfunction with eigenvalue 1 is harmonic with respect to P on V (see (3.3.2)). By Corollary 3.3.3, for a finite, irreducible chain the only harmonic functions are the constant functions. So the eigenspace corresponding to 1 is one-dimensional. We must have $\lambda_2 < 1$ by Lemma 5.2.2. ∎

When the chain is aperiodic, it cannot have an eigenvalue -1. Exercise 5.9 asks for a proof.

Lemma 5.2.4 *If P has an eigenvalue equal to -1, then P is not aperiodic.*

Lemma 5.2.5 *For all $j \ne 1$, $\pi f_j = 0$.*

Proof By orthonormality, $\langle f_1, f_j \rangle_\pi = 0$. Now use the fact that $f_1 = 1$. ∎

Let $\delta_x(y) := \mathbf{1}_{\{x=y\}}$.

Lemma 5.2.6 *For all x, y,*

$$\sum_{j=1}^n f_j(x) f_j(y) = \pi(x)^{-1} \delta_x(y).$$

Proof Using the notation of Theorem 5.2.1, the matrix Φ whose columns are the ϕ_js is orthogonal so $\Phi \Phi^T = I$. That is,

$$\sum_{j=1}^n \phi_j(x) \phi_j(y) = \delta_x(y),$$

or

$$\sum_{j=1}^n \sqrt{\pi(x)\pi(y)} f_j(x) f_j(y) = \delta_x(y).$$

Rearranging gives the result. ∎

Using the eigendecomposition of P, we get the following expression for its tth power P^t.

Theorem 5.2.7 (Spectral decomposition of P^t). *Let $\{f_j\}_{j=1}^n$ be the eigenfunctions of a reversible and irreducible transition matrix P with corresponding eigenvalues $\{\lambda_j\}_{j=1}^n$, as defined previously. Assume $\lambda_1 \geq \cdots \geq \lambda_n$. We have the decomposition*

$$\frac{P^t(x, y)}{\pi(y)} = 1 + \sum_{j=2}^{n} f_j(x) f_j(y) \lambda_j^t.$$

Proof Let F be the matrix whose columns are the eigenvectors $\{f_j\}_{j=1}^n$ and let D_λ be the diagonal matrix with $\{\lambda_j\}_{j=1}^n$ on the diagonal. Using the notation in the proof of Theorem 5.2.1,

$$D_\pi^{1/2} P^t D_\pi^{-1/2} = M^t = (D_\pi^{1/2} F) D_\lambda^t (D_\pi^{1/2} F)^T,$$

which after rearranging becomes

$$P^t D_\pi^{-1} = F D_\lambda^t F^T.$$

Expanding and using Lemma 5.2.3 gives the result. ∎ ■

Example 5.2.8 (Two-state chain). Let $V := \{0, 1\}$ and

$$P := \begin{pmatrix} 1 - \alpha & \alpha \\ \beta & 1 - \beta \end{pmatrix}$$

for $\alpha, \beta \in (0, 1)$. Observe that P is reversible with respect to the stationary distribution

$$\pi := \left(\frac{\beta}{\alpha + \beta}, \frac{\alpha}{\alpha + \beta} \right).$$

We know that $f_1 = 1$ is an eigenfunction with eigenvalue 1. As can be checked by direct computation, the other eigenfunction (in vector form) is

$$f_2 := \left(\sqrt{\frac{\alpha}{\beta}}, -\sqrt{\frac{\beta}{\alpha}} \right),$$

with eigenvalue $\lambda_2 := 1 - \alpha - \beta$. We normalized f_2 so that $\|f_2\|_\pi^2 = 1$.

By Theorem 5.2.7, the spectral decomposition at time t is therefore

$$P^t D_\pi^{-1} = \begin{pmatrix} 1 & 1 \\ 1 & 1 \end{pmatrix} + (1 - \alpha - \beta)^t \begin{pmatrix} \frac{\alpha}{\beta} & -1 \\ -1 & \frac{\beta}{\alpha} \end{pmatrix}.$$

Or, rearranging,

$$P^t = \begin{pmatrix} \frac{\beta}{\alpha+\beta} & \frac{\alpha}{\alpha+\beta} \\ \frac{\beta}{\alpha+\beta} & \frac{\alpha}{\alpha+\beta} \end{pmatrix} + (1 - \alpha - \beta)^t \begin{pmatrix} \frac{\alpha}{\alpha+\beta} & -\frac{\alpha}{\alpha+\beta} \\ -\frac{\beta}{\alpha+\beta} & \frac{\beta}{\alpha+\beta} \end{pmatrix}.$$

Note for instance that the case $\alpha + \beta = 1$ corresponds to a rank-one P, which immediately converges to stationarity.

Assume $\beta \geq \alpha$. Then, by (1.1.6) and Lemma 4.1.9,

$$d(t) = \max_x \frac{1}{2} \sum_y |P^t(x, y) - \pi(y)| = \frac{\beta}{\alpha + \beta} |1 - \alpha - \beta|^t.$$

As a result,

$$t_{mix}(\varepsilon) = \left\lceil \frac{\log\left(\varepsilon\frac{\alpha+\beta}{\beta}\right)}{\log|1-\alpha-\beta|} \right\rceil = \left\lceil \frac{\log\varepsilon^{-1} - \log\left(\frac{\alpha+\beta}{\beta}\right)}{\log|1-\alpha-\beta|^{-1}} \right\rceil.$$

◀

Spectral gap and mixing Assume further that P is aperiodic. Recall that by the convergence theorem (Theorem 1.1.33), for all x, y, $P^t(x,y) \to \pi(y)$ as $t \to +\infty$, and that the mixing time (Definition 1.1.35) is

$$t_{mix}(\varepsilon) := \min\{t \geq 0 : d(t) \leq \varepsilon\},$$

where $d(t) := \max_{x \in V} \|P^t(x,\cdot) - \pi(\cdot)\|_{TV}$. It will be convenient to work with a different notion of distance.

Definition 5.2.9 (Separation distance). *The* separation distance *is defined as*

SEPARATION DISTANCE

$$s_x(t) := \max_{y \in V} \left[1 - \frac{P^t(x,y)}{\pi(y)}\right],$$

and we let $s(t) := \max_{x \in V} s_x(t)$.

Lemma 5.2.10 (Separation distance and total variation distance).

$$d(t) \leq s(t).$$

Proof By Lemma 4.1.15,

$$\|P^t(x,\cdot) - \pi(\cdot)\|_{TV} = \sum_{y:P^t(x,y)<\pi(y)} \left[\pi(y) - P^t(x,y)\right]$$

$$= \sum_{y:P^t(x,y)<\pi(y)} \pi(y)\left[1 - \frac{P^t(x,y)}{\pi(y)}\right]$$

$$\leq s_x(t).$$

Since this holds for any x, the claim follows. ∎

It follows that, from the spectral decomposition (Theorem 5.2.7), the speed of convergence of $P^t(x,y)$ to $\pi(y)$ is dominated by the largest eigenvalue of P not equal to 1.

Definition 5.2.11 (Spectral gap). *The* absolute spectral gap *is* $\gamma_* := 1 - \lambda_*$ *where* $\lambda_* := |\lambda_2| \vee |\lambda_n|$. *The* spectral gap *is* $\gamma := 1 - \lambda_2$.

ABSOLUTE SPECTRAL GAP

By Lemmas 5.2.3 and 5.2.4, we have $\gamma_* > 0$ when P is irreducible and aperiodic. Note that the eigenvalues of the lazy version $\frac{1}{2}P + \frac{1}{2}I$ of P are $\left\{\frac{1}{2}(\lambda_j + 1)\right\}_{j=1}^n$, which are all non-negative. So, there, $\gamma_* = \gamma$.

Definition 5.2.12 (Relaxation time). *The* relaxation time *is defined as*

RELAXATION TIME

$$t_{rel} := \gamma_*^{-1}.$$

Example 5.2.13 (Two-state chain (continued)). Returning to Example 5.2.8, there are two cases:

- $\alpha + \beta \leq 1$: In that case the spectral gap is $\gamma = \gamma_* = \alpha + \beta$ and the relaxation time is $t_{rel} = 1/(\alpha + \beta)$.
- $\alpha + \beta > 1$: In that case the spectral gap is $\gamma = \gamma_* = 2 - \alpha - \beta$ and the relaxation time is $t_{rel} = 1/(2 - \alpha - \beta)$.

$\blacktriangleleft$

The following result clarifies the relationship between the mixing and relaxation times. Let $\pi_{min} = \min_x \pi(x)$.

Theorem 5.2.14 (Mixing time and relaxation time). *Let P be reversible, irreducible, and aperiodic with positive stationary distribution π. For all $\varepsilon > 0$,*

$$(t_{rel} - 1) \log\left(\frac{1}{2\varepsilon}\right) \leq t_{mix}(\varepsilon) \leq \log\left(\frac{1}{\varepsilon \pi_{min}}\right) t_{rel}.$$

Proof We start with the upper bound. By Lemma 5.2.10, it suffices to find t such that $s(t) \leq \varepsilon$. By the spectral decomposition and Cauchy–Schwarz,

$$\left|\frac{P^t(x,y)}{\pi(y)} - 1\right| \leq \lambda_*^t \sum_{j=2}^n |f_j(x)f_j(y)| \leq \lambda_*^t \sqrt{\sum_{j=2}^n f_j(x)^2 \sum_{j=2}^n f_j(y)^2}.$$

By Lemma 5.2.6, $\sum_{j=2}^n f_j(x)^2 \leq \pi(x)^{-1}$. Plugging this back above, we get

$$\left|\frac{P^t(x,y)}{\pi(y)} - 1\right| \leq \lambda_*^t \sqrt{\pi(x)^{-1}\pi(y)^{-1}} \leq \frac{\lambda_*^t}{\pi_{min}} = \frac{(1-\gamma_*)^t}{\pi_{min}} \leq \frac{e^{-\gamma_* t}}{\pi_{min}}, \tag{5.2.5}$$

where we used that $1 - z \leq e^{-z}$ for all $z \in \mathbb{R}$ (see Exercise 1.16). Observe that the right-hand side is less than ε when $t \geq \log\left(\frac{1}{\varepsilon \pi_{min}}\right) t_{rel}$.

For the lower bound, let f_* be an eigenfunction associated with an eigenvalue achieving $\lambda_* := |\lambda_2| \vee |\lambda_n|$. Let z be such that $|f_*(z)| = \|f_*\|_\infty$. By Lemma 5.2.5, $\pi f_* = 0$. Hence,

$$\lambda_*^t |f_*(z)| = |P^t f_*(z)|$$

$$= \left|\sum_y [P^t(z,y)f_*(y) - \pi(y)f_*(y)]\right|$$

$$\leq \|f_*\|_\infty \sum_y |P^t(z,y) - \pi(y)| \leq \|f_*\|_\infty 2d(t),$$

so $d(t) \geq \frac{1}{2}\lambda_*^t$. When $t = t_{mix}(\varepsilon)$, $\varepsilon \geq \frac{1}{2}\lambda_*^{t_{mix}(\varepsilon)}$. Therefore,

$$t_{mix}(\varepsilon)\left(\frac{1}{\lambda_*} - 1\right) \geq t_{mix}(\varepsilon) \log\left(\frac{1}{\lambda_*}\right) \geq \log\left(\frac{1}{2\varepsilon}\right).$$

The result follows from $\left(\frac{1}{\lambda_*} - 1\right)^{-1} = \left(\frac{1-\lambda_*}{\lambda_*}\right)^{-1} = \left(\frac{\gamma_*}{1-\gamma_*}\right)^{-1} = t_{rel} - 1$. $\blacksquare$

5.2.2 ▷ *Random Walks: A Spectral Look at Cycles and Hypercubes*

We illustrate the results in the previous subsection to random walk on cycles and hypercubes.

Random walk on a cycle

Consider simple random walk on the n-cycle (see Example 1.1.17). That is, $V := \{0, 1, \ldots, n-1\}$ and $P(x, y) = 1/2$ if and only if $|x - y| = 1 \mod n$. We assume that n is odd to avoid periodicity issues. Let $\pi \equiv n^{-1}$ be the stationary distribution (by symmetry and $|V| = n$). We showed in Section 4.3.2 that (for the lazy version of the chain) the mixing time is $t_{\text{mix}}(\varepsilon) = \Theta(n^2)$.

Here we use spectral techniques. We first compute the eigendecomposition, which in this case can be determined explicitly.

Lemma 5.2.15 (Cycle: eigenbasis). *For $j = 1, \ldots, n-1$, the function*

$$g_j(x) := \sqrt{2} \cos\left(\frac{2\pi j x}{n}\right), \qquad x = 0, 1, \ldots, n-1,$$

is an eigenfunction of P with eigenvalue

$$\mu_j := \cos\left(\frac{2\pi j}{n}\right),$$

and $g_0 = 1$ is an eigenfunction with eigenvalue 1. Moreover, the g_js are orthonormal in $\ell^2(V, \pi)$.

Proof We know from Lemma 5.2.3 that 1 is an eigenfunction with eigenvalue 1. Let $j \in \{1, \ldots, n-1\}$. Note that, for all x, switching momentarily to the complex representation (where we use i for the imaginary unit),

$$\sum_y P(x, y) g_j(y) = \frac{1}{2}\left[\sqrt{2}\cos\left(\frac{2\pi j(x-1)}{n}\right) + \sqrt{2}\cos\left(\frac{2\pi j(x+1)}{n}\right)\right]$$

$$= \frac{\sqrt{2}}{2}\left[\frac{e^{i\frac{2\pi j(x-1)}{n}} + e^{-i\frac{2\pi j(x-1)}{n}}}{2} + \frac{e^{i\frac{2\pi j(x+1)}{n}} + e^{-i\frac{2\pi j(x+1)}{n}}}{2}\right]$$

$$= \sqrt{2}\left[\frac{e^{i\frac{2\pi j x}{n}} + e^{-i\frac{2\pi j x}{n}}}{2}\right]\left[\frac{e^{i\frac{2\pi j}{n}} + e^{-i\frac{2\pi j}{n}}}{2}\right]$$

$$= \left[\sqrt{2}\cos\left(\frac{2\pi j x}{n}\right)\right]\left[\cos\left(\frac{2\pi j}{n}\right)\right]$$

$$= \cos\left(\frac{2\pi j}{n}\right) g_j(x).$$

The orthonormality follows from standard trigonometric identities. We prove only that they have unit norm. We use the Dirichlet kernel (see Exercise 5.8)

$$1 + 2\sum_{k=1}^{n} \cos k\theta = \frac{\sin((n+1/2)\theta)}{\sin(\theta/2)}$$

for $\theta \neq 0$, and the identity $\cos^2(\theta) = \frac{1}{2}(1 + \cos(2\theta))$. For $j = 0$, $g_j = 1$ and the norm squared is $\sum_x \pi(x) = 1$. For $j \neq 0$, we have $\|g_j\|_\pi^2$ is

$$\sum_{x \in V} \pi(x) g_j(x)^2 = \frac{1}{n} \sum_{x=0}^{n-1} 2 \cos^2 \left(\frac{2\pi j x}{n} \right)$$

$$= \frac{1}{n} \sum_{x=0}^{n-1} \left(1 + \cos \left(\frac{4\pi j x}{n} \right) \right)$$

$$= 1 + \frac{1}{n} \sum_{k=1}^{n} \cos \left(k \frac{4\pi j}{n} \right)$$

$$= 1 + \frac{1}{2} \left[\frac{\sin((n + 1/2)(4\pi j/n))}{\sin((4\pi j/n)/2)} - 1 \right],$$

which is indeed 1.   ∎

From the eigenvalues, we derive the relaxation time (Definition 5.2.12) analytically.

Theorem 5.2.16 (Cycle: relaxation time).   *The relaxation time for lazy simple random walk on the n-cycle is*

$$t_{rel} = \frac{1}{1 - \cos \left(\frac{2\pi}{n} \right)} = \Theta(n^2).$$

Proof   By Lemma 5.2.15, the absolute spectral gap (Definition 5.2.11) is $1 - \cos \left(\frac{2\pi}{n} \right)$, using that n is odd. By a Taylor expansion,

$$1 - \cos \left(\frac{2\pi}{n} \right) = \frac{4\pi^2}{n^2} + O(n^{-4}).$$ ∎

Since $\pi_{\min} = 1/n$, we get $t_{\mix}(\varepsilon) = O(n^2 \log n)$ and $t_{\mix}(\varepsilon) = \Omega(n^2)$ by Theorem 5.2.14.

It turns out our upper bound is off by a logarithmic factor. A sharper bound on the mixing time can be obtained by working directly with the spectral decomposition. By Lemma 4.1.9 and Cauchy–Schwarz (Theorem B.4.8), for any $x \in V$,

$$4 \| P^t(x, \cdot) - \pi(\cdot) \|_{TV}^2 = \left\{ \sum_y \pi(y) \left| \frac{P^t(x, y)}{\pi(y)} - 1 \right| \right\}^2$$

$$\leq \sum_y \pi(y) \left(\frac{P^t(x, y)}{\pi(y)} - 1 \right)^2$$

$$= \left\| \sum_{j=1}^{n-1} \mu_j^t g_j(x) g_j \right\|_\pi^2$$

$$= \sum_{j=1}^{n-1} \mu_j^{2t} g_j(x)^2,$$

where we used the spectral decomposition of P^t (Theorem 5.2.7) on the third line and Parseval's identity (i.e., (5.2.1)) on the fourth line.

Here comes the trick: the total variation distance does not depend on the starting point x by symmetry. Multiplying by $\pi(x)$ and summing over x – on the right-hand side only – gives

$$4\|P^t(x,\cdot) - \pi(\cdot)\|_{\text{TV}}^2 \leq \sum_x \pi(x) \sum_{j=1}^{n-1} \mu_j^{2t} g_j(x)^2$$

$$= \sum_{j=1}^{n-1} \mu_j^{2t} \sum_x \pi(x) g_j(x)^2$$

$$= \sum_{j=1}^{n-1} \mu_j^{2t},$$

where we used that $\|g_j\|_\pi^2 = 1$.

We get

$$4d(t)^2 \leq \sum_{j=1}^{n-1} \cos^{2t}\left(\frac{2\pi j}{n}\right) = 2 \sum_{j=1}^{(n-1)/2} \cos^{2t}\left(\frac{2\pi j}{n}\right).$$

For $x \in [0, \pi/2)$, $\cos x \leq e^{-x^2/2}$ (see Exercise 1.16). Then,

$$4d(t)^2 \leq 2 \sum_{j=1}^{(n-1)/2} \exp\left(-\frac{4\pi^2 j^2}{n^2} t\right)$$

$$\leq 2 \exp\left(-\frac{4\pi^2}{n^2} t\right) \sum_{j=1}^{\infty} \exp\left(-\frac{4\pi^2(j^2-1)}{n^2} t\right)$$

$$\leq 2 \exp\left(-\frac{4\pi^2}{n^2} t\right) \sum_{\ell=0}^{\infty} \exp\left(-\frac{4\pi^2 t}{n^2} \ell\right)$$

$$= \frac{2 \exp\left(-\frac{4\pi^2}{n^2} t\right)}{1 - \exp\left(-\frac{4\pi^2}{n^2} t\right)}.$$

So $t_{\text{mix}}(\varepsilon) = O(n^2)$.

Random walk on the hypercube

Consider simple random walk on the hypercube $V := \{-1, +1\}^n$, where $x \sim y$ if they differ at exactly one coordinate. We consider the lazy version to avoid issues of periodicity (see Example 1.1.31). Let P be the transition matrix and let $\pi \equiv 2^{-n}$ be the stationary distribution (by symmetry and $|V| = 2^n$). We showed in Section 4.3.2 that $t_{\text{mix}}(\varepsilon) = \Theta(n \log n)$. Here we use spectral techniques.

For $J \subseteq [n]$, we let

$$\chi_J(x) = \prod_{j \in J} x_j, \qquad x \in V.$$

These are called *parity functions*. We show that the parity functions form an eigenbasis of the transition matrix.

PARITY FUNCTION

Lemma 5.2.17 (Hypercube: eigenbasis). *For all $J \subseteq [n]$, the function χ_J is an eigenfunction of P with eigenvalue*

$$\mu_J := \frac{n - |J|}{n}.$$

Moreover, the χ_Js are orthonormal in $\ell^2(V, \pi)$.

Proof For $x \in V$ and $i \in [n]$, let $x^{[i]}$ be x where coordinate i is flipped. Note that, for all J, x,

$$\sum_y P(x, y)\chi_J(y) = \frac{1}{2}\chi_J(x) + \frac{1}{2}\sum_{i=1}^n \frac{1}{n}\chi_J(x^{[i]})$$

$$= \left\{\frac{1}{2} + \frac{1}{2}\frac{n - |J|}{n}\right\} \chi_J(x) - \frac{1}{2}\frac{|J|}{n}\chi_J(x)$$

$$= \frac{n - |J|}{n}\chi_J(x).$$

For the orthonormality, note that

$$\sum_{x \in V} \pi(x)\chi_J(x)^2 = \sum_{x \in V} \frac{1}{2^n}\prod_{j \in J} x_j^2 = 1.$$

For $J \neq J' \subseteq [n]$,

$$\sum_{x \in V} \pi(x)\chi_J(x)\chi_{J'}(x)$$

$$= \sum_{x \in V} \frac{1}{2^n} \prod_{j \in J \cap J'} x_j^2 \prod_{j \in J \setminus J'} x_j \prod_{j \in J' \setminus J} x_j$$

$$= \frac{2^{|J \cap J'|}}{2^n} \prod_{j \in J \setminus J'} \left(\sum_{x_j \in \{-1, +1\}} x_j\right) \prod_{j \in J' \setminus J} \left(\sum_{x_j \in \{-1, +1\}} x_j\right)$$

$$= 0,$$

since at least one of $J \setminus J'$ or $J' \setminus J$ is non-empty. ∎

From the eigenvalues, we obtain the relaxation time.

Theorem 5.2.18 (Hypercube: relaxation time). *The relaxation time for lazy simple random walk on the n-dimensional hypercube is*

$$t_{\text{rel}} = n.$$

Proof From Lemma 5.2.17, the absolute spectral gap is

$$\gamma_* = \gamma = 1 - \frac{n - 1}{n} = \frac{1}{n}.$$ ∎

Note that $\pi_{\min} = 1/2^n$. Hence, by Theorem 5.2.14, we have $t_{\text{mix}}(\varepsilon) = O(n^2)$ and $t_{\text{mix}}(\varepsilon) = \Omega(n)$. Those bounds, it turns out, are both off.

As we did for the cycle, we obtain a sharper upper bound by working directly with the spectral decomposition. By the same argument we used there,

$$4d(t)^2 \leq \sum_{J \neq \emptyset} \mu_J^{2t}.$$

Then,

$$
\begin{aligned}
4d(t)^2 &\leq \sum_{J \neq \emptyset} \left(\frac{n - |J|}{n} \right)^{2t} \\
&= \sum_{\ell=1}^{n} \binom{n}{\ell} \left(1 - \frac{\ell}{n} \right)^{2t} \\
&\leq \sum_{\ell=1}^{n} \binom{n}{\ell} \exp\left(-\frac{2t\ell}{n} \right) \\
&= \left(1 + \exp\left(-\frac{2t}{n} \right) \right)^{n} - 1,
\end{aligned}
$$

where we used that $1 - x \leq e^{-x}$ for all x (see Exercise 1.16). So, by definition, $t_{\mathrm{mix}}(\varepsilon) \leq \frac{1}{2} n \log n + O(n)$.

Remark 5.2.19 *In fact, lazy simple random walk on the n-dimensional hypercube has a "cutoff" at $(1/2)n \log n$. Roughly speaking, within a time window of size $O(n)$, the total variation distance to the stationary distribution goes from near 1 to near 0. See, for example, [LPW06, section 18.2.2].*

5.2.3 ▷ *Markov Chains: Varopoulos–Carne and Diameter-Based Bounds on the Mixing Time*

If (S_t) is simple random walk on $\mathbb{Z}$, then Lemma 2.4.3 guarantees that for any $x, y \in \mathbb{Z}$,

$$P^t(x, y) \leq e^{-|x-y|^2/2t}, \tag{5.2.6}$$

where P is the transition matrix of (S_t). Interestingly, a similar bound holds for *any* reversible Markov chain, and Lemma 2.4.3 plays an unexpected role in its proof. An application to mixing times is discussed in Claim 5.2.25.

Varopoulos–Carne bound

Our main bound is the following. Recall that a reversible Markov chain is equivalent to a random walk on the network corresponding to its positive transition probabilities (see Definition 1.2.7 and the discussion following it).

Theorem 5.2.20 (Varopoulos–Carne bound). *Let P be the transition matrix of an irreducible Markov chain (X_t) on the countable state space V. Assume further that P is reversible with respect to the stationary measure π and that the corresponding network $\mathcal{N}$ is locally finite. Then the following hold:*

$$\forall x, y \in V, \forall t \in \mathbb{N}, \qquad P^t(x, y) \leq 2 \sqrt{\frac{\pi(y)}{\pi(x)}} e^{-\rho(x,y)^2/2t},$$

where $\rho(x, y)$ is the graph distance between x and y on $\mathcal{N}$.

As a sanity check before proving the theorem, note that if the chain is aperiodic and π is the stationary distribution, then by the convergence theorem (Theorem 1.1.33),

$$P^t(x, y) \to \pi(y) \leq 2\sqrt{\frac{\pi(y)}{\pi(x)}} \qquad \text{as } t \to +\infty,$$

since $\pi(x), \pi(y) \leq 1$.

Proof of Theorem 5.2.20 The idea of the proof is to show that

$$P^t(x, y) \leq 2\sqrt{\frac{\pi(y)}{\pi(x)}} \mathbb{P}[S_t \geq \rho(x, y)],$$

where again (S_t) is simple random walk on $\mathbb{Z}$ started at 0, and then use the Chernoff bound (Lemma 2.4.3).

By the local finiteness assumption, only a finite number of states can be reached by time t. Hence we can reduce the problem to a finite state space. More precisely, let $\tilde{V} = \{z \in V : \rho(x, z) \leq t\}$ and for $z, w \in \tilde{V}$

$$\tilde{P}(z, w) = \begin{cases} P(z, w) & \text{if } z \neq w, \\ P(z, z) + P(z, V \setminus \tilde{V}) & \text{otherwise.} \end{cases}$$

By construction, $\tilde{P}$ is reversible with respect to $\tilde{\pi} = \pi / \pi(\tilde{V})$ on $\tilde{V}$. Because within time t one never reaches a state z where $P(z, V \setminus \tilde{V}) > 0$, by Chapman–Kolmogorov (Theorem 1.1.20) and using the fact that $\tilde{\pi}(y)/\tilde{\pi}(x) = \pi(y)/\pi(x)$, it suffices to prove the result for $\tilde{P}$. Hence we assume without loss of generality that V is finite with $|V| = n$.

To relate (X_t) to simple random walk on $\mathbb{Z}$, we use a special representation of P^t based on Chebyshev polynomials. For $\xi = \cos\theta \in [-1, 1]$,

$$T_k(\xi) = \cos k\theta$$

CHEBYSHEV POLYNOMIALS is a *Chebyshev polynomial of the first kind*. Note that $|T_k(\xi)| \leq 1$ on $[-1, 1]$ by definition. The classical trigonometric identity (to see this, write it in complex form)

$$\cos((k+1)\theta) + \cos((k-1)\theta) = 2\cos\theta\cos(k\theta)$$

implies the recursion

$$T_{k+1}(\xi) + T_{k-1}(\xi) = 2\xi\, T_k(\xi),$$

which in turn implies that T_k is indeed a polynomial. It has degree k from induction and the fact that $T_0(\xi) = 1$ and $T_1(\xi) = \xi$. The connection to simple random walk on $\mathbb{Z}$ comes from the following somewhat miraculous representation (which does not rely on reversibility). Let $T_k(P)$ denote the polynomial T_k evaluated at P as a matrix polynomial.

Lemma 5.2.21

$$P^t = \sum_{k=-t}^{t} \mathbb{P}[S_t = k]\, T_{|k|}(P).$$

Proof It suffices to prove

$$\xi^t = \sum_{k=-t}^{t} \mathbb{P}[S_t = k] \, T_{|k|}(\xi)$$

as an identity of polynomials. By the binomial theorem (Appendix A),

$$\xi^t = \left(\frac{e^{i\theta} + e^{-i\theta}}{2}\right)^t = \sum_{\ell=0}^{t} 2^{-t} \binom{t}{\ell} (e^{i\theta})^{\ell} (e^{-i\theta})^{t-\ell} = \sum_{k=-t}^{t} \mathbb{P}[S_t = k] e^{ik\theta},$$

where we used that the probability that

$$S_t = -t + 2\ell = (+1)\ell + (-1)(t - \ell)$$

is the event of making ℓ steps to the right and $t - \ell$ steps to the left. Now take real parts on both sides and use that $\cos(k\theta) = \cos(-k\theta)$ to get the claim. (Put differently, $(\cos\theta)^t$ is the characteristic function $\mathbb{E}[e^{i\theta S_t}]$ of S_t.) ∎

We bound $T_k(P)(x, y)$ as follows.

Lemma 5.2.22 *It holds that*

$$T_k(P)(x, y) = 0 \qquad \forall k < \rho(x, y)$$

and

$$T_k(P)(x, y) \leq \sqrt{\frac{\pi(y)}{\pi(x)}} \qquad \forall k \geq \rho(x, y).$$

Proof Note that $T_k(P)(x, y) = 0$ when $k < \rho(x, y)$ because $T_k(P)(x, y)$ is a function of the entries $P^{\ell}(x, y)$ for $\ell \leq k$, all of which are 0.

We work on $\ell^2(V, \pi)$. Let $f_1, \ldots, f_n$ be an eigendecomposition of P orthonormal with respect to the inner product $\langle \cdot, \cdot \rangle_\pi$ with eigenvalues $\lambda_1, \ldots, \lambda_n \in [-1, 1]$. Such a decomposition exists by Theorem 5.2.1. Then, $f_1, \ldots, f_n$ is also an eigendecomposition of the polynomial $T_k(P)$ with eigenvalues

$$T_k(\lambda_1), \ldots, T_k(\lambda_n) \in [-1, 1],$$

by the definition of the Chebyshev polynomials. By decomposing any function $f = \sum_{i=1}^{n} \alpha_i f_i$ over this eigenbasis, it implies that

$$\|T_k(P)f\|_\pi^2 = \left\| \sum_{i=1}^{n} \alpha_i T_k(\lambda_i) f_i \right\|_\pi^2$$

$$= \sum_{i=1}^{n} \alpha_i^2 \, T_k(\lambda_i)^2$$

$$\leq \sum_{i=1}^{n} \alpha_i^2$$

$$= \|f\|_\pi^2, \tag{5.2.7}$$

where we used Parseval's identity (5.2.1) twice and the fact that $T_k(\lambda_i)^2 \in [0, 1]$.

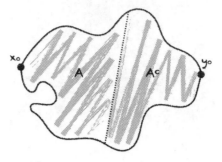

Figure 5.2 The supports of $P^{\lfloor(\Delta-1)/2\rfloor}(x_0,\cdot\,)$ and $P^{\lfloor(\Delta-1)/2\rfloor}(y_0,\cdot\,)$ are contained in A and A^c, respectively.

Let δ_z denote the point mass at z. By Cauchy–Schwarz (Theorem B.4.8) and (5.2.7),

$$T_k(P)(x,y) = \frac{\langle \delta_x, T_k(P)\delta_y\rangle_\pi}{\pi(x)} \leq \frac{\|\delta_x\|_\pi \|\delta_y\|_\pi}{\pi(x)} = \frac{\sqrt{\pi(x)}\sqrt{\pi(y)}}{\pi(x)} = \sqrt{\frac{\pi(y)}{\pi(x)}}$$

for any k (in particular for $k \geq \rho(x,y)$) and we have proved the claim.   ∎

Combining the two lemmas gives the result.   ∎

Remark 5.2.23 *The local finiteness assumption is made for simplicity only. The result holds for any countable-space, reversible chain. See [LP16, section 13.2].*

Lower bound on mixing Let (X_t) be an irreducible aperiodic (for now not necessarily reversible) Markov chain with finite state space V and stationary distribution π. Recall that, for a fixed $0 < \varepsilon < 1/2$, the mixing time is

$$\mathrm{t_{mix}}(\varepsilon) = \min\{t\colon d(t) \leq \varepsilon\},$$

where

$$d(t) = \max_{x \in V} \|P^t(x,\cdot\,) - \pi\|_{\mathrm{TV}}.$$

It is intuitively clear that $\mathrm{t_{mix}}(\varepsilon)$ is at least of the order of the "diameter" of the transition graph of P. For $x,y \in V$, let $\rho(x,y)$ be the graph distance between x and y on the undirected version of the transition graph, that is, ignoring the orientation of the edges. With this definition, a shortest directed path from x to y contains at least $\rho(x,y)$ edges. Here we define the *diameter* of the transition graph as $\Delta := \max_{x,y \in V} \rho(x,y)$. Let x_0, y_0 be a pair of vertices achieving the diameter. Then we claim that $P^{\lfloor(\Delta-1)/2\rfloor}(x_0,\cdot\,)$ and $P^{\lfloor(\Delta-1)/2\rfloor}(y_0,\cdot\,)$ are supported on disjoint sets. To see this, let

$$A = \{z \in V\colon \rho(x_0,z) < \rho(y_0,z)\}$$

DIAMETER

be the set of states closer to x_0 than y_0. See Figure 5.2. By the triangle inequality for ρ, any z such that $\rho(x_0,z) \leq \lfloor(\Delta-1)/2\rfloor$ is in A, otherwise we would have $\rho(y_0,z) \leq \rho(x_0,z) \leq \lfloor(\Delta-1)/2\rfloor$ and hence $\rho(x_0,y_0) \leq \rho(x_0,z)+\rho(y_0,z) \leq 2\lfloor(\Delta-1)/2\rfloor < \Delta$, a contradiction.

Similarly, if $\rho(y_0, z) \leq \lfloor (\Delta - 1)/2 \rfloor$, then $z \in A^c$. By the triangle inequality for the total variation distance,

$$
\begin{aligned}
d(\lfloor (\Delta - 1)/2 \rfloor) &\geq \frac{1}{2} \left\| P^{\lfloor (\Delta-1)/2 \rfloor}(x_0, \cdot) - P^{\lfloor (\Delta-1)/2 \rfloor}(y_0, \cdot) \right\|_{\mathrm{TV}} \\
&\geq \frac{1}{2} \left\{ P^{\lfloor (\Delta-1)/2 \rfloor}(x_0, A) - P^{\lfloor (\Delta-1)/2 \rfloor}(y_0, A) \right\} \\
&= \frac{1}{2} \{1 - 0\} = \frac{1}{2},
\end{aligned}
\tag{5.2.8}
$$

where we used (1.1.4) on the second line, so that:

Claim 5.2.24

$$
t_{\mathrm{mix}}(\varepsilon) \geq \frac{\Delta}{2}.
$$

This bound is often far from the truth. Consider for instance simple random walk on a cycle of size n. The diameter is $\Delta = n/2$. But Lemma 2.4.3 suggests that it takes time of order Δ^2 to even reach the antipode of the starting point, let alone achieve stationarity. More generally, when P is *reversible*, the "diffusive behavior" captured by the Varopoulos–Carne bound (Theorem 5.2.20) implies that the mixing time does indeed scale at least as the *square* of the diameter.

Assume that P is reversible with respect to π and has diameter Δ. Letting $n = |V|$ and $\pi_{\min} = \min_{x \in V} \pi(x)$, we then have the following.

Claim 5.2.25 *The following lower bound holds*

$$
t_{\mathrm{mix}}(\varepsilon) \geq \frac{\Delta^2}{12 \log n + 4 |\log \pi_{\min}|}
$$

provided $n \geq \frac{16}{(1-2\varepsilon)^2}$.

Proof The proof is based on the same argument we used to derive our first diameter-based bound, except that the Varopoulos–Carne bound gives a better dependence on the diameter. Namely, let x_0, y_0, and A be as above. By the Varopoulos–Carne bound,

$$
P^t(x_0, A^c) = \sum_{z \in A^c} P^t(x_0, z) \leq \sum_{z \in A^c} 2\sqrt{\frac{\pi(z)}{\pi(x_0)}} e^{-\frac{\rho^2(x_0, z)}{2t}} \leq 2n\pi_{\min}^{-1/2} e^{-\frac{\Delta^2}{8t}},
$$

where we used that $|A^c| \leq n$ and $\rho(x_0, z) \geq \frac{\Delta}{2}$ for $z \in A^c$. For any

$$
t < \frac{\Delta^2}{12 \log n + 4 |\log \pi_{\min}|},
\tag{5.2.9}
$$

we get that

$$
P^t(x_0, A^c) \leq 2n\pi_{\min}^{-1/2} \exp\left(-\frac{3 \log n + |\log \pi_{\min}|}{2} \right) = \frac{2}{\sqrt{n}},
$$

or $P^t(x_0, A) \geq 1 - \frac{2}{\sqrt{n}}$. Similarly, $P^t(y_0, A) \leq \frac{2}{\sqrt{n}}$ so that arguing as in (5.2.8)

$$
d(t) \geq \frac{1}{2} \left\{ 1 - \frac{2}{\sqrt{n}} - \frac{2}{\sqrt{n}} \right\} = \frac{1}{2} - \frac{2}{\sqrt{n}} \geq \varepsilon,
$$

for t as in (5.2.9) and n as in the statement.   ∎

Remark 5.2.26 *The dependence on Δ and $\pi_{\min}$ in Claim 5.2.25 cannot be improved. See [LP16, section 13.3].*

5.2.4 ▷ *Randomized Algorithms: Markov Chain Monte Carlo and a Quantitative Ergodic Theorem*

In Markov chain Monte Carlo methods, one generates samples from a probability distribution of interest π over some state space V in order to estimate some of its properties, for example, its mean, by designing and then running a Markov chain with stationary distribution π. The Metropolis algorithm from Example 1.1.30 is a standard way of constructing such a chain. These techniques play a central role in Bayesian statistics in particular where π is the so-called posterior distribution given the data.

We restrict ourselves here to finite V and, without loss of generality, we assume that $V = [n]$. Let P be an irreducible chain reversible with respect to a stationary distribution $\pi = (\pi_x)_{x \in V}$. As previously, we work on $\ell^2(V, \pi)$. Let $f: V \to \mathbb{R}$ be a function in $\ell^2(V, \pi)$. Recall that

$$\pi f = \sum_{x \in V} \pi_x f(x).$$

Our goal is to estimate πf from the sample path of the Markov chain $(X_t)_{t \geq 0}$ with transition matrix P. Indeed, the ergodic theorem guarantees that

$$\frac{1}{T} \sum_{t=1}^{T} f(X_t) \to \pi f$$

almost surely as $T \to +\infty$ for any starting point. We derive a simple, quantitative version of this statement that provides insights into how long the chain needs to be run to get an accurate estimate in terms of the spectral gap.

Theorem 5.2.27 (Ergodic theorem: reversible case). *Let $P = (P_{x,y})_{x,y \in V}$ be an irreducible aperiodic transition matrix over a finite state space V reversible with respect to the stationary distribution $\pi = (\pi_x)_{x \in V}$. Let $f: V \to \mathbb{R}$ be a function in $\ell^2(V, \pi)$. Then for any initial distribution $\mu = (\mu_x)_{x \in V}$,*

$$\frac{1}{T} \sum_{t=1}^{T} f(X_t) \to \pi f,$$

in probability as $T \to +\infty$. Moreover, for any $\varepsilon > 0$,

$$\mathbb{P}\left[\left| \frac{1}{T} \sum_{t=1}^{T} f(X_t) - \pi f \right| \geq \varepsilon \right] \leq \frac{9 \pi_{\min}^{-1} \|f\|_{\infty}^2 \gamma_*^{-1} \frac{1}{T}}{(\varepsilon - \pi_{\min}^{-1} \|f\|_{\infty} \gamma_*^{-1} \frac{1}{T})^2},$$

as $T \to +\infty$, where $\gamma_* > 0$ is the absolute spectral gap of P.

Recall that, by Lemmas 5.2.3 and 5.2.4, we have $\gamma_* > 0$ since P is irreducible and aperiodic. We will first need the following lemma.

Lemma 5.2.28 (Convergence of the expectation). *For any initial distribution* $\mu = (\mu_x)_{x \in V}$ *and any t,*

$$|\mathbb{E}[f(X_t)] - \pi f| \le (1 - \gamma_*)^t \pi_{\min}^{-1} \|f\|_\infty.$$

Proof We have

$$|\mathbb{E}[f(X_t)] - \pi f| = \left| \sum_x \sum_y \mu_x P_{x,y}^t f(y) - \sum_y \pi_y f(y) \right|.$$

Because $\sum_x \mu_x = 1$, the right-hand side is

$$= \left| \sum_x \sum_y \mu_x P_{x,y}^t f(y) - \sum_x \sum_y \mu_x \pi_y f(y) \right|$$

$$\le \sum_x \mu_x \sum_y |P_{x,y}^t - \pi_y| \, |f(y)|,$$

by the triangle inequality.

Now by (5.2.5) this is

$$\le \sum_x \mu_x \sum_y (1 - \gamma_*)^t \frac{\pi_y}{\pi_{\min}} |f(y)|$$

$$= (1 - \gamma_*)^t \frac{1}{\pi_{\min}} \sum_x \mu_x \sum_y \pi_y |f(y)|$$

$$\le (1 - \gamma_*)^t \pi_{\min}^{-1} \|f\|_\infty.$$

That proves the claim. ∎

Proof of Theorem 5.2.27 We use Chebyshev's inequality (Theorem 2.1.2), similarly to the proof of the L^2 weak law of large numbers (Theorem 2.1.6). In particular, we note that the X_ts are *not* independent.

By Lemma 5.2.28, the expectation of the time average can be bounded as follows

$$\left| \mathbb{E}\left[\frac{1}{T} \sum_{t=1}^T f(X_t) \right] - \pi f \right| \le \frac{1}{T} \sum_{t=1}^T |\mathbb{E}[f(X_t)] - \pi f|$$

$$\le \frac{1}{T} \sum_{t=1}^T (1 - \gamma_*)^t \pi_{\min}^{-1} \|f\|_\infty$$

$$\le \pi_{\min}^{-1} \|f\|_\infty \frac{1}{T} \sum_{t=0}^{+\infty} (1 - \gamma_*)^t$$

$$= \pi_{\min}^{-1} \|f\|_\infty \gamma_*^{-1} \frac{1}{T} \to 0,$$

as $T \to +\infty$, since $\gamma_* > 0$.

Next we bound the variance of the sum. We have

$$\mathrm{Var}\left[\frac{1}{T} \sum_{t=1}^T f(X_t) \right] = \frac{1}{T^2} \sum_{t=1}^T \mathrm{Var}[f(X_t)] + \frac{2}{T^2} \sum_{1 \le s < t \le T} \mathrm{Cov}[f(X_s), f(X_t)].$$

We bound the variance and covariance terms separately.

To obtain convergence, a trivial bound on the variance suffices

$$0 \leq \mathrm{Var}[f(X_t)] \leq \mathbb{E}[f(X_t)^2] \leq \|f\|_\infty^2.$$

Hence,

$$0 \leq \frac{1}{T^2} \sum_{t=1}^T \mathrm{Var}[f(X_t)] \leq \frac{T\|f\|_\infty^2}{T^2} \to 0$$

as $T \to +\infty$.

Bounding the covariance requires a more delicate argument. Fix $1 \leq s < t \leq T$. The trick is to condition on X_s and use the Markov Property (Theorem 1.1.18). By definition of the co-variance, the tower property (Lemma B.6.16) and taking out what is known (Lemma B.6.13),

$$\mathrm{Cov}[f(X_s), f(X_t)]$$
$$= \mathbb{E}\left[(f(X_s) - \mathbb{E}[f(X_s)])(f(X_t) - \mathbb{E}[f(X_t)])\right]$$
$$= \sum_x \mathbb{E}\left[(f(X_s) - \mathbb{E}[f(X_s)])(f(X_t) - \mathbb{E}[f(X_t)]) \mid X_s = x\right] \mathbb{P}[X_s = x]$$
$$= \sum_x \mathbb{E}\left[f(X_t) - \mathbb{E}[f(X_t)] \mid X_s = x\right] (f(x) - \mathbb{E}[f(X_s)]) \mathbb{P}[X_s = x].$$

We now use the time homogeneity of the chain to note that

$$\mathbb{E}\left[f(X_t) - \mathbb{E}[f(X_t)] \mid X_s = x\right]$$
$$= \mathbb{E}[f(X_t) \mid X_s = x] - \mathbb{E}[f(X_t)]$$
$$= \mathbb{E}[f(X_{t-s}) \mid X_0 = x] - \mathbb{E}[f(X_t)].$$

By Lemma 5.2.28,

$$|\mathbb{E}\left[f(X_t) - \mathbb{E}[f(X_t)] \mid X_s = x\right]|$$
$$= |\mathbb{E}[f(X_{t-s}) \mid X_0 = x] - \mathbb{E}[f(X_t)]|$$
$$= |(\mathbb{E}[f(X_{t-s}) \mid X_0 = x] - \pi f) - (\mathbb{E}[f(X_t)] - \pi f)|$$
$$\leq |\mathbb{E}[f(X_{t-s}) \mid X_0 = x] - \pi f| + |\mathbb{E}[f(X_t)] - \pi f|$$
$$\leq (1 - \gamma_*)^{t-s} \pi_{\min}^{-1} \|f\|_\infty + (1 - \gamma_*)^t \pi_{\min}^{-1} \|f\|_\infty$$
$$\leq 2(1 - \gamma_*)^{t-s} \pi_{\min}^{-1} \|f\|_\infty,$$

which does not depend on x. Plugging back above,

$$|\mathrm{Cov}[f(X_s), f(X_t)]|$$
$$\leq \sum_x |\mathbb{E}[f(X_t) - \mathbb{E}[f(X_t)] \mid X_s = x]| \, |f(x) - \mathbb{E}[f(X_s)]| \, \mathbb{P}[X_s = x]$$
$$\leq 2(1 - \gamma_*)^{t-s} \pi_{\min}^{-1} \|f\|_\infty \sum_x |f(x) - \mathbb{E}[f(X_s)]| \, \mathbb{P}[X_s = x]$$
$$\leq 2(1 - \gamma_*)^{t-s} \pi_{\min}^{-1} \|f\|_\infty \sum_x 2\|f\|_\infty \mathbb{P}[X_s = x]$$
$$\leq 4(1 - \gamma_*)^{t-s} \pi_{\min}^{-1} \|f\|_\infty^2.$$

Returning to the sum over the covariances, the previous bound gives

$$\left| \frac{2}{T^2} \sum_{1 \le s < t \le T} \mathrm{Cov}[f(X_s), f(X_t)] \right|$$

$$\le \frac{2}{T^2} \sum_{1 \le s < t \le T} |\mathrm{Cov}[f(X_s), f(X_t)]|$$

$$\le \frac{2}{T^2} \sum_{1 \le s < t \le T} 4(1 - \gamma_\star)^{t-s} \pi_{\min}^{-1} \|f\|_\infty^2.$$

To evaluate the sum we make the change of variable $h = t - s$ to get that the previous expression is

$$\le 4\pi_{\min}^{-1} \|f\|_\infty^2 \frac{2}{T^2} \sum_{1 \le s \le T} \sum_{h=1}^{T-s} (1 - \gamma_\star)^h$$

$$\le 4\pi_{\min}^{-1} \|f\|_\infty^2 \frac{2}{T^2} \sum_{1 \le s \le T} \sum_{h=0}^{+\infty} (1 - \gamma_\star)^h$$

$$= 4\pi_{\min}^{-1} \|f\|_\infty^2 \frac{2}{T^2} \sum_{1 \le s \le T} \frac{1}{\gamma_\star}$$

$$= 8\pi_{\min}^{-1} \|f\|_\infty^2 \gamma_\star^{-1} \frac{1}{T} \to 0$$

as $T \to +\infty$.

Combining the variance and covariance bounds, we have shown that

$$\mathrm{Var}\left[\frac{1}{T} \sum_{t=1}^{T} f(X_t) \right] \le \|f\|_\infty^2 \frac{1}{T} + 8\pi_{\min}^{-1} \|f\|_\infty^2 \gamma_\star^{-1} \frac{1}{T} \le 9\pi_{\min}^{-1} \|f\|_\infty^2 \gamma_\star^{-1} \frac{1}{T}.$$

For any $\varepsilon > 0$,

$$\mathbb{P}\left[\left| \frac{1}{T} \sum_{t=1}^{T} f(X_t) - \pi f \right| \ge \varepsilon \right]$$

$$= \mathbb{P}\left[\left| \frac{1}{T} \sum_{t=1}^{T} f(X_t) - \mathbb{E}\left[\frac{1}{T} \sum_{t=1}^{T} f(X_t) \right] + \left(\mathbb{E}\left[\frac{1}{T} \sum_{t=1}^{T} f(X_t) \right] - \pi f \right) \right| \ge \varepsilon \right]$$

$$\le \mathbb{P}\left[\left| \frac{1}{T} \sum_{t=1}^{T} f(X_t) - \mathbb{E}\left[\frac{1}{T} \sum_{t=1}^{T} f(X_t) \right] \right| + \left| \mathbb{E}\left[\frac{1}{T} \sum_{t=1}^{T} f(X_t) \right] - \pi f \right| \ge \varepsilon \right]$$

$$\le \mathbb{P}\left[\left| \frac{1}{T} \sum_{t=1}^{T} f(X_t) - \mathbb{E}\left[\frac{1}{T} \sum_{t=1}^{T} f(X_t) \right] \right| \ge \varepsilon - \pi_{\min}^{-1} \|f\|_\infty \gamma_\star^{-1} \frac{1}{T} \right].$$

We can now apply Chebyshev's inequality to get

$$\mathbb{P}\left[\left| \frac{1}{T} \sum_{t=1}^{T} f(X_t) - \pi f \right| \ge \varepsilon \right] \le \frac{9\pi_{\min}^{-1} \|f\|_\infty^2 \gamma_\star^{-1} \frac{1}{T}}{(\varepsilon - \pi_{\min}^{-1} \|f\|_\infty \gamma_\star^{-1} \frac{1}{T})^2} \to 0$$

as $T \to +\infty$. ∎

5.2.5 Spectral Radius

The results in this section have so far concerned finite state spaces. The countably infinite case presents a number of complications. We start with a few observations:

- Suppose P is irreducible, aperiodic, and positive recurrent. Then we know from the convergence theorem (Theorem 1.1.33) that if π is the stationary distribution, then for all x,

$$\|P^t(x,\cdot) - \pi(\cdot)\|_{\mathrm{TV}} \to 0$$

as $t \to +\infty$. The convergence rate depends on the starting point x. In the infinite state space case, one typically needs to make that dependence explicit to get meaningful results. In particular, the mixing time – as we have defined it – may not be a useful concept.

- In the transient and null recurrent cases, there is no stationary distribution to converge to by Theorem 3.1.20. Instead, we have the following by Theorem 3.1.21: if P is an irreducible chain which is either transient or null recurrent, then we have that

$$\lim_t P^t(x,y) = 0$$

for all $x, y \in V$.

- Conditions stronger than reversibility are needed for the spectral theorem – in a form similar to what we used – to apply. Specifically, one needs that P is a *compact operator*: whenever $(f_n)_n \in \ell^2(V, \pi)$ is a bounded sequence, there exists a subsequence $(f_{n_k})_k$ such that $(P f_{n_k})$ converges in the norm. Unfortunately, that is often not the case, as the next example illustrates, even in the reversible positive recurrent case.

Example 5.2.29 (A positive recurrent chain whose P is not compact). For $p < 1/2$, let (X_t) be the birth-death chain with $V := \{0, 1, 2, \ldots\}$, $P(0,0) := 1 - p$, $P(0,1) = p$, $P(x, x+1) := p$ and $P(x, x-1) := 1 - p$ for all $x \geq 1$, and $P(x,y) := 0$ if $|x - y| > 1$. As can be checked by direct computation, P is reversible with respect to the stationary distribution $\pi(x) = (1 - \gamma)\gamma^x$ for $x \geq 0$, where $\gamma := \frac{p}{1-p}$. For $j \geq 1$, define $g_j(x) := \pi(j)^{-1/2}\mathbf{1}_{\{x=j\}}$. Then, $\|g_j\|_\pi^2 = 1$ for all j so $\{g_j\}_j$ is bounded in $\ell^2(V, \pi)$. On the other hand,

$$Pg_j(x) = p\pi(j)^{-1/2}\mathbf{1}_{\{x=j-1\}} + (1-p)\pi(j)^{-1/2}\mathbf{1}_{\{x=j+1\}}.$$

So

$$\|Pg_j\|_\pi^2 = p^2\pi(j)^{-1}\pi(j-1) + (1-p)^2\pi(j)^{-1}\pi(j+1)$$
$$= p^2\frac{1-p}{p} + (1-p)^2\frac{p}{1-p}$$
$$= 2p(1-p).$$

Hence, $\{Pg_j\}_j$ is also bounded. However, for $j > \ell$,

$$\|Pg_j - Pg_\ell\|_\pi^2 \geq (1-p)^2\pi(j)^{-1}\pi(j+1) + p^2\pi(\ell)^{-1}\pi(\ell-1)$$
$$= 2p(1-p).$$

So $\{Pg_j\}_j$ does not have a converging subsequence. ◀

We will not say much about the spectral theory of infinite networks. In this subsection, we establish a relationship between the operator norm of P – which is related to its spectrum – and the decay of $P^t(x, y)$.

Let $\ell_0(V)$ be the set of real-valued functions on V with finite support. It is dense in $\ell^2(V, \pi)$. Indeed, let $v_1, v_2, \ldots$ be an enumeration of V and, for $f \in \ell^2(V, \pi)$, define $f|_n(v_i) := f(v_i)\mathbf{1}_{i \leq n}$ to be f restricted to $v_1, \ldots, v_n$. Then,

$$\|f - f|_n\|_\pi^2 = \sum_{i=n+1}^{\infty} \pi(v_i)f(v_i)^2 \to 0 \tag{5.2.10}$$

as $n \to \infty$, since $\|f\|_\pi^2 = \sum_x \pi(x)f(x)^2 < +\infty$. We will also need the following:

$$\|Pf - P(f|_n)\|_\pi^2 = \|P(f - f|_n)\|_\pi^2 \leq \|f - f|_n\|_\pi^2 \to 0, \tag{5.2.11}$$

where we used (5.2.2).

Definition 5.2.30 (Operator norm). *The* operator norm *of P is*

OPERATOR NORM

$$\|P\|_\pi = \sup\left\{ \frac{\|Pf\|_\pi}{\|f\|_\pi} : f \in \ell_0(V), f \neq 0 \right\}.$$

By definition, for any $f \in \ell_0(V)$,

$$\|Pf\|_\pi \leq \|P\|_\pi \|f\|_\pi. \tag{5.2.12}$$

The same can be seen to hold for any $f \in \ell^2(V, \pi)$ by considering the sequence $(f|_n)_n$ and noting that $\|f|_n\|_\pi \to \|f\|_\pi$ and $\|P(f|_n)\|_\pi \to \|Pf\|_\pi$ as $n \to \infty$ by (5.2.10), (5.2.11), and the triangle inequality. This latter observation explains why it suffices to restrict the supremum to ℓ_0 in the definition of the norm.

Note that, by (5.2.2), $\|P\|_\pi \leq 1$. Note further that if V is finite or, more generally, if π is summable, then we have in fact $\|P\|_\pi = 1$ by taking $f \equiv 1$ above. When P is self-adjoint, the norm $\|P\|_\pi$ is also equal to what is known as the *spectral radius*, that is, the radius of the smallest disk centered at 0 in the complex plane that contains the spectrum of P. We will not need to define what that means formally here. (But Exercise 5.5 asks for a proof in the setting of symmetric matrices.)

SPECTRAL RADIUS

Our main result is the following.

Theorem 5.2.31 (Spectral radius). *Let P be irreducible and reversible with respect to $\pi > 0$. Then,*

$$\rho(P) := \limsup_t P^t(x, y)^{1/t} = \|P\|_\pi.$$

In particular, the limit does not depend on x, y. Moreover, for all t,

$$P^t(x, y) \leq \sqrt{\frac{\pi(y)}{\pi(x)}} \|P\|_\pi^t.$$

In the positive recurrent case (for instance, if the chain is finite), we have $P^t(x, y) \to \pi(y) > 0$ and so $\rho(P) = 1 = \|P\|_\pi$. The theorem says that the equality between $\rho(P)$ and $\|P\|_\pi$ holds in general for reversible chains.

Proof of Theorem 5.2.31   To see that the limit does not depend on x, y, let $u, v, x, y \in V$ and $k, m \geq 0$ such that $P^m(u, x) > 0$ and $P^k(y, v) > 0$. Then,

$$P^{t+m+k}(u, v)^{1/(t+m+k)}$$
$$\geq (P^m(u, x)P^t(x, y)P^k(y, v))^{1/(t+m+k)}$$
$$\geq P^m(u, x)^{1/(t+m+k)}P^t(x, y)^{1/t}P^k(y, v)^{1/(t+m+k)},$$

which shows that $\limsup_t P^t(u, v)^{1/t} \geq \limsup_t P^t(x, y)^{1/t}$ for all u, v, x, y.

We first show that $\rho(P) \leq \|P\|_\pi$. Observe that applying (5.2.4) and (5.2.12) repeatedly gives that P^t is self-adjoint and satisfies the inequality $\|P^t\|_\pi \leq \|P\|_\pi^t$. Because $\|\delta_z\|_\pi^2 = \pi(z) \leq 1$, by Cauchy–Schwarz,

$$\pi(x)P^t(x, y) = \langle \delta_x, P^t\delta_y \rangle_\pi \leq \|P\|_\pi^t \|\delta_x\|_\pi \|\delta_y\|_\pi = \|P\|_\pi^t \sqrt{\pi(x)\pi(y)}.$$

Hence, $P^t(x, y) \leq \sqrt{\frac{\pi(y)}{\pi(x)}} \|P\|_\pi^t$ and

$$\rho(P) = \limsup_t P^t(x, y)^{1/t}$$
$$\leq \limsup_t \left(\sqrt{\frac{\pi(y)}{\pi(x)}} \|P\|_\pi^t \right)^{1/t}$$
$$= \|P\|_\pi.$$

To establish the inequality in the other direction, we make a series of observations. Fix a non-zero $f \in \ell_0(V)$.

- By self-adjointness and Cauchy–Schwarz,

$$\|P^{t+1}f\|_\pi^2 = \langle P^{t+1}f, P^{t+1}f \rangle_\pi = \langle P^{t+2}f, P^tf \rangle_\pi \leq \|P^{t+2}f\|_\pi \|P^tf\|_\pi,$$

   or

$$\frac{\|P^{t+1}f\|_\pi}{\|P^tf\|_\pi} \leq \frac{\|P^{t+2}f\|_\pi}{\|P^{t+1}f\|_\pi}.$$

So $\frac{\|P^{t+1}f\|_\pi}{\|P^tf\|_\pi}$ is non-decreasing and therefore has a limit $L \leq +\infty$. Moreover, for $t = 0$, we get

$$\frac{\|Pf\|_\pi}{\|f\|_\pi} \leq L, \tag{5.2.13}$$

so it suffices to prove $L \leq \rho(P)$.

- Observe that

$$\left(\frac{\|P^tf\|_\pi}{\|f\|_\pi} \right)^{1/t} = \left(\frac{\|Pf\|_\pi}{\|f\|_\pi} \times \cdots \times \frac{\|P^tf\|_\pi}{\|P^{t-1}f\|_\pi} \right)^{1/t} \to L,$$

   so in fact

$$L = \lim_t \|P^tf\|_\pi^{1/t}.$$

- By self-adjointness again,

$$\|P^t f\|_\pi^2 = \langle f, P^{2t} f \rangle_\pi = \sum_x \pi(x) f(x) \sum_y f(y) P^{2t}(x, y).$$

By definition of $\rho(P)$, for any $\varepsilon > 0$, there is a t large enough that

$$P^{2t}(x, y) \le (\rho(P) + \varepsilon)^{2t},$$

for all x, y in the support of f. For such a t, plugging back into the previous display,

$$\|P^t f\|_\pi^{1/t} \le (\rho(P) + \varepsilon) \left(\sum_x \pi(x) |f(x)| \sum_y |f(y)| \right)^{1/2t}.$$

The expression in parentheses on the right-hand side is finite because f has finite support. Since ε is arbitrary, we get

$$L = \lim_t \|P^t f\|_\pi^{1/t} \le \rho(P). \tag{5.2.14}$$

So, combining (5.2.13) and (5.2.14), we have shown that $\|P\|_\pi \le \rho(P)$ and that concludes the proof. ∎

Corollary 5.2.32 *Let P be irreducible and reversible with respect to π. If $\|P\|_\pi < 1$, then P is transient.*

Proof By Theorem 5.2.31, $P^t(x, x) \le \|P\|_\pi^t$ so

$$\sum_t P^t(x, x) \le \sum_t \|P\|_\pi^t < +\infty.$$

Let (X_t) be a chain with transition matrix P. Because

$$\sum_t P^t(x, x) = \mathbb{E}_x \left[\sum_t \mathbf{1}_{\{X_t = x\}} \right],$$

we have that $\sum_t \mathbf{1}_{\{X_t = x\}} < +\infty$, $\mathbb{P}_x$-a.s., and (X_t) is transient. ∎

This is not an if and only if. Random walk on $\mathbb{Z}^3$ is transient, yet $P^{2t}(0, 0) = \Theta(t^{-3/2})$ so there $\|P\|_\pi = \rho(P) = 1$.

In the non-reversible case, our definition of $\|P\|_\pi$ still makes sense with respect to any stationary measure π (although P is not self-adjoint). But the equality in Theorem 5.2.31 no longer holds in general.

Example 5.2.33 (Counter-example). Let (X_t) be asymmetric random walk on $\mathbb{Z}$ with probability $p \in (1/2, 1)$ of going to the right. Then both $\pi_0(x) := \left(\frac{p}{1-p} \right)^x$ and $\pi_1(x) := 1$ define stationary measures, but the transition matrix P is only reversible with respect to π_0.

Under π_1, we have $\|P\|_{\pi_1} = 1$. Indeed, let $g_n(x) := \mathbf{1}_{\{|x| \le n\}}$ and note that

$$(Pg_n)(x) = \mathbf{1}_{\{|x| \le n-1\}} + p \mathbf{1}_{\{x = -n-1 \text{ or } -n\}} + (1-p) \mathbf{1}_{\{x = n \text{ or } n+1\}},$$

so $\|g_n\|_{\pi_1}^2 = 2n + 1$ and $\|Pg_n\|_{\pi_1}^2 \ge 2(n-1) + 1$. Hence,

$$\limsup_n \frac{\|Pg_n\|_{\pi_1}}{\|g_n\|_{\pi_1}} \ge 1$$

and $\|P\|_{\pi_1} \geq 1$. But we already showed that $\|P\|_{\pi_1} \leq 1$ in (5.2.2), so the claim follows.

On the other hand, $\mathbb{E}_0[X_t] = (2p - 1)t$. So the martingale $Z_t := X_t - (2p - 1)t$ (see Example 3.1.29), as a sum of t independent centered random variables in $\{-1-(2p-1), 1-(2p-1)\}$, satisfies the assumptions of the Azuma–Hoeffding inequality (Theorem 3.2.1) with increment bound $c_t := 2$. So

$$
\begin{aligned}
P^t(0, 0)^{1/t} &\leq \mathbb{P}_0[X_t \leq 0]^{1/t} \\
&= \mathbb{P}_0[X_t - (2p - 1)t \leq -(2p - 1)t]^{1/t} \\
&\leq e^{-\frac{2(2p-1)^2 t^2}{2^2 t} \frac{1}{t}}.
\end{aligned}
$$

Therefore,

$$
\limsup_t P^t(0, 0)^{1/t} \leq e^{-(2p-1)^2/2} < 1.
$$

◄

5.3 Geometric Bounds

The goal of this section is to relate the spectral gap to certain geometric properties of the underlying network, more specifically isoperimetric properties, that is, relationships between the "volume" of sets and their "circumference." The classical *isoperimetric inequality* states that the area enclosed by any rectifiable simple closed curve in the plane is at most the length of the curve squared divided by 4π. Moreover, equality is achieved if and only if the curve is a circle.

Remark 5.3.1 *Here is an easy proof in the smooth case. Suppose $r(s) = (x(s), y(s))$, $s \in [0, 2\pi]$ is the parametrization of a positively oriented, smooth, simple closed curve in the plane centered at the origin with arc-length 2π, where $\|r'(s)\|_2 = 1$ for all s, $\int_0^{2\pi} r(s)\,\mathrm{d}s = 0$ and $x(0) = x(2\pi) = 0$. By Green's theorem, the area enclosed by the curve is*

$$
A = \int_0^{2\pi} x(s)y'(s)\,\mathrm{d}s = \frac{1}{2} \int_0^{2\pi} [x(s)^2 + y'(s)^2 - (x(s) - y'(s))^2]\,\mathrm{d}s,
$$

where we used that $2ab = a^2 + b^2 - (a - b)^2$. By the one-dimensional Poincaré inequality (Remark 3.2.7),

$$
A \leq \frac{1}{2} \int_0^{2\pi} [x(s)^2 + y'(s)^2]\,\mathrm{d}s \leq \frac{1}{2} \int_0^{2\pi} [x'(s)^2 + y'(s)^2]\,\mathrm{d}s = \pi,
$$

which is indeed the area of a circle of circumference 2π.

Edge expansion We define our isoperimetric quantity of interest. Let (X_t) be a finite, irreducible Markov chain on V reversible with respect to a stationary measure $\pi > 0$. (In this section, we do not necessarily assume that π is a probability distribution.) Let P be its transition matrix. We think of (X_t) as a random walk on the network $\mathcal{N} = (G, c)$, where G is the transition graph and $c(x, y) := \pi(x)P(x, y) = \pi(y)P(y, x)$.

For a subset $S \subseteq V$, we let the *edge boundary* of S be

$$
\partial_E S := \{e = (x, y) \in E \colon x \in S, y \in S^c\}.
$$

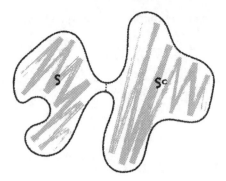

Figure 5.3 A bottleneck.

Let $g\colon E \to \mathbb{R}_+$ be an edge weight function. For $F \subseteq E$, we define

$$|f|_g := \sum_{e \in F} g(e),$$

and similarly for a vertex function. Finally, for $S \subseteq V$, we let

$$\Phi_E(S; g, h) := \frac{|\partial_E S|_g}{|S|_h}.$$

Roughly speaking, this is the ratio of the "size of the boundary" of a set to its "volume."

Our main definition, the edge expansion constant, quantifies the worst such ratio. First, one last piece of notation: for disjoint subsets $S_0, S_1 \subseteq V$, we let

$$c(S_0, S_1) := \sum_{x_0 \in S_0} \sum_{x_1 \in S_1} c(x_0, x_1).$$

Definition 5.3.2 (Edge expansion). *For a subset of states $S \subseteq V$, the* edge expansion constant *(or* bottleneck ratio*) of S is*

$$\Phi_E(S; c, \pi) = \frac{|\partial_E S|_c}{|S|_\pi} = \frac{c(S, S^c)}{\pi(S)}.$$

We refer to (S, S^c) as a cut. *The* edge expansion constant *(or* bottleneck ratio *or* Cheeger number *or* isoperimetric constant[1]*) of $\mathcal{N}$ is*

$$\Phi_* := \min\left\{\Phi_E(S; c, \pi)\colon S \subseteq V,\ 0 < \pi(S) \le \frac{1}{2}\right\}.$$

EDGE
EXPANSION
CONSTANT

Intuitively, a small value of Φ_* suggests the existence of a "bottleneck" in $\mathcal{N}$. Conversely, a large value of Φ_* indicates that all sets "expand out." See Figure 5.3. Note that the quantity $\Phi_E(S; c, \pi)$ has a natural probabilistic interpretation: pick a stationary state and make one step according to the transition matrix; then, $\Phi_E(S; c, \pi)$ is the conditional probability that, given that the first state is in S, the next one is in S^c.

---

[1] It is also called "conductance," but that terminology clashes with our use of the term.

Equivalently, the edge expansion constant can be expressed as

$$\Phi_* := \min\left\{ \frac{c(S, S^c)}{\pi(S) \wedge \pi(S^c)} : S \subseteq V, \, 0 < \pi(S) < 1 \right\}.$$

Example 5.3.3 (Edge expansion: complete graph). Let $G = K_n$ be the complete graph on n vertices and assume $c(x, y) = 1/n^2$ for all $x \neq y$. For simplicity, take n even. Then for a subset S of size $|S| = k$,

$$\Phi_E(S; c, \pi) = \frac{|\partial_E S|_c}{|S|_\pi} = \frac{k(n-k)/n^2}{k/n} = \frac{n-k}{n}.$$

Thus, the minimum is achieved for $k = n/2$ and

$$\Phi_* = \frac{n - n/2}{n} = \frac{1}{2}.$$ ◀

Dirichlet form, Rayleigh quotient, and normalized Laplacian We relate the edge expansion constant of $\mathcal{N}$ to the spectral gap of P. Recall that we denote by $\lambda_1, \ldots, \lambda_n$ the eigenvalues of P in decreasing order. First, we adapt the variational characterization of Theorem 5.1.3 to the network setting.

DIRICHLET
FORM

The *Dirichlet form* is defined over $\ell^2(V, \pi)$ as the bilinear form

$$\mathcal{D}(f, g) := \langle f, (I - P)g \rangle_\pi.$$

DIRICHLET
ENERGY

The associated quadratic form, also known as *Dirichlet energy*, is $\mathcal{D}(f) := \mathcal{D}(f, f)$. Note that, using stochasticity and reversibility,

$$\langle f, (I - P)f \rangle_\pi = \langle f, f \rangle_\pi - \langle f, Pf \rangle_\pi$$

$$= \frac{1}{2} \sum_{x,y} f(x)^2 \pi(x) P(x, y)$$

$$+ \frac{1}{2} \sum_{x,y} f(y)^2 \pi(y) P(y, x) - \sum_{x,y} \pi(x) f(x) f(y) P(x, y)$$

$$= \frac{1}{2} \sum_{x,y} \pi(x) P(x, y) f(x)^2$$

$$+ \frac{1}{2} \sum_{x,y} \pi(x) P(x, y) f(y)^2 - \sum_{x,y} \pi(x) P(x, y) f(x) f(y)$$

$$= \frac{1}{2} \sum_{x,y} c(x, y) [f(x) - f(y)]^2,$$

which is indeed consistent with the expression we encountered previously in Theorem 3.3.25.

RAYLEIGH
QUOTIENT

The *Rayleigh quotient* for $I - P$ over $\ell^2(V, \pi)$ is then

$$\frac{\langle f, (I - P)f \rangle_\pi}{\langle f, f \rangle_\pi} = \frac{\frac{1}{2} \sum_{x,y} c(x, y) [f(x) - f(y)]^2}{\sum_x \pi(x) f(x)^2}$$

$$= \frac{\mathbf{z}^T \mathcal{L} \mathbf{z}}{\mathbf{z}^T \mathbf{z}},$$

where $\mathcal{L}$ is the normalized Laplacian of the network $\mathcal{N}$ and we defined the vector $\mathbf{z} = (z_x)_{x \in V}$ with $z_x := \sqrt{\pi(x)} f(x)$. Consequently, the Courant–Fischer theorem (Theorem 5.1.3) in the form (5.1.5) gives the following. Here $\eta_2 = 1 - \lambda_2 = \gamma$ is the spectral gap of P, which can also be seen as the second smallest eigenvalue of $I - P$ (which has the same eigenfunctions as P itself). We have

$$\gamma = \inf \left\{ \frac{\langle f, (I - P)f \rangle_\pi}{\langle f, f \rangle_\pi} : \pi f = 0, f \neq 0 \right\}.$$

The infimum is achieved by the eigenfunction f_2 of P corresponding to its second largest eigenvalue λ_2. (Recall from Lemma 5.2.5 that $\pi f_2 = 0$.)

We note further that if $\pi f = 0$, then

$$\langle f, f \rangle_\pi = \langle f - \pi f, f - \pi f \rangle_\pi = \mathrm{Var}_\pi[f],$$

where the last expression denotes the variance under π. So the variational characterization of γ implies that

$$\mathrm{Var}_\pi[f] \leq \gamma^{-1} \mathcal{D}(f)$$

for all f such that $\pi f = 0$. In fact, it holds for *any* f by considering $f - \pi f$ and noticing that both sides are unaffected by subtracting a constant to f.

We have shown:

Theorem 5.3.4 (Poincaré inequality for $\mathcal{N}$). *Let P be finite, irreducible, and reversible with respect to π. Then,*

$$\mathrm{Var}_\pi[f] \leq \gamma^{-1} \mathcal{D}(f) \tag{5.3.1}$$

for all $f \in \ell^2(V \pi)$. Equality is achieved by the eigenfunction f_2 of P corresponding to the second largest eigenvalue λ_2.

An inequality of the type

$$\mathrm{Var}_\pi[f] \leq C \mathcal{D}(f) \qquad \forall f \tag{5.3.2}$$

is known as a *Poincaré inequality*, a simple version of which we encountered previously in Remark 3.2.7. To see the connection with that one-dimensional version, it will be conven- ient to work with directed edges. Let $\vec{E}$ be an orientation of E, that is, for each $e \in \{x, y\}$, $\vec{E}$ includes either (x, y) or (y, x) with associated weight $c(\vec{e}) := c(e) > 0$. For a function $f: V \to \mathbb{R}$ and an edge $\vec{e} = (x, y) \in \vec{E}$, we define the "discrete gradient"

$$\nabla f(\vec{e}) = f(y) - f(x).$$

With this notation, we can rewrite the Dirichlet energy as

$$\mathcal{D}(f) = \frac{1}{2} \sum_{x,y} c(x, y) [f(x) - f(y)]^2 = \sum_{\vec{e}} c(\vec{e}) [\nabla f(\vec{e})]^2, \tag{5.3.3}$$

hence (5.3.1) is a network analogue of (3.2.7).

POINCARÉ INEQUALITY

5.3.1 Cheeger's Inequality

The edge expansion constant and the spectral gap are related through the following isoperimetric inequalities. The lower bound is known as *Cheeger's inequality*.

Theorem 5.3.5 *Let P be a finite, irreducible, reversible Markov transition matrix and let* $\gamma = 1 - \lambda_2$ *be the spectral gap of P. Then,*

$$\frac{\Phi_*^2}{2} \leq \gamma \leq 2\Phi_*.$$

In terms of the relaxation time $t_{\mathrm{rel}} = \gamma^{-1}$, these inequalities have an intuitive meaning: the presence or absence of a bottleneck in the state space leads to slow or fast mixing, respectively. We detail some applications to mixing times in the next subsections.

Before giving a proof of the theorem, we start with a trivial – yet insightful – example.

Example 5.3.6 (Two-state chain). Let $V := \{0, 1\}$ and, for $\alpha, \beta \in (0, 1)$,

$$P := \begin{pmatrix} 1 - \alpha & \alpha \\ \beta & 1 - \beta \end{pmatrix},$$

which has stationary distribution

$$\pi := \left(\frac{\beta}{\alpha + \beta}, \frac{\alpha}{\alpha + \beta} \right).$$

Recall from Example 5.2.8 that the second right eigenvector is

$$f_2 := \left(\sqrt{\frac{\alpha}{\beta}}, -\sqrt{\frac{\beta}{\alpha}} \right) = \left(\sqrt{\frac{\pi_1}{\pi_0}}, -\sqrt{\frac{\pi_0}{\pi_1}} \right),$$

with eigenvalue $\lambda_2 := 1 - \alpha - \beta$, so the spectral gap is $\alpha + \beta$. Assume that $\beta \leq \alpha$. Then, the bottleneck ratio is

$$\Phi_* = \frac{c(0, 1)}{\pi(0)} = P(0, 1) = \alpha.$$

Then Theorem 5.3.5 reads

$$\frac{\alpha^2}{2} \leq \alpha + \beta \leq 2\alpha,$$

which is indeed satisfied for all $0 < \beta \leq \alpha < 1$. Note that the upper bound is tight when $\alpha = \beta$.   ◀

Proof of Theorem 5.3.5   We start with the upper bound. In view of the Poincaré inequality for $\mathcal{N}$ (Theorem 5.3.4), to get an upper bound on the spectral gap, it suffices to plug in a well-chosen function f in (5.3.1). Taking a hint from Example 5.3.6, for $S \subseteq V$ with $\pi(S) \in (0, 1/2]$, we let

$$f_S(x) := \begin{cases} -\sqrt{\dfrac{\pi(S^c)}{\pi(S)}} & x \in S, \\ \sqrt{\dfrac{\pi(S)}{\pi(S^c)}} & x \in S^c. \end{cases}$$

Then,

$$\sum_x \pi(x) f_S(x) = \pi(S) \left[-\sqrt{\frac{\pi(S^c)}{\pi(S)}} \right] + \pi(S^c) \left[\sqrt{\frac{\pi(S)}{\pi(S^c)}} \right] = 0$$

and

$$\sum_x \pi(x) f_S(x)^2 = \pi(S) \left[-\sqrt{\frac{\pi(S^c)}{\pi(S)}} \right]^2 + \pi(S^c) \left[\sqrt{\frac{\pi(S)}{\pi(S^c)}} \right]^2 = 1.$$

So $\mathrm{Var}_\pi[f_S] = 1$. Hence, from Theorem 5.3.4,

$$\begin{aligned}
\gamma &\le \frac{\mathscr{D}(f_S)}{\mathrm{Var}_\pi[f_S]} \\
&= \frac{1}{2} \sum_{x,y} c(x,y)[f_S(x) - f_S(y)]^2 \\
&= \sum_{x \in S, y \in S^c} c(x,y) \left[-\sqrt{\frac{\pi(S^c)}{\pi(S)}} - \sqrt{\frac{\pi(S)}{\pi(S^c)}} \right]^2 \\
&= \sum_{x \in S, y \in S^c} c(x,y) \left[-\frac{\pi(S^c) + \pi(S)}{\sqrt{\pi(S)\pi(S^c)}} \right]^2 \\
&= \frac{c(S, S^c)}{\pi(S)\pi(S^c)} \\
&\le 2 \frac{c(S, S^c)}{\pi(S)},
\end{aligned}$$

as claimed.

The other direction is trickier. Because we seek an upper bound on the edge expansion constant Φ_*, our goal is to find a cut (S, S^c) such that

$$\frac{c(S, S^c)}{\pi(S) \wedge \pi(S^c)} \le \sqrt{2\gamma}. \tag{5.3.4}$$

Because the eigenfunction f_2 achieves γ in Theorem 5.3.4, it is natural to look to it for "good cuts." Thinking of f_2 as a one-dimensional embedding of the network, it turns out to be enough to consider only "sweep cuts" of the form $S := \{v : f_2(v) \le \theta\}$ for a threshold θ. How to pick the right threshold is less obvious.

Here we use a probabilistic argument, that is, we construct a *random* cut (Z, Z^c). Observe that it suffices that

$$\mathbb{E}\left[c(Z, Z^c)\right] \le \sqrt{2\gamma} \, \mathbb{E}\left[\pi(Z) \wedge \pi(Z^c)\right], \tag{5.3.5}$$

since then $\mathbb{E}\left[\sqrt{2\gamma}\pi(Z) \wedge \pi(Z^c) - c(Z, Z^c)\right] \ge 0$, which in turn implies that we have $\mathbb{P}\left[\sqrt{2\gamma}\pi(Z) \wedge \pi(Z^c) - c(Z, Z^c) \ge 0\right] > 0$ by the first moment principle (Theorem 2.2.1); in other words, there exists a cut satisfying (5.3.4).

We now describe the random cut (Z, Z^c):

1. *(Cuts from f_2)* Let again f_2 be an eigenfunction corresponding to the eigenvalue λ_2 of P with $\|f_2\|_\pi^2 = 1$. Order the vertices $V := \{v_1, \ldots, v_n\}$ in such a way that

$$f_2(v_i) \le f_2(v_{i+1}) \qquad \forall i = 1, \ldots, n-1.$$

As we described above, the function f_2 naturally produces a series of cuts (S_i, S_i^c), where $S_i := \{v_1, \ldots, v_i\}$. By definition of the bottleneck ratio,

$$\Phi_* \le \frac{c(S_i, S_i^c)}{\pi(S_i) \wedge \pi(S_i^c)}. \tag{5.3.6}$$

2. *(Normalization)* Let

$$m := \min\{i : \pi(S_i) > 1/2\},$$

and define the translated function

$$f := f_2 - f_2(v_m).$$

We further set $g := \alpha f$, where $\alpha > 0$ is chosen so that

$$g(v_1)^2 + g(v_n)^2 = 1.$$

Note that, by construction, $g(v_m) = 0$ and $g(v_1) \le \cdots g(v_m) = 0 \le g(v_{m+1}) \le \cdots \le g(v_n)$. The function g is related to γ as follows:

Lemma 5.3.7

$$\frac{1}{2} \sum_{x,y} c(x,y)(g(x) - g(y))^2 \le \gamma \sum_x \pi(x)g(x)^2.$$

Proof By Theorem 5.3.4,

$$\gamma = \frac{\mathcal{D}(f_2)}{\mathrm{Var}_\pi[f_2]}.$$

Because neither the numerator nor the denominator is affected by adding a constant, we have also

$$\gamma = \frac{\mathcal{D}(f)}{\mathrm{Var}_\pi[f]}.$$

Furthermore, notice that a constant multiplying f cancels out in the ratio so

$$\gamma = \frac{\mathcal{D}(g)}{\mathrm{Var}_\pi[g]}.$$

Now use the fact that $\mathrm{Var}_\pi[g] \le \sum_x \pi(x)g(x)^2$. ∎

1. *(Random cut)* Pick Θ in $[g(v_1), g(v_n)]$ with density $2|\theta|$. Note that

$$\int_{g(v_1)}^{g(v_n)} 2|\theta| \, d\theta = g(v_1)^2 + g(v_n)^2 = 1.$$

Finally, define

$$Z := \{v_i : g(v_i) < \Theta\}.$$

The rest of the proof is calculations. We bound the expectations on both sides of (5.3.5).

Lemma 5.3.8 *The following hold:*

(i)

$$\mathbb{E}[\pi(Z) \wedge \pi(Z^c)] = \sum_x \pi(x)g(x)^2.$$

(ii)

$$\mathbb{E}[c(Z, Z^c)] \leq \left(\frac{1}{2} \sum_{x,y} c(x,y)(g(x) - g(y))^2 \right)^{1/2} \left(2 \sum_x \pi(x)g(x)^2 \right)^{1/2}.$$

Lemmas 5.3.7 and 5.3.8 immediately imply (5.3.5) and that concludes the proof of Theorem 5.3.5. So it remains to prove this last lemma.

Proof of Lemma 5.3.8 We start with (i). By definition of g, $\Theta \leq 0$ implies that $\pi(Z) \wedge \pi(Z^c) = \pi(Z)$ and vice versa. Thus,

$$
\begin{aligned}
\mathbb{E}[\pi(Z) \wedge \pi(Z^c)] &= \mathbb{E}\left[\sum_{\ell < m} \pi(v_\ell)\mathbf{1}_{\{v_\ell \in Z\}}\mathbf{1}_{\{\Theta \leq 0\}} + \sum_{\ell \geq m} \pi(v_\ell)\mathbf{1}_{\{v_\ell \in Z^c\}}\mathbf{1}_{\{\Theta > 0\}} \right] \\
&= \mathbb{E}\left[\sum_{\ell < m} \pi(v_\ell)\mathbf{1}_{\{g(v_\ell) < \Theta \leq 0\}} + \sum_{\ell \geq m} \pi(v_\ell)\mathbf{1}_{\{0 < \Theta \leq g(v_\ell)\}} \right] \\
&= \sum_{\ell < m} \pi(v_\ell)\mathbb{P}\left[g(v_\ell) < \Theta \leq 0 \right] + \sum_{\ell \geq m} \pi(v_\ell)\mathbb{P}\left[0 < \Theta \leq g(v_\ell) \right] \\
&= \sum_{\ell < m} \pi(v_\ell)g(v_\ell)^2 + \sum_{\ell \geq m} \pi(v_\ell)g(v_\ell)^2 \\
&= \sum_x \pi(x)g(x)^2, \quad\quad\quad (5.3.7)
\end{aligned}
$$

where we integrated over the density of Θ to obtain the fourth line.

We move on to (ii). To compute $\mathbb{E}[c(Z, Z^c)]$, we note that $x_k \in Z$ and $x_\ell \in Z^c$ if and only if $g(v_k) < \Theta \leq g(v_\ell)$. The probability of that event depends on the signs of $g(v_k)$ and $g(v_\ell)$. If $g(v_k)g(v_\ell) \geq 0$,

$$
\begin{aligned}
\mathbb{P}[g(v_k) < \Theta \leq g(v_\ell)] &= |g(v_k)^2 - g(v_\ell)^2| \\
&= |g(v_k) - g(v_\ell)||g(v_k) + g(v_\ell)| \\
&= |g(v_k) - g(v_\ell)|(|g(v_k)| + |g(v_\ell)|).
\end{aligned}
$$

If $g(v_k)g(v_\ell) < 0$,

$$
\begin{aligned}
\mathbb{P}[g(v_k) < \Theta \leq g(v_\ell)] &= g(v_k)^2 + g(v_\ell)^2 \\
&\leq g(v_k)^2 + g(v_\ell)^2 - 2g(v_k)g(v_\ell) \\
&= (g(v_k) - g(v_\ell))^2 \\
&= |g(v_k) - g(v_\ell)|(|g(v_k)| + |g(v_\ell)|).
\end{aligned}
$$

We apply Cauchy–Schwarz to get

$$\mathbb{E}[c(Z, Z^c)] = \sum_{k < \ell} c(v_k, v_\ell) \mathbb{P}[g(v_k) < \Theta \leq g(v_\ell)]$$

$$\leq \sum_{k < \ell} c(v_k, v_\ell) |g(v_k) - g(v_\ell)| (|g(v_k)| + |g(v_\ell)|)$$

$$\leq \left(\sum_{k < \ell} c(v_k, v_\ell)(g(v_k) - g(v_\ell))^2 \right)^{1/2}$$

$$\times \left(\sum_{k < \ell} c(v_k, v_\ell)(|g(v_k)| + |g(v_\ell)|)^2 \right)^{1/2}.$$

The expression in the first parentheses is equal to $\frac{1}{2} \sum_{x,y} c(x,y)(g(x) - g(y))^2$. So it remain to bound the expression in the second parentheses.

Note that

$$(|g(x)| + |g(y)|)^2 = 2g(x)^2 + 2g(y)^2 - (|g(x)| - |g(y)|)^2 \leq 2g(x)^2 + 2g(y)^2.$$

Therefore, since $\sum_y c(x,y) = \sum_y c(y,x) = \pi(x)$,

$$\sum_{k < \ell} c(v_k, v_\ell)(|g(v_k)| + |g(v_\ell)|)^2 \leq \frac{1}{2} \sum_{x,y} c(x,y)(|g(x)| + |g(y)|)^2$$

$$\leq \sum_x \pi(x) g(x)^2 + \sum_y \pi(y) g(y)^2$$

$$= 2 \sum_x \pi(x) g(x)^2.$$

That concludes the proof.   ∎

∎

5.3.2 ▷ *Random Walks: Trees, Cycles, and Hypercubes Revisited*

We use the techniques of the previous subsection to bound the mixing time of random walk on some simple graphs. In particular, we revisit the examples of Section 4.3.2.

b-**ary tree** Let (Z_t) be lazy simple random walk on the ℓ-level rooted b-ary tree, $\widehat{\mathbb{T}}_b^\ell$. The root, 0, is on level 0 and the leaves, L, are on level ℓ. All vertices have degree $b + 1$, except for the root which has degree b and the leaves which have degree 1. Recall that the stationary distribution is

$$\pi(x) := \frac{\delta(x)}{2(n-1)}, \tag{5.3.8}$$

where n is the number of vertices and $\delta(x)$ is the degree of x. We take $b = 2$ to simplify.

It is intuitively clear that each edge of this graph constitutes a bottleneck, with the root being the most "balanced" one. Let x_0 be a leaf of $\widehat{\mathbb{T}}_b^\ell$ and let A be the set of vertices "on the other side of the root (inclusively)," that is, vertices whose graph distance from x_0 is at least ℓ. See Figure 4.7. Let S be the remaining vertices. Then, by symmetry, $\pi(S) \le 1/2$. Note that there is a single edge connecting S and $S^c = A$, namely, the edge linking 0 and the root of the subtree T_S formed by the vertices in S. More precisely, let v_S be the root of T_S. From (5.3.8), $P(v_S, 0) = \frac{1}{2} \cdot \frac{1}{3} = \frac{1}{6}$ (where the $1/2$ accounts for the laziness), $\pi(v_S) = \frac{3}{2n-2}$, and, by symmetry,

$$\pi(S) = \frac{(2n - 2 - 2)/2}{2n - 2} = \frac{n - 2}{2n - 2},$$

where in the numerator we subtracted the degree of the root before dividing the sum of the remaining degrees by 2. Hence,

$$\Phi_* \le \frac{\frac{1}{6}\left(\frac{3}{2n-2}\right)}{\frac{n-2}{2n-2}} = \frac{1}{2(n - 2)}.$$

By Theorem 5.3.5,

$$\gamma \le 2\Phi_* \le \frac{1}{n - 2} \qquad \text{and} \qquad t_{\mathrm{rel}} = \gamma^{-1} \ge n - 2.$$

Thus, by Theorem 5.2.14 and the fact that the chain is lazy

$$t_{\mathrm{mix}}(\varepsilon) \ge (t_{\mathrm{rel}} - 1) \log\left(\frac{1}{2\varepsilon}\right) = \Omega(n).$$

We showed in Section 4.3.2, using other techniques, that $t_{\mathrm{mix}}(\varepsilon) = \Theta(n)$.

Cycle Let (Z_t) be lazy simple random walk on the cycle of size n, $\mathbb{Z}_n := \{0, 1, \ldots, n - 1\}$, where $i \sim j$ if $|j - i| = 1 \pmod n$. Assume n is even.

Consider a subset of vertices S. Note that by symmetry $\pi(S) = \frac{|S|}{n}$. Moreover, for all $i \sim j$, $c(i, j) = \pi(i)P(i, j) = \frac{1}{n} \cdot \frac{1}{2} \cdot \frac{1}{2} = \frac{1}{4n}$. Among all sets of size $|S|$, consecutive vertices minimize the size of the boundary. So

$$\Phi_* \le \frac{2\frac{1}{4n}}{\frac{\ell}{n}} = \frac{1}{2\ell}$$

for all $\ell \le n/2$. This expression is minimized for $\ell = n/2$ so

$$\Phi_* = \frac{1}{n}.$$

By Theorem 5.3.5,

$$\frac{1}{2n^2} = \frac{\Phi_*^2}{2} \le \gamma \le 2\Phi_* = \frac{2}{n}$$

and

$$\frac{n}{2} \le t_{\mathrm{rel}} = \gamma^{-1} \le 2n^2.$$

Thus, by Theorem 5.2.14,

$$t_{\mathrm{mix}}(\varepsilon) \ge (t_{\mathrm{rel}} - 1) \log\left(\frac{1}{2\varepsilon}\right) = \Omega(n)$$

and

$$t_{\text{mix}}(\varepsilon) \leq \log\left(\frac{1}{\varepsilon \pi_{\min}}\right) t_{\text{rel}} = O(n^2 \log n).$$

We know from exact eigenvalue computations (see Section 5.2.2 where technically we considered the non-lazy chain; laziness only affects the relaxation time by a factor of 2) that in fact $\gamma = \frac{2\pi^2}{n^2} + O(n^{-4})$. We also showed in that section that $t_{\text{mix}}(\varepsilon) = O(n^2)$. (Exercise 4.15 shows this is tight up to a constant factor.)

Hypercube Let (Z_t) be lazy simple random walk on the n-dimensional hypercube $\mathbb{Z}_2^n := \{0, 1\}^n$, where $i \sim j$ if $\|i - j\|_1 = 1$.

To get a bound on the edge expansion constant, consider the set $S = \{x \in \mathbb{Z}_2^n : x_1 = 0\}$. By symmetry $\pi(S) = \frac{1}{2}$. For each $i \sim j$, $c(i, j) = \frac{1}{2^n} \cdot \frac{1}{2} \cdot \frac{1}{n} = \frac{1}{n2^{n+1}}$. Hence,

$$\Phi_* \leq \frac{2^{n-1} \frac{1}{n2^{n+1}}}{\frac{1}{2}} = \frac{1}{2n},$$

where in the numerator we used that $|S| = 2^{n-1}$. By Theorem 5.3.5,

$$\gamma \leq 2\Phi_* \leq \frac{1}{n}.$$

Thus, by Theorem 5.2.14,

$$t_{\text{mix}}(\varepsilon) \geq (t_{\text{rel}} - 1) \log\left(\frac{1}{2\varepsilon}\right) = \Omega(n).$$

We know from exact eigenvalue computations (Section 5.2.2) that in fact $\gamma = \frac{1}{n}$.

We also showed in Section 4.3.2 that $t_{\text{mix}}(\varepsilon) = \Theta(n \log n)$.

5.3.3 ▷ *Random Graphs: Existence of an Expander Family and Application to Mixing*

In many applications, it is useful to construct "bottleneck-free" graphs. In particular, random walks mix rapidly on such graphs. Formally:

Definition 5.3.9 (Expander family). *Let $\{G_n\}_n$ be a collection of finite d-regular graphs with $\lim_n |V(G_n)| = +\infty$, where $V(G_n)$ is the vertex set of G_n. Let*

$$\Phi_*(G_n) := \min\left\{\frac{|\partial_E S|}{d|S|} : S \subseteq V(G_n), \, 0 < |S| \leq \frac{|V(G_n)|}{2}\right\}$$

denote the edge expansion constant of G_n with unit conductances. Let $\alpha > 0$. We say that $\{G_n\}_n$ is a (d, α)-expander family if for all n,

$$\Phi_*(G_n) \geq \alpha.$$

The key point of the definition is that the edge expansion constant of all graphs in an expander family is bounded away from 0 *uniformly in n*. Note that it is trivial to construct such a family *if we drop the bounded degree assumption*: the edge expansion constant of the complete graph K_n is $1/2$ by Example 5.3.3. On the other hand, it is far from obvious that one

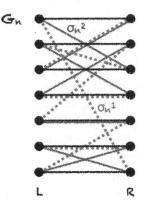

Figure 5.4 A draw from Pinsker's model.

can construct a family of *sparse* graphs (i.e., such that $|E(G_n)| = O(|V(G_n)|)$) with an edge expansion constant uniformly bounded away from 0. It turns out that a simple probabilistic construction does the trick.

We will need the following definition. For a subset $S \subseteq V$, we let the *vertex boundary* of S be

$$\partial_V S := \{y \in S^c \colon \exists x \in S \text{ s.t. } x \sim y\}.$$

VERTEX BOUNDARY

Existence of expander graphs For simplicity, we allow multigraphs (i.e., E is a multiset; or, put differently, there can be multiple edges between the same two vertices) and consider the case $d = 3$. We construct a random bipartite multigraph $G_n = (L_n, R_n, E_n)$ on $2n$ vertices known as *Pinsker's model*. Denote the vertices by $L_n = \{\ell_1, \ldots, \ell_n\}$ and $R_n = \{r_1, \ldots, r_n\}$. Let σ_n^1 and σ_n^2 be independent uniform random permutations of $[n]$. The edge set of G_n is given by

$$E_n := \{(\ell_i, r_i) \colon i \in [n]\} \cup \left\{(\ell_i, r_{\sigma_n^1(i)}) \colon i \in [n]\right\} \cup \left\{(\ell_i, r_{\sigma_n^2(i)}) \colon i \in [n]\right\}.$$

In other words, G_n is a union of three independent uniform perfect matchings (and its vertices are labeled so that one of the matchings is $\{(\ell_i, r_i)\}_i$). See Figure 5.4. Observe that, as a multigraph, all vertices of G_n have degree 3. We show that there exists $\alpha > 0$ such that, for all n large enough, with positive (in fact, high) probability G_n has an edge expansion constant bounded below by α. In particular, such a G_n exists for all n large enough and, thus, there exists a $(3, \alpha)$-expander family.

Claim 5.3.10 (Pinsker's model: edge expansion constant). *There exists $\alpha > 0$ such that*

$$\lim_n \mathbb{P}[\Phi_*(G_n) \geq \alpha] = 1.$$

Proof For convenience, assume n is even. We need to show that with probability going to 1, for any S with $|S| \leq |V(G_n)|/2 = n$, we have $|\partial_E S| \geq \alpha d |S|$ for some $\alpha > 0$. We first reduce the proof to a statement about sets of vertices lying *on one side* of G_n.

Lemma 5.3.11 *There is $\beta > 0$ such that*

$$\lim_n \mathbb{P}\left[|\partial_V K| \geq (1+\beta)|K|, \ \forall K \subseteq L, \ |K| \leq n/2\right] = 1.$$

The same holds for R.

Before proving Lemma 5.3.11, we argue that it implies Claim 5.3.10. Note that the lemma concerns the *vertex* boundary of K. To relate the latter to the edge boundary, let S with $|S| \leq n$, and let $S_L := S \cap L$ and $S_R := S \cap R$. For any subset $K \subseteq S_L$, the size of the edge boundary of S can be bounded below as follows:

$$|\partial_E S| \geq |\partial_V K| - |S_R|, \tag{5.3.9}$$

where we took into account that the vertices of $\partial_V K$ in S_R do not contribute to the edge boundary, but the others do as they are incident to at least one edge in $\partial_E S$. It remains to find a good K.

Assume without loss of generality that $|S_L| \geq |S_R|$ (in the other case, just interchange the roles of L and R), and suppose that the event in the lemma holds. In particular, $|S_R| \leq |S|/2$. We claim that there is a subset K of S_L such that

$$|S_R| \leq |S|/2 \leq |K| \leq n/2. \tag{5.3.10}$$

There are two cases:

- If $|S_L| < n/2$, then take $K = S_L$. It follows that $|K| = |S_L| \geq |S|/2$.
- If $|S_L| \geq n/2$, then let K be any subset of S_L of size $n/2$. Since $|S| \leq n$, it follows that $|K| = n/2 \geq |S|/2$.

Under the event in the lemma, $|\partial_V K| \geq (1+\beta)|K|$.

Going back to (5.3.9), using the lower bound on $|\partial_V K|$ and (5.3.10), we get

$$|\partial_E S| \geq (1+\beta)|K| - |K| = \beta|K| \geq \frac{\beta}{2}|S| = \alpha|S|,$$

where we set $\alpha = \beta/2$. Since this holds for any set S with $|S| \leq n$, we have proved Claim 5.3.10.

It remains to prove the lemma.

Proof of Lemma 5.3.11   Let $K \subseteq L$ with $k := |K| \leq n/2$. Without loss of generality assume $K = \{\ell_1, \ldots, \ell_k\}$. Observe that, by construction, $\partial_V K \supseteq K'$, where $K' = \{r_1, \ldots, r_k\}$. We analyze the "bad event"

$$\mathcal{B}_K := \{|\partial_V K| \leq k + \lfloor \beta k \rfloor\}$$

by considering all subsets of $\{r_{k+1}, \ldots, r_n\}$ of size $\lfloor \beta k \rfloor$ and bounding the probability that *all edges* out of K fall into *one of them and K'*. Note that there are $\binom{n-k}{\lfloor \beta k \rfloor}$ such subsets. See Figure 5.5.

Since σ_n^1 and σ_n^2 are uniform and independent, they each match K to a uniformly chosen subset of the same size in R and we have by a union bound

$$\mathbb{P}[\mathcal{B}_K] \leq \binom{n-k}{\lfloor \beta k \rfloor}\left[\frac{\binom{k+\lfloor \beta k \rfloor}{k}}{\binom{n}{k}}\right]^2 \leq \binom{n}{\lfloor \beta k \rfloor}\frac{\binom{k+\lfloor \beta k \rfloor}{\lfloor \beta k \rfloor}^2}{\binom{n}{k}^2},$$

where we used that $\binom{n}{s} = \binom{n}{n-s}$.

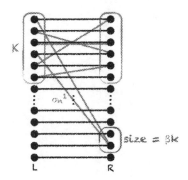

Figure 5.5 Illustration of the main step in proof of the lemma.

Taking a union bound again, this time over Ks, we have

$$\mathbb{P}[\exists K \subseteq L, \ |K| \leq n/2, \ |\partial_{\mathrm{V}} K| \leq (1+\beta)|K|]$$

$$\leq \sum_{K \subseteq L, \ |K| \leq n/2} \mathbb{P}[\mathcal{B}_K]$$

$$\leq \sum_{k=1}^{n/2} \binom{n}{k} \binom{n}{\lfloor \beta k \rfloor} \frac{\binom{k+\lfloor \beta k \rfloor}{\lfloor \beta k \rfloor}^2}{\binom{n}{k}^2}. \qquad (5.3.11)$$

We use the bound $\frac{n^s}{s^s} \leq \binom{n}{s} \leq \frac{e^s n^s}{s^s} \leq \frac{e^t n^t}{t^t}$ for $s \leq t < n$ (Appendix A). To see the last inequality, note that $\frac{\mathrm{d}}{\mathrm{d}t} \log(\frac{e^t n^t}{t^t}) = \log(\frac{n}{t}) > 0$ for $0 < t < n$. We obtain that the sum in the last display is bounded as

$$\sum_{k=1}^{n/2} \binom{n}{k} \binom{n}{\lfloor \beta k \rfloor} \frac{\binom{k+\lfloor \beta k \rfloor}{\lfloor \beta k \rfloor}^2}{\binom{n}{k}^2} = \sum_{k=1}^{n/2} \binom{n}{\lfloor \beta k \rfloor} \frac{\binom{k+\lfloor \beta k \rfloor}{\lfloor \beta k \rfloor}^2}{\binom{n}{k}}$$

$$\leq \sum_{k=1}^{n/2} \frac{e^{\beta k} n^{\beta k}}{(\beta k)^{\beta k}} \frac{\left(\frac{e^{\beta k}(k+\beta k)^{\beta k}}{(\beta k)^{\beta k}}\right)^2}{\frac{n^k}{k^k}}$$

$$\leq \sum_{k=1}^{n/2} \left(\frac{k}{n}\right)^{k(1-\beta)} \left(\frac{e^3(1+\beta)^2}{\beta^3}\right)^{\beta k}$$

$$= \sum_{k=1}^{\infty} f_n(k), \qquad (5.3.12)$$

where we defined

$$f_n(k) := \mathbf{1}_{\{k \leq n/2\}} \left(\frac{k}{n}\right)^{k(1-\beta)} \left(\frac{e^3(1+\beta)^2}{\beta^3}\right)^{\beta k}.$$

Let also

$$g(k) := \left[\left(\frac{1}{2}\right)^{1-\beta} \left(\frac{e^3(1+\beta)^2}{\beta^3}\right)^{\beta}\right]^k$$

and notice that for β small enough

$$|f_n(k)| \leq g(k) \qquad \forall k$$

since $\frac{k}{n} \leq \frac{1}{2}$ for $k \leq \frac{n}{2}$ and

$$\gamma_\beta := \left(\frac{1}{2}\right)^{1-\beta} \left(\frac{e^3(1+\beta)^2}{\beta^3}\right)^\beta < 1,$$

using that $\beta^\beta \to 1$ as $\beta \to 0$. Moreover, for each k,

$$f_n(k) \to 0,$$

as $n \to +\infty$, and

$$\sum_{k=1}^\infty g(k) \leq \frac{1}{1-\gamma_\beta} < +\infty.$$

Hence, by the dominated convergence theorem (Theorem B.4.7), combining (5.3.11) and (5.3.12), we get

$$\mathbb{P}[\exists K \subseteq L, \ |K| \leq n/2, \ |\partial_V K| \leq (1+\beta)|K|] = \sum_{k=1}^\infty f_n(k) \to 0.$$

That concludes the proof.   ∎

That concludes the proof of Claim 5.3.10.   ∎

Claim 5.3.10 implies:

Theorem 5.3.12 (Existence of expander family). *For $\alpha > 0$ small enough, there exists a $(3,\alpha)$-expander (multigraph) family.*

Proof By Claim 5.3.10, for all n large enough, there exists G_n with $\Phi_*(G_n) \geq \alpha$ for some fixed $\alpha > 0$.   ∎

Fast mixing on expander graphs As we mentioned at the beginning of this subsection, an important property of an expander graph is that random walk on such a graph mixes rapidly. We make this precise.

Claim 5.3.13 (Mixing on expanders). *Let $\{G_n\}$ be a (d,α)-expander family. Then, $t_{\mathrm{mix}}(\varepsilon) = \Theta(\log|V(G_n)|)$, where the constant depends on ε and α.*

Proof Because of the degree assumption, random walk on G_n is reversible with respect to the uniform distribution (see Example 1.1.29). So, by Theorems 5.2.14 and 5.3.5, the mixing time is upper bounded by

$$t_{\mathrm{mix}}(\varepsilon) \leq \log\left(\frac{1}{\varepsilon\pi_{\min}}\right) t_{\mathrm{rel}} \leq \log\left(\frac{|V(G_n)|}{\varepsilon}\right) 2\alpha^{-2} = O(\log|V(G_n)|).$$

By the diameter-based lower bound on the mixing time for reversible chains (Claim 5.2.25), for n large enough,

$$t_{\mathrm{mix}}(\varepsilon) \geq \frac{\Delta^2}{12\log|V(G_n)| + 4|\log\pi_{\min}|},$$

where Δ is the diameter of G_n. For a d-regular graph G_n, the diameter is at least $\log |V(G_n)|$. Indeed, by induction, the number of vertices within graph distance k of any vertex is at most d^k. For d^k to be greater than $|V(G_n)|$, we need $k \geq \log_d |V(G_n)|$. Finally,

$$t_{\text{mix}}(\varepsilon) \geq \frac{(\log_d |V(G_n)|)^2}{16 \log |V(G_n)|} = \Omega(\log |V(G_n)|).$$

That concludes the proof. ∎

5.3.4 ▷ *Ising Model: Glauber Dynamics on Complete Graphs and Expanders*

Let $G = (V, E)$ be a finite, connected graph with maximal degree $\bar{\delta}$. Define $\mathcal{X} := \{-1, +1\}^V$. Recall from Example 1.2.5 that the (ferromagnetic) Ising model on V with *inverse temperature* β is the probability distribution over *spin configurations* $\sigma \in \mathcal{X}$ given by

$$\mu_\beta(\sigma) := \frac{1}{\mathcal{Z}(\beta)} e^{-\beta \mathcal{H}(\sigma)},$$

where

$$\mathcal{H}(\sigma) := - \sum_{i \sim j} \sigma_i \sigma_j$$

is the *Hamiltonian* and

$$\mathcal{Z}(\beta) := \sum_{\sigma \in \mathcal{X}} e^{-\beta \mathcal{H}(\sigma)}$$

is the *partition function*. In this context, recall that vertices are often referred to as *sites*. The single-site Glauber dynamics of the Ising model (Definition 1.2.8) is the Markov chain on $\mathcal{X}$ which, at each time, selects a site $i \in V$ uniformly at random and updates the spin σ_i according to $\mu_\beta(\sigma)$ conditioned on agreeing with σ at all sites in $V \setminus \{i\}$. Specifically, for $\gamma \in \{-1, +1\}$, $i \in V$, and $\sigma \in \mathcal{X}$, let $\sigma^{i,\gamma}$ be the configuration σ with the state at i being set to γ. Then, letting $n = |V|$, the transition matrix of the Glauber dynamics is

$$Q_\beta(\sigma, \sigma^{i,\gamma}) := \frac{1}{n} \cdot \frac{e^{\gamma \beta S_i(\sigma)}}{e^{-\beta S_i(\sigma)} + e^{\beta S_i(\sigma)}} = \frac{1}{n} \left\{ \frac{1}{2} + \frac{1}{2} \tanh(\gamma \beta S_i(\sigma)) \right\},$$

where

$$S_i(\sigma) := \sum_{j \sim i} \sigma_j.$$

All other transitions have probability 0. Recall that this chain is irreducible and reversible with respect to μ_β. In particular, μ_β is the stationary distribution of Q_β. We showed in Claim 4.3.15 that the Glauber dynamics is fast mixing at high temperature. More precisely we proved that $t_{\text{mix}}(\varepsilon) = O(n \log n)$ when $\beta < \bar{\delta}^{-1}$. Here we prove a converse: at low temperature, graphs with good enough expansion properties produce exponentially slow mixing of the Glauber dynamics.

Curie–Weiss model

Let $G = K_n$ be the complete graph on n vertices. In this case, the Ising model is often referred to as the *Curie–Weiss model*. It is natural to scale β with n. We define $\alpha := \beta(n-1)$. Since $\bar{\delta} = n - 1$, we have that when $\alpha < 1$, $\beta = \frac{\alpha}{n-1} < \bar{\delta}^{-1}$ so $t_{\mathrm{mix}}(\varepsilon) = O(n \log n)$. In the other direction, we prove:

Claim 5.3.14 (Curie–Weiss model: slow mixing at low temperature). *For $\alpha > 1$, $t_{\mathrm{mix}}(\varepsilon) = \Omega(\exp(r(\alpha)n))$ for some function $r(\alpha) > 0$ not depending on n.*

Proof We first prove exponential mixing when α is large enough, an argument which will be useful in the generalization to expander graphs in Claim 5.3.15.

The idea of the proof is to bound the edge expansion constant and use Theorem 5.3.5. To simplify the proof, assume n is odd. We denote the edge expansion constant of the chain by $\Phi_*^{\mathcal{X}}$ to avoid confusion with that of the base graph G. Intuitively, because the spins tend to align strongly at low temperature, it takes a considerable amount of time to travel from a configuration with a majority of -1s to a configuration with a majority of $+1$s. Because the model tends to prefer agreeing spins but does not favor any particular spin, a natural place to look for a bottleneck is the set:

$$\mathcal{M} := \left\{ \sigma \in \mathcal{X} : \sum_i \sigma_i < 0 \right\},$$

where the quantity $m(\sigma) := \sum_i \sigma_i$ is called the *magnetization*. Note that the magnetization is positive if and only if a majority of spins are $+1$ and that it forms a Markov chain by itself. So the boundary of the set $\mathcal{M}$ must be crossed to travel from configurations with mostly -1 spins to configurations with mostly -1 spins.

Observe further that $\mu_\beta(\mathcal{M}) = 1/2$. The edge expansion constant is hence bounded by

$$\Phi_*^{\mathcal{X}} \leq \frac{\sum_{\sigma \in \mathcal{M}, \sigma' \notin \mathcal{M}} \mu_\beta(\sigma) Q_\beta(\sigma, \sigma')}{\mu_\beta(\mathcal{M})} = 2 \sum_{\sigma \in \mathcal{M}, \sigma' \notin \mathcal{M}} \mu_\beta(\sigma) Q_\beta(\sigma, \sigma'). \tag{5.3.13}$$

Because the Glauber dynamics changes a single spin at a time, in order for $\sigma \in \mathcal{M}$ to be adjacent to a configuration $\sigma' \notin \mathcal{M}$, it must be that

$$\sigma \in \mathcal{M}_{-1} := \{\sigma \in \mathcal{X} : m(\sigma) = -1\},$$

and that $\sigma' = \sigma^{j,+}$ for some site j such that

$$j \in \mathcal{J}_\sigma := \{j \in V : \sigma_j = -1\}.$$

Because the number of such sites is $(n + 1)/2$ on $\mathcal{M}_{-1}$, that is, $|\mathcal{J}_\sigma| = (n + 1)/2$ for all $\sigma \in \mathcal{M}_{-1}$, and the Glauber dynamics picks a site uniformly at random, it follows that for $\sigma \in \mathcal{M}_{-1}$,

$$\sum_{\sigma \in \mathcal{M}, \sigma' \notin \mathcal{M}} \mu_\beta(\sigma) Q_\beta(\sigma, \sigma') = \sum_{\sigma \in \mathcal{M}_{-1}} \mu_\beta(\sigma) \sum_{j \in \mathcal{J}_\sigma} Q_\beta(\sigma, \sigma^{j,+})$$

$$\leq \sum_{\sigma \in \mathcal{M}_{-1}} \mu_\beta(\sigma) \frac{(n+1)/2}{n} \tag{5.3.14}$$

$$= \frac{1}{2} \left(1 + \frac{1}{n} \right) \mu_\beta(\mathcal{M}_{-1}). \tag{5.3.15}$$

Thus, plugging this back in (5.3.13) gives

$$\Phi_*^{\mathcal{X}} \leq \left(1 + \frac{1}{n} \right) \mu_\beta(\mathcal{M}_{-1})$$

$$= (1 + o(1)) \sum_{\sigma \in \mathcal{M}_{-1}} \frac{e^{-\beta \mathcal{H}(\sigma)}}{\mathcal{Z}(\beta)} \tag{5.3.16}$$

$$= (1 + o(1)) \sum_{\sigma \in \mathcal{M}_{-1}} \frac{\exp\left(\frac{\alpha}{n-1} \left[\binom{|\mathcal{J}_\sigma|}{2} + \binom{|\mathcal{J}_\sigma^c|}{2} - |\mathcal{J}_\sigma| |\mathcal{J}_\sigma^c| \right] \right)}{\mathcal{Z}(\beta)}.$$

We bound the partition function $\mathcal{Z}(\beta) = \sum_{\sigma \in \mathcal{X}} e^{-\beta \mathcal{H}(\sigma)}$ with the term for the all-(-1) configuration, leading to

$$\Phi_*^{\mathcal{X}} \leq (1 + o(1)) \sum_{\sigma \in \mathcal{M}_{-1}} \frac{\exp\left(\frac{\alpha}{n-1} \left[\binom{|\mathcal{J}_\sigma|}{2} + \binom{|\mathcal{J}_\sigma^c|}{2} - |\mathcal{J}_\sigma| |\mathcal{J}_\sigma^c| \right] \right)}{\exp\left(\frac{\alpha}{n-1} \left[\binom{|\mathcal{J}_\sigma|}{2} + \binom{|\mathcal{J}_\sigma^c|}{2} + |\mathcal{J}_\sigma| |\mathcal{J}_\sigma^c| \right] \right)} \tag{5.3.17}$$

$$= (1 + o(1)) \sum_{\sigma \in \mathcal{M}_{-1}} \exp\left(-\frac{2\alpha}{n-1} |\mathcal{J}_\sigma| |\mathcal{J}_\sigma^c| \right)$$

$$= (1 + o(1)) \binom{n}{(n+1)/2} \exp\left(-\frac{2\alpha}{n-1} \left[\frac{n+1}{2} \right] \left[\frac{n-1}{2} \right] \right)$$

$$= (1 + o(1)) \sqrt{\frac{2}{\pi n}} 2^n (1 + o(1)) \exp\left(-\frac{\alpha(n+1)}{2} \right)$$

$$\leq C_\alpha \sqrt{\frac{2}{\pi n}} \exp\left(-n \left[\frac{\alpha}{2} - \log 2 \right] \right)$$

for some constant $C_\alpha > 0$ depending on α, where we used Stirling's formula (see Appendix A). Hence, by Theorems 5.2.14 and 5.3.5, for $\alpha > 2 \log 2$, there is $r(\alpha) > 0$

$$t_{\text{mix}}(\varepsilon) \geq (t_{\text{rel}} - 1) \log \left(\frac{1}{2\varepsilon} \right) \geq \exp(r(\alpha) n) \log \left(\frac{1}{2\varepsilon} \right).$$

That proves the weaker result.

We now show that $\alpha > 1$ in fact suffices. For this, we need to improve our bound on the partition function in (5.3.17). Writing

$$\mathcal{Z}(\beta) = \sum_{\sigma \in \mathcal{X}} e^{-\beta \mathcal{H}(\sigma)}$$

$$= \sum_{k=0}^{n} \binom{n}{k} \exp\left(\frac{\alpha}{n-1}\left[\binom{k}{2} + \binom{n-k}{2} - k(n-k)\right]\right)$$

$$= 2 \sum_{k=0}^{(n-1)/2} \binom{n}{k} \exp\left(\frac{\alpha}{n-1}\left[\binom{k}{2} + \binom{n-k}{2} - k(n-k)\right]\right)$$

$$=: 2 \sum_{k=0}^{(n-1)/2} \mathcal{Y}_{\alpha,k},$$

we see that the partition function is a sum of $O(n)$ exponentially large terms and is therefore dominated by the term corresponding to the largest exponent. Using Stirling's formula,

$$\log \binom{n}{k} = (1 + o(1)) n H(k/n),$$

where $H(p) = -p \log p - (1-p) \log(1-p)$ is the entropy, and therefore

$$\log \mathcal{Y}_{\alpha,k} = (1 + o(1)) n \underbrace{\left[H(k/n) + \alpha \frac{(k/n)^2 + (1-k/n)^2 - 2(k/n)(1-k/n)}{2}\right]}_{\mathcal{K}_\alpha(k/n)},$$

where, for $p \in [0, 1]$, we let

$$\mathcal{K}_\alpha(p) := H(p) + \alpha \frac{(1-2p)^2}{2}.$$

Note that the first term in $\mathcal{K}_\alpha(p)$ is increasing on $[0, 1/2]$, while the second term is decreasing on $[0, 1/2]$. In a sense, we are looking at the trade-off between the contribution from the entropy (i.e., how many ways are there to have k spins with value -1) and that from the Hamiltonian (i.e., how much such a configuration is favored). We seek to maximize $\mathcal{K}_\alpha(p)$ to determine the leading term in the partition function.

By a straightforward computation,

$$\mathcal{K}'_\alpha(p) = \log\left(\frac{1-p}{p}\right) - 2\alpha(1-2p)$$

and

$$\mathcal{K}''_\alpha(p) = -\frac{1}{p(1-p)} + 4\alpha.$$

Observe first that when $\alpha < 1$ (i.e., at high temperature), $\mathcal{K}'_\alpha(1/2) = 0$ and $\mathcal{K}''_\alpha(p) < 0$ for all $p \in [0, 1]$ since $p(1-p) \leq 1/4$. Hence, in that case, $\mathcal{K}_\alpha$ is maximized at $p = 1/2$.

In our case of interest, on the other hand, that is, when $\alpha > 1$, $\mathcal{K}''_\alpha(p) > 0$ in an interval around $1/2$ so there is $p_* < 1/2$ with $\mathcal{K}_\alpha(p_*) > \mathcal{K}_\alpha(1/2) = 1$. So the distribution significantly favors "unbalanced" configurations and crossing $\mathcal{M}_{-1}$ becomes a bottleneck for the Glauber dynamics. Going back to (5.3.17) and bounding $\mathcal{Z}(\beta) \geq 2\mathcal{Y}_{\alpha,\lfloor p_* n \rfloor}$, we get

$$\Phi_*^{\mathcal{X}} = O\left(\exp(-n[\mathcal{K}_\alpha(p_*) - \mathcal{K}_\alpha(1/2)])\right).$$

Applying Theorems 5.2.14 and 5.3.5 concludes the proof.   ∎

Expander graphs

In the proof of Claim 5.3.14, the bottleneck slowing down the chain arises as a result of the fact that, when $m(\sigma) = -1$, there is a large number of edges in the base graph K_n connecting $\mathcal{J}_\sigma$ and $\mathcal{J}_\sigma^c$. That produces a low probability for such configurations under the ferromagnetic Ising model, where agreeing spins are favored. The same argument easily extends to expander graphs. In words, we prove something that – at first – may seem a bit counter-intuitive: good expansion properties in the base graph produces a bottleneck in the Glauber dynamics at low temperature.

Claim 5.3.15 (Ising model on expander graphs: slow mixing of the Glauber dynamics). *Let $\{G_n\}_n$ be a (d, γ)-expander family. For large enough inverse temperature $\beta > 0$, the Glauber dynamics of the Ising model on G_n satisfies $t_{\mathrm{mix}}(\varepsilon) = \Omega(\exp(r(\beta)|V(G_n)|))$ for some function $r(\beta) > 0$ not depending on n.*

Proof Let μ_β be the probability distribution over spin configurations under the Ising model over $G_n = (V, E)$ with inverse temperature β. Let Q_β be the transition matrix of the Glauber dynamics. For not necessarily disjoint subsets of vertices $W_0, W_1 \subseteq V$ in the base graph G_n, let

$$E(W_0, W_1) := \{\{u, v\} : u \in W_0, v \in W_1, \{u, v\} \in E\}$$

be the set of edges with one endpoint in W_0 and one endpoint in W_1. Let $N = |V(G_n)|$ and assume it is odd for simplicity. We use the notation in the proof of Claim 5.3.14. Following the argument in that proof, we observe that (5.3.15) and (5.3.16) still hold. Thus,

$$\Phi_*^{\mathcal{X}} \leq (1 + o(1)) \sum_{\sigma \in \mathcal{M}_{-1}} \frac{\exp\left(\beta\left[|E(\mathcal{J}_\sigma, \mathcal{J}_\sigma)| + |E(\mathcal{J}_\sigma^c, \mathcal{J}_\sigma^c)| - |E(\mathcal{J}_\sigma, \mathcal{J}_\sigma^c)|\right]\right)}{\mathcal{Z}(\beta)}.$$

As we did in (5.3.17), we bound the partition function $\mathcal{Z}(\beta) = \sum_{\sigma \in \mathcal{X}} e^{-\beta \mathcal{H}(\sigma)}$ with the term for the all-(-1) configuration, leading to

$$\Phi_*^{\mathcal{X}} \leq (1 + o(1)) \sum_{\sigma \in \mathcal{M}_{-1}} \frac{\exp\left(\beta\left[|E(\mathcal{J}_\sigma, \mathcal{J}_\sigma)| + |E(\mathcal{J}_\sigma^c, \mathcal{J}_\sigma^c)| - |E(\mathcal{J}_\sigma, \mathcal{J}_\sigma^c)|\right]\right)}{\exp\left(\beta\left[|E(\mathcal{J}_\sigma, \mathcal{J}_\sigma)| + |E(\mathcal{J}_\sigma^c, \mathcal{J}_\sigma^c)| + |E(\mathcal{J}_\sigma, \mathcal{J}_\sigma^c)|\right]\right)}$$

$$= (1 + o(1)) \sum_{\sigma \in \mathcal{M}_{-1}} \exp\left(-2\beta|E(\mathcal{J}_\sigma, \mathcal{J}_\sigma^c)|\right)$$

$$= (1 + o(1)) \sum_{\sigma \in \mathcal{M}_{-1}} \exp\left(-2\beta|\partial_E \mathcal{J}_\sigma^c|\right)$$

$$\leq (1 + o(1)) \binom{N}{(N+1)/2} \exp\left(-2\beta\gamma d|\mathcal{J}_\sigma^c|\right)$$

$$= (1 + o(1)) \sqrt{\frac{2}{\pi N}} 2^N (1 + o(1)) \exp\left(-\beta\gamma d(N - 1)\right)$$

$$\leq C_{\beta,\gamma,d} \sqrt{\frac{2}{\pi N}} \exp\left(-N\left[\beta\gamma d - \log 2\right]\right),$$

for some constant $C_{\beta,\gamma,d} > 0$. We used the definition of an expander family (Definition 5.3.9) on the fourth line above. Taking $\beta > 0$ large enough gives the result. ∎

Spectral Methods

5.3.5 Congestion Ratio

Recall from (5.3.2) that an inequality of the type

$$\text{Var}_\pi[f] \leq C\mathscr{D}(f) \tag{5.3.18}$$

holding for all f is known as a Poincaré inequality. By Theorem 5.3.4, it implies the lower bound $\gamma \geq C^{-1}$ on the spectral gap $\gamma = 1 - \lambda_2$. In this section, we derive such an inequality using a formal measure of "congestion" in the network.

Let $\mathcal{N} = (G, c)$ be a finite, connected network with $G = (V, E)$. We assume that $c(x, y) = \pi(x)P(x, y)$ and therefore $c(x) = \sum_{y \sim x} c(x, y) = \pi(x)$, where π is the stationary distribution of random walk on $\mathcal{N}$. To state the bound, it will be convenient to work with directed edges – this time in both directions. Let $\widetilde{E}$ contain all edges from E with both orientations, that is, for each $e \in \{x, y\}$, $\widetilde{E}$ includes (x, y) and (y, x) with associated weight $c(x, y) = c(y, x) = c(e) > 0$. For a function $f \in \ell^2(V, \pi)$ and an edge $\vec{e} = (x, y) \in \widetilde{E}$, we define as before

$$\nabla f(\vec{e}) = f(y) - f(x).$$

With this notation, we can rewrite the Dirichlet energy as

$$\mathscr{D}(f) = \frac{1}{2} \sum_{x,y} c(x, y)[f(x) - f(y)]^2 = \frac{1}{2} \sum_{\vec{e} \in \widetilde{E}} c(\vec{e})[\nabla f(\vec{e})]^2. \tag{5.3.19}$$

For each pair of vertices x, y, let $\nu_{x,y}$ be a directed path between x and y in the digraph $\widetilde{G} = (V, \widetilde{E})$, as a collection of directed edges. Let $|\nu_{x,y}|$ be the number of edges in the path. CONGESTION RATIO The *congestion ratio* associated with the paths $\boldsymbol{\nu} = \{\nu_{x,y}\}_{x,y \in V}$ is

$$C_{\boldsymbol{\nu}} = \max_{\vec{e} \in \widetilde{E}} \frac{1}{c(\vec{e})} \sum_{x,y : \vec{e} \in \nu_{x,y}} |\nu_{x,y}| \pi(x)\pi(y).$$

CANONICAL PATHS Note that $C_{\boldsymbol{\nu}}$ tends to be large when many selected paths, called *canonical paths*, go through the same "congested" edge. To get a good bound in the theorem below, one must choose canonical paths that are well "spread out."

Theorem 5.3.16 (Canonical paths method). *For any choice of paths $\boldsymbol{\nu}$ as above, we have the following bound on the spectral gap*

$$\gamma \geq \frac{1}{C_{\boldsymbol{\nu}}}.$$

Proof We establish a Poincaré inequality (5.3.18) with $C := C_{\boldsymbol{\nu}}$. The proof strategy is to start with the variance and manipulate it to bring out canonical paths.

For any $f \in \ell^2(V, \pi)$, it can be checked by expanding that

$$\text{Var}_\pi[f] = \frac{1}{2} \sum_{x,y} \pi(x)\pi(y)(f(x) - f(y))^2. \tag{5.3.20}$$

To bring out terms similar to those in (5.3.19), we write $f(x) - f(y)$ as a telescoping sum over the canonical path between x and y. That is, letting $\vec{e}_1, \ldots, \vec{e}_{|\nu_{x,y}|}$ be the edges in $\nu_{x,y}$, observe that

$$f(y) - f(x) = \sum_{i=1}^{|\nu_{x,y}|} \nabla f(\vec{e}_i).$$

By Cauchy–Schwarz (Theorem B.4.8),

$$(f(y) - f(x))^2 = \left(\sum_{i=1}^{|v_{x,y}|} \nabla f(\vec{e}_i) \right)^2$$

$$\leq \left(\sum_{i=1}^{|v_{x,y}|} 1^2 \right) \left(\sum_{i=1}^{|v_{x,y}|} \nabla f(\vec{e}_i)^2 \right)$$

$$= |v_{x,y}| \sum_{\vec{e} \in v_{x,y}} \nabla f(\vec{e})^2.$$

Combining the last display with (5.3.20) and rearranging, we arrive at

$$\operatorname{Var}_\pi[f] \leq \frac{1}{2} \sum_{x,y} \pi(x)\pi(y)|v_{x,y}| \sum_{\vec{e} \in v_{x,y}} \nabla f(\vec{e})^2$$

$$= \frac{1}{2} \sum_{\vec{e} \in \widetilde{E}} \nabla f(\vec{e})^2 \sum_{x,y: \vec{e} \in v_{x,y}} |v_{x,y}| \pi(x)\pi(y)$$

$$= \frac{1}{2} \sum_{\vec{e} \in \widetilde{E}} c(\vec{e}) \nabla f(\vec{e})^2 \left(\frac{1}{c(\vec{e})} \sum_{x,y: \vec{e} \in v_{x,y}} |v_{x,y}| \pi(x)\pi(y) \right)$$

$$\leq C_v \, \mathscr{D}(f).$$

That concludes the proof. ∎

We give an example next.

Example 5.3.17 (Random walk inside a box). Consider random walk on the following d-dimensional box with sides of length n:

$$V = [n]^d = \{1, \ldots, n\}^d,$$

$$E = \{x, y \in [n]^d : \|x - y\|_1 = 1\},$$

$$P(x, y) = \frac{1}{|\{z : z \sim x\}|} \quad \forall x, y \in [n]^d, \, x \sim y,$$

$$\pi(x) = \frac{|\{z : z \sim x\}|}{2|E|},$$

and

$$c(e) = \frac{1}{2|E|}, \forall e \in E.$$

We define $\widetilde{E}$ as before.

We use Theorem 5.3.16 to bound the spectral gap. For $x = (x_1, \ldots, x_d), y = (y_1, \ldots, y_d) \in [n]^d$, we construct $v_{x,y}$ by matching each coordinate in turn. That is, for two vertices w,

$z \in [n]^d$ with a single distinct coordinate, let $[w, z]$ be the directed path from w to z in $\widetilde{G} = (V, \widetilde{E})$ corresponding to a straight line (or the empty path if $w = z$). Then,

$$v_{x,y} = \bigcup_{i=1}^{d} [(y_1, \ldots, y_{i-1}, x_i, x_{i+1}, \ldots, x_d), (y_1, \ldots, y_{i-1}, y_i, x_{i+1}, \ldots, x_d)]. \quad (5.3.21)$$

It remains to bound

$$C_v = \max_{\vec{e} \in \widetilde{E}} \frac{1}{c(\vec{e})} \sum_{x,y: \vec{e} \in v_{x,y}} |v_{x,y}| \pi(x) \pi(y)$$

from above.

Each term in the union defining $v_{x,y}$ contains at most n edges, and therefore

$$|v_{x,y}| \leq dn, \forall x, y.$$

Not attempting to get the best constant factors, the edge weights (i.e., conductances) satisfy

$$c(\vec{e}) = \frac{1}{2|E|} \geq \frac{1}{2 \cdot 2dn^d} = \frac{1}{4dn^d}$$

for all $\vec{e}$, since there are n^d vertices and each has at most $2d$ incident edges. Likewise, for any x,

$$\pi(x) = \frac{|\{z: z \sim x\}|}{2|E|} \leq \frac{2d}{2 \cdot (dn^d)/2} = \frac{2}{n^d},$$

where we divided by two in the denominator to account for the double-counting of edges. Hence we get

$$C_v \leq \max_{\vec{e} \in \widetilde{E}} \frac{1}{1/(4dn^d)} \sum_{x,y: \vec{e} \in v_{x,y}} (dn)(2/n^d)(2/n^d)$$

$$= \frac{16d^2}{n^{d-1}} \max_{\vec{e} \in \widetilde{E}} \left| \{x, y: \vec{e} \in v_{x,y}\} \right|.$$

To bound the cardinality of the set on the last line, we note that any edge $\vec{e} \in \widetilde{E}$ is of the form

$$\vec{e} = ((z_1, \ldots, z_{i-1}, z_i, z_{i+1}, \ldots, z_d), (z_1, \ldots, z_{i-1}, z_i \pm 1, z_{i+1}, \ldots, z_d)),$$

that is, the endvertices differ by exactly one unit along a single coordinate. By the construction of the path $v_{x,y}$ in (5.3.21), if $\vec{e} \in v_{x,y}$, then it must lie in the subpath

$$((z_1, \ldots, z_{i-1}, z_i, z_{i+1}, \ldots, z_d), (z_1, \ldots, z_{i-1}, z_i \pm 1, z_{i+1}, \ldots, z_d))$$

$$\in [(y_1, \ldots, y_{i-1}, x_i, x_{i+1}, \ldots, x_d), (y_1, \ldots, y_{i-1}, y_i, x_{i+1}, \ldots, x_d)].$$

But that imposes constraints on x and y. Namely, we must have

$$y_1 = z_1, \ldots, y_{i-1} = z_{i-1}, x_{i+1} = z_{i+1}, \ldots, x_d = z_d.$$

The remaining components of x and y (of which there are i of the former and $d - (i - 1)$ of the latter) have at most n possible values (although not all of them are allowed), so that

$$\left| \{x, y: \vec{e} \in v_{x,y}\} \right| \leq n^i n^{d-(i-1)} = n^{d+1}.$$

This upper bound is valid for any $\vec{e}$.

Putting everything together, we get the bound

$$C_v \leq \frac{16d^2}{n^{d-1}} n^{d+1} = 16d^2 n^2,$$

so that

$$\gamma \geq \frac{1}{16d^2 n^2}.$$

Observe that this lower bound on the spectral gap depends only mildly (i.e., polynomially) in the dimension. ◀

One advantage of the canonical paths method is that it is somewhat robust to modifying the underlying network through comparison arguments. See Exercise 5.17 for a simple illustration.

Exercises

Exercise 5.1 Let A be an $n \times n$ symmetric random matrix. We assume that the entries on and above the diagonal, $A_{i,j}, i \leq j$, are independent and uniform in $\{+1, -1\}$ (and each entry below the diagonal is equal to the corresponding entry above). Use Talagrand's inequality (Theorem 3.2.32) to prove concentration of the largest eigenvalue of A around its mean (which you do not need to compute).

Exercise 5.2 Let $G = (V, E, w)$ be a network.

(i) Prove formula (5.1.3) for the Laplacian quadratic form. (Hint: For an orientation $G^\sigma = (V, E^\sigma)$ of G (that is, give an arbitrary direction to each edge to turn it into a digraph), consider the matrix $B^\sigma \in \mathbb{R}^{n \times m}$ where the column corresponding to arc (i, j) has $-\sqrt{w_{ij}}$ in row i and $\sqrt{w_{ij}}$ in row j, and every other entry is 0.)
(ii) Show that the network Laplacian is positive semi-definite.

Exercise 5.3 Let $G = (V, E, w)$ be a weighted graph with normalized Laplacian $\mathcal{L}$. Show that

$$\mathbf{x}^T \mathcal{L} \mathbf{x} = \sum_{\{i,j\} \in E} w_{ij} \left(\frac{x_i}{\sqrt{\delta(i)}} - \frac{x_j}{\sqrt{\delta(j)}} \right)^2$$

for $\mathbf{x} = (x_1, \ldots, x_n) \in \mathbb{R}^n$.

Exercise 5.4 (2-norm). Prove that

$$\sup_{\mathbf{x} \in \mathbb{S}^{n-1}} \|A\mathbf{x}\|_2 = \sup_{\substack{\mathbf{x} \in \mathbb{S}^{n-1} \\ \mathbf{y} \in \mathbb{S}^{m-1}}} \langle A\mathbf{x}, \mathbf{y} \rangle.$$

(Hint: Use Cauchy–Schwarz (Theorem B.4.8) for one direction, and set $\mathbf{y} = A\mathbf{x}/\|A\mathbf{x}\|_2$ for the other one.)

Exercise 5.5 (Spectral radius of a symmetric matrix). Let $A \in \mathbb{R}^{n \times n}$ be a symmetric matrix. The set $\sigma(A)$ of eigenvalues of A is called the *spectrum* of A and SPECTRUM

$$\rho(A) = \max\{|\lambda| : \lambda \in \sigma(A)\}$$

is its *spectral radius*. Prove that

$$\rho(A) = \|A\|_2,$$

where we recall that

$$\|A\|_2 = \max_{0 \neq \mathbf{x} \in \mathbb{R}^m} \frac{\|A\mathbf{x}\|}{\|\mathbf{x}\|}.$$

Exercise 5.6 (Community recovery in sparse networks). Assume without proof the following theorem.

Theorem 5.3.18 (Remark 3.13 of [BH16]). *Consider a symmetric matrix* $\mathbf{Z} = [Z_{i,j}] \in \mathbb{R}^{n \times n}$ *whose entries are independent and obey,* $\mathbb{E}Z_{i,j} = 0$ *and* $Z_{i,j} \leq B$ $\forall 1 \leq i, j \leq n$, $\mathbb{E}Z_{i,j}^2 \leq \sigma^2$, *then with high probability we have* $\|\mathbf{Z}\| \lesssim \sigma\sqrt{n} + B\sqrt{\log n}$.

Let $(X, G) \sim \text{SBM}_{n,p_n,q_n}$. Show that, under the conditions $p_n \gtrsim \frac{\log n}{n}$ and $\sqrt{\frac{p_n}{n}} = o(p_n - q_n)$, spectral clustering achieves almost exact recovery.

Exercise 5.7 (Parseval's identity). Prove Parseval's identity (i.e., (5.2.1)) in the finite-dimensional case.

Exercise 5.8 (Dirichlet kernel). Prove that for $\theta \neq 0$,

$$1 + 2\sum_{k=1}^{n} \cos k\theta = \frac{\sin((n + 1/2)\theta)}{\sin(\theta/2)}.$$

(Hint: Switch to the complex representation and use the formula for a geometric series.)

Exercise 5.9 (Eigenvalues and periodicity). Let P be a finite irreducible transition matrix reversible with respect to π over V. Show that if P has a non-zero eigenfunction f with eigenvalue -1, then P is not aperiodic. (Hint: Look at x achieving $\|f\|_\infty$.)

Exercise 5.10 (Mixing time: necessary condition for cutoff). Consider a sequence of Markov chains indexed by $n = 1, 2, \ldots$. Assume that each chain has a finite state space and is irreducible, aperiodic, and reversible. Let $t_{\text{mix}}^{(n)}(\varepsilon)$ and $t_{\text{rel}}^{(n)}$ be, respectively, the mixing time and relaxation time of the nth chain. The sequence is said to have pre-cutoff if

$$\sup_{0 < \varepsilon < 1/2} \limsup_{n \to +\infty} \frac{t_{\text{mix}}^{(n)}(\varepsilon)}{t_{\text{mix}}^{(n)}(1 - \varepsilon)} < +\infty.$$

Show that if for some $\varepsilon > 0$,

$$\sup_{n \geq 1} \frac{t_{\text{mix}}^{(n)}(\varepsilon)}{t_{\text{rel}}^{(n)}} < +\infty,$$

then there is no pre-cutoff. In particular, there is no cutoff, as defined in Remark 4.3.8.

Exercise 5.11 (Relaxation time and variance). Let P be a finite irreducible transition matrix reversible with respect to π over V. Define

$$\text{Var}_\pi[g] = \sum_{x \in V} \pi(x)[g(x) - \pi g]^2.$$

Let γ_* be the absolute spectral gap of P. Show that

$$\text{Var}_\pi[P^t f] \leq (1 - \gamma_*)^{2t}\text{Var}_\pi[f].$$

Exercise 5.12 (Lumping). Let (X_t) be a Markov chain on a finite state space V with transition P. Suppose there is an equivalence relation $\sim$ on V with equivalence classes $V^\sharp$, denoting by $[x]$ the equivalence class of x, such that $[X_t]$ is a Markov chain with transition matrix $P^\sharp([x],[y]) = P(x,[y])$.

  (i) Let $f: V \to \mathbb{R}$ be an eigenfunction of P with eigenvalue λ and assume that f is constant on each equivalence class. Prove that $f^\sharp([x]) := f(x)$ defines an eigenfunction of $P^\sharp$. What is its eigenvalue?

 (ii) Suppose $g: V^\sharp \to \mathbb{R}$ is eigenfunction of $P^\sharp$ with eigenvalue λ. Prove that $g^\flat: V \to \mathbb{R}$ defined by $g^\flat(x) := g([x])$ is eigenfunction of P. What is its eigenvalue?

Exercise 5.13 (Random walk on path with reflecting boundaries). Let n be an even positive integer. Let (X_t) be simple random walk on the path $\{1,\dots,n\}$ with reflecting boundaries, that is, the transition matrix P is defined by $P(x,x-1) = P(x,x+1) = 1/2$ for $x \in \{2,\dots,n-1\}$, and $P(1,2) = P(n,n-1) = 1$. Use Exercise 5.12 to compute the eigenfunctions of P. (Hint: Use the results of Section 5.2.2.)

Exercise 5.14 (Product chain). For $j = 1,\dots,d$, let P_j be a transition matrix on the finite state space V_j reversible with respect to the stationary distribution π_j. Let $w = (w_j)_{j \in [d]}$ be a probability distribution over $[d]$. Consider the following Markov chain (X_t) on $V := V_1 \times \dots \times V_d$: at each step, pick j according to w, then take one step along the jth coordinate according to P_j.

  (i) Compute the transition matrix P and stationary distribution π of the chain (X_t). Show that P is reversible with respect to π.

 (ii) Construct an orthonormal basis of $\ell^2(V,\pi)$ made of eigenfunctions of P in terms of eigenfunctions of the P_js. What are the corresponding eigenvalues?

(iii) Compute the spectral gap γ of P in terms of the spectral gaps γ_j of the P_js.

Exercise 5.15 (Hypercube revisited). Use Exercise 5.14 to recover Lemma 5.2.17.

Exercise 5.16 (Norm and Rayleigh quotient). Let P be irreducible and reversible with respect to $\pi > 0$.

  (i) Prove the polarization identity

$$\langle Pf, g\rangle_\pi = \frac{1}{4}\left[\langle P(f+g),f+g\rangle_\pi - \langle P(f-g),f-g\rangle_\pi\right].$$

 (ii) Show that

$$\|P\|_\pi = \sup\left\{\frac{\langle f, Pf\rangle_\pi}{\langle f,f\rangle_\pi} : f \in \ell_0(V), f \neq 0\right\}.$$

Exercise 5.17 (Random walk on a box with holes). Consider the random walk in Example 5.3.17 with $d = 2$. Suppose we remove from the network an arbitrary collection of horizontal edges at even heights. Use the canonical paths method to derive a lower bound on the spectral gap of the form $\gamma \geq 1/(Cn^2)$. (Hint: Modify the argument in Example 5.3.17 and relate the congestion ratio before and after the removal.)

Bibliographic Remarks

Section 5.1 General references on the spectral theorem, the Courant–Fischer and perturbation results include the classics [HJ13, Ste98]. Much more on spectral graph theory can be gleaned from [Chu97, Nic18]. Section 5.1.4 is based largely on [Abb18], which gives a broad survey of theoretical results for community recovery, and [Ver18, section 4.5] as well as on scribe notes by Joowon Lee, Aidan Howells, Govind Gopakumar, and Shuyao Li for "MATH 888: Topics in Mathematical Data Science" taught at the University of Wisconsin–Madison in Fall 2021.

Section 5.2 For a great introduction to Hilbert space theory and its applications (including to the Dirichlet problem), consult [SS05, chapters 4,5]. Section 5.2.1 borrows from [LP17, chapter 12]. A representation-theoretic approach to computing eigenvalues and eigenfunctions, greatly generalizing the calculations in 5.2.2, is presented in [Dia88]. The presentation in Section 5.2.3 follows [KP, section 3] and [LP16, section 13.3]. The Varopoulos–Carne bound is due to Carne [Car85] and Varopoulos [Var85]. For a probabilistic approach to the Varopoulos–Carne bound, see Peyre's proof [Pey08]. The application to mixing times is from [LP16]. There are many textbooks dedicated to Markov chain Monte Carlo (MCMC) and its uses in data analysis, for example, [RC04, GL06, GCS$^+$14]. See also [Dia09]. A good overview of the techniques developed in the statistics literature to bound the rate of convergence of MCMC methods (a combination of coupling and Lyapounov arguments) is [JH01]. A deeper treatment of these ideas is developed in [MT09]. A formal definition of the spectral radius and its relationship to the operator norm can be found, for instance, in [Rud73, Part III].

Section 5.3 This section follows partly the presentation in [LP16, section 6.4], [LPW06, section 13.6], and [Spi12]. Various proofs of the isoperimetric inequality can be found in [SS03, SS05]. Theorem 5.3.5 is due to [SJ89, LS88]. The approach to its proof used here is due to Luca Trevisan. The original Cheeger inequality was proved, in the context of manifolds, in [Che70]. For a fascinating introduction to expander graphs and their applications, see [HLW06]. A detailed account of the Curie–Weiss model can be found in [FV18]. Section 5.3.5 is based partly on [Ber14, sections 3 and 4]. The method of canonical paths, and some related comparison techniques, were developed in [JS89, DS91, DSC93b, DSC93a]. For more advanced functional techniques for bounding the mixing time, see, for example, [MT06].

6

Branching Processes

Branching processes, which are the focus of this chapter, arise naturally in the study of sto-chastic processes on trees and locally tree-like graphs. Similarly to martingales, finding a hidden (or not-so-hidden) branching process within a probabilistic model can lead to useful bounds and insights into asymptotic behavior. After a review of the basic *extinction theory* of branching processes in Section 6.1 and of a fruitful *random-walk perspective* in Section 6.2, we give a couple of examples of applications in discrete probability in Section 6.3. In par-ticular, we analyze the height of a binary search tree, a standard data structure in computer science. We also give an introduction to phylogenetics, where a "multitype" variant of the Galton–Watson branching process plays an important role; we use the techniques derived in this chapter to establish a phase transition in the reconstruction of ancestral molecular sequences. We end this chapter in Section 6.4 with a detailed look into the *phase transition of the Erdős–Rényi graph model*. The random-walk perspective mentioned above allows one to analyze the "exploration" of a largest connected component, leading to information about the "evolution" of its size as edge density increases. Tools from all chapters come to bear on this final, marquee application.

6.1 Background

We begin with a review of the theory of Galton–Watson branching processes, a standard stochastic model for population growth. In particular, we discuss extinction theory. We also briefly introduce a multitype variant, where branching process and Markov chain aspects interact to produce interesting new behavior.

6.1.1 Basic Definitions

Recall the definition of a Galton–Watson process.

Definition 6.1.1 *A* Galton–Watson branching process *is a Markov chain of the following form:*

- *Let $Z_0 := 1$.*
- *Let $X(i,t)$, $i \geq 1$, $t \geq 1$, be an array of i.i.d. $\mathbb{Z}_+$-valued random variables with finite mean $m = \mathbb{E}[X(1,1)] < +\infty$, and define inductively*

$$Z_t := \sum_{1 \leq i \leq Z_{t-1}} X(i,t).$$

We denote by $\{p_k\}_{k \geq 0}$ the law of $X(1, 1)$. We also let $f(s) := \mathbb{E}[s^{X(1,1)}]$ be the corresponding probability generating function. To avoid trivialities we assume $\mathbb{P}[X(1, 1) = i] < 1$ for all $i \geq 0$. We further assume that $p_0 > 0$.

In words, Z_t models the size of a population at time (or generation) t. The random variable $X(i, t)$ corresponds to the number of offspring of the ith individual (if there is one) in generation $t - 1$. Generation t is formed of all offspring of the individuals in generation $t - 1$.

By tracking genealogical relationships, that is, who is whose child, we obtain a tree T rooted at the single individual in generation 0 with a vertex for each individual in the progeny and an edge for each parent–child relationship. We refer to T as a *Galton–Watson tree*.

A basic observation about Galton–Watson processes is that their growth (or decay) is exponential in t.

Lemma 6.1.2 (Exponential growth I). *Let*

$$W_t := m^{-t}Z_t. \tag{6.1.1}$$

Then (W_t) is a non-negative martingale with respect to the filtration

$$\mathcal{F}_t = \sigma(Z_0, \ldots, Z_t).$$

In particular, $\mathbb{E}[Z_t] = m^t$.

Proof   We use Lemma B.6.17. Observe that on $\{Z_{t-1} = k\}$,

$$\mathbb{E}[Z_t \mid \mathcal{F}_{t-1}] = \mathbb{E}\left[\sum_{1 \leq j \leq k} X(j, t) \,\middle|\, \mathcal{F}_{t-1} \right] = mk = mZ_{t-1}.$$

This is true for all k. Rearranging shows that (W_t) is a martingale. For the second claim, note that $\mathbb{E}[W_t] = \mathbb{E}[W_0] = 1$.   ∎

In fact, the martingale convergence theorem (Theorem 3.1.47) gives the following.

Lemma 6.1.3 (Exponential growth II). *We have $W_t \to W_\infty < +\infty$ almost surely for some non-negative random variable $W_\infty \in \sigma(\cup_t \mathcal{F}_t)$ with $\mathbb{E}[W_\infty] \leq 1$.*

Proof   This follows immediately from the martingale convergence theorem for non-negative martingales (Corollary 3.1.48).   ∎

6.1.2 Extinction

Observe that 0 is a fixed point of the process. The event

$$\{Z_t \to 0\} = \{\exists t \colon Z_t = 0\}$$

is called *extinction*. Establishing when extinction occurs is a central question in branching process theory. We let η be the probability of extinction. Recall that, to avoid trivialities, we assume $p_0 > 0$ and $p_1 < 1$. Here is a first observation about extinction.

Lemma 6.1.4 *Almost surely either $Z_t \to 0$ or $Z_t \to +\infty$.*

Proof The process (Z_t) is integer-valued and 0 is the only fixed point of the process under the assumption that $p_1 < 1$. From any state k, the probability of never coming back to $k > 0$ is at least $p_0^k > 0$, so every state $k > 0$ is transient. So the only possibilities left are $Z_t \to 0$ and $Z_t \to +\infty$, and the claim follows. ∎

In the critical case, that immediately implies almost sure extinction.

Theorem 6.1.5 (Extinction: critical case). *Assume $m = 1$. Then $Z_t \to 0$ almost surely, that is, $\eta = 1$.*

Proof When $m = 1$, (Z_t) itself is a martingale. Hence (Z_t) must converge to 0 by Lemma 6.1.3. ∎

We address the general case using probability generating functions. Let $f_t(s) = \mathbb{E}[s^{Z_t}]$, where by convention we set $f_t(0) := \mathbb{P}[Z_t = 0]$. Note that, by monotonicity,

$$\eta = \mathbb{P}[\exists t \geq 0 \colon Z_t = 0] = \lim_{t \to +\infty} \mathbb{P}[Z_t = 0] = \lim_{t \to +\infty} f_t(0). \tag{6.1.2}$$

Moreover, by the tower property (Lemma B.6.16) and the Markov property (Theorem 1.1.18), f_t has a natural recursive form

$$\begin{aligned}
f_t(s) &= \mathbb{E}[s^{Z_t}] \\
&= \mathbb{E}[\mathbb{E}[s^{Z_t} \mid \mathcal{F}_{t-1}]] \\
&= \mathbb{E}[f(s)^{Z_{t-1}}] \\
&= f_{t-1}(f(s)) = \cdots = f^{(t)}(s), \tag{6.1.3}
\end{aligned}$$

where $f^{(t)}$ is the tth iterate of f. The subcritical case below has an easier proof (see Exercise 6.1).

Theorem 6.1.6 (Extinction: subcriticial and supercritical cases). *The probability of extinction η is given by the smallest fixed point of f in $[0, 1]$. Moreover:*

(i) (Subcritical regime) If $m < 1$, then $\eta = 1$.
(ii) (Supercritical regime) If $m > 1$, then $\eta < 1$.

Proof The case $p_0 + p_1 = 1$ is straightforward: the process dies almost surely after a geometrically distributed time. So we assume $p_0 + p_1 < 1$ for the rest of the proof.

We first summarize without proof some properties of f which follow from standard power series facts.

Lemma 6.1.7 *On $[0, 1]$, the function f satisfies:*

(i) $f(0) = p_0, f(1) = 1$;
(ii) f is infinitely differentiable on $[0, 1)$;
(iii) f is strictly convex and increasing; and
(iv) $\lim_{s \uparrow 1} f'(s) = m < +\infty$.

We first characterize the fixed points of f. See Figure 6.1 for an illustration.

Lemma 6.1.8 *We have the following.*

(i) If $m > 1$, then f has a unique fixed point $\eta_0 \in [0, 1)$.
(ii) If $m < 1$, then $f(t) > t$ for $t \in [0, 1)$. Let $\eta_0 := 1$ in that case.

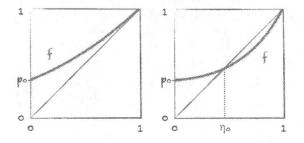

Figure 6.1 Fixed points of f in subcritical (a) and supercritical (b) cases.

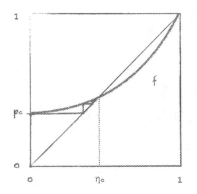

Figure 6.2 Convergence of iterates to a fixed point.

Proof Assume $m > 1$. Since $f'(1) = m > 1$, there is $\delta > 0$ such that $f(1 - \delta) < 1 - \delta$. On the other hand, $f(0) = p_0 > 0$ so by continuity of f there must be a fixed point in $(0, 1 - \delta)$. Moreover, by strict convexity and the fact that $f(1) = 1$, if $x \in (0, 1)$ is a fixed point, then $f(y) < y$ for $y \in (x, 1)$, proving uniqueness.

The second part follows by strict convexity and monotonicity.   ∎

It remains to prove convergence of the iterates to the appropriate fixed point. See Figure 6.2 for an illustration.

Lemma 6.1.9 *We have the following.*

  (i) *If $x \in [0, \eta_0)$, then $f^{(t)}(x) \uparrow \eta_0$.*
  (ii) *If $x \in (\eta_0, 1)$, then $f^{(t)}(x) \downarrow \eta_0$.*

Proof We only prove (i). The argument for (ii) is similar. By monotonicity, for $x \in [0, \eta_0)$, we have $x < f(x) < f(\eta_0) = \eta_0$. Iterating,

$$x < f^{(1)}(x) < \cdots < f^{(t)}(x) < f^{(t)}(\eta_0) = \eta_0.$$

So $f^{(t)}(x) \uparrow L \le \eta_0$ as $t \to \infty$. By continuity of f, we can take the limit $t \to \infty$ inside of f on the right-hand side of the equality

$$f^{(t)}(x) = f(f^{(t-1)}(x))$$

to get $L = f(L)$. So by definition of η_0 we must have $L = \eta_0$.   ∎

The result then follows from the above lemmas together with (6.1.2) and (6.1.3). ■

Example 6.1.10 (Poisson branching process). Consider the offspring distribution $X(1, 1) \sim$ Poi(λ) with mean $\lambda > 0$. We refer to this case as the *Poisson branching process*. Then,

$$f(s) = \mathbb{E}[s^{X(1,1)}] = \sum_{i \geq 0} e^{-\lambda} \frac{\lambda^i}{i!} s^i = e^{\lambda(s-1)}.$$

So the process goes extinct with probability 1 when $\lambda \leq 1$. For $\lambda > 1$, the probability of extinction η is the smallest solution in $[0, 1]$ to the equation

$$e^{-\lambda(1-x)} = x.$$

The survival probability $\zeta_\lambda := 1 - \eta$ satisfies $1 - e^{-\lambda \zeta_\lambda} = \zeta_\lambda$. ◄

We can use these extinction results to obtain more information on the limit in Lemma 6.1.3. Recall the definition of (W_t) in (6.1.1). Of course, conditioned on extinction, $W_\infty = 0$ almost surely. On the other hand:

Lemma 6.1.11 (Exponential growth III). *Conditioned on non-extinction, either $W_\infty = 0$ almost surely or $W_\infty > 0$ almost surely.*

As a result, $\mathbb{P}[W_\infty = 0] \in \{\eta, 1\}$.

Proof of Lemma 6.1.11 A property of rooted trees is said to be *inherited* if all finite trees satisfy the property and whenever a tree satisfies the property then so do all subtrees rooted at the children of the root. The property $\{W_\infty = 0\}$, as a property of the Galton–Watson tree T, is inherited, seeing that Z_t is a sum over the children of the root of the number of descendants at the corresponding generation $t - 1$. The result then follows from the following 0-1 law.

Lemma 6.1.12 (0-1 law for inherited properties). *For a Galton–Watson tree T, an inherited property A has, conditioned on non-extinction, probability 0 or 1.*

Proof Let $T^{(1)}, \ldots, T^{(Z_1)}$ be the descendant subtrees of the children of the root. We use the notation $T \in A$ to mean that tree T satisfies A. By the tower property, the definition of inherited, and conditional independence, we have

$$\begin{aligned}
\mathbb{P}[A] &= \mathbb{E}[\mathbb{P}[T \in A \mid Z_1]] \\
&\leq \mathbb{E}[\mathbb{P}[T^{(i)} \in A, \forall i \leq Z_1 \mid Z_1]] \\
&= \mathbb{E}[\mathbb{P}[A]^{Z_1}] \\
&= f(\mathbb{P}[A]).
\end{aligned}$$

So $\mathbb{P}[A] \in [0, \eta] \cup \{1\}$ by the proof of Lemma 6.1.8.

Moreover, since A holds for finite trees, we have $\mathbb{P}[A] \geq \eta$, where we recall that η is the probability of extinction. Hence, in fact, $\mathbb{P}[A] \in \{\eta, 1\}$. Conditioning on non-extinction gives the claim. ■

That concludes the proof. ■

A further moment assumption provides a more detailed picture.

Lemma 6.1.13 (Exponential growth IV). *Let (Z_t) be a Galton–Watson branching process with $m = \mathbb{E}[X(1,1)] > 1$ and $\sigma^2 = \mathrm{Var}[X(1,1)] < +\infty$. Then, (W_t) converges in L^2 and, in particular, $\mathbb{E}[W_\infty] = 1$. Furthermore, $\mathbb{P}[W_\infty = 0] = \eta$.*

Proof We bound $\mathbb{E}[W_t^2]$ by computing it explicitly by induction. From the orthogonality of increments (Lemma 3.1.50), it holds that

$$\mathbb{E}[W_t^2] = \mathbb{E}[W_{t-1}^2] + \mathbb{E}[(W_t - W_{t-1})^2].$$

Since $\mathbb{E}[W_t \mid \mathcal{F}_{t-1}] = W_{t-1}$ by the martingale property,

$$
\begin{aligned}
\mathbb{E}[(W_t - W_{t-1})^2 \mid \mathcal{F}_{t-1}] &= \mathrm{Var}[W_t \mid \mathcal{F}_{t-1}] \\
&= m^{-2t}\,\mathrm{Var}[Z_t \mid \mathcal{F}_{t-1}] \\
&= m^{-2t}\,\mathrm{Var}\left[\sum_{i=1}^{Z_{t-1}} X(i,t)\,\middle|\, \mathcal{F}_{t-1}\right] \\
&= m^{-2t} Z_{t-1}\sigma^2.
\end{aligned}
$$

Hence, taking expectations and using Lemma 6.1.2, we get

$$\mathbb{E}[W_t^2] = \mathbb{E}[W_{t-1}^2] + m^{-t-1}\sigma^2.$$

Since $\mathbb{E}[W_0^2] = 1$, induction gives

$$\mathbb{E}[W_t^2] = 1 + \sigma^2 \sum_{i=2}^{t+1} m^{-i},$$

which is uniformly bounded from above when $m > 1$.

By the convergence theorem for martingales bounded in L^2 (Theorem 3.1.51), (W_t) converges almost surely and in L^2 to a finite limit W_∞ and

$$1 = \mathbb{E}[W_t] \to \mathbb{E}[W_\infty].$$

The last statement follows from Lemma 6.1.11.   ∎

Remark 6.1.14 *A theorem of Kesten and Stigum gives a necessary and sufficient condition for $\mathbb{E}[W_\infty] = 1$ to hold [KS66b]. See, for example, [LP16, chapter 12].*

6.1.3 ▷ *Percolation: Galton–Watson Trees*

Let T be the Galton–Watson tree for an offspring distribution with mean $m > 1$. Now perform bond percolation on T with density p (see Definition 1.2.1). Let $\mathcal{C}_0$ be the open cluster of the root in T. Recall from Section 2.3.3 that the critical value is

$$p_c(T) = \sup\{p \in [0,1] \colon \theta(p) = 0\},$$

where the percolation function (conditioned on T) is $\theta(p) = \mathbb{P}_p[|\mathcal{C}_0| = +\infty \mid T]$.

Theorem 6.1.15 (Bond percolation on Galton–Watson trees). *Assume $m > 1$. Conditioned on non-extinction of T,*

$$p_c(T) = \frac{1}{m}$$

almost surely.

Proof We can think of $\mathcal{C}_0$ (or more precisely, its size on each level) as being itself generated by a Galton–Watson branching process, where this time the offspring distribution is the law of $\sum_{i=1}^{X(1,1)} I_i$, where the I_is are i.i.d. $\text{Ber}(p)$ and $X(1,1)$ is distributed according to the offspring distribution of T. In other words, we are "thinning" T. By conditioning on $X(1,1)$ and then using the tower property (Lemma B.6.16), the offspring mean under the process generating $\mathcal{C}_0$ is mp.

If $mp \leq 1$, then by the extinction theory (Theorems 6.1.5 and 6.1.6),

$$1 = \mathbb{P}_p[|\mathcal{C}_0| < +\infty] = \mathbb{E}[\mathbb{P}_p[|\mathcal{C}_0| < +\infty \mid T]],$$

and we must have $\mathbb{P}_p[|\mathcal{C}_0| < +\infty \mid T] = 1$ almost surely. Taking $p = 1/m$, we get $p_c(T) \geq \frac{1}{m}$ almost surely. That holds, in particular, on the non-extinction of T, which happens with positive probability.

For the other direction, fix p such that $mp > 1$. The property of trees $\{\mathbb{P}_p[|\mathcal{C}_0| < +\infty \mid T] = 1\}$ is inherited. So by Lemma 6.1.12, conditioned on non-extinction of T, it has probability 0 or 1. That probability is of course 1 on extinction. By Theorem 6.1.6,

$$1 > \mathbb{P}_p[|\mathcal{C}_0| < +\infty] = \mathbb{E}[\mathbb{P}_p[|\mathcal{C}_0| < +\infty \mid T]],$$

and, conditioned on non-extinction of T, we must have $\mathbb{P}_p[|\mathcal{C}_0| < +\infty \mid T] = 0$ – i.e., $p_c(T) < p$ – almost surely. Repeating this argument for a sequence $p_n \downarrow 1/m$ simultaneously (i.e., on the same T) and using the monotonicity of $\mathbb{P}_p[|\mathcal{C}_0| < +\infty \mid T]$, we get that $p_c(T) \leq 1/m$ almost surely conditioned on non-extinction of T. That proves the claim. $\blacksquare$

6.1.4 Multitype Branching Processes

Multitype branching processes are a useful generalization of Galton–Watson processes (Definition 6.1.1). Their behavior combines aspects of branching processes (exponential growth, extinction, etc.) and Markov chains (reducibility, mixing, etc.). We will not develop the full theory here. In this section, we define this class of processes and hint (largely without proofs) at their properties. In Section 6.3.2, we illustrate some of the more intricate interplay between the driving phenomena involved in a special example of practical importance.

MULTITYPE BRANCHING PROCESSES

Definition In a multitype branching process, each individual has one of τ types, which we will denote in this section by $1, \ldots, \tau$ for simplicity. Each type $\alpha \in [\tau] = \{1, \ldots, \tau\}$ has its own offspring distribution $\{p_{\mathbf{k}}^{(\alpha)} : \mathbf{k} \in \mathbb{Z}_+^\tau\}$, which specifies the distribution of the number of offspring of each type it has. Just to emphasize, this is a *collection of (typically distinct) multivariate distributions*.

For reasons that will become clear below, it will be convenient to work with row vectors. For each $\alpha \in [\tau]$, let

$$\mathbf{X}^{(\alpha)}(i,t) = \left(X_1^{(\alpha)}(i,t), \ldots, X_\tau^{(\alpha)}(i,t) \right) \quad \forall i, t \geq 1$$

be an array of i.i.d. $\mathbb{Z}_+^\tau$-valued random row vectors with distribution $\{p_{\mathbf{k}}^{(\alpha)}\}$. Let

$$\mathbf{Z}_0 = \mathbf{k}_0 \in \mathbb{Z}_+^\tau$$

be the initial population at time 0, again as a row vector. Recursively, the population vector

$$\mathbf{Z}_t = (Z_{t,1}, \ldots, Z_{t,\tau}) \in \mathbb{Z}_+^\tau$$

at time $t \geq 1$ is set to

$$\mathbf{Z}_t := \sum_{\alpha=1}^{\tau} \sum_{i=1}^{Z_{t-1,\alpha}} \mathbf{X}^{(\alpha)}(i, t). \tag{6.1.4}$$

In other words, the ith individual of type α at generation $t-1$ produces $X_\beta^{(\alpha)}(i, t)$ individuals of type β at generation t (before itself dying). Let $\mathcal{F}_t = \sigma(\mathbf{Z}_0, \ldots, \mathbf{Z}_t)$ be the corresponding filtration. We assume throughout that $\mathbb{P}[\|\mathbf{X}^{(\alpha)}(1, 1)\|_1 = 1] < 1$ for at least one α (which

NON-
SINGULAR
CASE

is referred to as the *non-singular case*); otherwise the process reduces to a simple finite Markov chain.

Martingales As in the single-type case, the means of the offspring distributions play a

MEAN MATRIX

key role in the theory. This time however they form a matrix, the so-called *mean matrix* $M = (m_{\alpha,\beta})$ with entries

$$m_{\alpha,\beta} = \mathbb{E}\left[X_\beta^{(\alpha)}(1, 1)\right] \quad \forall \alpha, \beta \in [\tau].$$

That is, $m_{\alpha,\beta}$ is the expected number of offspring of type β of an individual of type α. We assume throughout that $m_{\alpha,\beta} < +\infty$ for all α, β.

To see how M drives the growth of the process, we generalize the proof of Lemma 6.1.2. By the recursive formula (6.1.4),

$$\begin{aligned}
\mathbb{E}[\mathbf{Z}_t \mid \mathcal{F}_{t-1}] &= \mathbb{E}\left[\sum_{\alpha=1}^{\tau} \sum_{i=1}^{Z_{t-1,\alpha}} \mathbf{X}^{(\alpha)}(i, t) \,\middle|\, \mathcal{F}_{t-1}\right] \\
&= \sum_{\alpha=1}^{\tau} \sum_{i=1}^{Z_{t-1,\alpha}} \mathbb{E}\left[\mathbf{X}^{(\alpha)}(i, t) \mid \mathcal{F}_{t-1}\right] \\
&= \sum_{\alpha=1}^{\tau} Z_{t-1,\alpha}\, \mathbb{E}\left[\mathbf{X}^{(\alpha)}(1, 1)\right] \\
&= \mathbf{Z}_{t-1} M,
\end{aligned} \tag{6.1.5}$$

where we recall that $\mathbf{Z}_{t-1}$ and $\mathbf{Z}_t$ are row vectors. Inductively,

$$\mathbb{E}[\mathbf{Z}_t \mid \mathbf{Z}_0] = \mathbf{Z}_0 M^t. \tag{6.1.6}$$

Moreover, any real right eigenvector $\mathbf{u}$ (as a column vector) of M with real eigenvalue $\lambda \neq 0$ gives rise to a martingale

$$U_t := \lambda^{-t} \mathbf{Z}_t \mathbf{u}, \quad t \geq 0, \tag{6.1.7}$$

since

$$\begin{aligned}
\mathbb{E}[U_t \mid \mathcal{F}_{t-1}] &= \mathbb{E}[\lambda^{-t}\mathbf{Z}_t\mathbf{u} \mid \mathcal{F}_{t-1}] \\
&= \lambda^{-t}\,\mathbb{E}[\mathbf{Z}_t \mid \mathcal{F}_{t-1}]\,\mathbf{u} \\
&= \lambda^{-t}\mathbf{Z}_{t-1}M\mathbf{u} \\
&= \lambda^{-t}\mathbf{Z}_{t-1}\lambda\mathbf{u} \\
&= U_{t-1}.
\end{aligned}$$

Extinction The classical *Perron–Frobenius Theorem* characterizes the direction of largest growth of the matrix M. We state a version of it without proof in the case where all entries of M are strictly positive, which is referred to as the *positive regular case*. Note that, unlike POSITIVE the case of simple finite Markov chains, the matrix M is not in general stochastic, as it also REGULAR reflects the "growth" of the population in addition to the "transitions" between types. We CASE encountered the following concept in Section 5.2.5 and Exercise 5.5.

Definition 6.1.16 *The* spectral radius $\rho(A)$ *of a matrix A is the maximum of the eigenvalues* SPECTRAL *of A in absolute value.* RADIUS

Theorem 6.1.17 (Perron–Frobenius theorem: positive regular case). *Let M be a strictly positive, square matrix. Then $\rho := \rho(M)$ is an eigenvalue of M with algebraic and geometric multiplicities 1. It is also the only eigenvalue with absolute value ρ. The corresponding left and right eigenvectors, denoted by $\mathbf{v}$ (as a row vector) and $\mathbf{w}$ (as a column vector) respectively, are positive vectors. They are referred to as left and right* Perron *vector. We* PERRON *assume that they are normalized so that $\mathbf{1w} = 1$ and $\mathbf{vw} = 1$. Here $\mathbf{1}$ is the all-one row* VECTOR *vector.*

Because $\mathbf{w}$ is positive, the martingale

$$W_t := \rho^{-t}\mathbf{Z}_t\mathbf{w}, \quad t \geq 0$$

is non-negative. Therefore, it converges almost surely to a random limit with a finite mean by Corollary 3.1.48. When $\rho < 1$, an argument based on Markov's inequality (Theorem 2.1.1) implies that the process goes extinct almost surely. Formally, let $q^{(\alpha)}$ be the probability of extinction when started with a single individual of type α, that is,

$$q^{(\alpha)} := \mathbb{P}[\mathbf{Z}_t = \mathbf{0} \text{ for some } t \mid \mathbf{Z}_0 = \mathbf{e}_\alpha],$$

where $\mathbf{e}_\alpha \in \mathbb{Z}_+^\tau$ is the standard basis row vector with a one in the αth coordinate, and let $\mathbf{q} := (q^{(1)}, \dots, q^{(\tau)})$. Then,

$$\rho < 1 \implies \mathbf{q} = \mathbf{1}. \tag{6.1.8}$$

Exercise 6.1 asks for the proof. We state the following more general result without proof. We use the notation of Theorem 6.1.17. We will also refer to the generating functions

$$f^{(\alpha)}(\mathbf{s}) := \mathbb{E}\left[\prod_{\beta=1}^{\tau} s_\beta^{X_\beta^{(\alpha)}(1,1)}\right], \quad \mathbf{s} \in [0,1]^\tau$$

with $\mathbf{f} = (f^{(1)}, \dots, f^{(\tau)})$.

Theorem 6.1.18 (Extinction: multitype case). *Let* $(\mathbf{Z})_t$ *be a positive regular, nonsingular multitype branching process with a finite mean matrix* M.

(i) *If* $\rho \leq 1$, *then* $\mathbf{q} = \mathbf{1}$.

(ii) *If* $\rho > 1$, *then:*

- *It holds that* $\mathbf{q} < \mathbf{1}$.
- *The unique solution to* $\mathbf{f}(\mathbf{s}) = \mathbf{s}$ *in* $[0, 1)^\tau$ *is* $\mathbf{q}$.
- *Almost surely*

$$\lim_{t \to +\infty} \rho^{-t} \mathbf{Z_t} = \mathbf{v} W_\infty,$$

where W_∞ *is a non-negative random variable.*

- *If in addition* $\mathrm{Var}[X_\beta^{(\alpha)}(1, 1)] < +\infty$ *for all* α, β *then*

$$\mathbb{E}[W_\infty \mid \mathbf{Z}_0 = \mathbf{e}_\alpha] = w_\alpha$$

and

$$q^{(\alpha)} = \mathbb{P}[W_\infty = 0 \mid \mathbf{Z}_0 = \mathbf{e}_\alpha]$$

for all $\alpha \in [\tau]$.

Remark 6.1.19 *As in the single-type case, a theorem of Kesten and Stigum gives a necessary and sufficient condition for the last claim of Theorem 6.1.18 (ii) to hold* [KS66b].

Linear functionals Theorem 6.1.18 also characterizes the limit behavior of linear functionals of the form $\mathbf{Z_t u}$ for any vector that is *not* orthogonal to $\mathbf{v}$. In contrast, interesting new behavior arises when $\mathbf{u}$ is orthogonal to $\mathbf{v}$. We will not derive the general theory here. We only show through a second moment calculation that a phase transition takes place.

We restrict ourselves to the supercritical case $\rho > 1$ and to $\mathbf{u} = (u_1, \ldots, u_\tau)$ being a real right eigenvector of M with a real eigenvalue $\lambda \notin \{0, \rho\}$. Let U_t be the corresponding martingale from (6.1.7). The vector $\mathbf{u}$ is necessarily orthogonal to $\mathbf{v}$. Indeed, $\mathbf{v}M\mathbf{u}$ is equal to both $\rho \mathbf{v}\mathbf{u}$ and $\lambda \mathbf{v}\mathbf{u}$. Because $\rho \neq \lambda$ by assumption, this is only possible if all three expressions are 0. That implies $\mathbf{v}\mathbf{u} = 0$ since we also have $\rho \neq 0$ by assumption.

To compute the second moment of U_t, we mimic the computations in the proof of Lemma 6.1.13. We have

$$\mathbb{E}[U_t^2 \mid \mathbf{Z}_0] = \mathbb{E}[U_{t-1}^2 \mid \mathbf{Z}_0] + \mathbb{E}[(U_t - U_{t-1})^2 \mid \mathbf{Z}_0],$$

by the orthogonality of increments (Lemma 3.1.50). Since $\mathbb{E}[U_t \mid \mathcal{F}_{t-1}] = U_{t-1}$, by the martingale property, we get

$$\begin{aligned}
\mathbb{E}[(U_t - U_{t-1})^2 \mid \mathcal{F}_{t-1}] &= \mathrm{Var}[U_t \mid \mathcal{F}_{t-1}] \\
&= \mathrm{Var}[\lambda^{-t} \mathbf{Z_t u} \mid \mathcal{F}_{t-1}] \\
&= \lambda^{-2t} \mathrm{Var}\left[\left(\sum_{\alpha=1}^{\tau} \sum_{i=1}^{Z_{t-1,\alpha}} \mathbf{X}^{(\alpha)}(i, t) \right) \mathbf{u} \,\middle|\, \mathcal{F}_{t-1} \right] \\
&= \lambda^{-2t} \sum_{\alpha=1}^{\tau} Z_{t-1,\alpha} \, \mathrm{Var}\left[\mathbf{X}^{(\alpha)}(1, 1) \mathbf{u} \right] \\
&= \lambda^{-2t} \mathbf{Z}_{t-1} \mathbf{S}^{(\mathbf{u})},
\end{aligned}$$

where $\mathbf{S}^{(\mathbf{u})} = (\mathrm{Var}[\mathbf{X}^{(1)}(1,1)\,\mathbf{u}], \ldots, \mathrm{Var}[\mathbf{X}^{(\tau)}(1,1)\,\mathbf{u}])$ as a column vector. In the last display, we used (6.1.4) on the third line and the independence of the random vectors $\mathbf{X}^{(\alpha)}(i,t)$ on the fourth line. Hence, taking expectations and using (6.1.6), we get

$$\mathbb{E}[U_t^2 \mid \mathbf{Z}_0] = \mathbb{E}[U_{t-1}^2 \mid \mathbf{Z}_0] + \lambda^{-2t}\mathbf{Z}_0 M^{t-1}\mathbf{S}^{(\mathbf{u})},$$

and finally,

$$\mathbb{E}[U_t^2 \mid \mathbf{Z}_0] = (\mathbf{Z}_0\mathbf{u})^2 + \sum_{s=1}^{t} \lambda^{-2s}\mathbf{Z}_0 M^{s-1}\mathbf{S}^{(\mathbf{u})}. \tag{6.1.9}$$

The case $\mathbf{S}^{(\mathbf{u})} = \mathbf{0}$ is trivial (see Exercise 6.4), so we exclude it from the following lemma.

Lemma 6.1.20 (Second moment of U_t). *Assume* $\mathbf{S}^{(\mathbf{u})} \neq \mathbf{0}$ *and* $\mathbf{Z}_0 \neq \mathbf{0}$. *The sequence* $\mathbb{E}[U_t^2 \mid \mathbf{Z}_0]$, $t = 0, 1, 2, \ldots$, *is non-decreasing and satisfies*

$$\sup_{t \geq 0} \mathbb{E}[U_t^2 \mid \mathbf{Z}_0] \begin{cases} < +\infty & \text{if } \rho < \lambda^2, \\ = +\infty & \text{otherwise.} \end{cases}$$

Proof Because $\mathbf{S}^{(\mathbf{u})} \neq \mathbf{0}$ and non-negative and the matrix M is strictly positive by assumption, we have that

$$\widetilde{\mathbf{S}}^{(\mathbf{u})} := M\mathbf{S}^{(\mathbf{u})} > \mathbf{0}.$$

Since $\mathbf{w}$ is also strictly positive, there is $0 < C^- \leq C^+ < +\infty$ such that

$$C^-\mathbf{w} \leq \widetilde{\mathbf{S}}^{(\mathbf{u})} \leq C^+\mathbf{w}.$$

Moreover, since M is positive, each inequality is preserved when multiplying on both sides by M, that is, for any $s \geq 1$,

$$C^-\rho^s\mathbf{w} \leq M^s\widetilde{\mathbf{S}}^{(\mathbf{u})} \leq C^+\rho^s\mathbf{w}. \tag{6.1.10}$$

Now rewrite (6.1.9) as

$$\mathbb{E}[U_t^2 \mid \mathbf{Z}_0] = (\mathbf{Z}_0\mathbf{u})^2 + \lambda^{-2}\mathbf{Z}_0\mathbf{S}^{(\mathbf{u})} + \lambda^{-4}\sum_{s=2}^{t} \mathbf{Z}_0(1/\lambda^2)^{s-2}M^{s-2}\widetilde{\mathbf{S}}^{(\mathbf{u})}.$$

There are two cases:

- When $\rho < \lambda^2$, using (6.1.10), the sum on the right-hand side can be bounded *above* by

$$C^+\mathbf{Z}_0\mathbf{w}\sum_{s=2}^{t} \left(\frac{\rho}{\lambda^2}\right)^{s-2} \leq C^+\mathbf{Z}_0\mathbf{w}\frac{1}{1 - (\rho/\lambda^2)} < +\infty$$

  uniformly in t.
- When $\rho \geq \lambda^2$, the same sum can be bounded from *below* by

$$C^-\mathbf{Z}_0\mathbf{w}\sum_{s=2}^{t} \left(\frac{\rho}{\lambda^2}\right)^{s-2} \to +\infty$$

  as $t \to +\infty$. Indeed, $\mathbf{Z}_0 \neq \mathbf{0}$ implies that the inner product $\mathbf{Z}_0\mathbf{w}$ is strictly positive.

That proves the claim. ∎

In the case $\rho < \lambda^2$, the martingale (U_t) is bounded in L^2 and therefore converges almost surely to a limit U_∞ with $\mathbb{E}[U_\infty \mid \mathbf{Z}_0] = \mathbf{Z}_0\mathbf{u}$ by Theorem 3.1.51. On the other hand, when $\rho \geq \lambda^2$, it can be shown (we will not do this here) that $\mathbf{Z}_t\mathbf{u}/\sqrt{\mathbf{Z}_t\mathbf{w}}$ satisfies a central limit theorem with a limit independent of $\mathbf{Z}_0$. Implications of these claims are illustrated in Section 6.3.2.

6.2 Random-Walk Representation

In this section, we develop a random-walk representation of the Galton–Watson process. We give two applications: a characterization of the Galton–Watson process *conditioned on extinction* in terms of a dual branching process, and a formula for the *size of the total progeny*. We illustrate both in Section 6.2.4, where we revisit percolation on the infinite *b*-ary tree.

6.2.1 Exploration Process

We introduce an exploration process where a random-walk perspective will naturally arise.

Exploration of a graph

Because this will be useful again later, we describe it first in the context of a locally finite graph $G = (V, E)$. The exploration process starts at an arbitrary vertex $v \in V$ and has three types of vertices:

ACTIVE
EXPLORED
NEUTRAL

- $\mathcal{A}_t$: *active* vertices,
- $\mathcal{E}_t$: *explored* vertices,
- $\mathcal{N}_t$: *neutral* vertices.

At the beginning, we have $\mathcal{A}_0 := \{v\}$, $\mathcal{E}_0 := \emptyset$, and $\mathcal{N}_0$ contains all other vertices in G. At time t, if $\mathcal{A}_{t-1} = \emptyset$ (i.e., there are no active vertices), we let $(\mathcal{A}_t, \mathcal{E}_t, \mathcal{N}_t) := (\mathcal{A}_{t-1}, \mathcal{E}_{t-1}, \mathcal{N}_{t-1})$. Otherwise, we pick an element, a_t, from $\mathcal{A}_{t-1}$ (say in first-come, first-served basis to be explicit) and set:

- $\mathcal{A}_t := (\mathcal{A}_{t-1} \backslash \{a_t\}) \cup \{x \in \mathcal{N}_{t-1} : \{x, a_t\} \in E\}$,
- $\mathcal{E}_t := \mathcal{E}_{t-1} \cup \{a_t\}$,
- $\mathcal{N}_t := \mathcal{N}_{t-1} \backslash \{x \in \mathcal{N}_{t-1} : \{x, a_t\} \in E\}$.

We imagine revealing the edges of G as they are encountered in the exploration process. In words, starting with v, the connected component $\mathcal{C}_v$ of v is progressively grown by adding to it at each time a vertex adjacent to one of the previously explored vertices and uncovering its remaining neighbors in G. In this process, $\mathcal{E}_t$ is the set of previously explored vertices and $\mathcal{A}_t$ – the frontier of the process – is the set of vertices that are known to belong to $\mathcal{C}_v$ but whose full neighborhood is waiting to be uncovered. The rest of the vertices form the set $\mathcal{N}_t$. See Figure 6.3.

Let $A_t := |\mathcal{A}_t|$, $E_t := |\mathcal{E}_t|$, and $N_t := |\mathcal{N}_t|$. Note that (E_t) – not to be confused with the edge set – is non-decreasing, while (N_t) is non-increasing. Let

$$\tau_0 := \inf\{t \geq 0 : A_t = 0\}$$

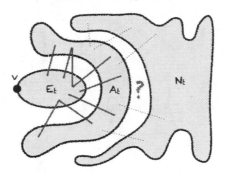

Figure 6.3 Exploration process for $\mathcal{C}_v$.

be the first time A_t is 0 (which by convention is $+\infty$ if there is no such t). The process is fixed for all $t > \tau_0$. Notice that $E_t = t$ for all $t \leq \tau_0$, as exactly one vertex is explored at each time until the set of active vertices is empty. The size of the connected component of v can be characterized as follows.

Lemma 6.2.1

$$\tau_0 = |\mathcal{C}_v|.$$

Proof Indeed, a single vertex of $\mathcal{C}_v$ is explored at each time until all of $\mathcal{C}_v$ has been visited. At that point, $\mathcal{A}_t$ is empty. ∎

Random-walk representation of a Galton–Watson tree

Let $(Z_i)_{i \geq 0}$ be a Galton–Watson branching process and let T be the corresponding Galton–Watson tree. We run the exploration process above on T started at the root 0. We will refer to the index i in Z_i as a "generation," and to the index t in the exploration process as "time" – they are *not* the same. Let $(\mathcal{A}_t, \mathcal{E}_t, \mathcal{N}_t)$ and $A_t := |\mathcal{A}_t|$, $E_t := |\mathcal{E}_t|$, and $N_t := |\mathcal{N}_t|$ be as above. Let $(\mathcal{F}_t)$ be the corresponding filtration. Because we explore the vertices on first-come, first-served basis, we exhaust all vertices in generation i before considering vertices in generation $i + 1$ (i.e., we perform breadth-first search).

The random-walk representation is the following. Observe that the process (A_t) admits a simple recursive form. We start with $A_0 := 1$. Then, conditioning on $\mathcal{F}_{t-1}$:

- If $A_{t-1} = 0$, the exploration process has finished its course and $A_t = 0$.
- Otherwise, (a) one active vertex becomes an explored vertex and (b) its offspring become active vertices. That is,

$$A_t := \begin{cases} A_{t-1} + \left(\underbrace{-1}_{(a)} + \underbrace{X_t}_{(b)} \right) & \text{if } t - 1 < \tau_0, \\ 0 & \text{otherwise,} \end{cases}$$

where X_t is distributed according to the offspring distribution.

We let $Y_t := X_t - 1$ and

$$S_t := 1 + \sum_{s=1}^{t} Y_s,$$

with $S_0 := 1$. Then,

$$\tau_0 = \inf\{t \geq 0 : S_t = 0\},$$

and

$$(A_t) = (S_{t \wedge \tau_0})$$

is a random walk started at 1 with i.i.d. increments (Y_t) stopped when it hits 0 for the first time.

We refer to

$$H = (X_1, \ldots, X_{\tau_0})$$

HISTORY

as the *history* of the process (Z_i). Observe that, under breadth-first search, the process (Z_i) can be reconstructed from H: $Z_0 = 1$, $Z_1 = X_1$, $Z_2 = X_2 + \ldots + X_{Z_1+1}$, and so forth. (Exercise 6.5 asks for a general formula.) As a result, (Z_i) can be recovered from (S_t) as well. We call $(x_1, \ldots, x_t)$ a *valid history* if

VALID
HISTORY

$$1 + (x_1 - 1) + \cdots + (x_s - 1) > 0$$

for all $s < t$ and

$$1 + (x_1 - 1) + \cdots + (x_t - 1) = 0.$$

Note that a valid history may have probability 0 under the offspring distribution.

6.2.2 Duality Principle

The random-walk representation above is useful to prove the following duality principle.

Theorem 6.2.2 (Duality principle). *Let (Z_i) be a branching process with offspring distribution $\{p_k\}_{k \geq 0}$ and extinction probability $\eta < 1$. Let (Z_i') be a branching process with offspring distribution $\{p_k'\}_{k \geq 0}$ where*

$$p_k' = \eta^{k-1} p_k.$$

Then, (Z_i) conditioned on extinction has the same distribution as (Z_i'), which is referred to as the dual branching process.

DUAL
BRANCHING
PROCESS

Let f be the probability generating function of the offspring distribution of (Z_i). Note that

$$\sum_{k \geq 0} p_k' = \sum_{k \geq 0} \eta^{k-1} p_k = \eta^{-1} f(\eta) = 1,$$

because η is a fixed point of f by Theorem 6.1.6. So $\{p_k'\}_{k \geq 0}$ is indeed a probability distribution. Note further that its expectation is

$$\sum_{k \geq 0} k p_k' = \sum_{k \geq 0} k \eta^{k-1} p_k = f'(\eta) < 1,$$

since, by Lemma 6.1.7, f' is strictly increasing, $f(\eta) = \eta < 1$ and $f(1) = 1$ (which would not be possible if $f'(\eta)$ were greater or equal to 1; see Figure 6.1 for an illustration). So the dual branching process is subcritical.

Proof of Theorem 6.2.2 We use the random-walk representation. Let $H = (X_1, \ldots, X_{\tau_0})$ and $H' = (X'_1, \ldots, X'_{\tau_0})$ be the histories of (Z_i) and (Z'_i), respectively. In the case of extinction of (Z_i), the history H has finite length.

By definition of the conditional probability, for a valid history $(x_1, \ldots, x_t)$ with a finite t,

$$\mathbb{P}[H = (x_1, \ldots, x_t) \mid \tau_0 < +\infty] = \frac{\mathbb{P}[H = (x_1, \ldots, x_t)]}{\mathbb{P}[\tau_0 < +\infty]} = \eta^{-1} \prod_{s=1}^{t} p_{x_s}.$$

Because $(x_1 - 1) + \cdots + (x_t - 1) = -1$,

$$\eta^{-1} \prod_{s=1}^{t} p_{x_s} = \eta^{-1} \prod_{s=1}^{t} \eta^{1-x_s} p'_{x_s} = \prod_{s=1}^{t} p'_{x_s} = \mathbb{P}[H' = (x_1, \ldots, x_t)].$$

Since this is true for all valid histories and the processes can be recovered from their histories, we have proved the claim. ∎

Example 6.2.3 (Poisson branching process). Let (Z_i) be a Galton–Watson branching process with offspring distribution Poi(λ) where $\lambda > 1$. Then, the dual probability distribution is given by

$$p'_k = \eta^{k-1} p_k = \eta^{k-1} e^{-\lambda} \frac{\lambda^k}{k!} = \eta^{-1} e^{-\lambda} \frac{(\lambda\eta)^k}{k!},$$

where recall from Example 6.1.10 that $e^{-\lambda(1-\eta)} = \eta$, so

$$p'_k = e^{\lambda(1-\eta)} e^{-\lambda} \frac{(\lambda\eta)^k}{k!} = e^{-\lambda\eta} \frac{(\lambda\eta)^k}{k!}.$$

That is, the dual branching process has offspring distribution Poi($\lambda\eta$). ◀

6.2.3 Hitting-Time Theorem

The random-walk representation also gives a formula for the distribution of the size of the progeny.

Law of total progeny The key is the following claim.

Lemma 6.2.4 (Total progeny and random-walk representation). *Let W be the total progeny of the Galton–Watson branching process (Z_i). Then,*

$$W = \tau_0.$$

Proof Recall that

$$\tau_0 := \inf\{t \geq 0 : A_t = 0\}.$$

If the process does not go extinct, then $\tau_0 = +\infty$ as there are always more vertices to explore.

Suppose the process goes extinct and that $W = n$. Notice that $E_t = t$ for all $t \le \tau_0$, as exactly one vertex is explored at each time until the set of active vertices is empty. Moreover, for all t, $(\mathcal{A}_t, \mathcal{E}_t, \mathcal{N}_t)$ forms a partition of $[n]$, so

$$A_t + t + N_t = n \qquad \forall t \le \tau_0.$$

At $t = \tau_0$, $A_t = N_t = 0$ and we get

$$\tau_0 = n.$$

That proves the claim. ∎

To compute the distribution of $W = \tau_0$, we use the following hitting-time theorem, which is proved later in this subsection.

Theorem 6.2.5 (Hitting-time theorem). *Let (R_t) be a random walk started at 0 with i.i.d. increments (U_t) satisfying*

$$\mathbb{P}[U_t \le 1] = 1.$$

Fix a positive integer ℓ. Let σ_ℓ be the first time t such that $R_t = \ell$. Then,

$$\mathbb{P}[\sigma_\ell = t] = \frac{\ell}{t} \mathbb{P}[R_t = \ell].$$

Finally we get:

Theorem 6.2.6 (Law of total progeny). *Let (Z_t) be a Galton–Watson branching process with total progeny W. In the random-walk representation of (Z_t),*

$$\mathbb{P}[W = t] = \frac{1}{t} \mathbb{P}[X_1 + \cdots + X_t = t - 1]$$

for all $t \ge 1$.

Proof Recall that $Y_t := X_t - 1 \ge -1$ and

$$S_t = 1 + \sum_{s=1}^{t} Y_s,$$

with $S_0 = 1$, and that

$$\begin{aligned}
\tau_0 &= \inf\{t \ge 0 \colon S_t = 0\} \\
&= \inf\{t \ge 0 \colon 1 + (X_1 - 1) + \cdots + (X_t - 1) = 0\} \\
&= \inf\{t \ge 0 \colon X_1 + \cdots + X_t = t - 1\}.
\end{aligned}$$

Define $R_t := 1 - S_t$ and $U_t := -Y_t$ for all t. Then $R_0 := 0$,

$$\{X_1 + \cdots + X_t = t - 1\} = \{R_t = 1\},$$

and

$$\tau_0 = \inf\{t \ge 0 \colon R_t = 1\}.$$

The process (R_t) satisfies the assumptions of the hitting-time theorem (Theorem 6.2.5) with $\ell = 1$ and $\sigma_\ell = \tau_0 = W$. Applying the theorem gives the claim. ∎

Example 6.2.7 (Poisson branching process (continued)). Let (Z_i) be a Galton–Watson branching process with offspring distribution $\text{Poi}(\lambda)$ where $\lambda > 0$. Let W be its total progeny. By the hitting-time theorem, for $t \geq 1$,

$$\mathbb{P}[W = t] = \frac{1}{t}\mathbb{P}[X_1 + \cdots + X_t = t - 1]$$

$$= \frac{1}{t}e^{-\lambda t}\frac{(\lambda t)^{t-1}}{(t-1)!}$$

$$= e^{-\lambda t}\frac{(\lambda t)^{t-1}}{t!},$$

where we used that a sum of independent Poisson is Poisson. ◄

Spitzer's combinatorial lemma Before proving the hitting-time theorem, we begin with a combinatorial lemma of independent interest. Let $u_1, \ldots, u_t \in \mathbb{R}$ and define $r_0 := 0$ and $r_j := u_1 + \cdots + u_j$ for $1 \leq j \leq t$. We say that j is a *ladder index* if $r_j > r_0 \vee \cdots \vee r_{j-1}$. Consider the cyclic permutations of $\mathbf{u} = (u_1, \ldots, u_t)$, that is, $\mathbf{u}^{(0)} = \mathbf{u}, \mathbf{u}^{(1)} = (u_2, \ldots, u_t, u_1)$, $\ldots, \mathbf{u}^{(t-1)} = (u_t, u_1, \ldots, u_{t-1})$. Define the corresponding partial sums $r_j^{(\beta)} := u_1^{(\beta)} + \cdots + u_j^{(\beta)}$ for $j = 1, \ldots, t$ and $\beta = 0, \ldots, t - 1$.

Lemma 6.2.8 (Spitzer's combinatorial lemma). *Assume $r_t > 0$. Let ℓ be the number of cyclic permutations such that t is a ladder index. Then $\ell \geq 1$ and each such cyclic permutation has exactly ℓ ladder indices.*

Proof We will need the following observation:

$$(r_1^{(\beta)}, \ldots, r_t^{(\beta)})$$
$$= (r_{\beta+1} - r_\beta, r_{\beta+2} - r_\beta, \ldots, r_t - r_\beta,$$
$$[r_t - r_\beta] + r_1, [r_t - r_\beta] + r_2, \ldots, [r_t - r_\beta] + r_\beta)$$
$$= (r_{\beta+1} - r_\beta, r_{\beta+2} - r_\beta, \ldots, r_t - r_\beta,$$
$$r_t - [r_\beta - r_1], r_t - [r_\beta - r_2], \ldots, r_t - [r_\beta - r_{\beta-1}], r_t). \qquad (6.2.1)$$

We first show that $\ell \geq 1$, that is, there is at least one cyclic permutation where t is a ladder index. Let $\beta \geq 1$ be the smallest index achieving the maximum of $r_1, \ldots, r_t$, that is,

$$r_\beta > r_1 \vee \cdots \vee r_{\beta-1} \quad \text{and} \quad r_\beta \geq r_{\beta+1} \vee \cdots \vee r_t.$$

Moreover, $r_t > 0 = r_0$ by assumption. Hence,

$$r_{\beta+j} - r_\beta \leq 0 < r_t \qquad \forall j = 1, \ldots, t - \beta$$

and

$$r_t - [r_\beta - r_j] < r_t \qquad \forall j = 1, \ldots, \beta - 1.$$

From (6.2.1), in $\mathbf{u}^{(\beta)}$, t is a ladder index.

For the second claim, since $\ell \geq 1$, we can assume without loss of generality that $\mathbf{u}$ is such that t is a ladder index. (Note that $r_t^{(\beta)} = r_t$ for all β.) We show that β is a ladder index in $\mathbf{u}$ if and only if t is a ladder index in $\mathbf{u}^{(\beta)}$. That does indeed imply the claim as there are ℓ cyclic permutations where t is a ladder index by assumption. We use (6.2.1) again. Observe that β is a ladder index in $\boldsymbol{u}$ if and only if

$$r_\beta > r_0 \vee \cdots \vee r_{\beta-1},$$

which holds if and only if

$$r_\beta > r_0 = 0 \quad \text{and} \quad r_t - [r_\beta - r_j] < r_t \; \forall j = 1, \ldots, \beta - 1. \tag{6.2.2}$$

Moreover, because $r_t > r_j$ for all j by the assumption that t is ladder index, the last display holds if and only if

$$r_{\beta+j} - r_\beta < r_t \; \forall j = 1, \ldots, t - \beta \tag{6.2.3}$$

and

$$r_t - [r_\beta - r_j] < r_t, \; \forall j = 1, \ldots, \beta - 1, \tag{6.2.4}$$

that is, if and only if t is a ladder index in $\boldsymbol{u}^{(\beta)}$ by (6.2.1). Indeed, the second condition (i.e., (6.2.4)) is intact from (6.2.2), while the first one (i.e., (6.2.3)) can be rewritten as $r_\beta > -(r_t - r_{\beta+j})$, where the right-hand side is < 0 for $j = 1, \ldots, t - \beta - 1$ and $= 0$ for $j = t - \beta$. ∎

Proof of hitting-time theorem We are now ready to prove the hitting-time theorem. We only handle the case $\ell = 1$ (which is the one we used for the law of the total progeny). Exercise 6.6 asks for the full proof.

Proof of Theorem 6.2.5 Recall that $R_t = \sum_{s=1}^t U_s$ and $\sigma_1 = \inf\{j \geq 0 : R_j = 1\}$. By the assumption that $U_s \leq 1$ almost surely for all s,

$$\{\sigma_1 = t\} = \{t \text{ is the first ladder index in } R_1, \ldots, R_t\}.$$

By symmetry, for all $\beta = 0, \ldots, t - 1$,

$$\mathbb{P}[t \text{ is the first ladder index in } R_1, \ldots, R_t]$$
$$= \mathbb{P}[t \text{ is the first ladder index in } R_1^{(\beta)}, \ldots, R_t^{(\beta)}].$$

Let $\mathcal{E}_\beta$ be the event on the last line. Then,

$$\mathbb{P}[\sigma_1 = t] = \mathbb{E}[\mathbf{1}_{\mathcal{E}_0}] = \frac{1}{t} \mathbb{E}\left[\sum_{\beta=0}^{t-1} \mathbf{1}_{\mathcal{E}_\beta} \right].$$

By Spitzer's combinatorial lemma (Lemma 6.2.8), there is at most one cyclic permutation where t is the *first* ladder index. (There is at least one cyclic permutation where t is a ladder index – but it may not be the *first* one, that is, there may be multiple ladder indices.) In particular, $\sum_{\beta=0}^{t-1} \mathbf{1}_{\mathcal{E}_\beta} \in \{0, 1\}$. So, by the previous display,

$$\mathbb{P}[\sigma_1 = t] = \frac{1}{t} \mathbb{P}\left[\cup_{\beta=0}^{t-1} \mathcal{E}_\beta \right].$$

Finally, we claim that $\{R_t = 1\} = \cup_{\beta=0}^{t-1} \mathcal{E}_\beta$. Indeed, because $R_0 = 0$ and $U_s \leq 1$ for all s, the partial sum at the jth ladder index must take value j. So the event $\cup_{\beta=0}^{t-1} \mathcal{E}_\beta$ implies $\{R_t = 1\}$ since the last partial sum of all cyclic permutations is R_t. Similarly, because there is at least one cyclic permutation such that t is a ladder index, the event $\{R_t = 1\}$ implies that t is in fact the first ladder index in that cyclic permutation, and therefore it implies $\cup_{\beta=1}^{t} \mathcal{E}_\beta$. Hence,

$$\mathbb{P}[\sigma_1 = t] = \frac{1}{t} \mathbb{P}[R_t = 1],$$

which concludes the proof (for the case $\ell = 1$). ∎

6.2.4 ▷ *Percolation: Critical Exponents on the Infinite b-ary Tree*

In this section, we use branching processes to study bond percolation (Definition 1.2.1) on the infinite b-ary tree $\widehat{\mathbb{T}}_b$ and derive explicit expressions for quantities of interest. Close to the critical value, we prove the existence of "critical exponents." We illustrate the use of both the duality principle (Theorem 6.2.2) and the hitting-time theorem (Theorem 6.2.6).

Critical value We denote the root by 0. Similarly to what we did in Section 6.1.3, we think of the open cluster of the root, $\mathcal{C}_0$, as the progeny of a branching process as follows. Denote by ∂_n the nth level of $\widehat{\mathbb{T}}_b$, that is, the vertices of $\widehat{\mathbb{T}}_b$ at graph distance n from the root. In the branching process interpretation, we think of the immediate descendants in $\mathcal{C}_0$ of a vertex v as the offspring of v. By construction, v has at most b children, independently of all other vertices in the same generation. In this branching process, the offspring distribution $\{q_k\}_{k=0}^{b}$ is binomial with parameters b and p; $Z_n := |\mathcal{C}_0 \cap \partial_n|$ represents the size of the progeny at generation n; and $W := |\mathcal{C}_0|$ is the total progeny of the process. In particular, $|\mathcal{C}_0| < +\infty$ if and only if the process goes extinct. Because the mean number of offspring is bp, by Theorem 6.1.6, this leads immediately to a second proof of (a rooted variant of) Claim 2.3.9:

Claim 6.2.9

$$p_{\mathrm{c}}\left(\widehat{\mathbb{T}}_b\right) = \frac{1}{b}.$$

Percolation function The generating function of the offspring distribution is $\phi(s) := ((1-p) + ps)^b$. So, by Theorems 6.1.5 and 6.1.6, the percolation function

$$\theta(p) = \mathbb{P}_p[|\mathcal{C}_0| = +\infty]$$

is 0 on $[0, 1/b]$, while on $(1/b, 1]$ the quantity $\eta(p) := 1 - \theta(p)$ is the unique solution in $[0, 1)$ of the fixed point equation

$$s = ((1-p) + ps)^b. \tag{6.2.5}$$

For $b = 2$, for instance, we can compute the fixed point explicitly by noting that

$$\begin{aligned}
0 &= ((1-p) + ps)^2 - s \\
&= p^2 s^2 + [2p(1-p) - 1]s + (1-p)^2,
\end{aligned}$$

whose solution for $p \in (1/2, 1]$ is

$$s^* = \frac{-[2p(1-p) - 1] \pm \sqrt{[2p(1-p) - 1]^2 - 4p^2(1-p)^2}}{2p^2}$$

$$= \frac{-[2p(1-p) - 1] \pm \sqrt{1 - 4p(1-p)}}{2p^2}$$

$$= \frac{-[2p(1-p) - 1] \pm (2p - 1)}{2p^2}$$

$$= \frac{2p^2 + [(1 - 2p) \pm (2p - 1)]}{2p^2}.$$

So, rejecting the fixed point 1,

$$\theta(p) = 1 - \frac{2p^2 + 2(1 - 2p)}{2p^2} = \frac{2p - 1}{p^2}.$$

We have proved:

Claim 6.2.10 *For $b = 2$,*

$$\theta(p) = \begin{cases} 0, & 0 \le p \le \frac{1}{2}, \\ \frac{2(p - \frac{1}{2})}{p^2}, & \frac{1}{2} < p \le 1. \end{cases}$$

Since $\eta(p) = (1 - \theta(p))$, we have in that case

$$\eta(p) = \begin{cases} 1, & 0 \le p \le \frac{1}{2}, \\ \frac{(1-p)^2}{p^2}, & \frac{1}{2} < p \le 1. \end{cases}$$

Conditioning on a finite cluster The expected size of the population at generation n is $(bp)^n$ by Lemma 6.1.2, so for $p \in [0, \frac{1}{b})$,

$$\mathbb{E}_p |\mathcal{C}_0| = \sum_{n \ge 0} (bp)^n = \frac{1}{1 - bp}. \tag{6.2.6}$$

For $p \in (\frac{1}{b}, 1)$, the total progeny is infinite with positive probability (and in particular the expectation is infinite), but we can compute the expected cluster size *on the event that* $|\mathcal{C}_0| < +\infty$. For this purpose we use the duality principle.

Recall that $q_k = \binom{b}{k} p^k (1-p)^{b-k}$, $k = 0, \ldots, b$, is the offspring distribution. For $0 \le k \le b$, we let the dual offspring distribution be

$$\hat{q}_k := [\eta(p)]^{k-1} q_k$$

$$= [\eta(p)]^{k-1} \binom{b}{k} p^k (1-p)^{b-k}$$

$$= \frac{[\eta(p)]^k}{((1-p) + p\,\eta(p))^b} \binom{b}{k} p^k (1-p)^{b-k}$$

$$= \binom{b}{k} \left(\frac{p\,\eta(p)}{(1-p) + p\,\eta(p)} \right)^k \left(\frac{1-p}{(1-p) + p\,\eta(p)} \right)^{b-k}$$

$$=: \binom{b}{k} \hat{p}^k (1-\hat{p})^{b-k},$$

where we used (6.2.5) and implicitly defined the dual density

$$\hat{p} := \frac{p\,\eta(p)}{(1-p) + p\,\eta(p)}. \tag{6.2.7}$$

In particular, $\{\hat{q}_k\}$ is a probability distribution as expected under Theorem 6.2.2 – it is in fact binomial with parameters b and $\hat{p}$. Summarizing the implications of Theorem 6.2.2:

Claim 6.2.11 *Conditioned on* $|\mathcal{C}_0| < +\infty$, *(supercritical) percolation on* $\widehat{\mathbb{T}}_b$ *with density* $p \in (\frac{1}{b}, 1)$ *has the same distribution as (subcritical) percolation on* $\widehat{\mathbb{T}}_b$ *with density defined by* (6.2.7).

Hence, using (6.2.6) with both p and $\hat{p}$ as well as the fact that $\mathbb{P}_p[|\mathcal{C}_0| < +\infty] = \eta(p)$, we have the following.

Claim 6.2.12

$$\chi^{\mathrm{f}}(p) := \mathbb{E}_p \left[|\mathcal{C}_0| \mathbf{1}_{\{|\mathcal{C}_0| < +\infty\}} \right] = \begin{cases} \frac{1}{1-bp}, & p \in [0, \frac{1}{b}), \\ \frac{\eta(p)}{1-b\hat{p}}, & p \in (\frac{1}{b}, 1). \end{cases}$$

For $b = 2$, $\eta(p) = 1 - \theta(p) = \left(\frac{1-p}{p} \right)^2$ so

$$\hat{p} = \frac{p \left(\frac{1-p}{p} \right)^2}{(1-p) + p \left(\frac{1-p}{p} \right)^2} = \frac{(1-p)^2}{p(1-p) + (1-p)^2} = 1 - p,$$

and

Claim 6.2.13 *For* $b = 2$,

$$\chi^{\mathrm{f}}(p) = \begin{cases} \frac{1/2}{\frac{1}{2}-p}, & p \in [0, \frac{1}{2}), \\ \frac{\frac{1}{2} \left(\frac{1-p}{p} \right)^2}{p - \frac{1}{2}}, & p \in (\frac{1}{2}, 1). \end{cases}$$

Distribution of the open cluster size In fact, the hitting-time theorem gives an explicit formula for the distribution of $|\mathcal{C}_0|$. Namely, recall that $|\mathcal{C}_0| \overset{\mathrm{d}}{=} \tau_0$, where

$$\tau_0 = \inf\{t \geq 0 : S_t = 0\}.$$

for $S_t = \sum_{\ell \leq t} X_\ell - (t-1)$, where $S_0 = 1$ and the X_ℓs are i.i.d. binomial with parameters b and p. By Theorem 6.2.6,

$$\mathbb{P}[\tau_0 = t] = \frac{1}{t}\,\mathbb{P}[S_t = 0],$$

and we have

$$\mathbb{P}_p[|\mathcal{C}_0| = \ell] = \frac{1}{\ell}\,\mathbb{P}\left[\sum_{i \leq \ell} X_\ell = \ell - 1\right] = \frac{1}{\ell}\binom{b\ell}{\ell - 1}p^{\ell-1}(1-p)^{b\ell-(\ell-1)}, \qquad (6.2.8)$$

where we used that a sum of independent binomials with the same p is itself binomial. In particular, at criticality (where $|\mathcal{C}_0| < +\infty$ almost surely; see Claim 3.1.52), using Stirling's formula (see Appendix A) it can be checked that

$$\mathbb{P}_{p_c}[|\mathcal{C}_0| = \ell] \sim \frac{1}{\ell}\frac{1}{\sqrt{2\pi p_c(1-p_c)b\ell}} = \frac{1}{\sqrt{2\pi(1-p_c)\ell^3}}$$

as $\ell \to +\infty$.

Critical exponents Close to criticality, physicists predict that many quantities behave according to power laws of the form $|p - p_c|^\beta$, where the exponent is referred to as a *critical exponent*. The critical exponents are believed to satisfy certain "universality" properties. But even proving the existence of such exponents in general remains a major open problem. On trees, though, we can simply read off the critical exponents from the above formulas. For $b = 2$, Claims 6.2.10 and 6.2.13 imply for instance that as $p \to p_c$,

CRITICAL
EXPONENT

$$\theta(p) \sim 8(p - p_c)\mathbf{1}_{\{p > 1/2\}}$$

and

$$\chi^f(p) \sim \frac{1}{2}|p - p_c|^{-1}.$$

In fact, as can be seen from Claim 6.2.12, the critical exponent of $\chi^f(p)$ does not depend on b. The same holds for $\theta(p)$ (see Exercise 6.9). Using (6.2.8), the higher moments of $|\mathcal{C}_0|$ can also be studied around criticality (see Exercise 6.10).

6.3 Applications

We develop two applications of branching processes in discrete probability. First, we prove a result about the height of a random binary search tree. Then we describe a phase transition in an Ising model on a tree with applications to evolutionary biology. In the next section, we also use branching processes to study the phase transition of an Erdős–Rényi random graph model.

6.3.1 ▷ *Probabilistic Analysis of Algorithms: Binary Search Tree*

BINARY
SEARCH TREE
A *binary search tree* is a commonly used data structure in computer science. It consists of a rooted binary tree $T_n = (V_n, E_n)$. Each vertex has a "left" and "right" subtree (possibly empty) and a "key" from an input sequence $x_1, \ldots, x_n \in \mathbb{R}$ (which we assume are distinct)

that satisfies the BST property: the key at vertex $v \in V$ is greater than all keys in the left subtree below it and less than all keys in the right subtree below it. Such a data structure can be used for a variety of algorithmic tasks, such as searching for keys or sorting them.

The tree is constructed recursively as follows. Assume that the keys $x_1, \ldots, x_i$ have already been inserted and that the current tree T_i satisfies the BST property. To insert x_{i+1},

- start at the root;
- if the root's key is strictly larger than x_{i+1}, then move to its left descendant, otherwise move to its right descendant;
- if such a descendant does not exist, then create it and assign it x_{i+1} as its key;
- otherwise repeat.

Inserting keys (and other operations such as deleting keys, which we do not describe) takes time proportional to the height H_n of the tree T_n, that is, the length of the *longest* path from the root to a leaf. While, in general, the height can be as large as n (if keys are inserted in order for instance), the typical behavior can be much smaller.

Indeed, here we study the case of n keys $X_1, \ldots, X_n$ i.i.d. from a continuous distribution on $\mathbb{R}$ and establish a much better behavior for the random height. Let γ be the unique solution greater than 1 of

$$\left(\frac{1}{e}\right)\left(\frac{2e}{\gamma}\right)^{\gamma} = 1. \tag{6.3.1}$$

See Exercise 6.14 for a proof that γ is well defined and that the left-hand side is strictly decreasing at γ. We show:

Claim 6.3.1 *$H_n/\log n \to_p \gamma$ as $n \to \infty$.*

Alternative representation of the height

The main idea of the proof is to relate the height H_n of the tree T_n to a product of independent uniform random variables. We make a series of observations about the structure of the tree. First:

Observation 1 Keys affect the construction of the binary search tree *only through their ordering*. Let σ be the corresponding (random) permutation, that is,

$$X_{\sigma(1)} < X_{\sigma(2)} < \cdots < X_{\sigma(n)}.$$

Let $t[\sigma]$ be the binary search tree generated by the permutation σ.

Second, by symmetry:

Observation 2 The permutation σ is *uniformly distributed*.

Denote by S_v the size of the subtree rooted at v (including v itself) in $t[\sigma]$. At the root ρ, we have $S_\rho = n$. What is the size of the subtree rooted at the left descendant ρ' of ρ? Eventually all keys with a rank lower than $\sigma^{-1}(1)$, that is, those keys with indices in $\left\{\sigma(i): i < \sigma^{-1}(1)\right\}$, find their way into the left subtree of the root. In other words,

$$S_{\rho'} = \sigma^{-1}(1) - 1.$$

Similarly, denoting by ρ'' the right descendant of ρ, we see that

$$S_{\rho''} = n - \sigma^{-1}(1).$$

We refer to $\sigma^{-1}(1)$ as the *rank* of the root. By Observation 2:

Observation 3 The rank $\sigma^{-1}(1)$ of the root is *uniformly distributed* in $[n]$. Moreover, it is identically distributed to $\lfloor S_\rho W_\rho \rfloor + 1$, where W_ρ is uniform in $[0, 1]$.

The second part of this last observation can be checked by direct computation. Rename $X_1', \ldots, X_{S_{\rho'}}'$ the keys in the subtree rooted at ρ' *in the order that they are inserted* and let σ' be the (random) permutation corresponding to their ordering, that is,

$$X_{\sigma'(1)}' < X_{\sigma'(2)}' < \cdots < X_{\sigma'(S_{\rho'})}'.$$

Define σ'' similarly for ρ''. Again by symmetry:

Observation 4 Conditioned on $\sigma^{-1}(1)$ (and therefore on $S_{\rho'}$ and $S_{\rho''}$), the permutations σ' and σ'' are *independent* and *uniformly distributed*.

Finally, recursively:

Observation 5 The binary search tree $t[\sigma]$ is obtained by appending the left subtree $t[\sigma']$ and right subtree $t[\sigma'']$ to the root ρ.

If $S_{\rho'} = 0$, then $t[\sigma'] = \emptyset$ (and there is in fact no ρ'); while, if $S_{\rho'} = 1$, the tree $t[\sigma']$ is comprised of the single vertex ρ'. Similarly for σ''. Hence, this recursive process stops whenever we reach a vertex v with $S_v \in \{0, 1\}$. But it will be convenient to extend it indefinitely to produce an *infinite binary tree* $\mathcal{T} = \widehat{\mathbb{T}}_2$, where all additional vertices v are assigned $S_v = 0$.

The upshot of all these observations is that we obtain the following alternative characterization of the height H_n:

- assign an independent $U[0, 1]$ (i.e., uniform in $[0, 1]$) random variable W_v to each vertex v in the infinite binary tree $\mathcal{T}$;
- at the root ρ, set

$$S_\rho = n;$$

- then recursively from the root down, set

$$S_{v'} := \lfloor S_v W_v \rfloor \quad \text{and} \quad S_{v''} := \lfloor S_v (1 - W_v) \rfloor, \tag{6.3.2}$$

where v' and v'' are the left and right descendants of v in $\mathcal{T}$.

It can be checked that $S_{v'} + S_{v''} = S_v - 1$ almost surely, provided $S_v \geq 1$ (see Exercise 6.15). Moreover, notice that when $S_v = 1$, then $S_{v'} = S_{v''} = 0$ almost surely; while if $S_v = 0$, then $S_{v'} = S_{v''} = 0$. Finally, the height H_n is the highest level containing a vertex with subtree size at least 1, that is,

$$H_n = \sup \{h : \exists v \in \mathcal{L}_h, \ S_v \geq 1\}, \tag{6.3.3}$$

where $\mathcal{L}_h$ is the set of vertices of $\mathcal{T}$ at graph distance h from the root.

Key technical bound

Because $W \sim U[0, 1]$ implies also that $(1 - W) \sim U[0, 1]$, we immediately get from (6.3.2) that:

Lemma 6.3.2 (Distribution of subtree size). *Let v be a vertex at topological distance ℓ from the root of $\mathcal{T}$. Let $U_1, \ldots, U_\ell$ be i.i.d. $U[0, 1]$. Then we have the equality in distribution*

$$S_v \overset{d}{=} \lfloor \cdots \lfloor \lfloor nU_1 \rfloor U_2 \rfloor \cdots U_\ell \rfloor.$$

From Lemma 6.3.2 and the characterization of the height in (6.3.3), we need to control how fast products of independent uniforms decrease. But that is only half of the story: the number of paths of length ℓ from the root *grows exponentially with* ℓ. The following lemma, which takes both effects into account, will play a key role in the analysis. It also explains the definition of γ in (6.3.1). Note that we ignore – for the time being – the repeated rounding in Lemma 6.3.2; it will turn out to have a minor effect.

Lemma 6.3.3 (Product of uniforms). *Let $U_1, U_2, \ldots$ be i.i.d. $U[0, 1]$. Then*

$$\lim_{\ell \to +\infty} 2^\ell \, \mathbb{P}\left[U_1 \cdots U_\ell \geq e^{-\ell/c}\right] = \begin{cases} +\infty & \text{if } c < \gamma, \\ 0 & \text{if } c > \gamma. \end{cases}$$

Proof Taking logarithms turns the product on the left-hand side into a sum of i.i.d. random variables

$$2^\ell \, \mathbb{P}\left[U_1 \cdots U_\ell \geq e^{-\ell/c}\right] = 2^\ell \, \mathbb{P}\left[\sum_{i=1}^{\ell}(-\log U_i) \leq \ell/c\right]. \tag{6.3.4}$$

Now it is elementary to bound the right-hand side.

Lemma 6.3.4 (A tail bound). *Let $U_1, \ldots, U_\ell$ be i.i.d. $U[0, 1]$. Then, for any $y > 0$,*

$$\frac{y^\ell e^{-y}}{\ell!} \leq \mathbb{P}\left[\sum_{i=1}^{\ell}(-\log U_i) \leq y\right] \leq \frac{y^\ell e^{-y}}{\ell!}\left(\frac{1}{1 - \frac{y}{\ell+1}}\right). \tag{6.3.5}$$

Proof We prove a more general claim, specifically

$$\mathbb{P}\left[\sum_{i=1}^{\ell}(-\log U_i) \leq y\right] = e^{-y}\left\{\sum_{i=\ell}^{+\infty}\frac{y^i}{i!}\right\},$$

from which (6.3.5) follows: the lower bound is obtained by keeping only the first term in the sum; the upper bound is obtained by factoring out $y^\ell e^{-y}/\ell!$ and relating the remaining sum to a geometric series.

So it remains to prove the general claim. First note that $-\log U_1$ is exponentially distributed. Indeed, for any $y \geq 0$,

$$\mathbb{P}\left[-\log U_1 > y\right] = \mathbb{P}\left[U_1 < e^{-y}\right] = e^{-y}.$$

So

$$\mathbb{P}\left[-\log U_1 \leq y\right] = 1 - e^{-y} = e^{-y}\left\{\sum_{i=0}^{+\infty}\frac{y^i}{i!} - 1\right\} = e^{-y}\left\{\sum_{i=1}^{+\infty}\frac{y^i}{i!}\right\},$$

as claimed in the base case $\ell = 1$.

Proceeding by induction, suppose the claim holds up to $\ell - 1$. Then,

$$
\mathbb{P}\left[\sum_{i=1}^{\ell}(-\log U_i) > y\right] = \int_0^{+\infty} e^{-z} \mathbb{P}\left[\sum_{i=1}^{\ell-1}(-\log U_i) > y - z\right] dz
$$

$$
= e^{-y} + \int_0^y e^{-z} \mathbb{P}\left[\sum_{i=1}^{\ell-1}(-\log U_i) > y - z\right] dz
$$

$$
= e^{-y} + \int_0^y e^{-z} e^{-(y-z)} \left\{\sum_{i=0}^{\ell-2} \frac{(y-z)^i}{i!}\right\} dz
$$

$$
= e^{-y} + e^{-y} \sum_{i=0}^{\ell-2} \frac{y^{i+1}}{i!(i+1)}
$$

$$
= e^{-y} \sum_{j=0}^{\ell-1} \frac{y^j}{j!}.
$$

That proves the claim. ∎

We return to the proof of Lemma 6.3.3. Plugging (6.3.5) into (6.3.4), we get

$$
\frac{2^{\ell}(\ell/c)^{\ell} e^{-(\ell/c)}}{\ell!} \leq 2^{\ell} \mathbb{P}\left[U_1 \cdots U_{\ell} \geq e^{-\ell/c}\right]
$$

$$
\leq \frac{2^{\ell}(\ell/c)^{\ell} e^{-(\ell/c)}}{\ell!} \left(\frac{1}{1 - \frac{(\ell/c)}{\ell+1}}\right). \tag{6.3.6}
$$

As $\ell \to +\infty$,

$$
\frac{1}{1 - \frac{(\ell/c)}{\ell+1}} \to \frac{1}{1 - \frac{1}{c}}, \tag{6.3.7}
$$

which is positive when $c > 1$. We will use the standard bound (see Exercise 1.3 for a proof)

$$
\frac{\ell^{\ell}}{e^{\ell-1}} \leq \ell! \leq \frac{\ell^{\ell+1}}{e^{\ell-1}}.
$$

It implies immediately that

$$
\frac{2^{\ell}(\ell/c)^{\ell} e^{-(\ell/c)} e^{\ell-1}}{\ell^{\ell+1}} \leq \frac{2^{\ell}(\ell/c)^{\ell} e^{-(\ell/c)}}{\ell!} \leq \frac{2^{\ell}(\ell/c)^{\ell} e^{-(\ell/c)} e^{\ell-1}}{\ell^{\ell}},
$$

which after simplifying gives

$$
(e\ell)^{-1}\left[\left(\frac{1}{e}\right)\left(\frac{2e}{c}\right)^c\right]^{\ell/c} \leq 2^{\ell} \frac{(\ell/c)^{\ell} e^{-(\ell/c)}}{\ell!} \leq e^{-1}\left[\left(\frac{1}{e}\right)\left(\frac{2e}{c}\right)^c\right]^{\ell/c}. \tag{6.3.8}
$$

By (6.3.1) and the remark following it, the expression in square brackets is > 1 or < 1 depending on whether $c < \gamma$ or $c > \gamma$. Combining (6.3.6), (6.3.7), and (6.3.8) and taking a limit as $\ell \to +\infty$ gives the claim. ∎

As an immediate consequence of Lemma 6.3.3, we bound the height from above. Fix any $\varepsilon > 0$ and let $h := (\gamma + \varepsilon) \log n$. We use a union bound as follows:

$$\mathbb{P}[H_n \geq h] = \mathbb{P}\left[\bigcup_{v \in \mathcal{L}_h} \{S_v \geq 1\}\right] \leq \sum_{v \in \mathcal{L}_h} \mathbb{P}[S_v \geq 1] = 2^h \, \mathbb{P}[S_v \geq 1] \tag{6.3.9}$$

for any $v \in \mathcal{L}_h$, where the first equality follows from (6.3.3). Since

$$\lfloor \cdots \lfloor \lfloor nU_1 \rfloor U_2 \rfloor \cdots U_h \rfloor \leq nU_1 U_2 \cdots U_h,$$

Lemmas 6.3.2 and 6.3.3 imply that

$$\begin{aligned}
2^h \, \mathbb{P}[S_v \geq 1] &\leq 2^h \, \mathbb{P}[nU_1 U_2 \cdots U_h \geq 1] \\
&= 2^h \, \mathbb{P}\left[U_1 U_2 \cdots U_h \geq e^{-h/(\gamma+\varepsilon)}\right] \\
&\to 0
\end{aligned} \tag{6.3.10}$$

as $h \to +\infty$. From (6.3.9) and (6.3.10), we obtain finally that for any $\varepsilon > 0$,

$$\mathbb{P}[H_n / \log n \geq \gamma + \varepsilon] \to 0$$

as $n \to +\infty$, which establishes one direction of Claim 6.3.1.

Lower bounding the height: a branching process

Establishing the other direction is where branching processes enter the scene. We will need some additional notation. Fix $c < \gamma$ and let ℓ be a positive integer that will be set later on. For any pair of vertex $v, w \in \mathcal{T}$ with w a descendant of v, let $\mathcal{Q}[v, w]$ be the set of vertices on the path between v and w, including v but excluding w. Further, recalling (6.3.2), define

$$\mathcal{U}[v, w] = \prod_{z \in \mathcal{Q}[u,v]} U_z^{v,w},$$

where $U_z^{v,w} = W_z$ (respectively $1 - W_z$) if the path from v to w takes the left (respectively right) edge upon exiting z. Denote by $\mathcal{L}_\ell[v]$ the set of descendant vertices of v in $\mathcal{T}$ at graph distance ℓ from v and consider the random subset

$$\mathcal{L}_\ell^*[v] = \left\{w \in \mathcal{L}_\ell[v] : \mathcal{U}[v, w] \geq e^{-\ell/c}\right\}.$$

Fix a vertex $u \in \mathcal{T}$. We define the following Galton–Watson branching process.

- Initialize $Z_0^{u,\ell} := 1$ and $u_{0,1} := u$.
- For $t \geq 1$, set

$$Z_t^{u,\ell} = \sum_{r=1}^{Z_{t-1}^{u,\ell}} \left|\mathcal{L}_\ell^*[u_{t-1,r}]\right|,$$

and let $u_{t,1}, \ldots, u_{t,Z_t^{u,\ell}}$ be the vertices in $\cup_{r=1}^{Z_{t-1}^{u,\ell}} \mathcal{L}_\ell^*[u_{t-1,r}]$ from left to right.

In other words, $Z_1^{u,\ell}$ counts the number of vertices ℓ levels below u whose subtree sizes (ignoring rounding) have not decreased "too much" compared to that of u (in the sense of Lemma 6.3.3). We let such vertices (if any) be $u_{1,1}, \ldots, u_{1,Z_1^{u,\ell}}$. Similarly, $Z_2^{u,\ell}$ counts the same quantity over all vertices ℓ levels below the vertices $u_{1,1}, \ldots, u_{1,Z_1^{u,\ell}}$, and so forth.

Because the W_vs are i.i.d., this process is indeed a Galton–Watson branching process. The expectation of the offspring distribution (which by symmetry does not depend on the choice of u) is

$$m = \mathbb{E}\left[Z_1^{u,\ell}\right] = 2^{\ell}\,\mathbb{P}\left[U_1 \cdots U_{\ell} \geq e^{-\ell/c}\right],$$

where we used the notation of Lemma 6.3.3. By that lemma, we can choose ℓ large enough that $m > 1$. Fix such an ℓ for the rest of the proof. In that case, by Theorem 6.1.6, the process survives with probability $1 - \eta$ for some $0 \leq \eta < 1$.

The relevance of this observation can be seen from taking $u = \rho$.

Claim 6.3.5 *Let $c' < c$. Conditioned on survival of $(Z_t^{\rho,\ell})$, for n large enough $H_n \geq c' \log n - \theta_n \ell$ almost surely for some $\theta_n \in [0, 1)$.*

Proof To account for the rounding, we will need the inequality

$$\lfloor \cdots \lfloor \lfloor nU_1 \rfloor U_2 \rfloor \cdots U_s \rfloor \geq nU_1 U_2 \cdots U_s - s, \tag{6.3.11}$$

which holds for all $n, s \geq 1$, as can be checked by induction. Write $s = k\ell$ for some positive integer k to be determined. Conditioned on survival of $(Z_t^{\rho,\ell})$, the population at generation k satisfies

$$Z_k^{\rho,\ell} \geq 1,$$

which implies that, for some $v^* \in \mathcal{L}_s[\rho]$, it holds that

$$n\mathcal{U}[\rho, v^*] \geq n(e^{-\ell/c})^k.$$

Now take $s = c' \log n - \theta_n \ell$ with $c' < c$ and $\theta_n \in [0, 1)$ such that s is a multiple of ℓ. Then,

$$n(e^{-\ell/c})^k = n(e^{-s/c}) = n(n^{-c'/c}e^{-\theta_n \ell/c}) = n^{1-c'/c}e^{-\theta_n \ell/c}$$
$$\geq c' \log n - \theta_n \ell + 1 = s + 1 \tag{6.3.12}$$

for all n large enough, where we used that $1 - c'/c > 0$, $\theta_n \in [0, 1)$ and ℓ is fixed. So, using the characterization of the height in (6.3.2) and (6.3.3) together with inequality (6.3.11), we derive

$$S_{v^*} \geq n\mathcal{U}[\rho, v^*] - s \geq n(e^{-\ell/c})^k - s \geq 1. \tag{6.3.13}$$

That is, $H_n \geq c' \log n - \theta_n \ell$. ∎

But this is not quite what we want: this last claim holds only *conditioned on survival*; or put differently, it holds with probability $1 - \eta$, a value which could be significantly smaller than 1 in general. To handle this last issue, we consider a large number of *independent copies* of the Galton–Watson process above in order to "boost" the probability that *at least one of them survives* to a value arbitrarily close to 1.

Claim 6.3.6 *For any $\delta > 0$, there is a J so that $H_n \geq c' \log n - \theta_n \ell + J\ell$ with probability at least $1 - \delta$ for all n large enough.*

Proof Let $J\ell$ be a multiple of ℓ and let

$$u_1^*, \ldots, u_{2^{J\ell}}^*$$

be the vertices on level $J\ell$ from left to right. Each process

$$\left(Z_t^{u_i^*,\ell}\right)_{t\geq 0}, \quad i = 1,\ldots,2^{J\ell}$$

is an independent copy of $(Z_t^{\rho,\ell})_{t\geq 0}$.

We define two "bad events":

- *(No survival)* Let $\mathcal{B}_1$ be the event that all $(Z_t^{u_i^*,\ell})$s go extinct and choose J large enough that this event has probability $< \delta/2$, that is,

$$\mathbb{P}[\mathcal{B}_1] = \eta^{2^{J\ell}} < \delta/2.$$

Under $\mathcal{B}_1^c$, at least one of the branching processes survives; let I be the lowest index among them.

- *(Fast decay at the top)* To bound the height, we also need to control the effect of the first $J\ell$ levels on the subtree sizes. Let $\mathcal{B}_2$ be the event that at least one of the W-values associated with the $2^{J\ell} - 1$ vertices ancestral to the u_i^*s is outside the interval $(\alpha, 1 - \alpha)$. Choose α small enough that this event has probability $< \delta/2$, that is,

$$\mathbb{P}[\mathcal{B}_2] \leq (2\alpha)(2^{J\ell} - 1) < \delta/2.$$

Under $\mathcal{B}_2^c$, we have almost surely the lower bound

$$\mathcal{U}[\rho, u_I^*] \geq \alpha^{J\ell}, \tag{6.3.14}$$

since it in fact holds for all u_i^*s simultaneously.

We are now ready to conclude. Assume $\mathcal{B}_1^c$ and $\mathcal{B}_2^c$ hold. Taking

$$s = k\ell = c' \log n - \theta_n \ell$$

as before, we have $Z_k^{u_I^*,\ell} \geq 1$ so there is $v^* \in \mathcal{L}_s[u_I^*]$ such that

$$n\mathcal{U}[\rho, v^*] = n\mathcal{U}[\rho, u_I^*]\mathcal{U}[u_I^*, v^*] \geq n\alpha^{J\ell}(e^{-\ell/c})^k,$$

where we used (6.3.14). Observe that (6.3.12) remains valid (for potentially larger n) even after multiplying all expressions on the left-hand side of the inequality by $\alpha^{J\ell}$. Arguing as in (6.3.13), we get that $H_n \geq c' \log n - \theta_n \ell + J\ell$. This event holds with probability at least

$$\mathbb{P}[(\mathcal{B}_1 \cup \mathcal{B}_2)^c] = 1 - \mathbb{P}[\mathcal{B}_1] - \mathbb{P}[\mathcal{B}_2] \geq 1 - \delta.$$

We have proved the claim. ∎

For any $\varepsilon > 0$, we can choose $c' = \gamma - \varepsilon$ and $c' < c < \gamma$. Furthermore, δ can be made arbitrarily small (provided n is large enough). Put differently, we have proved that for any $\varepsilon > 0$,

$$\mathbb{P}[H_n/\log n \geq \gamma - \varepsilon] \to 1$$

as $n \to +\infty$, which establishes the other direction of Claim 6.3.1.

6.3.2 ▷ Data Science: The Reconstruction Problem, the Kesten–Stigum Bound and a Phase Transition in Phylogenetics

In this section, we explore an application of multitype branching processes in statistical phylogenetics, the reconstruction of evolutionary trees from molecular data. Informally, we consider a ferromagnetic Ising model (Example 1.2.5) on an infinite binary tree and we ask: When do the states at level h "remember" the state at the root? We establish the existence of a phase transition. Before defining the problem formally and explaining its connection to evolutionary biology, we describe an equivalent definition of the model. This alternative "Markov chain on a tree" perspective will make it easier to derive recursions for quantities of interest. Equivalence between the two models is proved in Exercise 6.16.

The reconstruction problem

MUTATION PROBABILITY

CFN MODEL

Consider a rooted infinite binary tree $\mathcal{T} = \widehat{\mathbb{T}}_2$, where the root is denoted by 0. Fix a parameter $0 < p < 1/2$, which we will refer to as the *mutation probability* for reasons that will be explained below. We assign a state σ_v in $\mathcal{C} = \{+1, -1\}$ to each vertex v as follows. At the root 0, the state σ_0 is picked uniformly at random in $\{+1, -1\}$. Moving away from the root, the state σ_v at a vertex v, conditioned on the state at its immediate ancestor u, is equal to σ_u with probability $1 - p$ and to $-\sigma_u$ with probability p. In the computational biology literature, this model is referred to as the *Cavender–Farris–Neyman (CFN) model*.

RECONSTRUCTION PROBLEM

For $h \geq 0$, let $\mathcal{L}_h$ be the set of vertices in $\mathcal{T}$ at graph distance h from the root. We denote by $\boldsymbol{\sigma}_h = (\sigma_\ell)_{\ell \in \mathcal{L}_h}$ the vector of states at level h and we denote by μ_h the distribution of $\boldsymbol{\sigma}_h$. The *reconstruction problem* consists in trying to "guess" the state at the root σ_0 given the states $\boldsymbol{\sigma}_h$ at level h. We first note that in general we cannot expect an arbitrarily good estimator. Indeed, rewriting the Markov transition matrix along the edges (i.e., the matrix encoding the probability of the state at a vertex given the state at its immediate ancestor) in its *random cluster* form

$$P := \begin{pmatrix} 1-p & p \\ p & 1-p \end{pmatrix} = (1-2p) \begin{pmatrix} 1 & 0 \\ 0 & 1 \end{pmatrix} + (2p) \begin{pmatrix} 1/2 & 1/2 \\ 1/2 & 1/2 \end{pmatrix}, \tag{6.3.15}$$

we see that the states $\boldsymbol{\sigma}_1$ at the first level are completely randomized (i.e., independent of σ_0) with probability $(2p)^2$ – in which case we cannot hope to reconstruct the root state better than a coin flip. Intuitively, the reconstruction problem is solvable if we can find an estimator of the root state which outperforms a random coin flip as h grows to $+\infty$. Let μ_h^+ be the distribution μ_h conditioned on the root state σ_0 being $+1$, and similarly for μ_h^-. Observe that $\mu_h = \frac{1}{2}\mu_h^+ + \frac{1}{2}\mu_h^-$. Recall also that

$$\|\mu_h^+ - \mu_h^-\|_{\mathrm{TV}} = \frac{1}{2} \sum_{\mathbf{s}_h \in \{+1,-1\}^{2^h}} |\mu_h^+(\mathbf{s}_h) - \mu_h^-(\mathbf{s}_h)|.$$

RECONSTRUCTION SOLVABILITY

Definition 6.3.7 (Reconstruction solvability). *We say that the reconstruction problem for $0 < p < 1/2$ is solvable if*

$$\liminf_{h \to +\infty} \|\mu_h^+ - \mu_h^-\|_{\mathrm{TV}} > 0,$$

otherwise the problem is unsolvable.

(Exercise 6.17 asks for a proof that $\|\mu_h^+ - \mu_h^-\|_{\text{TV}}$ is monotone in h and therefore has a limit.)

To see the connection with the description above, consider an arbitrary root estimator $\hat{\sigma}_0(\mathbf{s}_h)$. Then the probability of a mistake is

$$\mathbb{P}[\hat{\sigma}_0(\boldsymbol{\sigma}_h) \neq \sigma_0] = \frac{1}{2} \sum_{\mathbf{s}_h \in \{+1,-1\}^{2^h}} \mu_h^-(\mathbf{s}_h)\mathbf{1}\{\hat{\sigma}_0(\mathbf{s}_h) = +1\}$$

$$+ \frac{1}{2} \sum_{\mathbf{s}_h \in \{+1,-1\}^{2^h}} \mu_h^+(\mathbf{s}_h)\mathbf{1}\{\hat{\sigma}_0(\mathbf{s}_h) = -1\}.$$

This expression is minimized by choosing for each $\mathbf{s}_h$ separately

$$\hat{\sigma}_0(\mathbf{s}_h) = \begin{cases} +1 & \text{if } \mu_h^+(\mathbf{s}_h) \geq \mu_h^-(\mathbf{s}_h), \\ -1 & \text{otherwise.} \end{cases}$$

Let $\mu_h(s_0|\mathbf{s}_h)$ be the posterior probability of the root state, that is, the conditional probability of the root state s_0 given the states $\mathbf{s}_h$ at level h. By Bayes' rule,

$$\mu_h(+1|\mathbf{s}_h) = \frac{(1/2)\mu_h^+(\mathbf{s}_h)}{\mu_h(\mathbf{s}_h)},$$

and similarly for $\mu_h(+1|\mathbf{s}_h)$. Hence, the choice above is equivalent to

$$\hat{\sigma}_0(\mathbf{s}_h) = \begin{cases} +1 & \text{if } \mu_h(+1|\mathbf{s}_h) \geq \mu_h(-|\mathbf{s}_h), \\ -1 & \text{otherwise,} \end{cases}$$

which is known as the *maximum a posteriori (MAP) estimator*. (We encountered it in a MAP
different context in Section 5.1.4.) For short, we will denote it by $\hat{\sigma}_0^{\text{MAP}}$.   ESTIMATOR

Now note that

$$\mathbb{P}[\hat{\sigma}_0^{\text{MAP}}(\boldsymbol{\sigma}_h) = \sigma_0] - \mathbb{P}[\hat{\sigma}_0^{\text{MAP}}(\boldsymbol{\sigma}_h) \neq \sigma_0]$$

$$= \frac{1}{2} \sum_{\mathbf{s}_h \in \{+1,-1\}^{2^h}} \mu_h^+(\mathbf{s}_h) [\mathbf{1}\{\hat{\sigma}_0^{\text{MAP}}(\mathbf{s}_h) = +1\} - \mathbf{1}\{\hat{\sigma}_0^{\text{MAP}}(\mathbf{s}_h) = -1\}]$$

$$+ \frac{1}{2} \sum_{\mathbf{s}_h \in \{+1,-1\}^{2^h}} \mu_h^-(\mathbf{s}_h) [\mathbf{1}\{\hat{\sigma}_0^{\text{MAP}}(\mathbf{s}_h) = -1\} - \mathbf{1}\{\hat{\sigma}_0^{\text{MAP}}(\mathbf{s}_h) = +1\}]$$

$$= \frac{1}{2} \sum_{\mathbf{s}_h \in \{+1,-1\}^{2^h}} \mu_h^+(\mathbf{s}_h) \hat{\sigma}_0^{\text{MAP}}(\mathbf{s}_h)$$

$$- \frac{1}{2} \sum_{\mathbf{s}_h \in \{+1,-1\}^{2^h}} \mu_h^-(\mathbf{s}_h) \hat{\sigma}_0^{\text{MAP}}(\mathbf{s}_h)$$

$$= \frac{1}{2} \sum_{\mathbf{s}_h \in \{+1,-1\}^{2^h}} |\mu_h^+(\mathbf{s}_h) - \mu_h^-(\mathbf{s}_h)|$$

$$= \|\mu_h^+ - \mu_h^-\|_{\text{TV}},$$

where the third equality comes from

$$|a - b| = (a - b)\mathbf{1}\{a \geq b\} + (b - a)\mathbf{1}\{a < b\}.$$

Since $\mathbb{P}[\hat{\sigma}_0(\boldsymbol{\sigma}_h) = \sigma_0] + \mathbb{P}[\hat{\sigma}_0(\boldsymbol{\sigma}_h) \neq \sigma_0] = 1$, the display above can be rewritten as

$$\mathbb{P}[\hat{\sigma}_0^{\mathrm{MAP}}(\boldsymbol{\sigma}_h) \neq \sigma_0] = \frac{1}{2} - \frac{1}{2}\|\mu_h^+ - \mu_h^-\|_{\mathrm{TV}}.$$

Given that $\hat{\sigma}_0^{\mathrm{MAP}}$ was chosen to minimize the error probability, we also have that *for any root estimator* $\hat{\sigma}_0$,

$$\mathbb{P}[\hat{\sigma}_0(\boldsymbol{\sigma}_h) \neq \sigma_0] \geq \frac{1}{2} - \frac{1}{2}\|\mu_h^+ - \mu_h^-\|_{\mathrm{TV}}.$$

Since this last inequality also applies to the estimator $-\hat{\sigma}_0$, we have also that

$$\mathbb{P}[\hat{\sigma}_0(\boldsymbol{\sigma}_h) \neq \sigma_0] \leq \frac{1}{2}.$$

The next lemma summarizes the discussion above.

Lemma 6.3.8 (Probability of erroneous reconstruction). *The probability of an erroneous root reconstruction behaves as follows.*

(i) *If the reconstruction problem is solvable, then*

$$\lim_{h \to +\infty} \mathbb{P}[\hat{\sigma}_0^{\mathrm{MAP}}(\boldsymbol{\sigma}_h) \neq \sigma_0] < \frac{1}{2}.$$

(ii) *If the reconstruction problem is unsolvable, then for any root estimator* $\hat{\sigma}_0$

$$\lim_{h \to +\infty} \mathbb{P}[\hat{\sigma}_0(\boldsymbol{\sigma}_h) \neq \sigma_0] = \frac{1}{2}.$$

It turns out that the accuracy of the MAP estimator undergoes a phase transition at a critical mutation probability p_*. Our main theorem is the following.

Theorem 6.3.9 (Solvability). *Let θ_* be the unique positive solution to*

$$2\theta_*^2 = 1,$$

and set $p^* = \frac{1-\theta_*}{2}$. Then the reconstruction problem is:

(i) *solvable if $0 < p < p_*$;*
(ii) *unsolvable if $p_* \leq p < 1/2$.*

We will prove this theorem in the rest of the section.

But first, what does all of this have to do with evolutionary biology? Truncate $\mathcal{T}$ at level h to obtain a finite tree $\mathcal{T}_h$ with leaf set $\mathcal{L}_h$. In phylogenetics, one uses such a tree to depict evolutionary relationships between extant species that are represented by its leaves. Each internal branching corresponds to a past speciation event. Extinctions have been pruned from the tree. The genomes of ancestral species, starting from the most recent common ancestor at the root, are posited to have evolved along the (deterministic) tree $\mathcal{T}_h$ according to a random process of single-site substitutions. To simplify, each position in the genome is assumed to take one of two values, $+1$ or -1, and it evolves independently from all other positions under a CFN model on $\mathcal{T}_h$. That is, on each edge of the tree a mutation occurs with probability p, changing the state of the immediate descendant species at that position. This is of course only a toy model, but it is not far from what evolutionary biologists actually use in practice with great success. One practical problem of interest is to reconstruct the genome of ancestors given access to contemporary genomes. This is, in a nutshell, the reconstruction problem.

Kesten–Stigum bound

The condition in Theorem 6.3.9 is referred to as the *Kesten–Stigum bound*. We explain why next. We showed in Lemma 6.3.8 that the MAP estimator has an error probability bounded away from $1/2$ if and only if the reconstruction problem is solvable. Of course, other estimators may also achieve that same desirable outcome. In fact, from the lemma, to establish reconstruction solvability it *suffices* to exhibit one such "better-than-random" estimator. So, rather than analyzing $\hat{\sigma}_0^{\mathrm{MAP}}$, we look at a simpler estimator first and prove half of Theorem 6.3.9. The other half will be proven below using different ideas.

The key is to notice that a multitype branching process (see Section 6.1.4) hides in the background. For $h \geq 0$, consider the random row vector $\mathbf{Z}_h = (Z_{h,+}, Z_{h,-})$ where the first component records the number of $+1$ states (which we refer to as belonging to the $+$ type) in σ_h and, likewise, the second component counts the -1 states (referred to as of $-$ type). Then, $(\mathbf{Z}_h)_{h \geq 0}$ is a two-type Galton–Watson process where each individual has exactly two children. Their types depend on the type of the parent. A type $+$ individual has the following offspring distribution:

$$
p_{\mathbf{k}}^{(+)} = \begin{cases} (1-p)^2 & \text{if } \mathbf{k} = (2,0), \\ 2p(1-p) & \text{if } \mathbf{k} = (1,1), \\ p^2 & \text{if } \mathbf{k} = (0,2), \\ 0 & \text{otherwise.} \end{cases}
$$

Similar expressions hold for $p_{\mathbf{k}}^{(-)}$. The mean matrix is given by

$$
\begin{aligned}
M &= \begin{pmatrix} 2(1-p)^2 + 2p(1-p) & 2p(1-p) + 2p^2 \\ 2p(1-p) + 2p^2 & 2(1-p)^2 + 2p(1-p) \end{pmatrix} \\
&= 2 \begin{pmatrix} (1-p)(1-p+p) & p(1-p+p) \\ p(1-p+p) & (1-p)(1-p+p) \end{pmatrix} \\
&= 2 \begin{pmatrix} 1-p & p \\ p & 1-p \end{pmatrix} \\
&= 2P,
\end{aligned}
$$

where (not coincidentally) we have already encountered the matrix P in (6.3.15). As a symmetric matrix, by the spectral theorem (Theorem 5.1.1), P has a real eigenvector decomposition

$$P = \begin{pmatrix} 1-p & p \\ p & 1-p \end{pmatrix}$$

$$= \begin{pmatrix} 1/2 & 1/2 \\ 1/2 & 1/2 \end{pmatrix} + (1-2p) \begin{pmatrix} 1/2 & -1/2 \\ -1/2 & 1/2 \end{pmatrix}$$

$$= \lambda_1 \mathbf{x}_1 \mathbf{x}_1^T + \lambda_2 \mathbf{x}_2 \mathbf{x}_2^T,$$

where the eigenvalues and eigenvectors are

$$\lambda_1 = 1, \quad \lambda_2 = 1 - 2p, \quad \mathbf{x}_1 = \begin{pmatrix} 1/\sqrt{2} \\ 1/\sqrt{2} \end{pmatrix}, \quad \mathbf{x}_2 = \begin{pmatrix} 1/\sqrt{2} \\ -1/\sqrt{2} \end{pmatrix}.$$

The eigenvalues of M are twice those of P, while the eigenvectors are the same. In particular, using the notation and convention of the Perron–Frobenius Theorem (Theorem 6.1.17), we have

$$\rho = 2, \quad \mathbf{w} = \begin{pmatrix} 1/2 \\ 1/2 \end{pmatrix}.$$

These should not come entirely as a surprise. In particular, recall from Theorem 6.1.18 that ρ can be interpreted as an "overall rate of growth" of the population, which here is two since each individual has exactly two children (ignoring the types).

Let $\mathbf{u} = (1, -1)$ be a column vector proportional to the second right eigenvector of M. We know from Section 6.1.4 that

$$U_h = (2\lambda_2)^{-h} \mathbf{Z}_h \mathbf{u} = \frac{1}{2^h \theta^h} \sum_{\ell \in \mathcal{L}_h} \sigma_\ell, \quad h \geq 0$$

is a martingale, where we used the notation

$$\theta := \lambda_2 = 1 - 2p.$$

MAJORITY
ESTIMATOR Upon looking more closely, the quantity U_h has a natural interpretation: its sign is the *majority estimator*, that is, $\mathrm{sgn}(U_h) = +1$ if a majority of individuals at level h are of type $+$ (breaking ties in favor of $+$), and is -1 otherwise. We indicated previously that we only need to find one estimator with an error probability bounded away from $1/2$ to establish reconstruction solvability for a given value of p. The majority estimator

$$\hat{\sigma}_0^{\mathrm{Maj}} := \mathrm{sgn}(U_h)$$

is an obvious one to try. What is less obvious is that it works – all the way to the threshold. This essentially follows from the results of Section 6.1.4, as we detail next.

We begin with an informal discussion. When can $\hat{\sigma}_0^{\mathrm{Maj}}$ be expected to work? We will not in fact bound the error probability of $\hat{\sigma}_0^{\mathrm{Maj}}$, but instead analyze directly the properties of (U_h). By our modeling assumptions, $\mathbf{Z}_0$ is either $(1, 0)$ or $(0, 1)$ with equal probability. Hence, by the martingale property, we obtain that

$$\mathbb{E}[U_h \mid \mathbf{Z}_0] = \mathbf{Z}_0 \mathbf{u} = \sigma_0. \tag{6.3.16}$$

In other words, U_h is "centered" around the root state. Intuitively, its second moment therefore captures how informative it is about σ_0. Lemma 6.1.20 exhibits a phase transition for $\mathbb{E}[U_h^2 \mid \mathbf{Z}_0]$. The condition for that lemma to hold is

$$(\text{Var}[\mathbf{X}^{(+)}(1,1)\,\mathbf{u}], \text{Var}[\mathbf{X}^{(-)}(1,1)\,\mathbf{u}]) \neq \mathbf{0},$$

where $\mathbf{X}^{(+)}(1,1) \sim \{p_{\mathbf{k}}^{(+)}\}$ and $\mathbf{X}^{(-)}(1,1) \sim \{p_{\mathbf{k}}^{(-)}\}$. This is indeed satisfied. The lemma then states that $\mathbb{E}[U_h^2 \mid \mathbf{Z}_0]$ is uniformly bounded if and only if $\rho < (2\lambda_2)^2$, or after rearranging,

$$2\theta^2 > 1. \tag{6.3.17}$$

Note that this is the condition in Theorem 6.3.9. It arises as a trade-off between the rate of growth $\rho = 2$ and the second largest eigenvalue $\lambda_2 = \theta$ of the Markov transition matrix P. One way to make sense of it is to observe the following:

- On any infinite path out of the root, the process performs a finite Markov chain with transition matrix P. We know from Theorem 5.2.14 (see in particular Example 5.2.8) that the chain mixes – and therefore "forgets" its starting state σ_0 – at a rate governed by the spectral gap $1 - \lambda_2$.
- On the other hand, the tree itself is growing at rate $\rho = 2$, which produces an exponentially large number of (overlapping) paths out of the root. That growth helps preserve the information about σ_0 down the tree through the duplication of the state (with mutation) at each branching.
- The condition $\rho < (2\lambda_2)^2$ says in essence that when mixing is slow enough – corresponding to larger values of λ_2 – compared to the growth, then the reconstruction problem is solvable. Lemma 6.1.20 was first proved by Kesten and Stigum, and (6.3.17) is thereby known as the Kesten–Stigum bound.

It remains to turn these observations into a formal proof.

Denote by $\mathbb{E}^+$ the expectation conditioned on $\sigma_0 = +1$, and similarly for $\mathbb{E}^-$. The following lemma is a consequence of (6.3.16). We give a quick alternative proof.

Lemma 6.3.10 (Unbiasedness of U_h). *We have*

$$\mathbb{E}^+[U_h] = +1, \qquad \mathbb{E}^-[U_h] = -1.$$

Proof By applying the Markov transition matrix P on the first level and using the symmetries of the model, for any $\ell \in \mathcal{L}_h$ and $\ell' \in \mathcal{L}_{h-1}$, we have

$$\begin{aligned}
\mathbb{E}^+[\sigma_\ell] &= (1-p)\mathbb{E}^+[\sigma_{\ell'}] + p\,\mathbb{E}^-[\sigma_{\ell'}] \\
&= (1-p)\mathbb{E}^+[\sigma_{\ell'}] + p\,\mathbb{E}^+[-\sigma_{\ell'}] \\
&= (1-2p)\mathbb{E}^+[\sigma_{\ell'}] \\
&= \theta\,\mathbb{E}^+[\sigma_{\ell'}].
\end{aligned}$$

Iterating, we get $\mathbb{E}^+[\sigma_\ell] = \theta^h$. The claim follows by linearity of expectation. ∎

Although we do not strictly need it, we also derive an explicit formula for the variance. The proof is typical of how conditional independence properties of this kind of Markov model on trees can be used to derive recursions for quantities of interest.

Lemma 6.3.11 (Variance of U_h). *We have*

$$\text{Var}[U_h] \to \begin{cases} \dfrac{1/2}{1-(2\theta^2)^{-1}} & \text{if } 2\theta^2 > 1, \\ +\infty & \text{otherwise.} \end{cases}$$

Proof By the conditional variance formula

$$\text{Var}[U_h] = \text{Var}[\mathbb{E}[U_h \mid \sigma_0]] + \mathbb{E}[\text{Var}[U_h \mid \sigma_0]]$$
$$= \text{Var}[\sigma_0] + \mathbb{E}[\text{Var}[U_h \mid \sigma_0]]$$
$$= 1 + \text{Var}^+[U_{\bar{h}}], \tag{6.3.18}$$

where the last line follows from symmetry, with Var^+ indicating the conditional variance given that the root state σ_0 is $+1$. Write $U_h = \dot{U}_h + \ddot{U}_h$ as a sum over the left and right subtrees below the root, respectively. Using the conditional independence of those two subtrees given the root state, we get from (6.3.18) that

$$\text{Var}[U_h] = 1 + \text{Var}^+[U_h]$$
$$= 1 + \text{Var}^+\left[\dot{U}_h + \ddot{U}_h\right]$$
$$= 1 + 2\text{Var}^+\left[\dot{U}_h\right]$$
$$= 1 + 2\left(\mathbb{E}^+\left[\dot{U}_h^2\right] - \mathbb{E}^+\left[\dot{U}_h\right]^2\right). \tag{6.3.19}$$

We now use the Markov transition matrix on the first level to derive a recursion in h. Let $\dot{\sigma}_0$ be the state at the left child of the root. We use the fact that the random variables $2\theta \dot{U}_h$ conditioned on $\dot{\sigma}_0 = +1$ and U_{h-1} conditioned on $\sigma_0 = +1$ are identically distributed. Using $\mathbb{E}^+[\dot{U}_h] = 1/2$ (by Lemma 6.3.10 and symmetry), we get from (6.3.19) that

$$\text{Var}[U_h] = 1 - 2\mathbb{E}^+\left[\dot{U}_h\right]^2 + 2\,\mathbb{E}^+\left[\dot{U}_h^2\right]$$
$$= 1 - 2(1/2)^2 + 2\left[(1-p)\mathbb{E}^+\left[(2\theta)^{-2}U_{h-1}^2\right] + p\,\mathbb{E}^-\left[(2\theta)^{-2}U_{h-1}^2\right]\right]$$
$$= 1/2 + (2\theta^2)^{-1}\mathbb{E}^+[U_{h-1}^2]$$
$$= 1/2 + (2\theta^2)^{-1}\text{Var}[U_{h-1}], \tag{6.3.20}$$

where we used that

$$\text{Var}[U_{h-1}] = \mathbb{E}[U_{h-1}^2] = \mathbb{E}^+[U_{h-1}^2] = \mathbb{E}^-[U_{h-1}^2]$$

by symmetry and the fact that $\mathbb{E}[U_{h-1}] = 0$. Solving the affine recursion (6.3.20) gives

$$\text{Var}[U_h] = (2\theta^2)^{-h} + (1/2)\sum_{i=0}^{h-1}(2\theta^2)^{-i},$$

where we used that $\text{Var}[U_0] = \text{Var}[\sigma_0] = 1$. The result follows. ∎

We can now prove the first part of Theorem 6.3.9.

Proof of Theorem 6.3.9 (i) Let $\bar{\mu}_h$ be the distribution of U_h and define $\bar{\mu}_h^+$ and $\bar{\mu}_h^-$ similarly. We give a bound on $\|\mu_h^+ - \mu_h^-\|_{\text{TV}}$ through a bound on $\|\bar{\mu}_h^+ - \bar{\mu}_h^-\|_{\text{TV}}$. Let $\bar{\mathbf{s}}_h$ be the U_h-value associated to $\mathbf{s}_h = (s_{h,\ell})_{\ell \in \mathcal{L}_h} \in \{+1, -1\}^{2^h}$, that is,

$$\bar{\mathbf{s}}_h = \frac{1}{2^h\theta^h}\sum_{\ell \in \mathcal{L}_h} s_{h,\ell}.$$

Then, by marginalizing and the triangle inequality,

$$
\sum_z |\bar{\mu}_h^+(z) - \bar{\mu}_h^-(z)| = \sum_z \left| \sum_{\mathbf{s}_h : \bar{\mathbf{s}}_h = z} (\mu_h^+(\mathbf{s}_h) - \mu_h^-(\mathbf{s}_h)) \right|
$$

$$
\leq \sum_z \sum_{\mathbf{s}_h : \bar{\mathbf{s}}_h = z} |\mu_h^+(\mathbf{s}_h) - \mu_h^-(\mathbf{s}_h)|
$$

$$
= \sum_{\mathbf{s}_h \in \{+1,-1\}^{2^h}} |\mu_h^+(\mathbf{s}_h) - \mu_h^-(\mathbf{s}_h)|,
$$

where the first sum is over the support of $\bar{\mu}_h$. So it suffices to bound from below the left-hand side on the first line.

For that purpose, we apply Cauchy–Schwarz and use the variance bound in Lemma 6.3.11. First note that $\frac{1}{2}\bar{\mu}_h^+ + \frac{1}{2}\bar{\mu}_h^- = \bar{\mu}_h$ so that, by the triangle inequality,

$$
\frac{|\bar{\mu}_h^+(z) - \bar{\mu}_h^-(z)|}{2\bar{\mu}_h(z)} \leq \frac{\bar{\mu}_h^+(z) + \bar{\mu}_h^-(z)}{2\bar{\mu}_h(z)} = 1. \tag{6.3.21}
$$

Hence, we get

$$
\sum_z |\bar{\mu}_h^+(z) - \bar{\mu}_h^-(z)| = \sum_z \frac{|\bar{\mu}_h^+(z) - \bar{\mu}_h^-(z)|}{2\bar{\mu}_h(z)} 2\bar{\mu}_h(z)
$$

$$
\geq 2 \sum_z \left(\frac{\bar{\mu}_h^+(z) - \bar{\mu}_h^-(z)}{2\bar{\mu}_h(z)} \right)^2 \bar{\mu}_h(z)
$$

$$
\geq 2 \frac{\left(\sum_z z \left(\frac{\bar{\mu}_h^+(z) - \bar{\mu}_h^-(z)}{2\bar{\mu}_h(z)} \right) \bar{\mu}_h(z) \right)^2}{\sum_z z^2 \bar{\mu}_h(z)}
$$

$$
\geq \frac{1}{2} \frac{\left(\sum_z z \left(\bar{\mu}_h^+(z) - \bar{\mu}_h^-(z) \right) \right)^2}{\sum_z z^2 \bar{\mu}_h(z)}
$$

$$
= \frac{1}{2} \frac{(\mathbb{E}^+[U_h] - \mathbb{E}^-[U_h])^2}{\text{Var}[U_h]}
$$

$$
\geq 4(1 - (2\theta^2)^{-1}) > 0,
$$

where we used (6.3.21) on the second line, Cauchy–Schwarz on the third line (after rearranging), and Lemmas 6.3.10 and 6.3.11 on the last line. ∎

Remark 6.3.12 *The proof above and a correlation inequality of [EKPS00, Theorem 1.4] give a lower bound on the probability of reconstruction of the majority estimator.*

Impossibility of reconstruction

The previous result was based on showing that majority voting, that is, $\hat{\sigma}_0^{\text{Maj}}$, produces a good root-state estimator – up to $p = p_*$. Here we establish that this result is best possible. Majority is not in fact the best root-state estimator: in general, its error probability can be higher than $\hat{\sigma}_0^{\text{MAP}}$ as the latter also takes into account the *configuration of the states at level*

h. However, perhaps surprisingly, it turns out that the critical threshold for $\hat{\sigma}_0^{\text{Maj}}$ coincides with that of $\hat{\sigma}_0^{\text{MAP}}$ in the CFN model.

To prove the second part of Theorem 6.3.9 we analyze the MAP estimator. Recall that $\mu_h(s_0|\mathbf{s}_h)$ is the conditional probability of the root state s_0 given the states $\mathbf{s}_h$ at level h. It will be more convenient to work with the following "root magnetization":

$$R_h := \mu_h(+1|\boldsymbol{\sigma}_h) - \mu_h(-1|\boldsymbol{\sigma}_h),$$

which, as a function of $\boldsymbol{\sigma}_h$, is a *random variable*. Note that $\mathbb{E}[R_h] = 0$ by symmetry. By Bayes' rule and the fact that $\mu_h(+1|\boldsymbol{\sigma}_h) + \mu_h(-1|\boldsymbol{\sigma}_h) = 1$, we have the following alternative formulas which will prove useful

$$R_h = \frac{1}{2\mu_h(\boldsymbol{\sigma}_h)}[\mu_h^+(\boldsymbol{\sigma}_h) - \mu_h^-(\boldsymbol{\sigma}_h)], \qquad (6.3.22)$$

$$R_h = 2\mu_h(+1|\boldsymbol{\sigma}_h) - 1 = \frac{\mu_h^+(\boldsymbol{\sigma}_h)}{\mu_h(\boldsymbol{\sigma}_h)} - 1, \qquad (6.3.23)$$

$$R_h = 1 - 2\mu_h(-1|\boldsymbol{\sigma}_h) = 1 - \frac{\mu_h^-(\boldsymbol{\sigma}_h)}{\mu_h(\boldsymbol{\sigma}_h)}. \qquad (6.3.24)$$

It turns out to be enough to prove an upper bound on the variance of R_h.

Lemma 6.3.13 (Second moment bound). *It holds that*

$$\|\mu_h^+ - \mu_h^-\|_{\text{TV}} \le \sqrt{\mathbb{E}[R_h^2]}.$$

Proof   By (6.3.22),

$$\frac{1}{2} \sum_{\mathbf{s}_h \in \{+1, -1\}^{2^h}} |\mu_h^+(\mathbf{s}_h) - \mu_h^-(\mathbf{s}_h)|$$

$$= \sum_{\mathbf{s}_h \in \{+1, -1\}^{2^h}} \mu_h(\mathbf{s}_h) |\mu_h(+1|\mathbf{s}_h) - \mu_h(-1|\mathbf{s}_h)|$$

$$= \mathbb{E}|R_h|$$

$$\le \sqrt{\mathbb{E}[R_h^2]},$$

where we used Cauchy–Schwarz on the last line.   ∎

Let $\bar{z}_h = \mathbb{E}[R_h^2]$. In view of Lemma 6.3.13, the proof of Theorem 6.3.9 (ii) will follow from establishing the limit

$$\lim_{h \to +\infty} \bar{z}_h = 0.$$

We apply the same kind of recursive argument we used for the analysis of majority (see in particular Lemma 6.3.11): we condition on the root to exploit conditional independence; we use the Markov transition matrix on the top edges.

We first derive a recursion for R_h itself – as a random variable. We proceed in two steps:

- *Step 1:* We break up the first h levels of the tree into two identical $(h-1)$-level trees with an additional edge at their respective roots through conditional independence.
- *Step 2:* We account for that edge through the Markov transition matrix.

We will need some notation. Let $\dot{\sigma}_h$ be the states at level h (from the root) below the left child of the root and let $\dot{\mu}_h$ be the distribution of $\dot{\sigma}_h$. Define

$$\dot{Y}_h = \dot{\mu}_h(+1|\dot{\sigma}_h) - \dot{\mu}_h(-1|\dot{\sigma}_h),$$

where $\dot{\mu}_h(s_0|\dot{s}_h)$ is the conditional probability that the root is s_0 given that $\dot{\sigma}_h = \dot{s}_h$. Similarly, denote with a double dot the same quantities with respect to the subtree below the right child of the root. Expressions similar to (6.3.22), (6.3.23), and (6.3.24) also hold.

Lemma 6.3.14 (Recursion: Step 1). *It holds almost surely that*

$$R_h = \frac{\dot{Y}_h + \ddot{Y}_h}{1 + \dot{Y}_h \ddot{Y}_h}.$$

Proof Using $\mu_h^+(s_h) = \dot{\mu}_h^+(\dot{s}_h)\ddot{\mu}_h^+(\ddot{s}_h)$ by conditional independence (where the superscript indicates conditioning on the root), (6.3.22) applied to R_h, and (6.3.23) and (6.3.24) applied to $\dot{Y}_h$ and $\ddot{Y}_h$, we get

$$
\begin{aligned}
R_h &= \frac{1}{2} \sum_{\gamma=+,-} \gamma \frac{\mu_h^\gamma(\sigma_h)}{\mu_h(\sigma_h)} \\
&= \frac{1}{2} \frac{\dot{\mu}_h(\dot{\sigma}_h)\ddot{\mu}_h(\ddot{\sigma}_h)}{\mu_h(\sigma_h)} \sum_{\gamma=+,-} \gamma \frac{\dot{\mu}_h^\gamma(\dot{\sigma}_h)\ddot{\mu}_h^\gamma(\ddot{\sigma}_h)}{\dot{\mu}_h(\dot{\sigma}_h)\ddot{\mu}_h(\ddot{\sigma}_h)} \\
&= \frac{1}{2} \frac{\dot{\mu}_h(\dot{\sigma}_h)\ddot{\mu}_h(\ddot{\sigma}_h)}{\mu_h(\sigma_h)} \sum_{\gamma=+,-} \gamma \left(1 + \gamma \dot{Y}_h\right)\left(1 + \gamma \ddot{Y}_h\right) \\
&= \frac{\dot{\mu}_h(\dot{\sigma}_h)\ddot{\mu}_h(\ddot{\sigma}_h)}{\mu_h(\sigma_h)}(\dot{Y}_h + \ddot{Y}_h).
\end{aligned}
$$

The factor in front can be computed as follows:

$$
\begin{aligned}
\frac{\mu_h(\sigma_h)}{\dot{\mu}_h(\dot{\sigma}_h)\ddot{\mu}_h(\ddot{\sigma}_h)} &= \sum_{\gamma=+,-} \frac{1}{2} \frac{\mu_h^\gamma(\sigma_h)}{\dot{\mu}_h(\dot{\sigma}_h)\ddot{\mu}_h(\ddot{\sigma}_h)} \\
&= \sum_{\gamma=+,-} \frac{1}{2} \frac{\dot{\mu}_h^\gamma(\dot{\sigma}_h)\ddot{\mu}_h^\gamma(\ddot{\sigma}_h)}{\dot{\mu}_h(\dot{\sigma}_h)\ddot{\mu}_h(\ddot{\sigma}_h)} \\
&= \frac{1}{2} \sum_{\gamma=+,-} \left(1 + \gamma \dot{Y}_h\right)\left(1 + \gamma \ddot{Y}_h\right) \\
&= 1 + \dot{Y}_h \ddot{Y}_h.
\end{aligned}
$$

That proves the claim. ∎

For the second step of the recursion, we define

$$\dot{D}_h = \dot{v}_h(+1|\dot{\sigma}_h) - \dot{v}_h(-1|\dot{\sigma}_h),$$

where $\dot{v}_h(\dot{s}_0|\dot{s}_h)$ is the conditional probability that the left child of the root is $\dot{s}_0$ given that the states at level h (from the root) below the left child are $\dot{\sigma}_h = \dot{s}_h$; and similarly for the right child of the root. Again expressions similar to (6.3.22), (6.3.23), and (6.3.24) hold. The following lemma is left as an exercise (see Exercise 6.18).

Lemma 6.3.15 (Recursion: Step 2). *It holds almost surely that*

$$\dot{Y}_h = \theta \dot{D}_h.$$

We are now ready to prove the second half of our main theorem.

Proof of Theorem 6.3.9 (ii)   Putting Lemmas 6.3.14 and 6.3.15 together, we get

$$R_h = \frac{\theta(\dot{D}_h + \ddot{D}_h)}{1 + \theta^2 \dot{D}_h \ddot{D}_h}. \tag{6.3.25}$$

We now take expectations. Recall that we seek to compute the second moment of R_h. However, an important simplification arises from the following observation

$$
\begin{aligned}
\mathbb{E}^+[R_h] &= \sum_{\mathbf{s}_h \in \{+1,-1\}^{2^h}} \mu_h^+(\mathbf{s}_h) R_h(\mathbf{s}_h) \\
&= \sum_{\mathbf{s}_h \in \{+1,-1\}^{2^h}} \mu_h(\mathbf{s}_h) \frac{\mu_h^+(\mathbf{s}_h)}{\mu_h(\mathbf{s}_h)} R_h(\mathbf{s}_h) \\
&= \sum_{\mathbf{s}_h \in \{+1,-1\}^{2^h}} \mu_h(\mathbf{s}_h)(1 + R_h(\mathbf{s}_h)) R_h(\mathbf{s}_h) \\
&= \mathbb{E}[(1 + R_h)R_h] \\
&= \mathbb{E}[R_h^2],
\end{aligned}
$$

where we used (6.3.23) on the third line and $\mathbb{E}[R_h] = 0$ on the fifth line. *So it suffices to compute the conditional first moment.*

Using the expansion

$$\frac{1}{1+r} = 1 - r + \frac{r^2}{1+r},$$

with $r = \theta^2 \dot{D}_h \ddot{D}_h$, we have by (6.3.25) that

$$
\begin{aligned}
R_h &= \theta(\dot{D}_h + \ddot{D}_h) - \theta^3(\dot{D}_h + \ddot{D}_h)\dot{D}_h\ddot{D}_h + \theta^4 \dot{D}_h^2 \ddot{D}_h^2 R_h \\
&\leq \theta(\dot{D}_h + \ddot{D}_h) - \theta^3(\dot{D}_h + \ddot{D}_h)\dot{D}_h\ddot{D}_h + \theta^4 \dot{D}_h^2 \ddot{D}_h^2,
\end{aligned}
\tag{6.3.26}
$$

where we used $|R_h| \leq 1$.

We will need the conditional first and second moments of $\dot{D}_h$. For the first moment, note that by symmetry (more precisely, by the fact that R_{h-1} conditioned on $\sigma_0 = -1$ is equal in distribution to $-R_{h-1}$ conditioned on $\sigma_0 = +1$),

$$
\begin{aligned}
\mathbb{E}^+[\dot{D}_h] &= (1-p)\mathbb{E}^+[R_{h-1}] + p\,\mathbb{E}^-[R_{h-1}] \\
&= (1-p)\mathbb{E}^+[R_{h-1}] + p\,\mathbb{E}^+[-R_{h-1}] \\
&= (1-2p)\mathbb{E}^+[R_{h-1}] \\
&= \theta\,\mathbb{E}^+[R_{h-1}].
\end{aligned}
$$

Similarly, for the second moment, we have

$$
\begin{aligned}
\mathbb{E}^+[\dot{D}_h^2] &= (1-p)\mathbb{E}^+[R_{h-1}^2] + p\,\mathbb{E}^-[R_{h-1}^2] \\
&= \mathbb{E}[R_{h-1}^2] \\
&= \mathbb{E}^+[R_{h-1}],
\end{aligned}
$$

where we used that $\mathbb{E}^+[R_{h-1}^2] = \mathbb{E}^-[R_{h-1}^2]$ by symmetry so that $\mathbb{E}[R_{h-1}^2] = (1/2)\mathbb{E}^+[R_{h-1}^2] + (1/2)\mathbb{E}^-[R_{h-1}^2] = \mathbb{E}^+[R_{h-1}^2]$.

Taking expectations in (6.3.26), using conditional independence, and plugging in the formulas for $\mathbb{E}^+[\dot{D}_h]$ and $\mathbb{E}^+[\dot{D}_h^2]$ above, we obtain

$$
\begin{aligned}
\bar{z}_h &= \mathbb{E}^+[R_h] \\
&\le \theta(\mathbb{E}^+[\dot{D}_h] + \mathbb{E}^+[\ddot{D}_h]) - \theta^3(\mathbb{E}^+[\dot{D}_h^2]\,\mathbb{E}^+[\ddot{D}_h] + \mathbb{E}^+[\ddot{D}_h^2]\,\mathbb{E}^+[\dot{D}_h]) \\
&\quad + \theta^4 \mathbb{E}^+[\dot{D}_h^2]\,\mathbb{E}^+[\ddot{D}_h^2] \\
&= 2\theta^2 \mathbb{E}^+[R_{h-1}] - 2\theta^4 \mathbb{E}^+[R_{h-1}]^2 + \theta^4 \mathbb{E}^+[R_{h-1}]^2 \\
&= 2\theta^2 \bar{z}_{h-1} - \theta^4 \bar{z}_{h-1}^2.
\end{aligned}
\tag{6.3.27}
$$

We analyze this recursion next. At $h = 0$, we have $\bar{z}_0 = \mathbb{E}^+[R_0] = 1$.

- When $2\theta^2 < 1$, the sequence $\bar{z}_h$ decreases to 0 exponentially fast

$$\bar{z}_h \le (2\theta^2)^h, \qquad h \ge 0.$$

- When $2\theta^2 = 1$ on the other hand, convergence to 0 occurs at a slower rate. We show by induction that

$$\bar{z}_h \le \frac{4}{h}, \quad h \ge 0.$$

Note that $\bar{z}_1 \le \bar{z}_0 - \theta^4 \bar{z}_0^2 = 3/4 \le 4$ since $\theta^4 = 1/4$, which proves the base case. Assuming the bound holds for $h - 1$, we have from (6.3.27) that

$$
\begin{aligned}
\bar{z}_h &\le \bar{z}_{h-1} - \frac{1}{4}\bar{z}_{h-1}^2 \\
&\le \frac{4}{h-1} - \frac{4}{(h-1)^2} \\
&= 4\frac{h-2}{(h-1)^2} \\
&\le \frac{4}{h},
\end{aligned}
$$

where the last line follows from checking that $h(h-2) \le (h-1)^2$.

Since

$$\lim_{h \to +\infty} \bar{z}_h = 0,$$

the claim follows from Lemma 6.3.13. ∎

Remark 6.3.16 *While Theorem 6.3.9 part (i) can be generalized beyond the CFN model (see, for example, [MP03]), part (ii) cannot. A striking construction of [Mos01] shows that, under more general models, certain root-state estimators taking into account the configuration of the states at level h can "beat" the Kesten–Stigum bound.*

6.4 ▷ Finale: The Phase Transition of the Erdős–Rényi Model

A compelling way to view an Erdős–Rényi random graph – as its density varies – is the following coupling or "evolution." For each pair $\{i,j\}$, let $U_{\{i,j\}}$ be independent uniform random variables in $[0,1]$ and set $\mathcal{G}(p) := ([n], \mathcal{E}(p))$, where $\{i,j\} \in \mathcal{E}(p)$ if and only if $U_{\{i,j\}} \leq p$. Then, $\mathcal{G}(p)$ is distributed according to $\mathbb{G}_{n,p}$. As p varies from 0 to 1, we start with an empty graph and progressively add edges until the complete graph is obtained.

We showed in Section 2.3.2 that $\frac{\log n}{n}$ is a threshold function for connectivity. Before connectivity occurs in the evolution of the random graph, a quantity of interest is the size of the largest connected component. As we show in the current section, this quantity itself undergoes a remarkable phase transition: when $p = \frac{\lambda}{n}$ with $\lambda < 1$, the largest component has size $\Theta(\log n)$; as λ crosses 1, many components quickly merge to form a so-called giant component of size $\Theta(n)$.

This celebrated result is often referred to as "the" phase transition of the Erdős–Rényi graph model. Although the proof is quite long, it is well worth studying in details. It employs most tools we have seen up to this point: first and second moment methods, Chernoff–Cramér bounds, martingale techniques, coupling and stochastic domination, and branching processes. It is quintessential discrete probability.

6.4.1 Statement and Proof Sketch

Before stating the main theorems, we recall a basic result from Chapter 2.

- *(Poisson tail)* Let S_n be a sum of n i.i.d. Poi(λ) variables. Recall from (2.4.10) and (2.4.11) that for $a > \lambda$,

$$-\frac{1}{n} \log \mathbb{P}[S_n \geq an] \geq a \log \left(\frac{a}{\lambda}\right) - a + \lambda =: I_\lambda^{\mathrm{Poi}}(a), \qquad (6.4.1)$$

and similarly for $a < \lambda$,

$$-\frac{1}{n} \log \mathbb{P}[S_n \leq an] \geq I_\lambda^{\mathrm{Poi}}(a). \qquad (6.4.2)$$

To simplify the notation, we let

$$I_\lambda := I_\lambda^{\mathrm{Poi}}(1) = \lambda - 1 - \log \lambda \geq 0, \qquad (6.4.3)$$

where the inequality follows from the convexity of I_λ and the fact that it attains its minimum at $\lambda = 1$ where it is 0.

We let $p = \frac{\lambda}{n}$ and denote by $\mathcal{C}_{\max}$ a largest connected component. In the subcritical case, that is, when $\lambda < 1$, we show that the largest connected component has logarithmic size in n.

Theorem 6.4.1 (Subcritical case: upper bound on the largest cluster). *Let $G_n \sim \mathbb{G}_{n,p_n}$, where $p_n = \frac{\lambda}{n}$ with $\lambda \in (0,1)$. For all $\kappa > 0$,*

$$\mathbb{P}_{n,p_n}\left[|\mathcal{C}_{\max}| \geq (1+\kappa)I_\lambda^{-1} \log n\right] = o(1),$$

where I_λ is defined in (6.4.3).

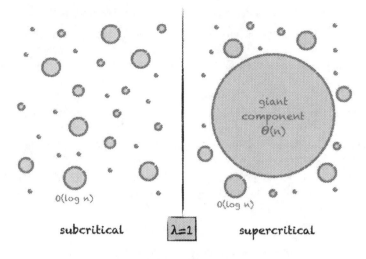

Figure 6.4 Illustration of the phase transition.

We also give a matching logarithmic lower bound on the size of $\mathcal{C}_{\max}$ in Theorem 6.4.11.

In the supercritical case, that is, when $\lambda > 1$, we prove the existence of a unique connected component of size linear in n, which is referred to as the *giant component*.

GIANT
COMPONENT

Theorem 6.4.2 (Supercritical regime: giant component). *Let* $G_n \sim \mathbb{G}_{n,p_n}$, *where* $p_n = \frac{\lambda}{n}$ *with* $\lambda > 1$. *For any* $\gamma \in (1/2, 1)$ *and* $\delta < 2\gamma - 1$,

$$\mathbb{P}_{n,p_n}\left[||\mathcal{C}_{\max}| - \zeta_\lambda n| \geq n^\gamma\right] = O(n^{-\delta}),$$

where ζ_λ *is the unique solution in* $(0, 1)$ *to the fixed point equation*

$$1 - e^{-\lambda\zeta} = \zeta.$$

In fact, with probability $1 - O(n^{-\delta})$, *there is a unique largest component and the second largest connected component has size* $O(\log n)$.

See Figure 6.4 for an illustration.

At a high level, the proof goes as follows:

- *(Subcritical regime)* In the subcritical case, we use an exploration process and a domination argument to approximate the size of the connected components with the progeny of a branching process. The result then follows from the hitting-time theorem and the Poisson tail.

- *(Supercritical regime)* In the supercritical case, a similar argument gives a bound on the expected size of the giant component, which is related to the survival of the branching process. Chebyshev's inequality gives concentration. The hard part there is to bound the variance.

6.4.2 Bounding Cluster Size: Domination by Branching Processes

For a vertex $v \in [n]$, let C_v be the connected component containing v, which we also refer to as the *cluster* of v. To analyze the size of C_v, we use the exploration process introduced in Section 6.2.1 and show that it is dominated above and below by branching processes.

Exploration process

Recall that the exploration process started at v has three types of vertices: the active vertices $\mathcal{A}_t$, the explored vertices $\mathcal{E}_t$, and the neutral vertices $\mathcal{N}_t$. We start with $\mathcal{A}_0 := \{v\}$, $\mathcal{E}_0 := \emptyset$, and $\mathcal{N}_0$ contains all other vertices in G_n. We imagine revealing the edges of G_n as they are encountered in this process and we let $(\mathcal{F}_t)$ be the corresponding filtration. In words, starting with v, the cluster of v is progressively grown by adding to it at each time a vertex adjacent to one of the previously explored vertices and uncovering its remaining neighbors in G_n.

Let as before $A_t := |\mathcal{A}_t|$, $E_t := |\mathcal{E}_t|$, and $N_t := |\mathcal{N}_t|$, and

$$\tau_0 := \inf\{t \geq 0 : A_t = 0\} = |C_v|,$$

where the rightmost equality is from Lemma 6.2.1. Recall that (E_t) is non-decreasing, while (N_t) is non-increasing, and that the process is fixed for all $t > \tau_0$. Since $E_t = t$ for all $t \leq \tau_0$ (as exactly one vertex is explored at each time until the set of active vertices is empty) and $(\mathcal{A}_t, \mathcal{E}_t, \mathcal{N}_t)$ forms a partition of $[n]$ for all t, we have

$$A_t + t + N_t = n \qquad \forall t \leq \tau_0. \tag{6.4.4}$$

Hence, in tracking the size of the exploration process, we can work with A_t or N_t. Moreover, at $t = \tau_0$ we have

$$|C_v| = \tau_0 = n - N_{\tau_0}. \tag{6.4.5}$$

Similarly to the case of a Galton–Watson tree, the processes (A_t) and (N_t) admit a simple recursive form. Conditioning on $\mathcal{F}_{t-1}$:

- *(Active vertices)* If $A_{t-1} = 0$, the exploration process has finished its course and $A_t = 0$. Otherwise, (a) one active vertex becomes explored and (b) its neutral neighbors become active vertices. That is,

$$A_t = A_{t-1} + \mathbf{1}_{\{A_{t-1}>0\}}\Big[\underbrace{-1}_{(a)} + \underbrace{X_t}_{(b)}\Big], \tag{6.4.6}$$

where X_t is binomial with parameters N_{t-1} and p_n. By (6.4.4), N_{t-1} can be written in terms of A_{t-1} as $N_{t-1} = n - (t-1) - A_{t-1}$. For the coupling arguments below, it will be useful to think of X_t as a sum of independent Bernoulli variables. That is, let $(I_{t,j} : t \geq 1, j \geq 1)$ be an array of independent, identically distributed $\{0,1\}$-variables with $\mathbb{P}[I_{1,1} = 1] = p_n$. We write

$$X_t = \sum_{i=1}^{N_{t-1}} I_{t,i}. \tag{6.4.7}$$

- *(Neutral vertices)* Similarly, if $A_{t-1} > 0$, that is, $N_{t-1} < n - (t-1)$, X_t neutral vertices become active. That is,

$$N_t = N_{t-1} - \mathbf{1}_{\{N_{t-1}<n-(t-1)\}} X_t. \tag{6.4.8}$$

Poisson branching process approximation

With these observations, we now relate the size of the cluster of v to the total progeny of a Poisson branching process with an appropriately chosen offspring mean. The intuition is simple: when $p_n = \lambda/n$, the number of neighbors of a vertex is well approximated by a Poisson distribution; therefore, exploration of the cluster of v is similar to that of the corresponding branching process. We will see that this holds long enough to prove accurate results about the subcritical regime (see Lemma 6.4.6). It will also be useful in the supercritical regime, but additional arguments will be required there (see Lemmas 6.4.8 and 6.4.7).

Lemma 6.4.3 (Cluster size: Poisson branching process approximation). *Let $G_n \sim \mathbb{G}_{n,p_n}$, where $p_n = \frac{\lambda}{n}$ with $\lambda > 0$ and let $\mathcal{C}_v$ be the connected component of $v \in [n]$. Let W_λ be the total progeny of a branching process with offspring distribution $\mathrm{Poi}(\lambda)$. Then, for $1 \leq k_n = o(\sqrt{n})$,*

$$\mathbb{P}[W_\lambda \geq k_n] - O\left(\frac{k_n^2}{n}\right) \leq \mathbb{P}_{n,p_n}[|\mathcal{C}_v| \geq k_n] \leq \mathbb{P}[W_\lambda \geq k_n].$$

From Example 6.2.7, we have an explicit formula for the distribution of W_λ.

Before proving the lemma, recall the following simple domination results from Chapter 4:

- *(Binomial domination)* We have

$$n \geq m \implies \mathrm{Bin}(n, p) \succeq \mathrm{Bin}(m, p). \tag{6.4.9}$$

The binomial distribution is also dominated by the Poisson distribution in the following way:

$$\lambda \in (0, 1) \implies \mathrm{Poi}(\lambda) \succeq \mathrm{Bin}\left(n - 1, \frac{\lambda}{n}\right). \tag{6.4.10}$$

For the proofs, see Examples 4.2.4 and 4.2.8.

We use these domination results to relate the size of a connected component to the progeny of a branching process.

Proof of Lemma 6.4.3 We start with the upper bound.

Upper bound: Because $N_{t-1} = n - (t-1) - A_{t-1} \leq n - 1$, conditioned on $\mathcal{F}_{t-1}$, the following stochastic domination relations hold

$$\mathrm{Bin}\left(N_{t-1}, \frac{\lambda}{n}\right) \preceq \mathrm{Bin}\left(n - 1, \frac{\lambda}{n}\right) \preceq \mathrm{Poi}(\lambda),$$

by (6.4.9) and (6.4.10). Observe that the center and rightmost distributions do not depend on N_{t-1}. Let $(X_t^{>})$ be a sequence of independent $\mathrm{Poi}(\lambda)$.

Using the coupling in Example 4.2.8, we can couple the processes $(I_{t,j})_j$ and $(X_t^{>})$ in such way that $X_t^{>} \geq \sum_{j=1}^{n-1} I_{t,j}$ almost surely for all t. Then by induction on t,

$$A_t \leq A_t^{>}$$

almost surely for all t, where we define (recalling (6.4.6))

$$A_t^{>} := A_{t-1}^{>} + \mathbf{1}_{\{A_{t-1}^{>} > 0\}}\left[-1 + X_t^{>}\right], \tag{6.4.11}$$

with $A_0^> := 1$. In words, $(A_t^>)$ is the size of the active set of a *Galton–Watson branching process with offspring distribution* Poi(λ), as defined in Section 6.2.1.

As a result, letting

$$W_\lambda = \tau_0^> := \inf\{t \geq 0 \colon A_t^> = 0\}$$

be the total progeny of this branching process, we immediately get

$$\mathbb{P}_{n,p_n}[|\mathcal{C}_v| \geq k_n] = \mathbb{P}_{n,p_n}[\tau_0 \geq k_n] \leq \mathbb{P}[\tau_0^> \geq k_n] = \mathbb{P}[W_\lambda \geq k_n].$$

Lower bound: In the other direction, we proceed in two steps. We first show that, up to a certain time, the process is bounded from below by a branching process with binomial offspring distribution. In a second step, we show that this binomial branching process can be approximated by a Poisson branching process.

1. *(Domination from below)* Let $A_t^<$ be defined as (again recalling (6.4.6))

$$A_t^< := A_{t-1}^< + \mathbf{1}_{\{A_{t-1}^< > 0\}}\big[-1 + X_t^<\big], \tag{6.4.12}$$

with $A_0^< := 1$, where

$$X_t^< := \sum_{i=1}^{n-k_n} I_{t,j}. \tag{6.4.13}$$

Note that we use the same $I_{t,j}$s as in the definition of X_t, that is, we couple the two processes. This time $(A_t^<)$ is the size of the active set in the exploration process of a Galton–Watson branching process with offspring distribution Bin($n - k_n, p_n$). Let

$$\tau_0^< := \inf\{t \geq 0 \colon A_t^< = 0\}$$

be the total progeny of this branching process. We prove the following relationship between τ_0 and $\tau_0^<$.

Lemma 6.4.4 *We have*

$$\mathbb{P}[\tau_0^< \geq k_n] \leq \mathbb{P}_{n,p_n}[\tau_0 \geq k_n].$$

Proof We claim that A_t is bounded from below by $A_t^<$ *up to the stopping time*

$$\sigma_{n-k_n} := \inf\{t \geq 0 \colon N_t \leq n - k_n\},$$

which by convention is $+\infty$ if the event is not reached (i.e., if the cluster is "small"; see below). Indeed, $N_0 = n - 1$ and for all $t \leq \sigma_{n-k_n}$, $N_{t-1} > n - k_n$ by definition. Hence, by the coupling (6.4.7) and (6.4.13), $X_t \geq X_t^<$ for all $t \leq \sigma_{n-k_n}$ and as a result, by induction on t,

$$A_t \geq A_t^< \qquad \forall t \leq \sigma_{n-k_n},$$

where we used the recursions (6.4.6) and (6.4.13).

Because the inequality between A_t and $A_t^<$ holds only up to time σ_{n-k_n}, we cannot compare directly τ_0 and $\tau_0^<$. However, we will use the following observation: the size of the cluster of v is at least the total number of active and explored vertices at any time t. In particular, when $\sigma_{n-k_n} < +\infty$,

$$\tau_0 = |\mathcal{C}_v| \geq A_{\sigma_{n-k_n}} + E_{\sigma_{n-k_n}} = n - N_{\sigma_{n-k_n}} \geq k_n.$$

On the other hand, when $\sigma_{n-k_n} = +\infty$, $N_t > n - k_n$ for all t – in particular for $t = \tau_0$ – and therefore $|\mathcal{C}_v| = \tau_0 = n - N_{\tau_0} < k_n$ by (6.4.5). Moreover, in that case, because $A_t \geq A_t^{<}$ for all $t \leq \sigma_{n-k_n} = +\infty$, it holds in addition that $\tau_0^{<} \leq \tau_0 < k_n$. To sum up, we have proved the implications

$$\tau_0^{<} \geq k_n \implies \sigma_{n-k_n} < +\infty \implies \tau_0 \geq k_n.$$

In particular, we have proved the lemma. ∎

2. *(Poisson approximation)* Our next step is approximate the tail of $\tau_0^{<}$ by that of $\tau_0^{>}$.

Lemma 6.4.5 *We have*

$$\mathbb{P}[\tau_0^{<} \geq k_n] = \mathbb{P}[\tau_0^{>} \geq k_n] + O\left(\frac{k_n^2}{n}\right).$$

Proof By Theorem 6.2.6,

$$\mathbb{P}[\tau_0^{<} = t] = \frac{1}{t}\mathbb{P}\left[\sum_{i=1}^{t} X_i^{<} = t - 1\right], \qquad (6.4.14)$$

where the $X_i^{<}$s are independent $\mathrm{Bin}(n - k_n, p_n)$. Note further that, because the sum of independent binomials with the same success probability is binomial,

$$\sum_{i=1}^{t} X_i^{<} \sim \mathrm{Bin}(t(n - k_n), p_n).$$

Recall on the other hand that $(X_t^{>})$ is $\mathrm{Poi}(\lambda)$ and, because a sum of independent Poisson is Poisson (see Exercise 6.7), we have

$$\mathbb{P}[\tau_0^{>} = t] = \frac{1}{t}\mathbb{P}\left[\sum_{i=1}^{t} X_i^{>} = t - 1\right], \qquad (6.4.15)$$

where

$$\sum_{i=1}^{t} X_i^{>} \sim \mathrm{Poi}(t\lambda).$$

We use the Poisson approximation result in Theorem 4.1.18 to compare the probabilities on the right-hand sides of (6.4.14) and (6.4.15). In fact, because the Poisson approximation is in terms of the total variation distance – which bounds any event – one might be tempted to apply it directly to the tails of $\tau_0^{<}$ and $\tau_0^{>}$ by summing over t. However, note that the factor of $1/t$ in (6.4.14) and (6.4.15) prevents us from doing so.

Instead, we argue for each t separately and use that

$$\left|\mathbb{P}\left[\sum_{i=1}^{t} X_i^{<} = t - 1\right] - \mathbb{P}\left[\sum_{i=1}^{t} X_i^{>} = t - 1\right]\right|$$

$$\leq \|\mathrm{Bin}(t(n - k_n), p_n) - \mathrm{Poi}(t\lambda)\|_{\mathrm{TV}},$$

by the observations in the previous paragraph. Theorem 4.1.18 tells us that

$$\|\mathrm{Bin}(t(n - k_n), p_n) - \mathrm{Poi}\,(t(n - k_n)[-\log(1 - p_n)])\|_{\mathrm{TV}}$$
$$\leq \frac{1}{2} t(n - k_n)[-\log(1 - p_n)]^2.$$

We must adjust the mean of the Poisson distribution. To do so, we argue as in Example 4.1.12 to get

$$\|\mathrm{Poi}\,(t(n - k_n)[-\log(1 - p_n)]) - \mathrm{Poi}\,(t\lambda)\|_{\mathrm{TV}}$$
$$\leq |t\lambda - t(n - k_n)(-\log(1 - p_n))|.$$

Finally, recalling that $p_n = \lambda/n$, combining the last three displays and using the triangle inequality for the total variation distance,

$$\left| \mathbb{P}\left[\sum_{i=1}^{t} X_i^{\prec} = t - 1\right] - \mathbb{P}\left[\sum_{i=1}^{t} X_i^{\succ} = t - 1\right] \right|$$
$$\leq \frac{1}{2} t(n - k_n)[-\log(1 - p_n)]^2 + |t\lambda - t(n - k_n)(-\log(1 - p_n))|$$
$$\leq \frac{1}{2} tn\left(\frac{\lambda}{n} + O\left(\frac{\lambda^2}{n^2}\right)\right)^2 + \left| t\lambda - t(n - k_n)\left(\frac{\lambda}{n} + O\left(\frac{\lambda^2}{n^2}\right)\right) \right|$$
$$= O\left(\frac{tk_n}{n}\right),$$

where we used that $k_n \geq 1$ and λ is fixed.

So, by (6.4.14) and (6.4.15), dividing by t and then summing over $t < k_n$ gives

$$\left| \mathbb{P}[\tau_0^{\prec} < k_n] - \mathbb{P}[\tau_0^{\succ} < k_n] \right| = O\left(\frac{k_n^2}{n}\right).$$

Rearranging proves the lemma. ∎

Putting together Lemmas 6.4.4 and 6.4.5 gives

$$\mathbb{P}_{n,p_n}[|\mathcal{C}_v| \geq k_n] = \mathbb{P}_{n,p_n}[\tau_0 \geq k_n]$$
$$\geq \mathbb{P}[\tau_0^{\succ} \geq k_n] - O\left(\frac{k_n^2}{n}\right)$$
$$= \mathbb{P}[W_\lambda \geq k_n] - O\left(\frac{k_n^2}{n}\right),$$

as claimed. ∎

Subcritical regime: largest cluster

We are now ready to analyze the subcritical regime, that is, the case $\lambda < 1$.

Lemma 6.4.6 (Subcritical regime: upper bound on cluster size). *Let $G_n \sim \mathbb{G}_{n,p_n}$ where $p_n = \frac{\lambda}{n}$ with $\lambda \in (0, 1)$ and let $\mathcal{C}_v$ be the connected component of $v \in [n]$. For all $\kappa > 0$,*

$$\mathbb{P}_{n,p_n}\left[|\mathcal{C}_v| \geq (1 + \kappa)I_\lambda^{-1} \log n\right] = O(n^{-(1+\kappa)}).$$

Proof We use the Poisson branching process approximation (Lemma 6.4.3). To apply the lemma we need to bound the tail of the progeny W_λ of a Poisson branching process. Using the notation of Lemma 6.4.3, by Theorem 6.2.6,

$$\mathbb{P}[W_\lambda \geq k_n] = \mathbb{P}[W_\lambda = +\infty] + \sum_{t \geq k_n} \frac{1}{t} \mathbb{P}\left[\sum_{i=1}^{t} X_i^{\frown} = t - 1\right], \qquad (6.4.16)$$

where the $X_i^{\frown}$s are i.i.d. Poi(λ). Both terms on the right-hand side depend on whether or not the mean λ is smaller or larger than 1. When $\lambda < 1$, the Poisson branching process goes extinct with probability 1 by the extinction theory (Theorem 6.1.6). Hence $\mathbb{P}[W_\lambda = +\infty] = 0$.

As to the second term, the sum of the $X_i^{\frown}$s is Poi(λt). Using the Poisson tail (6.4.1) for $\lambda < 1$ and $k_n = \omega(1)$,

$$\sum_{t \geq k_n} \frac{1}{t} \mathbb{P}\left[\sum_{i=1}^{t} X_i^{\frown} = t - 1\right] \leq \sum_{t \geq k_n} \mathbb{P}\left[\sum_{i=1}^{t} X_i^{\frown} \geq t - 1\right]$$

$$\leq \sum_{t \geq k_n} \exp\left(-t I_\lambda^{\mathrm{Poi}}\left(\frac{t-1}{t}\right)\right)$$

$$\leq \sum_{t \geq k_n} \exp\left(-t(I_\lambda - O(t^{-1}))\right)$$

$$\leq \sum_{t \geq k_n} C \exp\left(-t I_\lambda\right)$$

$$= O\left(\exp\left(-I_\lambda k_n\right)\right) \qquad (6.4.17)$$

for some constant $C > 0$.

Let $c = (1 + \kappa) I_\lambda^{-1}$ for $\kappa > 0$. By Lemma 6.4.3,

$$\mathbb{P}_{n,p_n}[|\mathcal{C}_v| \geq c \log n] \leq \mathbb{P}[W_\lambda \geq c \log n].$$

By (6.4.16) and (6.4.17),

$$\mathbb{P}[W_\lambda \geq c \log n] = O\left(\exp\left(-I_\lambda c \log n\right)\right), \qquad (6.4.18)$$

which proves the claim. ∎

As before, let $\mathcal{C}_{\max}$ be a largest connected component of G_n (choosing the component containing the lowest label if there is more than one such component). A union bound and the previous lemma immediately imply an upper bound on the size of $\mathcal{C}_{\max}$ in the subcritical case.

Proof of Theorem 6.4.1 Let again $c = (1 + \kappa) I_\lambda^{-1}$ for $\kappa > 0$. By a union bound and symmetry,

$$\mathbb{P}_{n,p_n}[|\mathcal{C}_{\max}| \geq c \log n] = \mathbb{P}_{n,p_n}[\exists v, \ |\mathcal{C}_v| > c \log n]$$

$$\leq n \, \mathbb{P}_{n,p_n}[|\mathcal{C}_1| \geq c \log n]. \qquad (6.4.19)$$

By Lemma 6.4.6,

$$\mathbb{P}_{n,p_n}[|\mathcal{C}_{\max}| \geq c \log n] = O(n \cdot n^{-(1+\kappa)}) = O(n^{-\kappa}) \to 0$$

as $n \to +\infty$. ∎

In fact we prove below that the largest component is indeed of size roughly $I_\lambda^{-1} \log n$. But first we turn to the supercritical regime.

Supercritical regime: two phases

Applying the Poisson branching process approximation in the supercritical regime gives the following.

Lemma 6.4.7 (Supercritical regime: extinction). *Let $G_n \sim \mathbb{G}_{n,p_n}$ where $p_n = \frac{\lambda}{n}$ with $\lambda > 1$. and let $\mathcal{C}_v$ be the connected component of $v \in [n]$. Let ζ_λ be the unique solution in $(0, 1)$ to the fixed point equation*

$$1 - e^{-\lambda\zeta} = \zeta.$$

For any $\kappa > 0$,

$$\mathbb{P}_{n,p_n}\left[|\mathcal{C}_v| \geq (1 + \kappa)I_\lambda^{-1} \log n\right] = \zeta_\lambda + O\left(\frac{\log^2 n}{n}\right).$$

Note the small but critical difference with Lemma 6.4.6: this time the branching process can survive. This happens with probability ζ_λ by extinction theory (Theorem 6.1.6). In that case, we will need further arguments to nail down the cluster size. Observe also that the result holds for a *fixed* vertex v – and therefore does not yet tell us about the *largest* cluster. We come back to the latter in the next subsection.

Proof of Lemma 6.4.7 We adapt the proof of Lemma 6.4.6, beginning with (6.4.16), which recall states

$$\mathbb{P}[W_\lambda \geq k_n] = \mathbb{P}[W_\lambda = +\infty] + \sum_{t \geq k_n} \frac{1}{t}\mathbb{P}\left[\sum_{i=1}^{t} X_i^> = t - 1\right],$$

where the $X_i^>$s are i.i.d. Poi(λ). When $\lambda > 1$, $\mathbb{P}[W_\lambda = +\infty] = \zeta_\lambda$, where $\zeta_\lambda > 0$ is the survival probability of the branching process by Example 6.1.10. As to the second term, using (6.4.2) for $\lambda > 1$,

$$\sum_{t \geq k_n} \frac{1}{t}\mathbb{P}\left[\sum_{i=1}^{t} X_i^> = t - 1\right] \leq \sum_{t \geq k_n} \mathbb{P}\left[\sum_{i=1}^{t} X_i^> \leq t\right]$$

$$\leq \sum_{t \geq k_n} \exp(-tI_\lambda)$$

$$\leq C \exp(-I_\lambda k_n) \qquad (6.4.20)$$

for a constant $C > 0$.

Now let $c = (1 + \kappa)I_\lambda^{-1}$ for $\kappa > 0$. By Lemma 6.4.3,

$$\mathbb{P}_{n,p_n}[|\mathcal{C}_v| \geq c \log n] = \mathbb{P}[W_\lambda \geq c \log n] + O\left(\frac{\log^2 n}{n}\right). \qquad (6.4.21)$$

By (6.4.16) and (6.4.20),

$$\mathbb{P}[W_\lambda \geq c \log n] = \zeta_\lambda + O\left(\exp(-cI_\lambda \log n)\right)$$

$$= \zeta_\lambda + O(n^{-(1+\kappa)}). \qquad (6.4.22)$$

Combining (6.4.21) and (6.4.22), for any $\kappa > 0$,

$$\mathbb{P}_{n,p_n}[|\mathcal{C}_v| \geq c \log n] = \zeta_\lambda + O\left(\frac{\log^2 n}{n}\right), \tag{6.4.23}$$

as claimed. ∎

Recall that the Poisson branching process approximation was based on the fact that the degree of a vertex is well approximated by a Poisson distribution. When the exploration process goes on for too long however (i.e., when k_n is large), this approximation is not as accurate because of a saturation effect: at each step of the exploration, we uncover edges *to the neutral vertices* (which then become active); and, because an Erdős–Rényi graph has a finite pool of vertices from which to draw these edges, as the number of neutral vertices decreases so does the expected number of uncovered edges. Instead, we use the following lemma which explicitly accounts for the dwindling size of $\mathcal{N}_t$. Roughly speaking, we model the set of neutral vertices as a process that discards a fraction p_n of its current set at each time step (i.e., those neutral vertices with an edge to the current explored vertex).

Lemma 6.4.8 *Let $G_n \sim \mathbb{G}_{n,p_n}$, where $p_n = \frac{\lambda}{n}$ with $\lambda > 0$ and let $\mathcal{C}_v$ be the connected component of $v \in [n]$. Let $Y_t \sim \text{Bin}(n - 1, 1 - (1 - p_n)^t)$. Then, for any t,*

$$\mathbb{P}_{n,p_n}[|\mathcal{C}_v| = t] \leq \mathbb{P}[Y_t = t - 1].$$

Proof We work with neutral vertices. By (6.4.4) and Lemma 6.2.1, for any t,

$$\mathbb{P}_{n,p_n}[|\mathcal{C}_v| = t] = \mathbb{P}_{n,p_n}[\tau_0 = t] \leq \mathbb{P}_{n,p_n}[N_t = n - t]. \tag{6.4.24}$$

Recall that $N_0 = n - 1$ and

$$N_t = N_{t-1} - \mathbf{1}_{\{N_{t-1} < n - (t-1)\}} \sum_{i=1}^{N_{t-1}} I_{t,i}.$$

It is easier to consider the process *without the indicator* as it has a simple distribution. Define $N_0^0 := n - 1$ and

$$N_t^0 := N_{t-1}^0 - \sum_{i=1}^{N_{t-1}^0} I_{t,i},$$

and observe that $N_t \geq N_t^0$ for all t, as the two processes agree up to time τ_0 at which point N_t stays fixed. The interpretation of N_t^0 is straightforward: starting with $n - 1$ vertices, at each time each remaining vertex is discarded with probability p_n. Hence, the number of surviving vertices at time t has distribution

$$N_t^0 \sim \text{Bin}(n - 1, (1 - p_n)^t),$$

by the independence of the steps. Arguing as in (6.4.24),

$$\begin{aligned}
\mathbb{P}_{n,p_n}[|\mathcal{C}_v| = t] &\leq \mathbb{P}_{n,p_n}[N_t^0 = n - t] \\
&= \mathbb{P}_{n,p_n}[(n - 1) - N_t^0 = t - 1] \\
&= \mathbb{P}[Y_t = t - 1],
\end{aligned}$$

which concludes the proof. ∎

The previous lemma gives the following additional bound on the cluster size in the super-critical regime. Together with Lemma 6.4.7 it shows that, when $|\mathcal{C}_v| > c \log n$, the cluster size is in fact linear in n with high probability. We will have more to say about the largest cluster in the next subsection.

Lemma 6.4.9 (Supercritical regime: saturation). *Let $G_n \sim \mathbb{G}_{n,p_n}$, where $p_n = \frac{\lambda}{n}$ with $\lambda > 1$ and let $\mathcal{C}_v$ be the connected component of $v \in [n]$. Let ζ_λ be the unique solution in $(0,1)$ to the fixed point equation*

$$1 - e^{-\lambda \zeta} = \zeta.$$

For any $\alpha < \zeta_\lambda$ and any $\delta > 0$, there exists $\kappa_{\delta,\alpha} > 0$ large enough so that

$$\mathbb{P}_{n,p_n}\left[(1 + \kappa_{\delta,\alpha})I_\lambda^{-1}\log n \leq |\mathcal{C}_v| \leq \alpha n\right] = O(n^{-(1+\delta)}). \tag{6.4.25}$$

Proof By Lemma 6.4.8,

$$\mathbb{P}_{n,p_n}[|\mathcal{C}_v| = t] \leq \mathbb{P}[Y_t = t - 1] \leq \mathbb{P}[Y_t \leq t],$$

where $Y_t \sim \text{Bin}(n - 1, 1 - (1 - p_n)^t)$. Roughly, the right-hand side is negligible until the mean $\mu_t := (n - 1)(1 - (1 - \lambda/n)^t)$ is of the order of t. Let ζ_λ be as above, and recall that it is a solution to

$$1 - e^{-\lambda \zeta} - \zeta = 0.$$

Note in particular that when $t = \zeta_\lambda n$,

$$\mu_t = (n - 1)(1 - (1 - \lambda/n)^{\zeta_\lambda n}) \approx n(1 - e^{-\lambda \zeta_\lambda}) = \zeta_\lambda n = t.$$

Let $\alpha < \zeta_\lambda$.

For any $t \in [c \log n, \alpha n]$, by the Chernoff bound for Poisson trials (Theorem 2.4.7 (ii)(b)),

$$\mathbb{P}[Y_t \leq t] \leq \exp\left(-\frac{\mu_t}{2}\left(1 - \frac{t}{\mu_t}\right)^2\right). \tag{6.4.26}$$

For $t/n \leq \alpha < \zeta_\lambda$, using $1 - x \leq e^{-x}$ for $x \in (0, 1)$ (see Exercise 1.16), there is $\gamma_\alpha > 1$ such that

$$\mu_t \geq (n - 1)(1 - e^{-\lambda(t/n)})$$
$$= t\left(\frac{n - 1}{n}\right)\frac{1 - e^{-\lambda(t/n)}}{t/n}$$
$$\geq t\left(\frac{n - 1}{n}\right)\frac{1 - e^{-\lambda\alpha}}{\alpha}$$
$$\geq \gamma_\alpha t,$$

for n large enough, where we used that $1 - e^{-\lambda x}$ is increasing in x on the third line and that $1 - e^{-\lambda x} - x > 0$ for $0 < x < \zeta_\lambda$ on the fourth line (as can be checked by computing the first and second derivatives). Plugging this back into (6.4.26), we get

$$\mathbb{P}[Y_t \leq t] \leq \exp\left(-t\left\{\frac{\gamma_\alpha}{2}\left(1 - \frac{1}{\gamma_\alpha}\right)^2\right\}\right).$$

Therefore,

$$\sum_{t=c\log n}^{\alpha n} \mathbb{P}_{n,p_n}[|\mathcal{C}_v| = t] \leq \sum_{t=c\log n}^{\alpha n} \mathbb{P}[Y_t \leq t]$$

$$\leq \sum_{t=c\log n}^{+\infty} \exp\left(-t\left\{\frac{\gamma_\alpha}{2}\left(1 - \frac{1}{\gamma_\alpha}\right)^2\right\}\right)$$

$$= O\left(\exp\left(-c\log n\left\{\frac{\gamma_\alpha}{2}\left(1 - \frac{1}{\gamma_\alpha}\right)^2\right\}\right)\right).$$

Taking $\kappa > 0$ large enough proves (6.4.25).   ∎

6.4.3 Concentration of Cluster Size: Second Moment Bounds

To characterize the size of the largest cluster in the supercritical case, we use Chebyshev's inequality. We also use a related second moment argument to give a lower bound on the largest cluster in the subcritical regime.

Supercritical regime: giant component

Assume $\lambda > 1$. Our goal is to characterize the size of the largest component. We do this by bounding what is *not* in it (i.e., intuitively those vertices whose exploration process goes extinct). For $\delta > 0$ and $\alpha < \zeta_\lambda$, let $\kappa_{\delta,\alpha}$ be as defined in Lemma 6.4.9. Set

$$\underline{k}_n := (1 + \kappa_{\delta,\alpha})I_\lambda^{-1}\log n \qquad \text{and} \qquad \bar{k}_n := \alpha n.$$

We call a vertex v such that $|\mathcal{C}_v| \leq \underline{k}_n$ a *small vertex*.   SMALL

Let   VERTEX

$$S_k := \sum_{v\in[n]} \mathbf{1}_{\{|\mathcal{C}_v|\leq k\}}.$$

It will also be useful to work with

$$B_k = n - S_k = \sum_{v\in[n]} \mathbf{1}_{\{|\mathcal{C}_v|>k\}}.$$

The quantity $S_{\underline{k}_n}$ is the number of small vertices. By Lemma 6.4.7, its expectation is

$$\mathbb{E}_{n,p_n}[S_{\underline{k}_n}] = n(1 - \mathbb{P}_{n,p_n}[|\mathcal{C}_v| > \underline{k}_n]) = (1 - \zeta_\lambda)n + O\left(\log^2 n\right). \tag{6.4.27}$$

Using Chebyshev's inequality (Theorem 2.1.2), we prove that $S_{\underline{k}_n}$ is concentrated.

Lemma 6.4.10 (Concentration of $S_{\underline{k}_n}$). *For any $\gamma \in (1/2, 1)$ and $\delta < 2\gamma - 1$,*

$$\mathbb{P}_{n,p_n}[|S_{\underline{k}_n} - (1 - \zeta_\lambda)n| \geq n^\gamma] = O(n^{-\delta}).$$

Lemma 6.4.10, which is proved below, leads to our main result in the supercritical case: the   GIANT
existence of the *giant component*, a unique cluster $\mathcal{C}_{\max}$ of size linear in n.   COMPONENT

Proof of Theorem 6.4.2   Take $\alpha \in (\zeta_\lambda/2, \zeta_\lambda)$ and let $\underline{k}_n$, $\bar{k}_n$, and γ be as above. Let $\mathcal{B}_{1,n} :=$ $\{|B_{\underline{k}_n} - \zeta_\lambda n| \geq n^\gamma\}$. Because $\gamma < 1$, the event $\mathcal{B}_{1,n}^c$ implies that

$$\sum_{v \in [n]} \mathbf{1}_{\{|\mathcal{C}_v| > \underline{k}_n\}} = B_{\underline{k}_n} > \zeta_\lambda n - n^\gamma \geq 1,$$

for n large enough. That is, there is at least one "large" cluster of size $> \underline{k}_n$. In turn, that implies

$$|\mathcal{C}_{\max}| \leq B_{\underline{k}_n},$$

since there are at most $B_{\underline{k}_n}$ vertices in that large cluster.

Let $\mathcal{B}_{2,n} := \{\exists v, |\mathcal{C}_v| \in [\underline{k}_n, \bar{k}_n]\}$. If $\mathcal{B}_{2,n}^c$ holds, in addition to $\mathcal{B}_{1,n}^c$, then

$$|\mathcal{C}_{\max}| \leq B_{\underline{k}_n} = B_{\bar{k}_n},$$

since there is no cluster whose size falls in $[\underline{k}_n, \bar{k}_n]$. Moreover, there is equality across the last display if there is a unique cluster of size greater than $\bar{k}_n$.

This is indeed the case under $\mathcal{B}_{1,n}^c \cap \mathcal{B}_{2,n}^c$: if there were two distinct clusters of size $\bar{k}_n$, then since $2\alpha > \zeta_\lambda$ we would have for n large enough

$$B_{\underline{k}_n} = B_{\bar{k}_n} > 2\bar{k}_n = 2\alpha n > \zeta_\lambda n + n^\gamma,$$

a contradiction. Hence we have proved that under $\mathcal{B}_{1,n}^c \cap \mathcal{B}_{2,n}^c$,

$$|\mathcal{C}_{\max}| = B_{\underline{k}_n} = B_{\bar{k}_n}.$$

Take $\delta < 2\gamma - 1$. Applying Lemmas 6.4.9 and 6.4.10,

$$\mathbb{P}[\mathcal{B}_{1,n} \cup \mathcal{B}_{2,n}] \leq O(n^{-\delta}) + n \cdot O(n^{-(1+\delta)}) = O(n^{-\delta}),$$

which concludes the proof.   ∎

It remains to prove Lemma 6.4.10.

Proof of Lemma 6.4.10   As mentioned above, we use Chebyshev's inequality. Hence our main task is to bound the variance of $S_{\underline{k}_n}$.

Our starting point is the following expression for the second moment

$$\mathbb{E}_{n,p_n}[S_k^2] = \sum_{u,v \in [n]} \mathbb{P}_{n,p_n}[|\mathcal{C}_u| \leq k, |\mathcal{C}_v| \leq k]$$

$$= \sum_{u,v \in [n]} \{\mathbb{P}_{n,p_n}[|\mathcal{C}_u| \leq k, |\mathcal{C}_v| \leq k, u \leftrightarrow v]$$

$$+ \mathbb{P}_{n,p_n}[|\mathcal{C}_u| \leq k, |\mathcal{C}_v| \leq k, u \nleftrightarrow v]\}, \tag{6.4.28}$$

where $u \leftrightarrow v$ indicates that u and v are in the same connected component.

To bound the first term in (6.4.28), we note that $u \leftrightarrow v$ implies that $\mathcal{C}_u = \mathcal{C}_v$. Hence,

$$
\begin{aligned}
\sum_{u,v \in [n]} \mathbb{P}_{n,p_n}[|\mathcal{C}_u| \le k, |\mathcal{C}_v| \le k, u \leftrightarrow v] &= \sum_{u,v \in [n]} \mathbb{P}_{n,p_n}[|\mathcal{C}_u| \le k, v \in \mathcal{C}_u] \\
&= \sum_{u,v \in [n]} \mathbb{E}_{n,p_n}[\mathbf{1}_{\{|\mathcal{C}_u| \le k\}} \mathbf{1}_{\{v \in \mathcal{C}_u\}}] \\
&= \sum_{u \in [n]} \mathbb{E}_{n,p_n}\left[\mathbf{1}_{\{|\mathcal{C}_u| \le k\}} \sum_{v \in [n]} \mathbf{1}_{\{v \in \mathcal{C}_u\}} \right] \\
&= \sum_{u \in [n]} \mathbb{E}_{n,p_n}[|\mathcal{C}_u| \mathbf{1}_{\{|\mathcal{C}_u| \le k\}}] \\
&= n\, \mathbb{E}_{n,p_n}[|\mathcal{C}_1| \mathbf{1}_{\{|\mathcal{C}_1| \le k\}}] \\
&\le nk. \tag{6.4.29}
\end{aligned}
$$

To bound the second term in (6.4.28), we sum over the size of $\mathcal{C}_u$ and note that, conditioned on $\{|\mathcal{C}_u| = \ell, u \nleftrightarrow v\}$, the size of $\mathcal{C}_v$ has the same distribution as the unconditional size of $\mathcal{C}_1$ in a $\mathbb{G}_{n-\ell,p_n}$ random graph, that is,

$$
\mathbb{P}_{n,p_n}[|\mathcal{C}_v| \le k \mid |\mathcal{C}_u| = \ell, u \nleftrightarrow v] = \mathbb{P}_{n-\ell,p_n}[|\mathcal{C}_1| \le k].
$$

Observe that the probability on the right-hand side is increasing in ℓ (as can be seen, for example, by coupling; see below for a related argument). Hence,

$$
\begin{aligned}
&\sum_{u,v \in [n]} \sum_{\ell \le k} \mathbb{P}_{n,p_n}[|\mathcal{C}_u| = \ell, |\mathcal{C}_v| \le k, u \nleftrightarrow v] \\
&= \sum_{u,v \in [n]} \sum_{\ell \le k} \mathbb{P}_{n,p_n}[|\mathcal{C}_u| = \ell, u \nleftrightarrow v]\, \mathbb{P}_{n,p_n}[|\mathcal{C}_v| \le k \mid |\mathcal{C}_u| = \ell, u \nleftrightarrow v] \\
&= \sum_{u,v \in [n]} \sum_{\ell \le k} \mathbb{P}_{n,p_n}[|\mathcal{C}_u| = \ell, u \nleftrightarrow v]\, \mathbb{P}_{n-\ell,p_n}[|\mathcal{C}_v| \le k] \\
&\le \sum_{u,v \in [n]} \sum_{\ell \le k} \mathbb{P}_{n,p_n}[|\mathcal{C}_u| = \ell]\, \mathbb{P}_{n-k,p_n}[|\mathcal{C}_v| \le k] \\
&= \sum_{u,v \in [n]} \mathbb{P}_{n,p_n}[|\mathcal{C}_u| \le k]\, \mathbb{P}_{n-k,p_n}[|\mathcal{C}_v| \le k].
\end{aligned}
$$

To get a bound on the variance of S_k, we need to relate this last expression to $(\mathbb{E}_{n,p_n}[S_k])^2$, where we will use that

$$
\mathbb{E}_{n,p_n}[S_k] = \mathbb{E}_{n,p_n}\left[\sum_{v \in [n]} \mathbf{1}_{\{|\mathcal{C}_v| \le k\}} \right] = \sum_{v \in [n]} \mathbb{P}_{n,p_n}[|\mathcal{C}_v| \le k]. \tag{6.4.30}
$$

We define

$$
\Delta_k := \mathbb{P}_{n-k,p_n}[|\mathcal{C}_1| \le k] - \mathbb{P}_{n,p_n}[|\mathcal{C}_1| \le k].
$$

Then, plugging this back above, we get

$$\sum_{u,v\in[n]}\sum_{\ell\leq k}\mathbb{P}_{n,p_n}[|\mathcal{C}_u| = \ell, |\mathcal{C}_v| \leq k, u \leftrightarrow v]$$

$$\leq \sum_{u,v\in[n]}\mathbb{P}_{n,p_n}[|\mathcal{C}_u| \leq k](\mathbb{P}_{n,p_n}[|\mathcal{C}_v| \leq k] + \Delta_k)$$

$$\leq (\mathbb{E}_{n,p_n}[S_k])^2 + n^2|\Delta_k|$$

by (6.4.30). It remains to bound Δ_k.

We use a coupling argument. Let $H \sim \mathbb{G}_{n-k,p_n}$ and construct $H' \sim \mathbb{G}_{n,p_n}$ in the following manner: let H' coincide with H on the first $n - k$ vertices, and then pick the rest of the edges independently. Then clearly $\Delta_k \geq 0$ since the cluster of 1 in H' includes the cluster of 1 in H. In fact, Δ_k is the probability that under this coupling the cluster of 1 has at most k vertices in H but not in H'. That implies in particular that at least one of the vertices in the cluster of 1 in H is connected to a vertex in $\{n-k+1,\ldots,n\}$. Hence, by a union bound over those k^2 potential edges,

$$\Delta_k \leq k^2 p_n$$

and

$$\sum_{u,v\in[n]}\mathbb{P}_{n,p_n}[|\mathcal{C}_u| \leq k, |\mathcal{C}_v| \leq k, u \leftrightarrow v] \leq (\mathbb{E}_{n,p_n}[S_k])^2 + \lambda nk^2. \qquad (6.4.31)$$

Combining (6.4.29) and (6.4.31), we get

$$\mathrm{Var}[S_k] \leq 2\lambda nk^2.$$

The result follows from (6.4.27) and Chebyshev's inequality

$$\mathbb{P}[|S_{\underline{k}_n} - (1 - \zeta_\lambda)n| \geq n^\gamma]$$

$$\leq \mathbb{P}[|S_{\underline{k}_n} - \mathbb{E}_{n,p_n}[S_{\underline{k}_n}]| \geq n^\gamma - C\log^2 n]$$

$$\leq \frac{2\lambda n\underline{k}_n^2}{(n^\gamma - C\log^2 n)^2}$$

$$\leq \frac{2\lambda n(1 + \kappa_{\delta,\alpha})^2 I_\lambda^{-2}\log^2 n}{C'n^{2\gamma}}$$

$$\leq C''n^{-\delta}$$

for constants $C, C', C'' > 0$ and n large enough, where we used that $2\gamma > 1$ and $\delta < 2\gamma - 1$. ∎

Subcritical regime: second moment argument

A second moment argument also gives a lower bound on the size of the largest component in the subcritical case. We proved in Theorem 6.4.1 that, when $\lambda < 1$, the probability of observing a connected component of size larger than $I_\lambda^{-1}\log n$ is vanishingly small. In the other direction, we get:

Theorem 6.4.11 (Subcritical regime: lower bound on the largest cluster). *Let $G_n \sim \mathbb{G}_{n,p_n}$, where $p_n = \frac{\lambda}{n}$ with $\lambda \in (0, 1)$. For all $\kappa \in (0, 1)$,*

$$\mathbb{P}_{n,p_n}\left[|\mathcal{C}_{\max}| \leq (1-\kappa)I_\lambda^{-1} \log n\right] = o(1).$$

Proof Recall that

$$B_k = \sum_{v \in [n]} \mathbf{1}_{\{|\mathcal{C}_v| > k\}}.$$

It suffices to prove that with probability $1 - o(1)$ we have $B_k > 0$ when $k = (1 - \kappa)I_\lambda^{-1} \log n$. To apply the second moment method (Theorem 2.3.2), we need an upper bound on the second moment of B_k and a lower bound on its first moment. The following lemma is closely related to Lemma 6.4.10. Exercise 6.12 asks for a proof.

Lemma 6.4.12 (Second moment of X_k). *Assume $\lambda < 1$. There is a constant $C > 0$ such that*

$$\mathbb{E}_{n,p_n}[B_k^2] \leq (\mathbb{E}_{n,p_n}[B_k])^2 + Cnke^{-kI_\lambda} \quad \forall k \geq 0.$$

Lemma 6.4.13 (First moment of X_k). *Let $k_n = (1 - \kappa)I_\lambda^{-1} \log n$. Then, for any $\beta \in (0, \kappa)$ we have that*

$$\mathbb{E}_{n,p_n}[B_{k_n}] = \Omega(n^\beta)$$

for n large enough.

Proof By Lemma 6.4.3,

$$\mathbb{E}_{n,p_n}[B_{k_n}] = n\,\mathbb{P}_{n,p_n}[|\mathcal{C}_1| > k_n]$$
$$\geq n\,\mathbb{P}[W_\lambda > k_n] - O\left(\lceil k_n \rceil^2\right). \tag{6.4.32}$$

Once again, we use the random-walk representation of the total progeny of a branching process (Theorem 6.2.6). In contrast to the proof of Lemma 6.4.6, we need a lower bound this time. For this purpose, we use the explicit expression for the law of the total progeny W_λ from Example 6.2.7:

$$\mathbb{P}[W_\lambda > k_n] = \sum_{t > k_n} \frac{1}{t} e^{-\lambda t} \frac{(\lambda t)^{t-1}}{(t-1)!}.$$

Using Stirling's formula (see Appendix A) and (6.4.3), we note that

$$\frac{1}{t} e^{-\lambda t} \frac{(\lambda t)^{t-1}}{(t-1)!} = e^{-\lambda t} \frac{(\lambda t)^{t-1}}{t!}$$
$$= e^{-\lambda t} \frac{(\lambda t)^t}{\lambda t (t/e)^t \sqrt{2\pi t}(1 + o(1))}$$
$$= \frac{1 - o(1)}{\lambda \sqrt{2\pi t^3}} \exp\left(-t\lambda + t \log \lambda + t\right)$$
$$= \frac{1 - o(1)}{\lambda \sqrt{2\pi t^3}} \exp\left(-t I_\lambda\right).$$

Hence, for any $\varepsilon > 0$,

$$\mathbb{P}[W_\lambda > k_n] \geq \lambda^{-1} \sum_{t > k_n} \exp\left(-t(I_\lambda + \varepsilon)\right)$$

$$= \Omega\left(\exp\left(-k_n(I_\lambda + \varepsilon)\right)\right)$$

for n large enough. For any $\beta \in (0, \kappa)$, taking ε small enough we have

$$n\,\mathbb{P}[W_\lambda > k_n] = \Omega\left(n\exp\left(-k_n(I_\lambda + \varepsilon)\right)\right)$$

$$= \Omega\left(\exp\left(\{1 - (1 - \kappa)I_\lambda^{-1}(I_\lambda + \varepsilon)\}\log n\right)\right)$$

$$= \Omega(n^\beta).$$

Plugging this back into (6.4.32) gives

$$\mathbb{E}_{n,p_n}[B_{k_n}] = \Omega(n^\beta),$$

which proves the claim.   ∎

We return to the proof of Theorem 6.4.11. Let again $k_n = (1 - \kappa)I_\lambda^{-1}\log n$. By the second moment method and Lemmas 6.4.12 and 6.4.13,

$$\mathbb{P}_{n,p_n}[B_{k_n} > 0] \geq \frac{(\mathbb{E}B_{k_n})^2}{\mathbb{E}[B_{k_n}^2]}$$

$$\geq \left(1 + \frac{O(nk_ne^{-k_nI_\lambda})}{\Omega(n^{2\beta})}\right)^{-1}$$

$$= \left(1 + \frac{O(nk_ne^{(\kappa-1)\log n})}{\Omega(n^{2\beta})}\right)^{-1}$$

$$= \left(1 + \frac{O(k_nn^\kappa)}{\Omega(n^{2\beta})}\right)^{-1}$$

$$\to 1$$

for β close enough to κ. That proves the claim.   ∎

6.4.4 Critical Case via Martingales

It remains to consider the critical case, that is, when $\lambda = 1$. As it turns out, the model goes through a "double jump": as λ crosses 1, the largest cluster size goes from order $\log n$ to order $n^{2/3}$ to order n. Here we use martingale methods to show the following.

Theorem 6.4.14 (Critical case: upper bound on the largest cluster). *Let $G_n \sim \mathbb{G}_{n,p_n}$, where $p_n = \frac{1}{n}$. For all $\kappa > 1$,*

$$\mathbb{P}_{n,p_n}\left[|\mathcal{C}_{\max}| > \kappa n^{2/3}\right] \leq \frac{C}{\kappa^{3/2}}$$

for some constant $C > 0$.

Remark 6.4.15 *One can also derive a lower bound on the probability that $|\mathcal{C}_{\max}| > \kappa n^{2/3}$ for some $\kappa > 0$ [ER60]. Exercise 6.20 provides a sketch based on counting tree components;*

the combinatorial approach has the advantage of giving insights into the structure of the graph (see [Bol01] for more on this). See also [NP10] for a martingale proof of the lower bound as well as a better upper bound.

The key technical bound is the following.

Lemma 6.4.16 *Let $G_n \sim \mathbb{G}_{n,p_n}$, where $p_n = \frac{1}{n}$ and let $\mathcal{C}_v$ be the connected component of $v \in [n]$. There are constants $c, c' > 0$ such that for all $k \geq c$,*

$$\mathbb{P}_{n,p_n}[|\mathcal{C}_v| > k] \leq \frac{c'}{\sqrt{k}}.$$

Before we establish the lemma, we prove the theorem assuming it.

Proof of Theorem 6.4.14   Recall that

$$B_k = \sum_{v \in [n]} \mathbf{1}_{\{|\mathcal{C}_v| > k\}}.$$

Take

$$k_n := \kappa n^{2/3}.$$

By Markov's inequality (Theorem 2.1.1) and Lemma 6.4.16,

$$\begin{aligned}
\mathbb{P}_{n,p_n}[|\mathcal{C}_{\max}| > k_n] &\leq \mathbb{P}_{n,p_n}\left[B_{k_n} > k_n\right] \\
&\leq \frac{\mathbb{E}_{n,p_n}\left[B_{k_n}\right]}{k_n} \\
&= \frac{n\,\mathbb{P}_{n,p_n}[|\mathcal{C}_v| > k_n]}{k_n} \\
&\leq \frac{nc'}{k_n^{3/2}} \\
&\leq \frac{C}{\kappa^{3/2}}
\end{aligned}$$

for some constant $C > 0$.   ∎

It remains to prove the lemma.

Proof of Lemma 6.4.16   Once again, we use the exploration process defined in Section 6.4.2 started at v. Let $(\mathcal{F}_t)$ be the corresponding filtration and let $A_t = |\mathcal{A}_t|$ be the size of the active set.

Domination by a martingale   Recalling (6.4.6), we define

$$M_t := M_{t-1} + \left[-1 + \widetilde{X}_t\right], \tag{6.4.33}$$

with $M_0 := 1$ and $(\widetilde{X}_t)$ are i.i.d. Bin$(n, 1/n)$. We couple (A_t) and (M_t) through (6.4.7) by letting

$$\widetilde{X}_t = \sum_{i=1}^{n} I_{t,i}.$$

In particular, $M_t \geq A_t$ for all t.

Furthermore, we have

$$\mathbb{E}[M_t \mid \mathcal{F}_{t-1}] = M_{t-1} - 1 + n\frac{1}{n} = M_{t-1}.$$

So (M_t) is a martingale. We define the stopping time

$$\tilde{\tau}_0 := \inf\{t \geq 0 : M_t = 0\}.$$

Recalling that

$$\tau_0 = \inf\{t \geq 0 : A_t = 0\} = |\mathcal{C}_v|,$$

by Lemma 6.2.1, we have $\tilde{\tau}_0 \geq \tau_0 = |\mathcal{C}_v|$ almost surely. So

$$\mathbb{P}_{n,p_n}[|\mathcal{C}_v| > k] \leq \mathbb{P}[\tilde{\tau}_0 > k].$$

The tail of $\tilde{\tau}_0$ To bound the tail of $\tilde{\tau}_0$, we introduce a modified stopping time. For $h > 0$, let

$$\tau'_h := \inf\{t \geq 0 : M_t = 0 \text{ or } M_t \geq h\}.$$

We will use the inequality

$$\mathbb{P}[\tilde{\tau}_0 > k] = \mathbb{P}[M_t > 0, \forall t \leq k] \leq \mathbb{P}[\tau'_h > k] + \mathbb{P}[M_{\tau'_h} \geq h],$$

and we will choose h below to minimize the rightmost expression (or, more specifically, an upper bound on it). The rest of the analysis is similar to the gambler's ruin problem in Example 3.1.41, with some slight complications arising from the fact that the process is not nearest-neighbor.

We note that by the exponential tail of hitting times on finite state spaces (Lemma 3.1.25), the stopping time τ'_h is almost surely finite and, in fact, has a finite expectation. By two applications of Markov's inequality,

$$\mathbb{P}[M_{\tau'_h} \geq h] \leq \frac{\mathbb{E}[M_{\tau'_h}]}{h}$$

and

$$\mathbb{P}[\tau'_h > k] \leq \frac{\mathbb{E}\tau'_h}{k}.$$

So it remains to bound the expectations on the right-hand sides.

Bounding $\mathbb{E}M_{\tau'_h}$ and $\mathbb{E}\tau'_h$ To compute $\mathbb{E}M_{\tau'_h}$, we use the optional stopping theorem in the uniformly bounded case (Theorem 3.1.38 (ii)) to the stopped process $(M_{t \wedge \tau'_h})$ (which is also a martingale by Lemma 3.1.37) to get that

$$\mathbb{E}[M_{\tau'_h}] = \mathbb{E}[M_0] = 1.$$

We conclude that

$$\mathbb{P}[M_{\tau'_h} \geq h] \leq \frac{1}{h}. \tag{6.4.34}$$

To compute $\mathbb{E}\tau'_h$, we use a different martingale (adapted from Example 3.1.31), specifically

$$L_t := M_t^2 - \sigma^2 t,$$

where we let $\sigma^2 := n\frac{1}{n}\left(1 - \frac{1}{n}\right) = \left(1 - \frac{1}{n}\right)$, which is $\geq \frac{1}{2}$ when $n \geq 2$. To see that (L_t) is a martingale, note that by taking out what is known (Lemma B.6.13) and using the fact that (M_t) is itself a martingale

$$
\begin{aligned}
\mathbb{E}[L_t \mid \mathcal{F}_{t-1}] &= \mathbb{E}[(M_{t-1} + (M_t - M_{t-1}))^2 - \sigma^2 t \mid \mathcal{F}_{t-1}] \\
&= \mathbb{E}[M_{t-1}^2 + 2M_{t-1}(M_t - M_{t-1}) + (M_t - M_{t-1})^2 - \sigma^2 t \mid \mathcal{F}_{t-1}] \\
&= M_{t-1}^2 + 2M_{t-1} \cdot 0 + \sigma^2 - \sigma^2 t \\
&= L_{t-1}.
\end{aligned}
$$

By Lemma 3.1.37, the stopped process $(L_{t\wedge\tau'_h})$ is also a martingale; and it has bounded increments since

$$
\begin{aligned}
|L_{(t+1)\wedge\tau'_h} - L_{t\wedge\tau'_h}| &\leq |M_{(t+1)\wedge\tau'_h}^2 - M_{t\wedge\tau'_h}^2| + \sigma^2 \\
&\leq \left||(-1 + \widetilde{X}_{t+1})^2 + 2h| - |1 + \widetilde{X}_{t+1}|\right| + \sigma^2 \\
&\leq n^2 + 2hn + 1.
\end{aligned}
$$

We use the optional stopping theorem in the bounded increments case (Theorem 3.1.38 (iii)) on $(L_{t\wedge\tau'_h})$ to get

$$\mathbb{E}[M_{\tau'_h}^2 - \sigma^2\tau'_h] = \mathbb{E}[M_{\tau'_h}^2] - \sigma^2\mathbb{E}\tau'_h = 1.$$

After rearranging (6.4.35),

$$\mathbb{E}\tau'_h \leq \frac{1}{\sigma^2}\mathbb{E}[M_{\tau'_h}^2] \leq 2\,\mathbb{E}[M_{\tau'_h}^2], \tag{6.4.35}$$

where we used the fact that $\sigma^2 \geq 1/2$.

To bound $\mathbb{E}[M_{\tau'_h}^2]$, we need to control by how much the process "overshoots" h. A stochastic domination argument gives the desired bound; Exercise 6.21 asks for a proof.

Lemma 6.4.17 (Overshoot bound). *Let f be an increasing function and $W \sim \mathrm{Bin}(n, 1/n)$. Then,*

$$\mathbb{E}[f(M_{\tau'_h} - h) \mid M_{\tau'_h} \geq h] \leq \mathbb{E}[f(W)].$$

The lemma implies that

$$
\begin{aligned}
\mathbb{E}[M_{\tau'_h}^2 \mid M_{\tau'_h} \geq h] &= \mathbb{E}[(M_{\tau'_h} - h)^2 + 2(M_{\tau'_h} - h)h + h^2 \mid M_{\tau'_h} \geq h] \\
&\leq (\sigma^2 + 1) + 2h + h^2 \\
&\leq 4h^2.
\end{aligned}
$$

Plugging back into (6.4.35) gives

$$\mathbb{E}\tau'_h \leq 2\left\{\frac{1}{h}\mathbb{E}[M_{\tau'_h}^2 \mid M_{\tau'_h} \geq h]\right\} \leq 8h,$$

where we used (6.4.34).

Putting everything together Finally, take $h := \sqrt{\frac{k}{8}}$. Putting everything together,

$$\mathbb{P}_{n,p_n}[|\mathcal{C}_v| > k] \leq \mathbb{P}[\tilde{\tau}_0 > k] \leq \mathbb{P}[\tau'_h > k] + \mathbb{P}[M_{\tau'_h} \geq h] \leq \frac{8h}{k} + \frac{1}{h} = 2\sqrt{\frac{8}{k}}.$$

That concludes the proof. ∎

6.4.5 ▷ *Encore: Random Walk on the Erdős–Rényi Graph*

So far in this section we have used techniques from all chapters of the book – with the exception of Chapter 5. Not to be outdone, we discuss one last result that will make use of spectral techniques. We venture a little further down the evolution of the Erdős–Rényi graph model to the connected regime. Specifically, recall from Section 2.3.2 that $G_n = (V_n, E_n) \sim \mathbb{G}_{n,p_n}$ is connected with probability $1 - o(1)$ when $np_n = \omega(\log n)$.

We show in that regime that lazy simple random walk (X_t) on G_n "mixes fast." Recall from Example 1.1.29 that, when the graph is connected, the corresponding transition matrix P is reversible with respect to the stationary distribution

$$\pi(v) := \frac{\delta(v)}{2|E_n|},$$

where $\delta(v)$ is the degree of v. For a fixed $\varepsilon > 0$, the mixing time (see Definition 1.1.35) is

$$t_{\text{mix}}(\varepsilon) = \inf\{t \geq 0 \colon d(t) \leq \varepsilon\},$$

where

$$d(t) = \sup_{x \in V_n} \|P^t(x, \cdot) - \pi(\cdot)\|_{\text{TV}}.$$

By convention, we let $t_{\text{mix}}(\varepsilon) = +\infty$ if the graph is not connected. Our main result is the following.

Theorem 6.4.18 (Mixing on a connected Erdős–Rényi graph). *Let* $G_n \sim \mathbb{G}_{n,p_n}$ *with* $np_n = \omega(\log n)$. *With probability* $1 - o(1)$, *the mixing time is* $O(\log n)$.

Edge expansion We use Cheeger's inequality (Theorem 5.3.5) which, recall, states that

$$\gamma \geq \frac{\Phi_*^2}{2},$$

where γ is the spectral gap of P (see Definition 5.2.11) and

$$\Phi_* = \min\left\{\Phi_E(S; c, \pi) \colon S \subseteq V, \ 0 < \pi(S) \leq \frac{1}{2}\right\}$$

is the edge expansion constant (see Definition 5.3.2) with

$$\Phi_E(S; c, \pi) = \frac{c(S, S^c)}{\pi(S)}$$

for a subset of vertices $S \subseteq V_n$. Here, for a pair of vertices x, y connected by an edge,

$$c(x, y) = \pi(x)P(x, y) = \frac{\delta(x)}{2|E_n|} \frac{1}{\delta(x)} = \frac{1}{2|E_n|}.$$

Hence

$$c(S, S^c) = \frac{|E(S, S^c)|}{2|E_n|},$$

where $E(S, S^c)$ is the set of edges between S and S^c. Similarly,

$$\pi(S) = \frac{\sum_{x \in S} \delta(x)}{2|E_n|}.$$

The numerator is referred to as the volume of S and we use the notation $\mathrm{vol}(S) = \sum_{x \in S} \delta(x)$. So

$$\frac{c(S, S^c)}{\pi(S)} = \frac{|E(S, S^c)|}{\mathrm{vol}(S)}. \tag{6.4.36}$$

Because the random walk is lazy, the spectral gap is equal to the absolute spectral gap (see Definition 5.2.11), and as a consequence the relaxation time (see Definition 5.2.12) is

$$t_{\mathrm{rel}} = \gamma^{-1}.$$

Using Theorem 5.2.14, we get

$$t_{\mathrm{mix}}(\varepsilon) \leq \log\left(\frac{1}{\varepsilon \pi_{\min}}\right) t_{\mathrm{rel}} \leq \log\left(\frac{1}{\varepsilon \pi_{\min}}\right) \frac{2}{\Phi_*^2}, \tag{6.4.37}$$

where

$$\pi_{\min} = \min_x \pi(x) = \min_x \frac{\delta(x)}{2|E_n|} = \frac{\min_x \delta(x)}{\sum_y \delta(y)}.$$

So our main task is to bound $\delta(x)$ and $|E(S, S^c)|$ with high probability. We do this next.

Bounding the degrees In fact, we have already done half the work. Indeed, in Example 2.4.18 we studied the maximum degree of G_n

$$D_n = \max_{v \in V_n} \delta(v)$$

in the regime $np_n = \omega(\log n)$. We showed that for any $\zeta > 0$ as $n \to +\infty$,

$$\mathbb{P}\left[|D_n - np_n| \leq 2\sqrt{(1 + \zeta)np_n \log n}\right] \to 1.$$

The proof of that result actually shows something stronger: all degrees satisfy the inequality simultaneously, that is,

$$\mathbb{P}\left[\forall v \in V_n, \, |\delta(v) - np_n| \leq 2\sqrt{(1 + \zeta)np_n \log n}\right] = 1 - o(1). \tag{6.4.38}$$

We will use the fact that $2\sqrt{(1 + \zeta)np_n \log n} = o(np_n)$ when $np_n = \omega(\log n)$. In essence, all degrees are roughly np_n. That implies the following claims.

Lemma 6.4.19 (Bounds on stationary distribution and volume). *The following hold with probability $1 - o(1)$.*

(i) The smallest stationary probability satisfies

$$\pi_{\min} \geq \frac{1 - o(1)}{n}.$$

(ii) *For any set of vertices $S \subseteq V_n$ with $|S| > 2n/3$, we have*

$$\pi(S) > \frac{1}{2}.$$

(iii) *For any set of vertices $S \subseteq V_n$ with $s := |S|$,*

$$\text{vol}(S) = snp_n(1 + o(1)).$$

Proof We assume that the event in (6.4.38) holds.

For (i), that means

$$\pi_{\min} \geq \frac{np_n - 2\sqrt{(1+\zeta)np_n \log n}}{n(np_n + 2\sqrt{(1+\zeta)np_n \log n})} = \frac{1}{n}(1 - o(1)),$$

when $np_n = \omega(\log n)$.

For (ii), we get

$$\pi(S) = \frac{\sum_{x \in S} \delta(x)}{\sum_{x \in V_n} \delta(x)} \geq \frac{|S|(np_n - 2\sqrt{(1+\zeta)np_n \log n})}{n(np_n + 2\sqrt{(1+\zeta)np_n \log n})} > \frac{2}{3}(1 - o(1)).$$

Finally, (iii) follows similarly. ∎

Bounding the cut size An application of Bernstein's inequality (Theorem 2.4.17) gives the following bound.

Lemma 6.4.20 (Bound on the edge expansion). *With probability $1 - o(1)$,*

$$\Phi_* = \Omega(1).$$

Proof By the definition of Φ_* and Lemma 6.4.19 (ii), we can restrict ourselves to sets S of size at most $2n/3$. Let S be such a set with $s = |S|$. Then, $|E(S, S^c)|$ is $\text{Bin}(s(n - s), p_n)$. By Bernstein's inequality with $c = 1$ and $v_i = p_n(1 - p_n)$,

$$\mathbb{P}_{n,p_n}[|E(S, S^c)| \leq s(n - s)p_n - \beta] \leq \exp\left(-\frac{\beta^2}{4s(n - s)p_n(1 - p_n)}\right),$$

for $\beta \leq s(n - s)p_n(1 - p_n)$. We take $\beta = \frac{1}{2}s(n - s)p_n$ and get

$$\mathbb{P}_{n,p_n}\left[|E(S, S^c)| \leq \frac{1}{2}s(n - s)p_n\right] \leq \exp\left(-\frac{s(n - s)p_n}{16(1 - p_n)}\right).$$

By a union bound over all sets of size s and using the fact that $\binom{n}{s} \leq (\frac{ne}{s})^s$ (see Appendix A), there is a constant $C > 0$ such that

$$\mathbb{P}_{n,p_n}\left[\exists S, \, |S| = s, \, |E(S, S^c)| \leq \frac{1}{2}s(n - s)p_n\right]$$

$$\leq \binom{n}{s} \exp\left(-\frac{s(n - s)p_n}{16(1 - p_n)}\right)$$

$$\leq \exp\left(-s\frac{np_n}{48} + s\log(ne/s)\right)$$

$$\leq \exp\left(-Csnp_n\right),$$

for n large enough, where we also used that $n - s \geq n/3$ and $np_n = \omega(\log n)$. Summing over s gives, for a constant $C' > 0$,

$$\mathbb{P}_{n,p_n}\left[\exists S, \ 1 \leq |S| \leq 2n/3, \ |E(S, S^c)| \leq \frac{1}{2}|S|(n - |S|)p_n\right] \leq C' \exp\left(-Cnp_n\right),$$

which goes to 0 as $n \to +\infty$.

Using (6.4.36) and Lemma 6.4.19 (iii), any set S such that $|E(S, S^c)| > \frac{1}{2}|S|(n - |S|)p_n$ has edge expansion

$$\Phi_{\mathrm{E}}(S; c, \pi) \geq \frac{\frac{1}{2}|S|(n - |S|)p_n}{|S|np_n(1 + o(1))} \geq \frac{1}{6}(1 - o(1)).$$

That proves the claim. ∎

Proof of the theorem Finally, we are ready to prove the main result.

Proof of Theorem 6.4.18 Plugging Lemma 6.4.19 (i) and Lemma 6.4.20 into (6.4.37) gives

$$t_{\mathrm{mix}}(\varepsilon) \leq \log\left(\frac{1}{\varepsilon\pi_{\min}}\right)\frac{2}{\Phi_*^2} \leq C'' \log(\varepsilon^{-1}n(1 + o(1))) = O(\log n)$$

for some constant $C'' > 0$. ∎

Remark 6.4.21 *A mixing time of $O(\log n)$ in fact holds for lazy simple random walk on $\mathbb{G}_{n,p_n}$ when $p_n = \lambda \log(n)/n$ with $\lambda > 1$ [CF07]. See also [Dur06, section 6.5]. Mixing time on the giant component has also been studied. See, for example, [FR08, BKW14, DKLP11].*

Exercises

Exercise 6.1 (Galton–Watson process: subcritical case). We use Markov's inequality to analyze the subcritical case.

   (i) Let (Z_t) be a Galton–Watson process with offspring distribution mean $m < 1$. Use Markov's inequality (Theorem 2.1.1) to prove that extinction occurs almost surely.
  (ii) Prove the equivalent result in the multitype case, that is, prove (6.1.8).

Exercise 6.2 (Galton–Watson process: geometric offspring). Let (Z_t) be a Galton–Watson branching process with geometric offspring distribution (started at 0), that is, $p_k = p(1 - p)^k$ for all $k \geq 0$, for some $p \in (0, 1)$. Let $q := 1 - p$, let m be the mean of the offspring distribution, and let $W_t = m^{-t}Z_t$.

   (i) Compute the probability generating function f of $\{p_k\}_{k \geq 0}$ and the extinction probability $\eta := \eta_p$ as a function of p.
  (ii) If G is a 2×2 matrix, define

$$G(s) := \frac{G_{11}s + G_{12}}{G_{21}s + G_{22}}.$$

Show that $G(H(s)) = (GH)(s)$.

(iii) Assume $m \neq 1$. Use (ii) to derive

$$f_t(s) = \frac{pm^t(1-s) + qs - p}{qm^t(1-s) + qs - p}.$$

Deduce that when $m > 1$,

$$\mathbb{E}[\exp(-\lambda W_\infty)] = \eta + (1-\eta)\frac{(1-\eta)}{\lambda + (1-\eta)}.$$

(iv) Assume $m = 1$. Show that

$$f_t(s) = \frac{t - (t-1)s}{t + 1 - ts},$$

and deduce that

$$\mathbb{E}[e^{-\lambda Z_t/t} \mid Z_t > 0] \to \frac{1}{1+\lambda}.$$

Exercise 6.3 (Supercritical branching process: infinite line of descent). Let (Z_t) be a supercritical Galton–Watson branching process with offspring distribution $\{p_k\}_{k \geq 0}$. Let η be the extinction probability and define $\zeta := 1 - \eta$. Let Z_t^∞ be the number of individuals in the tth generation with an infinite line of descent, that is, whose descendant subtree is infinite. Denote by $\mathcal{S}$ the event of non-extinction of (Z_t). Define $p_0^\infty := 0$ and

$$p_k^\infty := \zeta^{-1} \sum_{j \geq k} \binom{j}{k} \eta^{j-k} \zeta^k p_j.$$

   (i) Show that $\{p_k^\infty\}_{k \geq 0}$ is a probability distribution and compute its expectation.
   (ii) Show that for any $k \geq 0$,

$$\mathbb{P}[Z_1^\infty = k \mid \mathcal{S}] = p_k^\infty.$$

   (Hint: Condition on Z_1.)
   (iii) Show by induction on t that, conditioned on non-extinction, the process (Z_t^∞) has the same distribution as a Galton–Watson branching process with offspring distribution $\{p_k^\infty\}_{k \geq 0}$.

Exercise 6.4 (Multitype branching processes: a special case). Extend Lemma 6.1.20 to the case $\mathbf{S}^{(\mathbf{u})} = \mathbf{0}$. (Hint: Show that $U_t = \mathbf{Z}_0 \mathbf{u}$ for all t almost surely.)

Exercise 6.5 (Galton–Watson: inverting history). Let

$$H = (X_1, \ldots, X_{\tau_0})$$

be the history (see Section 6.2) of the Galton–Watson process (Z_i). Write Z_i as a function of H, for all i.

Exercise 6.6 (Spitzer's lemma). Prove Theorem 6.2.5.

Exercise 6.7 (Sum of Poisson). Let Q_1 and Q_2 be independent Poisson random variables with respective means λ_1 and λ_2. Show by direct computation of the convolution that the sum $Q_1 + Q_2$ is Poisson with mean $\lambda_1 + \lambda_2$. (Hint: Recall that $\mathbb{P}[Q_1 = k] = e^{-\lambda_1}\lambda_1^k/k!$ for all $k \in \mathbb{Z}_+$.)

Exercise 6.8 (Percolation on bounded-degree graphs). Let $G = (V, E)$ be a countable graph such that all vertices have degree bounded by $b+1$ for $b \geq 2$. Let 0 be a distinguished vertex in G. For bond percolation on G, prove that

$$p_c(G) \geq p_c(\widehat{\mathbb{T}}_b),$$

by bounding the expected size of the cluster of 0. (Hint: Consider self-avoiding paths started at 0.)

Exercise 6.9 (Percolation on $\widehat{\mathbb{T}}_b$: critical exponent of $\theta(p)$). Consider bond percolation on the rooted infinite b-ary tree $\widehat{\mathbb{T}}_b$ with $b > 2$. For $\varepsilon \in [0, 1 - \frac{1}{b}]$ and $u \in [0, 1]$, define

$$h(\varepsilon, u) := u - \left(\left(1 - \tfrac{1}{b} - \varepsilon \right) (1 - u) + \tfrac{1}{b} + \varepsilon \right)^b.$$

(i) Show that there is a constant $C > 0$ not depending on ε, u such that

$$\left| h(\varepsilon, u) - b\varepsilon u + \frac{b-1}{2b} u^2 \right| \leq C(u^3 \vee \varepsilon u^2).$$

(ii) Use (i) to prove that

$$\lim_{p \downarrow p_c(\widehat{\mathbb{T}}_b)} \frac{\theta(p)}{(p - p_c(\widehat{\mathbb{T}}_b))} = \frac{2b^2}{b-1}.$$

Exercise 6.10 (Percolation on $\widehat{\mathbb{T}}_2$: higher moments of $|\mathcal{C}_0|$). Consider bond percolation on the rooted infinite binary tree $\widehat{\mathbb{T}}_2$. For density $p < \frac{1}{2}$, let Z_p be an integer-valued random variable with distribution

$$\mathbb{P}_p[Z_p = \ell] = \frac{\ell \, \mathbb{P}_p[|\mathcal{C}_0| = \ell]}{\mathbb{E}_p|\mathcal{C}_0|} \qquad \forall \ell \geq 1.$$

(i) Using the explicit formula for $\mathbb{P}_p[|\mathcal{C}_0| = \ell]$ derived in Section 6.2.4, show that for all $0 < a < b < +\infty$,

$$\mathbb{P}_p \left[\frac{Z_p}{(1/4)(\frac{1}{2} - p)^{-2}} \in [a, b] \right] \to C \int_a^b x^{-1/2} e^{-x} dx,$$

as $p \uparrow \frac{1}{2}$, for some constant $C > 0$.

(ii) Show that for all $k \geq 2$, there is $C_k > 0$ such that

$$\lim_{p \uparrow p_c(\widehat{\mathbb{T}}_2)} \frac{\mathbb{E}_p|\mathcal{C}_0|^k}{(p_c(\widehat{\mathbb{T}}_2) - p)^{-1-2(k-1)}} = C_k.$$

(iii) What happens when $p \downarrow p_c(\widehat{\mathbb{T}}_2)$?

Exercise 6.11 (Branching process approximation: improved bound). Let $p_n = \frac{\lambda}{n}$ with $\lambda > 0$. Let W_{n,p_n}, respectively W_λ, be the total progeny of a branching process with offspring distribution $\mathrm{Bin}(n, p_n)$, respectively $\mathrm{Poi}(\lambda)$.

(i) Show that

$$|\mathbb{P}[W_{n,p_n} \geq k] - \mathbb{P}[W_\lambda \geq k]|$$
$$\leq \max\{\mathbb{P}[W_{n,p_n} \geq k, W_\lambda < k], \mathbb{P}[W_{n,p_n} < k, W_\lambda \geq k]\}.$$

(ii) Couple the two processes step-by-step and use (i) to show that

$$|\mathbb{P}[W_{n,p_n} \geq k] - \mathbb{P}[W_\lambda \geq k]| \leq \frac{\lambda^2}{n} \sum_{i=1}^{k-1} \mathbb{P}[W_\lambda \geq i].$$

Exercise 6.12 (Subcritical Erdős–Rényi: second moment). Prove Lemma 6.4.12.

Exercise 6.13 (Random binary search tree: property (BST)). Show that the (BST) property is preserved by the algorithm described at the beginning of Section 6.3.1.

Exercise 6.14 (Random binary search tree: limit). Consider equation (6.3.1).

(i) Show that there exists a unique solution greater than 1.
(ii) Prove that the expression on the left-hand side is strictly decreasing at that solution.

Exercise 6.15 (Random binary search tree: height is well defined). Let $\mathcal{T}$ be an infinite binary tree. Assign an independent $U[0,1]$ random variable Z_v to each vertex v in $\mathcal{T}$, set $S_\rho = n$ and then recursively from the root down

$$S_{v'} := \lfloor S_v Z_v \rfloor \quad \text{and} \quad S_{v''} := \lfloor S_v (1 - Z_v) \rfloor,$$

where v' and v'' are the left and right descendants of v in $\mathcal{T}$.

(i) Show that, for any v, it holds that $S_{v'} + S_{v''} = S_v - 1$ almost surely provided $S_v \geq 1$.
(ii) Show that, for any v, there is almost surely a descendant w of v (not necessarily immediate) such that $S_w = 1$.
(iii) Let

$$H_n = \max\{h \colon \exists v \in \mathcal{V}_h,\ S_v = 1\},$$

where $\mathcal{V}_h$ is the set of vertices of $\mathcal{T}$ at topological distance h from the root. Show that $H_n \leq n$.

Exercise 6.16 (Ising vs. CFN). Let $\mathcal{T}_h$ be a rooted complete binary tree with h levels. Fix $0 < p < 1/2$. Assign to each vertex v a state $\sigma_v \in \{+1, -1\}$ at random according to the CFN model described in Section 6.3.2. Show that this distribution is equivalent to a ferromagnetic Ising model on $\mathcal{T}_h$ and determine the inverse temperature β in terms of p. (Hint: Write the distribution of the states under the CFN model as a product over the edges.)

Exercise 6.17 (Monotonicity of $\|\mu_h^+ - \mu_h^-\|_{\text{TV}}$). Let μ_h^+, μ_h^- be as in Section 6.3.2. Show that

$$\|\mu_{h+1}^+ - \mu_{h+1}^-\|_{\text{TV}} \leq \|\mu_h^+ - \mu_h^-\|_{\text{TV}}.$$

(Hint: Use the Markovian nature of the process.)

Exercise 6.18 (Unsolvability: recursion). Prove Lemma 6.3.15.

Exercise 6.19 (Cayley's formula). Let (Z_t) be a Poisson branching process with offspring mean 1 started at $Z_0 = 1$ and let T be the corresponding Galton–Watson tree. Let W be the total of size of the progeny, that is, the number of vertices in T. Recall from Example 6.2.7 that

$$\mathbb{P}[W = n] = \frac{n^{n-1} e^{-n}}{n!}.$$

(i) Given $W = n$, label the vertices of T uniformly at random with the integers $1, \ldots, n$. Show that every rooted labeled tree on n vertices arises with probability $e^{-n}/n!$. (Hint: Label the vertices as you grow the tree and observe that a lot of terms cancel out or simplify.)

(ii) Derive Cayley's formula: the number of labeled trees on n vertices is n^{n-2}.

Exercise 6.20 (Critical regime: tree components). Let $G_n \sim \mathbb{G}_{n,p_n}$, where $p_n = \frac{1}{n}$.

(i) Let $\gamma_{n,k}$ be the expected number of isolated tree components of size k in G_n. Justify the formula

$$\gamma_{n,k} = \binom{n}{k} k^{k-2} \left(\frac{1}{n}\right)^{k-1} \left(1 - \frac{1}{n}\right)^{k(n-k) + \binom{k}{2} - (k-1)}.$$

(Hint: We did a related calculation in Section 2.3.2.)

(ii) Show that if $k = \omega(1)$ and $k = o(n^{3/4})$, then

$$\gamma_{n,k} \sim n \frac{k^{-5/2}}{\sqrt{2\pi}} \exp\left(-\frac{k^3}{6n^2}\right).$$

(iii) Conclude that for $0 < \delta < 1$ the expectation of U, the number of isolated tree components of size in $[(\delta n)^{2/3}, n^{2/3}]$ is $\Omega(\delta^{-1})$ as $\delta \to 0$.

(iv) For $1 \leq k_1 \leq k_2 \leq n - k_1$, let σ_{n,k_1,k_2} be the expected number of pairs of isolated tree components where the first one has size k_1 and the second one has size k_2. Justify the formula

$$\sigma_{n,k_1,k_2} = \binom{n}{k_1} k_1^{k_1-2} \left(\frac{1}{n}\right)^{k_1-1} \left(1 - \frac{1}{n}\right)^{k_1(n-k_1) + \binom{k_1}{2} - (k_1-1)}$$
$$\times \binom{n-k_1}{k_2} k_2^{k_2-2} \left(\frac{1}{n}\right)^{k_2-1} \left(1 - \frac{1}{n}\right)^{k_2(n-(k_1+k_2)) + \binom{k_2}{2} - (k_2-1)},$$

and show that

$$\sigma_{n,k_1,k_2} \leq \gamma_{n,k_1} \gamma_{n,k_2}.$$

(Hint: You may need to prove that, for $0 < a \leq 1 \leq b$, it holds that $1 - ab \leq (1-a)^b$.)

(v) Prove that $\mathrm{Var}[U] = O(\mathbb{E}[U])$. (Hint: Use (2.1.6), (iv), and (ii).)

Exercise 6.21 (Critical regime: overshoot bound). The goal of this exercise is to prove Lemma 6.4.17. We use the notation of Section 6.4.4.

(i) Let $W, Z \sim \mathrm{Bin}(n, 1/n)$ and $0 \leq r \leq n$. Show that $W - r$ conditioned on $W \geq r$ is stochastically dominated by Z. (Hint: Use the representation of W as a sum of indicators. Thinking of the partial sums as a Markov chain, consider the first time it reaches r.)

(ii) Show that $M_{\tau_h'} - h$ conditioned on $M_{\tau_h'} \geq h$ is stochastically dominated by Z from (i). (Hint: By the tower property, it suffices to show that

$$\mathbb{P}[M_{\tau_h'} - h \geq z \mid \tau_h' = \ell, M_{\ell-1} = h - r, M_\ell \geq h] \leq \mathbb{P}[Z \geq z],$$

for the relevant ℓ, r, z.)

(iii) Use (ii) to prove Lemma 6.4.17.

Bibliographic Remarks

Section 6.1 See [Dur10, section 5.3.4] for a quick introduction to branching processes. A more detailed overview relating to its use in discrete probability can be found in [vdH17, chapter 3]. The classical reference on branching processes is [AN04]. The Kesten–Stigum Theorem is due to Kesten and Stigum [KS66b]. Our proof of a weaker version with the second moment condition follows [Dur10, Example 5.4.3]. Section 6.1.4 is based loosely on [AN04, chapter V]. A proof of Theorem 6.1.18 can be found in [Har63]. A good reference for the Perron–Frobenius Theorem (Theorem 6.1.17 as well as more general versions) is [HJ13, chapter 8]. The central limit theorem for $\rho \geq \lambda^2$ referred to at the end of Section 6.1.4 is due to Kesten and Stigum [KS66a, KS67]. The critical percolation threshold for percolation on Galton–Watson trees is due to R. Lyons [Lyo90].

Section 6.2 The exploration process in Section 6.2.1 dates back to [ML86] and [Kar90]. The hitting-time theorem (Theorem 6.2.5) in the case $\ell = 1$ was first proved in [Ott49]. For alternative proofs, see, for example, [vdHK08] or [Wen75]. Spitzer's combinatorial lemma (Lemma 6.2.8) is from [Spi56]. See also [Fel71, section XII.6]. The presentation in Section 6.2.4 follows [vdH10]. See also [Dur85].

Section 6.3 Section 6.3.1 follows [Dev98, section 2.1] from the excellent volume [HMRAR98]. Section 6.3.2 is partly a simplified version of [BCMR06]. Further applications in phylogenetics, specifically to the sample complexity of phylogeny inference algorithms, can be found in, for example, [Mos04, Mos03, Roc10, DMR11, RS17]. The reconstruction problem also has applications in community detection [MNS15b]. See [Abb18] for a survey.

Section 6.4 The phase transtion of the Erdős–Rényi graph model was first studied in [ER60]. For much more, see, for example, [vdH17, chapter 4], [JLR11, chapter 5], and [Bol01, chapter 6]. In particular, a central limit theorem for the giant component, proved by several authors, including Martin–Löf [ML98], Pittel [Pit90], and Barraez, Boucheron, and de la Vega [BBFdlV00], is established in [vdH17, section 4.5]. Section 6.4.4 is based on [NP10]. See also [Per09, sections 2 and 3]. Much more is known about the critical regime; see, for example, [Ald97, Bol84, Lu90, LuPW94]. Section 6.4.5 is based partly on [Dur06, section 6.5]. For a lot more on random walk on random graphs (not just Erdős–Rényi), see [Dur06, chapter 6]. For more on the spectral properties of random graphs, see [CL06].

Appendix A

Useful Combinatorial Formulas

Recall the following facts about factorials and binomial coefficients:

$$\frac{n^n}{e^{n-1}} \leq n! \leq \frac{n^{n+1}}{e^{n-1}},$$

$$\frac{n^k}{k^k} \leq \binom{n}{k} \leq \frac{e^k n^k}{k^k},$$

$$(x+y)^n = \sum_{k=0}^{n} \binom{n}{k} x^k y^{n-k},$$

$$\sum_{k=0}^{d} \binom{n}{k} \leq \left(\frac{en}{d}\right)^d,$$

$$n! \sim \sqrt{2\pi n} \left(\frac{n}{e}\right)^n,$$

$$\binom{2n}{n} = (1 + o(1)) \frac{4^n}{\sqrt{\pi n}},$$

and

$$\log \binom{n}{k} = (1 + o(1)) n H(k/n),$$

where $H(p) := -p \log p - (1 - p) \log(1 - p)$. The third one is the *binomial theorem*. The fifth one is *Stirling's formula*.

Appendix B

Measure-Theoretic Foundations

This appendix contains relevant background on measure-theoretic probability. We follow closely the highly recommended [Wil91]. Missing proofs (and a lot more details and examples) can be found there. Another excellent textbook on this topic is [Dur10].

B.1 Probability Spaces

Let S be a set. In general it turns out that we cannot assign a probability to every subset of S. Here we discuss "well-behaved" collections of subsets. First an algebra on S is a collection of subsets stable under finitely many set operations.

Definition B.1.1 (Algebra on S). *A collection Σ_0 of subsets of S is an* algebra on S *if the following conditions hold:*

(i) $S \in \Sigma_0$;
(ii) $F \in \Sigma_0$ implies $F^c \in \Sigma_0$;
(iii) $F, G \in \Sigma_0$ implies $F \cup G \in \Sigma_0$.

This, of course, implies that the empty set as well as all pairwise intersections are also in Σ_0. The collection Σ_0 is an actual algebra (i.e., a vector space with a bilinear product) with the symmetric difference as its "sum," the intersection as its "product" and the underlying field being the field with two elements.

Example B.1.2 On $\mathbb{R}$, sets of the form

$$\bigcup_{i=1}^{k} (a_i, b_i],$$

where the union is disjoint with $k < +\infty$ and $-\infty \leq a_i \leq b_i \leq +\infty$ form an algebra. ◀

Finite set operations are not enough for our purposes. For instance, we want to be able to take limits. A σ-algebra is stable under countably many set operations.

Definition B.1.3 (σ-algebra on S). *A collection Σ of subsets of S is a σ-algebra on S (or σ-field on S) if*

(i) $S \in \Sigma$;
(ii) $F \in \Sigma$ implies $F^c \in \Sigma$;
(iii) $F_n \in \Sigma \; \forall n$ implies $\cup_n F_n \in \Sigma$.

Example B.1.4 2^S is a trivial example. ◀

To give a non-trivial example, we need the following definition. We begin with a lemma.

Lemma B.1.5 (Intersection of σ-algebras). *Let $\mathcal{F}_i$, $i \in I$, be σ-algebras on S, where I is arbitrary. Then, $\cap_i \mathcal{F}_i$ is a σ-algebra.*

Proof We prove only one of the conditions. The other ones are similar. Suppose $A \in \mathcal{F}_i$ for all i. Then, A^c is in $\mathcal{F}_i$ for all i since each $\mathcal{F}_i$ is itself a σ-algebra. ∎

Definition B.1.6 (σ-algebra generated by $\mathcal{C}$). *Let $\mathcal{C}$ be a collection of subsets of S. Then we let $\sigma(\mathcal{C})$ be the smallest σ-algebra containing $\mathcal{C}$, defined as the intersection of all such σ-algebras (including in particular 2^S).*

Example B.1.7 The smallest σ-algebra containing all open sets in $\mathbb{R}$, denoted $\mathcal{B}(\mathbb{R})$, is called the *Borel σ-algebra*. This is a non-trivial σ-algebra in the sense that it can be proved that there exist subsets of $\mathbb{R}$ that are *not* in $\mathcal{B}$, but that any "reasonable" set is in $\mathcal{B}$. In particular, it contains the algebra in Example B.1.2. ◀

Example B.1.8 The σ-algebra generated by the algebra in Example B.1.2 is $\mathcal{B}(\mathbb{R})$. This follows from the fact that all open sets of $\mathbb{R}$ can be written as a countable union of open intervals. (Indeed, for $x \in O$ an open set, let I_x be the largest open interval contained in O and containing x. If $I_x \cap I_y \neq \emptyset$, then $I_x = I_y$ by maximality (i.e., take the union). Then, $O = \cup_x I_x$ and there are only countably many disjoint ones because each one contains a rational.) ◀

We now define measures.

Definition B.1.9 (Additivity and σ-additivity). *A non-negative set function on an algebra Σ_0*

$$\mu_0 : \Sigma_0 \to [0, +\infty]$$

is additive if

  (i) $\mu_0(\emptyset) = 0$;
  (ii) $F, G \in \Sigma_0$ with $F \cap G = \emptyset$ implies $\mu_0(F \cup G) = \mu_0(F) + \mu_0(G)$.

Moreover, μ_0 is said to be σ-additive if condition (ii) is true for any countable collection of disjoint sets whose union is in Σ_0, that is, if $F_n \in \Sigma_0$, $n \geq 0$, all pairwise disjoint with $\cup_n F_n \in \Sigma_0$, then $\mu_0(\cup_n F_n) = \sum_n \mu_0(F_n)$.

Example B.1.10 For the algebra in the Example B.1.2, the set function

$$\lambda_0 \left(\bigcup_{i=1}^{k} (a_i, b_i] \right) = \sum_{i=1}^{k} (b_i - a_i)$$

is additive. (In fact, it is also σ-additive. We will show this later.) ◀

Definition B.1.11 (Measure space). *Let Σ be a σ-algebra on S. Then (S, Σ) is a* measurable space. *A σ-additive function μ on Σ is called a* measure *and (S, Σ, μ) is called a* measure space.

Definition B.1.12 (Probability space). *If $(\Omega, \mathcal{F}, \mathbb{P})$ is a measure space with $\mathbb{P}(\Omega) = 1$, then $\mathbb{P}$ is called a* probability measure *and $(\Omega, \mathcal{F}, \mathbb{P})$ is called a* probability space *(or* probability triple*).*

The sets in $\mathcal{F}$ are referred to as *events*.

To define a measure on $\mathcal{B}(\mathbb{R})$ we need the following tools from abstract measure theory.

Theorem B.1.13 (Caratheodory's extension theorem). *Let Σ_0 be an algebra on S and let $\Sigma = \sigma(\Sigma_0)$. If μ_0 is σ-additive on Σ_0, then there exists a measure μ on Σ that agrees with μ_0 on Σ_0.*

If in addition μ_0 is finite, the next lemma implies that the extension is unique.

Lemma B.1.14 (Uniqueness of extensions). *Let $\mathcal{I}$ be a π-system on S, that is, a family of subsets closed under finite intersections, and let $\Sigma = \sigma(\mathcal{I})$. If μ_1, μ_2 are finite measures on (S, Σ) that agree on $\mathcal{I}$, then they agree on Σ.*

Example B.1.15 The sets $(-\infty, x]$ for $x \in \mathbb{R}$ form a π-system generating $\mathcal{B}(\mathbb{R})$. That is, $\mathcal{B}(\mathbb{R})$ is the smallest σ-algebra containing that π-system. ◀

Finally, we can define Lebesgue measure. We start with $(0, 1]$ and extend to $\mathbb{R}$ in the obvious way. We need the following lemma.

Lemma B.1.16 (σ-additivity of λ_0). *Let λ_0 be the set function defined above, restricted to $(0, 1]$. Then λ_0 is σ-additive.*

Definition B.1.17 (Lebesgue measure on unit interval). *The unique extension of λ_0 to $(0, 1]$ is denoted λ and is called* Lebesgue measure.

B.2 Random Variables

Let (S, Σ, μ) be a measure space and let $\mathcal{B} = \mathcal{B}(\mathbb{R})$.

Definition B.2.1 (Measurable function). *Suppose $h : S \to \mathbb{R}$ and define*

$$h^{-1}(A) = \{s \in S : h(s) \in A\}.$$

The function h is Σ-measurable if $h^{-1}(B) \in \Sigma$ for all $B \in \mathcal{B}$. We denote by $m\Sigma$ (resp., $(m\Sigma)^+$, $b\Sigma$) the Σ-measurable functions (resp., that are non-negative, bounded).

In the probabilistic case:

Definition B.2.2 *A* random variable *is a measurable function on a probability space $(\Omega, \mathcal{F}, \mathbb{P})$.*

The behavior of a random variable is characterized by its distribution function.

Definition B.2.3 (Distribution function). *Let X be a random variable on a probability space $(\Omega, \mathcal{F}, \mathbb{P})$. The* law *of X is*

$$\mathcal{L}_X = \mathbb{P} \circ X^{-1},$$

which is a probability measure on $(\mathbb{R}, \mathcal{B})$. By Lemma B.1.14, $\mathcal{L}_X$ is determined by the distribution function *(DF) of X:*

$$F_X(x) = \mathbb{P}[X \leq x], \quad x \in \mathbb{R}.$$

Example B.2.4 The distribution function of a constant random variable is a jump of size 1 at the value it takes almost surely. The distribution function of a random variable with law equal to Lebesgue measure on $(0, 1]$ is

$$F_X(x) = \begin{cases} x, & x \in (0, 1], \\ 0, & x \leq 0, \\ 1, & x > 1. \end{cases}$$

We refer to such as random variable as a *uniform random variable* over $(0, 1]$. ◀

Distribution functions are characterized by a few simple properties.

Proposition B.2.5 *Suppose $F = F_X$ is the distribution function of a random variable X on $(\Omega, \mathcal{F}, \mathbb{P})$. Then the following hold:*

(i) *F is non-decreasing;*
(ii) *$\lim_{x \to +\infty} F(x) = 1$, $\lim_{x \to -\infty} F(x) = 0$;*
(iii) *F is right-continuous.*

Proof The first property follows from the monotonicity of probability measure (which itself follows immediately from σ-additivity).

For the second property, note that the limit exists by the first property. The value of the limit follows from the following important lemma.

Lemma B.2.6 (Monotone convergence properties of measures). *Let (S, Σ, μ) be a measure space.*

(i) *If $F_n \in \Sigma$, $n \geq 1$, with $F_n \uparrow F$, then $\mu(F_n) \uparrow \mu(F)$.*
(ii) *If $G_n \in \Sigma$, $n \geq 1$, with $G_n \downarrow G$ and $\mu(G_k) < +\infty$ for some k, then $\mu(G_n) \downarrow \mu(G)$.*

Proof Clearly, $F = \cup_n F_n \in \Sigma$. For $n \geq 1$, write $H_n = F_n \backslash F_{n-1}$ (with $F_0 = \emptyset$). Then by disjointness,

$$\mu(F_n) = \sum_{k \leq n} \mu(H_k) \uparrow \sum_{k < +\infty} \mu(H_k) = \mu(F).$$

The second statement is similar. ∎

Similarly, for the third property, by Lemma B.2.6 again,

$$\mathbb{P}[X \leq x_n] \downarrow \mathbb{P}[X \leq x]$$

if $x_n \downarrow x$. ∎

It turns out that the properties above characterize distribution functions in the following sense.

Theorem B.2.7 (Skorokhod representation). *Let F satisfy the three properties above. Then there is a random variable X on*

$$(\Omega, \mathcal{F}, \mathbb{P}) = ((0, 1], \mathcal{B}(0, 1], \lambda)$$

with distribution function F. The law of X is called the Lebesgue–Stieltjes *measure associated to F.*

The result says that all real random variables can be generated from uniform random variables over $(0, 1]$.

Proof   Assume first that F is continuous and strictly increasing. Define $X(\omega) = F^{-1}(\omega)$ for all $\omega \in \Omega$. Then, $\forall x \in \mathbb{R}$,

$$\mathbb{P}[X \leq x] = \mathbb{P}[\{\omega \colon F^{-1}(\omega) \leq x\}] = \mathbb{P}[\{\omega \colon \omega \leq F(x)\}] = F(x).$$

In general, let

$$X(\omega) = \inf\{x \colon F(x) \geq \omega\}.$$

It suffices to prove that

$$X(\omega) \leq x \quad \Longleftrightarrow \quad \omega \leq F(x).$$

The $\Leftarrow$ direction is clear by definition of X. On the other hand, by the right-continuity of F, we have that $\omega \leq F(X(\omega))$. Therefore, by monotonicity of F,

$$X(\omega) \leq x \quad \Rightarrow \quad \omega \leq F(X(\omega)) \leq F(x).$$

That proves the claim.   ∎

Turning measurability on its head, we get the following important definition.

Definition B.2.8 *Let $(\Omega, \mathcal{F}, \mathbb{P})$ be a probability space. Let Y_γ, $\gamma \in \Gamma$, be a collection of maps from Ω to $\mathbb{R}$. We let*

$$\sigma(Y_\gamma, \gamma \in \Gamma)$$

be the smallest σ-algebra on which the Y_γs are measurable.

In a sense, the above σ-algebra corresponds to "the partial information available when the Y_γs are observed."

Example B.2.9 Suppose we flip two unbiased coins and let X be the number of heads observed. Then, denoting heads by H and tails by T,

$$\sigma(X) = \sigma(\{\{\text{HH}\}, \{\text{HT}, \text{TH}\}, \{\text{TT}\}\}),$$

which is coarser than the full σ-algebra 2^Ω.   ◀

Note that h^{-1} preserves all set operations. For example, $h^{-1}(A \cup B) = h^{-1}(A) \cup h^{-1}(B)$. This gives the following important lemma.

Lemma B.2.10 (Sufficient condition for measurability). *Suppose $\mathcal{C} \subseteq \mathcal{B}$ with $\sigma(\mathcal{C}) = \mathcal{B}$. Then, $h^{-1} \colon \mathcal{C} \to \Sigma$ implies $h \in m\Sigma$. That is, it suffices to check measurability on a collection generating $\mathcal{B}$.*

Proof   Let $\mathcal{E}$ be the sets such that $h^{-1}(B) \in \Sigma$. By the observation before the statement, $\mathcal{E}$ is a σ-algebra. But $\mathcal{C} \subseteq \mathcal{E}$, which implies $\sigma(\mathcal{C}) \subseteq \mathcal{E}$ by minimality.   ∎

As a consequence we get the following properties of measurable functions.

Proposition B.2.11 (Properties of measurable functions). *Let h, h_n, $n \geq 1$, be in $m\Sigma$ and $f \in m\mathcal{B}$.*

(i) $f \circ h \in m\Sigma$.

(ii) *If S is a topological space and h is continuous, then h is $\mathcal{B}(S)$-measurable, where $\mathcal{B}(S)$ is generated by the open sets of S.*

(iii) *The function $g \colon S \to \mathbb{R}$ is in $m\Sigma$ if for all $c \in \mathbb{R}$,*

$$\{g \le c\} \in \Sigma.$$

(iv) $\forall \alpha \in \mathbb{R}$, $h_1 + h_2, h_1 h_2, \alpha h \in m\Sigma$.

(v) $\inf h_n$, $\sup h_n$, $\liminf h_n$, $\limsup h_n$ *are in $m\Sigma$.*

(vi) *The set*

$$\{s \colon \lim h_n(s) \text{ exists in } \mathbb{R}\}$$

is measurable.

Proof We sketch the proof of a few of them.

(ii) This follows from Lemma B.2.10 by taking $\mathcal{C}$ as the open sets of $\mathbb{R}$.

(iii) Similarly, take $\mathcal{C}$ to be the sets of the form $(-\infty, c]$.

(iv) This follows from (iii). For example note that, writing the left-hand side as $h_1 > c - h_2$,

$$\{h_1 + h_2 > c\} = \cup_{q \in \mathbb{Q}}[\{h_1 > q\} \cap \{q > c - h_2\}],$$

which is a countable union of measurable sets by assumption.

(v) Note that

$$\{\sup h_n \le c\} = \cap_n \{h_n \le c\}.$$

furthermore, note that $\liminf$ is the sup of an inf. ∎

B.3 Independence

Let $(\Omega, \mathcal{F}, \mathbb{P})$ be a probability space.

Definition B.3.1 (Independence). *Sub-σ-algebras $\mathcal{G}_1, \mathcal{G}_2, \ldots$ of $\mathcal{F}$ are independent if, for* INDEPENDENCE *all $G_i \in \mathcal{G}_i$, $i \ge 1$, and distinct $i_1, \ldots, i_n$, we have*

$$\mathbb{P}[G_{i_1} \cap \cdots \cap G_{i_n}] = \prod_{j=1}^{n} \mathbb{P}[G_{i_j}].$$

Specializing to events and random variables:

Definition B.3.2 (Independent random variables). *Random variables $X_1, X_2, \ldots$ are independent if the σ-algebras $\sigma(X_1), \sigma(X_2), \ldots$ are independent.*

Definition B.3.3 (Independent events). *Events $E_1, E_2, \ldots$ are independent if the σ-algebras*

$$\mathcal{E}_i = \{\emptyset, E_i, E_i^c, \Omega\}, \quad i \ge 1,$$

are independent.

Recall the more familiar definitions.

Theorem B.3.4 (Independent random variables: familiar definition). *Random variables X, Y are independent if and only if for all $x, y \in \mathbb{R}$,*

$$\mathbb{P}[X \leq x, Y \leq y] = \mathbb{P}[X \leq x]\,\mathbb{P}[Y \leq y].$$

Theorem B.3.5 (Independent events: familiar definition). *Events E_1, E_2 are independent if and only if*

$$\mathbb{P}[E_1 \cap E_2] = \mathbb{P}[E_1]\,\mathbb{P}[E_2].$$

The proofs of these characterizations follow immediately from the following lemma.

Lemma B.3.6 (Independence and π-systems). *Suppose that $\mathcal{G}$ and $\mathcal{H}$ are sub-σ-algebras and that $\mathcal{I}$ and $\mathcal{J}$ are π-systems such that*

$$\sigma(\mathcal{I}) = \mathcal{G}, \quad \sigma(\mathcal{J}) = \mathcal{H}.$$

Then $\mathcal{G}$ and $\mathcal{H}$ are independent if and only if $\mathcal{I}$ and $\mathcal{J}$ are as well, that is,

$$\mathbb{P}[I \cap J] = \mathbb{P}[I]\,\mathbb{P}[J] \quad \forall I \in \mathcal{I}, J \in \mathcal{J}.$$

Proof Suppose $\mathcal{I}$ and $\mathcal{J}$ are independent. For fixed $I \in \mathcal{I}$, the measures $\mathbb{P}[I \cap H]$ and $\mathbb{P}[I]\,\mathbb{P}[H]$ are equal for $H \in \mathcal{J}$ and have total mass $\mathbb{P}[I] < +\infty$. By the Uniqueness of Extensions Lemma (Lemma B.1.14) the above measures agree on $\sigma(\mathcal{J}) = \mathcal{H}$.

Repeat the argument. Fix $H \in \mathcal{H}$. Then the measures $\mathbb{P}[G \cap H]$ and $\mathbb{P}[G]\,\mathbb{P}[H]$ agree on $\mathcal{I}$ and have total mass $\mathbb{P}[H] < +\infty$. Therefore, they must agree on $\sigma(\mathcal{I}) = \mathcal{G}$. ∎

We give a standard construction of an infinite sequence of independent random variables with prescribed distributions.

Let $(\Omega, \mathcal{F}, \mathbb{P}) = ((0, 1], \mathcal{B}(0, 1], \lambda)$ and for $\omega \in \Omega$ consider the binary expansion

$$\omega = 0.\omega_1 \omega_2 \ldots$$

(For dyadic rationals, use the all-1 ending and note that the dyadic rationals have measure 0 by countability.) This construction produces a sequence of independent so-called *Bernoulli trials*. That is, under λ, each bit is Bernoulli$(1/2)$ and any finite collection is independent.

To get two independent uniform random variables, consider the following construction:

$$U_1 = 0.\omega_1 \omega_3 \omega_5 \ldots,$$
$$U_2 = 0.\omega_2 \omega_4 \omega_6 \ldots$$

Let $\mathcal{A}_1$ (resp. $\mathcal{A}_2$) be the π-system consisting of all finite intersections of events of the form $\{\omega_i \in H_i\}$ for odd i (resp. even i). By Lemma B.3.6, the σ-fields $\sigma(\mathcal{A}_1)$ and $\sigma(\mathcal{A}_2)$ are independent.

More generally, let

$$V_1 = 0.\omega_1 \omega_3 \omega_6 \ldots,$$
$$V_2 = 0.\omega_2 \omega_5 \omega_9 \ldots,$$
$$V_3 = 0.\omega_4 \omega_8 \omega_{13} \ldots$$

$$\vdots = \ddots .$$

that is, fill up the array diagonally. By the argument above, the V_is are independent and Bernoulli($1/2$).

Finally, let μ_n, $n \geq 1$, be a sequence of probability measures with distribution functions F_n, $n \geq 1$. For each n, define

$$X_n(\omega) = \inf\{x \colon F_n(x) \geq V_n(\omega)\}.$$

By the (proof of the) Skorokhod Representation (Theorem B.2.7), X_n has distribution function F_n.

Definition B.3.7 (I.i.d. random variables). *A sequence of independent random variables* (X_n) *as above is independent and identically distributed (i.i.d.) if* $F_n = F$ *for some n.*

Alternatively, we have the following more general result.

Theorem B.3.8 (Kolmogorov's extension theorem). *Suppose we are given probability measures* μ_n *on* $(\mathbb{R}^n, \mathcal{B}(\mathbb{R}^n))$ *that are* consistent, *that is,*

$$\mu_{n+1}((a_1, b_1] \times \cdots \times (a_n, b_n] \times \mathbb{R}) = \mu_n((a_1, b_1] \times \cdots \times (a_n, b_n]).$$

Then there exists a unique probability measure $\mathbb{P}$ *on* $(\mathbb{R}^{\mathbb{N}}, \mathcal{R}^{\mathbb{N}})$ *with*

$$\mathbb{P}[\omega \colon \omega_i \in (a_i, b_i], 1 \leq i \leq n] = \mu_n((a_1, b_1] \times \cdots \times (a_n, b_n]).$$

Here $\mathcal{R}^{\mathbb{N}}$ *is the* product σ-algebra, *that is, the* σ-algebra generated by finite-dimensional rectangles.

Next, we discuss a first non-trivial result about independent sequences.

Definition B.3.9 (Tail σ-algebra). *Let* $X_1, X_2, \ldots$ *be random variables on a probability space* $(\Omega, \mathcal{F}, \mathbb{P})$. *Define*

$$\mathcal{T} = \bigcap_{n \geq 1} \mathcal{T}_n,$$

where

$$\mathcal{T}_n = \sigma(X_{n+1}, X_{n+2}, \ldots).$$

As an intersection of σ-algebras, $\mathcal{T}$ *is a* σ-algebra. *It is called the* tail σ-algebra *of the sequence* (X_n).

Intuitively, an event is in the tail if changing a finite number of values does not affect its occurence.

Example B.3.10 If $S_n = \sum_{k \leq n} X_k$, then

$$\{\lim_n S_n \text{ exists}\} \in \mathcal{T},$$

$$\{\limsup_n n^{-1} S_n > 0\} \in \mathcal{T},$$

but

$$\{\limsup_n S_n > 0\} \notin \mathcal{T}. \qquad \blacktriangleleft$$

Theorem B.3.11 (Kolmogorov's 0-1 law). *Let (X_n) be a sequence of independent random variables with tail σ-algebra $\mathcal{T}$. Then $\mathcal{T}$ is $\mathbb{P}$-trivial, that is, for all $A \in \mathcal{T}$ we have either $\mathbb{P}[A] = 0$ or 1.*

Proof Let $\mathcal{X}_n = \sigma(X_1, \ldots, X_n)$. Note that $\mathcal{X}_n$ and $\mathcal{T}_n$ are independent. Moreover, since $\mathcal{T} \subseteq \mathcal{T}_n$ we have that $\mathcal{X}_n$ is independent of $\mathcal{T}$. Now let

$$\mathcal{X}_\infty = \sigma(X_n, n \geq 1).$$

Note that

$$\mathcal{K}_\infty = \bigcup_{n \geq 1} \mathcal{X}_n$$

is a π-system generating $\mathcal{X}_\infty$. Therefore, by Lemma B.3.6, $\mathcal{X}_\infty$ is independent of $\mathcal{T}$. But $\mathcal{T} \subseteq \mathcal{X}_\infty$ and therefore $\mathcal{T}$ is independent of itself! Hence, if $A \in \mathcal{T}$,

$$\mathbb{P}[A] = \mathbb{P}[A \cap A] = \mathbb{P}[A]^2,$$

which can occur only if $\mathbb{P}[A] \in \{0, 1\}$.   ∎

B.4 Expectation

Let (S, Σ, μ) be a measure space. We denote by $\mathbf{1}_A$ the indicator of a set A, that is,

$$\mathbf{1}_A(s) = \begin{cases} 1 & \text{if } s \in A, \\ 0 & \text{o.w.} \end{cases}$$

Definition B.4.1 (Simple functions). *A simple function is a function of the form*

$$f = \sum_{k=1}^{m} a_k \mathbf{1}_{A_k},$$

where $a_k \in [0, +\infty]$ and $A_k \in \Sigma$ for all k. We denote the set of all such functions by SF^+. We define the integral *of f by*

$$\mu(f) := \sum_{k=1}^{m} a_k \mu(A_k) \leq +\infty.$$

We also write $\mu f = \mu(f)$.

The following is left as a (somewhat tedious but) immediate exercise.

Proposition B.4.2 *Let $f, g \in \mathrm{SF}^+$.*

(i) If $\mu(f \neq g) = 0$, then $\mu f = \mu g$. (Hint: Rewrite f and g over the same disjoint sets.)
(ii) For all $c \geq 0$, $f + g, cf \in \mathrm{SF}^+$ and

$$\mu(f + g) = \mu f + \mu g, \quad \mu(cf) = c\mu f.$$

(Hint: This one is obvious by definition.)
(iii) If $f \leq g$, then $\mu f \leq \mu g$. (Hint: Show that $g - f \in \mathrm{SF}^+$ and use linearity.)

The main definition and theorem of integration theory follows.

Definition B.4.3 (Non-negative functions). *Let $f \in (m\Sigma)^+$. Then the* integral *of f is defined by*

$$\mu(f) = \sup\{\mu(h) \colon h \in SF^+,\ h \leq f\}.$$

Again we also write $\mu f = \mu(f)$.

Theorem B.4.4 (Monotone convergence theorem). *If $f_n, f \in (m\Sigma)^+$, $n \geq 1$, with $f_n \uparrow f$, then*

$$\mu f_n \uparrow \mu f.$$

Many theorems in integration follow from the monotone convergence theorem. In that context, the following approximation is useful.

Definition B.4.5 (Staircase function). *For $f \in (m\Sigma)^+$ and $r \geq 1$, the rth staircase function $\alpha^{(r)}$ is*

$$\alpha^{(r)}(x) = \begin{cases} 0 & \text{if } x = 0, \\ (i-1)2^{-r} & \text{if } (i-1)2^{-r} < x \leq i2^{-r} \leq r, \\ r & \text{if } x > r. \end{cases}$$

We let $f^{(r)} = \alpha^{(r)}(f)$. Note that $f^{(r)} \in SF^+$ and $f^{(r)} \uparrow f$ as $r \to +\infty$.

Using the previous definition, we get for example the following properties.

Proposition B.4.6 *Let $f, g \in (m\Sigma)^+$.*

(i) *If $\mu(f \neq g) = 0$, then $\mu(f) = \mu(g)$.*
(ii) *For all $c \geq 0$, $f + g, cf \in (m\Sigma)^+$ and*

$$\mu(f + g) = \mu f + \mu g, \quad \mu(cf) = c\mu f.$$

(iii) *If $f \leq g$, then $\mu f \leq \mu g$.*

For a function f, let f^+ and f^- be the positive and negative parts of f, that is,

$$f^+(s) = f(s) \vee 0, \quad f^-(s) = (-f(s)) \vee 0.$$

Note that $|f| = f^+ + f^-$. Finally, we define

$$\mu(f) := \mu(f^+) - \mu(f^-),$$

provided $\mu(f^+) + \mu(f^-) < +\infty$, in which case we write $f \in L^1(S, \Sigma, \mu)$. Proposition B.4.6 can be generalized naturally to this definition. Moreover, we have the following.

Theorem B.4.7 (Dominated convergence theorem). *If $f_n, f \in m\Sigma$, $n \geq 1$, with $f_n(s) \to f(s)$ for all $s \in S$, and there is a non-negative function $g \in L^1(S, \Sigma, \mu)$ such that $|f_n| \leq g$, then*

$$\mu(|f_n - f|) \to 0,$$

and in particular

$$\mu f_n \to \mu f$$

as $n \to \infty$.

More generally, for $0 < p < +\infty$, the space $L^p(S, \Sigma, \mu)$ contains all functions $f : S \to \mathbb{R}$ such that $\|f\|_p < +\infty$, where

$$\|f\|_p := \mu(|f|^p)^{1/p}$$

up to equality almost everywhere. We state the following results without proof.

Theorem B.4.8 (Hölder's inequality). *Let* $1 < p, q < +\infty$ *such that* $p^{-1} + q^{-1} = 1$. *Then, for any* $f \in L^p(S, \Sigma, \mu)$ *and* $g \in L^q(S, \Sigma, \mu)$, *it holds that* $fg \in L^1(S, \Sigma, \mu)$ *and further*

$$\|fg\|_1 \leq \|f\|_p \|g\|_q.$$

CAUCHY–
SCHWARZ
INEQUALITY
The case $p = q = 2$ *is known as the* Cauchy–Schwarz inequality *(or* Schwarz inequality*)*.

Theorem B.4.9 (Minkowski's inequality). *Let* $1 < p < +\infty$. *Then, for any* $f, g \in L^p(S, \Sigma, \mu)$, *it holds that* $f + g \in L^p(S, \Sigma, \mu)$ *and furthermore,*

$$\|f + g\|_p \leq \|f\|_p + \|g\|_p.$$

Theorem B.4.10 (L^p completeness). *Let* $1 \leq p < +\infty$. *If* $(f_n)_n$ *in* $L^p(S, \Sigma, \mu)$ *is Cauchy, that is,*

$$\sup_{n,m \geq k} \|f_n - f_m\|_p \to 0,$$

as $k \to +\infty$, *then there exists* $f \in L^p(S, \Sigma, \mu)$ *such that*

$$\|f_n - f\|_p \to 0$$

as $n \to +\infty$.

We can now define the expectation. Let $(\Omega, \mathcal{F}, \mathbb{P})$ be a probability space.

Definition B.4.11 (Expectation). *If* $X \geq 0$ *is a random variable then we define the* expectation *of* X, *denoted by* $\mathbb{E}[X]$, *as the integral of* X *over* $\mathbb{P}$. *More generally, if*

$$\mathbb{E}|X| = \mathbb{E}[X^+] + \mathbb{E}[X^-] < +\infty,$$

we let

$$\mathbb{E}[X] = \mathbb{E}[X^+] - \mathbb{E}[X^-].$$

INTEGRABLE
We denote the set of all such integrable *random variables (up to equality almost surely) by* $L^1(\Omega, \mathcal{F}, \mathbb{P})$.

The properties of the integral for non-negative functions (see Proposition B.4.6) extend to the expectation.

Proposition B.4.12 *Let* X, X_1, X_2 *be random variables in* $L^1(\Omega, \mathcal{F}, \mathbb{P})$.

(LIN) *If* $a_1, a_2 \in \mathbb{R}$, *then* $\mathbb{E}[a_1 X_1 + a_2 X_2] = a_1 \mathbb{E}[X_1] + a_2 \mathbb{E}[X_2]$.
(POS) *If* $X \geq 0$, *then* $\mathbb{E}[X] \geq 0$.

One useful implication of (POS) is that $|X| - X \geq 0$ so that $\mathbb{E}[X] \leq \mathbb{E}|X|$ and, by applying the same argument to $-X$, we have further $|\mathbb{E}[X]| \leq \mathbb{E}|X|$.

The monotone convergence theorem (Theorem B.4.4) implies the following results. We first need a definition.

Definition B.4.13 (Convergence almost sure). *We say that $X_n \to X$ almost surely (a.s.) if*

$$\mathbb{P}[X_n \to X] = 1.$$

Proposition B.4.14 *Let X, Y, X_n, $n \geq 1$, be random variables in $L^1(\Omega, \mathcal{F}, \mathbb{P})$.*

(MON) *If $0 \leq X_n \uparrow X$, then $\mathbb{E}[X_n] \uparrow \mathbb{E}[X] \leq +\infty$.*
(FATOU) *If $X_n \geq 0$, then $\mathbb{E}[\liminf_n X_n] \leq \liminf_n \mathbb{E}[X_n]$.*
(DOM) *If $|X_n| \leq Y$, $n \geq 1$, with $\mathbb{E}[Y] < +\infty$ and $X_n \to X$ a.s., then*

$$\mathbb{E}|X_n - X| \to 0,$$

and, hence,

$$\mathbb{E}[X_n] \to \mathbb{E}[X].$$

(Indeed,

$$
\begin{aligned}
|\mathbb{E}[X_n] - \mathbb{E}[X]| &= |\mathbb{E}[X_n - X]| \\
&= |\mathbb{E}[(X_n - X)^+] - \mathbb{E}[(X_n - X)^-]| \\
&\leq \mathbb{E}[(X_n - X)^+] + \mathbb{E}[(X_n - X)^-] \\
&= \mathbb{E}|X_n - X|.)
\end{aligned}
$$

(SCHEFFE) *If $X_n \to X$ a.s. and $\mathbb{E}|X_n| \to \mathbb{E}|X|$, then*

$$\mathbb{E}|X_n - X| \to 0.$$

(BDD) *If $X_n \to X$ a.s. and $|X_n| \leq K < +\infty$ for all n, then*

$$\mathbb{E}|X_n - X| \to 0.$$

Proof We only prove (FATOU). To use (MON) we write the lim inf as an increasing limit. Letting $Z_k = \inf_{n \geq k} X_n$, we have

$$\liminf_n X_n = \uparrow \lim_k Z_k,$$

so that by (MON)

$$\mathbb{E}[\liminf_n X_n] = \uparrow \lim_k \mathbb{E}[Z_k].$$

For $n \geq k$ we have $X_n \geq Z_k$ so that $\mathbb{E}[X_n] \geq \mathbb{E}[Z_k]$, hence

$$\mathbb{E}[Z_k] \leq \inf_{n \geq k} \mathbb{E}[X_n].$$

Finally, we get

$$\mathbb{E}[\liminf_n X_n] \leq \uparrow \liminf_{k \ n \geq k} \mathbb{E}[X_n].$$

■

The following inequality is often useful. We give an example below.

Theorem B.4.15 (Jensen's inequality). *Let $h: G \to \mathbb{R}$ be a convex function on an open interval G such that $\mathbb{P}[X \in G] = 1$ and $X, h(X) \in L^1(\Omega, \mathcal{F}, \mathbb{P})$, then*

$$\mathbb{E}[h(X)] \geq h(\mathbb{E}[X]).$$

JENSEN'S INEQUALITY

The L^p norm defined earlier applies to random variables as well. That is, for $p \geq 1$, we let $\|X\|_p = \mathbb{E}[|X|^p]^{1/p}$ and denote by $L^p(\Omega, \mathcal{F}, \mathbb{P})$ the collection of random variables X (up to almost sure equality) such that $\|X\|_p < +\infty$. Jensen's inequality (Theorem B.4.15) implies the following relationship.

Lemma B.4.16 (Monotonicity of norms). *For $1 \leq p \leq r < +\infty$, we have $\|X\|_p \leq \|X\|_r$.*

Proof For $n \geq 0$, let

$$X_n = (|X| \wedge n)^p.$$

Take $h(x) = x^{r/p}$, which is convex on $(0, +\infty)$. Then, by Jensen's inequality,

$$(\mathbb{E}[X_n])^{r/p} \leq \mathbb{E}[(X_n)^{r/p}] = \mathbb{E}[(|X| \wedge n)^r] \leq \mathbb{E}[|X|^r].$$

Take $n \to \infty$ and use (MON). ∎

This latter inequality is useful among other things to argue about the convergence of expectations. We say that X_n converges to X_∞ in L^p if $\|X_n - X_\infty\|_p \to 0$. By the previous lemma, convergence on L^r implies convergence in L^p for $r \geq p \geq 1$. Further we have:

Lemma B.4.17 (Convergence of expectations). *Assume $X_n, X_\infty \in L^1$. Then,*

$$\|X_n - X_\infty\|_1 \to 0$$

implies

$$\mathbb{E}[X_n] \to \mathbb{E}[X_\infty].$$

Proof Note that

$$|\mathbb{E}[X_n] - \mathbb{E}[X_\infty]| \leq \mathbb{E}|X_n - X_\infty| \to 0.$$

∎

So, a fortiori, convergence in L^p implies convergence of expectations.

Square integrable random variables have a nice geometry by virtue of forming a Hilbert space.

Definition B.4.18 (Square integrable variables). *Recall that $L^2(\Omega, \mathcal{F}, \mathbb{P})$ denotes the set of* square integrable *all square integrable random variables (up to equality almost surely), that is, those X with $\mathbb{E}[X^2] < +\infty$. For $X, Y \in L^2(\Omega, \mathcal{F}, \mathbb{P})$, define the inner product $\langle X, Y \rangle := \mathbb{E}[XY]$. Then the L^2 norm is $\|X\|_2 = \sqrt{\langle X, X \rangle}$.*

Theorem B.4.19 (Cauchy–Schwarz inequality). *If $X, Y \in L^2(\Omega, \mathcal{F}, \mathbb{P})$, then $XY \in L^1(\Omega, \mathcal{F}, \mathbb{P})$ and*

$$\mathbb{E}|XY| \leq \sqrt{\mathbb{E}[X^2]\mathbb{E}[Y^2]},$$

or put differently,

$$|\langle X, Y \rangle| \leq \|X\|_2 \|Y\|_2^2.$$

parallelogram law **Theorem B.4.20** (Parallelogram law). *If $X, Y \in L^2(\Omega, \mathcal{F}, \mathbb{P})$, then*

$$\|X + Y\|_2^2 + \|X - Y\|_2^2 = 2\|X\|_2^2 + 2\|Y\|_2^2.$$

B.5 Fubini's Theorem

We now define product measures and state (without proof) Fubini's Theorem.

Definition B.5.1 (Product σ-algebra). *Let (S_1, Σ_1) and (S_2, Σ_2) be measure spaces. Let $S = S_1 \times S_2$ be the Cartesian product of S_1 and S_2. For $i = 1, 2$, let $\pi_i \colon S \to S_i$ be the projection on the ith coordinate, that is,*

$$\pi_i(s_1, s_2) = s_i.$$

The product σ-algebra $\Sigma = \Sigma_1 \times \Sigma_2$ is defined as

$$\Sigma = \sigma(\pi_1, \pi_2).$$

In other words, it is the smallest σ-algebra that makes coordinate maps measurable. It is generated by sets of the form

$$\pi_1^{-1}(B_1) = B_1 \times S_2, \quad \pi_2^{-1}(B_2) = S_1 \times B_2, \quad B_1 \in \Sigma_1, B_2 \in \Sigma_2.$$

Theorem B.5.2 (Fubini's Theorem). *For $F \in \Sigma$, let $f = \mathbf{1}_F$ and define*

$$\mu(f) := \int_{S_1} I_1^f(s_1)\mu_1(\mathrm{d}s_1) = \int_{S_2} I_2^f(s_2)\mu_2(\mathrm{d}s_2),$$

where

$$I_1^f(s_1) := \int_{S_2} f(s_1, s_2)\mu_2(\mathrm{d}s_2) \in \mathrm{b}\Sigma_1$$

and

$$I_2^f(s_2) := \int_{S_1} f(s_1, s_2)\mu_1(\mathrm{d}s_1) \in \mathrm{b}\Sigma_2.$$

(The equality and inclusions above are part of the statement.) The set function μ is a measure on (S, Σ) called the product measure *of μ_1 and μ_2 and we write $\mu = \mu_1 \times \mu_2$ and*

$$(S, \Sigma, \mu) = (S_1, \Sigma_1, \mu_1) \times (S_2, \Sigma_2, \mu_2).$$

Moreover, μ is the unique measure on (S, Σ) for which

$$\mu(A_1 \times A_2) = \mu(A_1)\mu(A_2), \quad A_i \in \Sigma_i.$$

If $f \in (\mathrm{m}\Sigma)^+$, then

$$\mu(f) = \int_{S_1} I_1^f(s_1)\mu_1(\mathrm{d}s_1) = \int_{S_2} I_2^f(s_2)\mu_2(\mathrm{d}s_2),$$

where I_1^f, I_2^f are defined as before (i.e., as the sup *over bounded functions from below). The same is valid if $f \in \mathrm{m}\Sigma$ and $\mu(|f|) < +\infty$.*

Some applications of Fubini's Theorem (Theorem B.5.2) follow. We first recall the following useful formula.

Theorem B.5.3 (Change-of-variables formula). *Let X be a random variable with law $\mathcal{L}$. If $f \colon \mathbb{R} \to \mathbb{R}$ is such that either $f \geq 0$ or $\mathbb{E}|f(X)| < +\infty$, then*

$$\mathbb{E}[f(X)] = \int_{\mathbb{R}} f(y)\mathcal{L}(\mathrm{d}y).$$

Proof   We use the standard machinery.

1. For $f = \mathbf{1}_B$ with $B \in \mathcal{B}$,

$$\mathbb{E}[\mathbf{1}_B(X)] = \mathcal{L}(B) = \int_{\mathbb{R}} \mathbf{1}_B(y)\mathcal{L}(\mathrm{d}y).$$

2. If $f = \sum_{k=1}^{m} a_k \mathbf{1}_{A_k}$ is a simple function, then by (LIN)

$$\mathbb{E}[f(X)] = \sum_{k=1}^{m} a_k \mathbb{E}[\mathbf{1}_{A_k}(X)] = \sum_{k=1}^{m} a_k \int_{\mathbb{R}} \mathbf{1}_{A_k}(y)\mathcal{L}(\mathrm{d}y) = \int_{\mathbb{R}} f(y)\mathcal{L}(\mathrm{d}y).$$

3. Let $f \geq 0$ and approximate f by a sequence $\{f_n\}$ of increasing simple functions. By (MON),

$$\mathbb{E}[f(X)] = \lim_n \mathbb{E}[f_n(X)] = \lim_n \int_{\mathbb{R}} f_n(y)\mathcal{L}(\mathrm{d}y) = \int_{\mathbb{R}} f(y)\mathcal{L}(\mathrm{d}y).$$

4. Finally, assume that f is such that $\mathbb{E}|f(X)| < +\infty$. Then, by (LIN),

$$\begin{aligned}
\mathbb{E}[f(X)] &= \mathbb{E}[f^+(X)] - \mathbb{E}[f^-(X)] \\
&= \int_{\mathbb{R}} f^+(y)\mathcal{L}(\mathrm{d}y) - \int_{\mathbb{R}} f^-(y)\mathcal{L}(\mathrm{d}y) \\
&= \int_{\mathbb{R}} f(y)\mathcal{L}(\mathrm{d}y). \qquad \blacksquare
\end{aligned}$$

Theorem B.5.4 *Let X and Y be independent random variables with respective laws μ and ν. Let f and g be measurable functions such that either $f, g \geq 0$ or $\mathbb{E}|f(X)|, \mathbb{E}|g(y)| < +\infty$. Then,*

$$\mathbb{E}[f(X)g(y)] = \mathbb{E}[f(X)]\mathbb{E}[g(y)].$$

Proof   From the change-of-variables formula (Theorem B.5.3) and Fubini's Theorem (Theorem B.5.2), we get

$$\begin{aligned}
\mathbb{E}[f(X)g(y)] &= \int_{\mathbb{R}^2} f(x)g(y)(\mu \times \nu)(\mathrm{d}x \times \mathrm{d}y) \\
&= \int_{\mathbb{R}} \left(\int_{\mathbb{R}} f(x)g(y)\mu(\mathrm{d}x) \right) \nu(\mathrm{d}y) \\
&= \int_{\mathbb{R}} (g(y)\mathbb{E}[f(X)])\, \nu(\mathrm{d}y) \\
&= \mathbb{E}[f(X)]\mathbb{E}[g(y)]. \qquad \blacksquare
\end{aligned}$$

Definition B.5.5 (Density). *Let X be a random variable with law μ. We say that X has density f_X if for all $B \in \mathcal{B}(\mathbb{R})$,*

$$\mu(B) = \mathbb{P}[X \in B] = \int_B f_X(x)\lambda(\mathrm{d}x).$$

Theorem B.5.6 (Convolution). *Let X and Y be independent random variables with distribution functions F and G, respectively. Then the distribution function, H, of $X + Y$ is*

$$H(z) = \int F(z - y)\mathrm{d}G(y).$$

This is called the convolution *of F and G. Moreover, if X and Y have densities f and g, respectively, then X + Y has density*

$$h(z) = \int f(z - y)g(y)\mathrm{d}y.$$

Proof From Fubini's Theorem (Theorem B.5.3), denoting the laws of X and Y by μ and ν, respectively,

$$\begin{aligned}
\mathbb{P}[X + Y \leq z] &= \int \int \mathbf{1}_{\{x+y\leq z\}}\mu(\mathrm{d}x)\nu(\mathrm{d}y) \\
&= \int F(z - y)\nu(\mathrm{d}y) \\
&= \int F(z - y)\mathrm{d}G(y) \\
&= \int \left(\int_{-\infty}^{z} f(x - y)\mathrm{d}x \right) \mathrm{d}G(y) \\
&= \int_{-\infty}^{z} \left(\int f(x - y)\mathrm{d}G(y) \right) \mathrm{d}x \\
&= \int_{-\infty}^{z} \left(\int f(x - y)g(y)\mathrm{d}y \right) \mathrm{d}x.
\end{aligned}$$

∎

See Exercise 2.1 for a proof of the following standard formula.

Theorem B.5.7 (Moments of non-negative random variables). *For any non-negative random variable X and positive integer k,*

$$\mathbb{E}[X^k] = \int_0^{+\infty} kx^{k-1}\mathbb{P}[X > x]\,\mathrm{d}x. \tag{B.5.1}$$

B.6 Conditional Expectation

Before defining the conditional expectation, we recall some elementary concepts. For two events A, B, the conditional probability of A given B is defined as

$$\mathbb{P}[A \mid B] = \frac{\mathbb{P}[A \cap B]}{\mathbb{P}[B]},$$

where we assume $\mathbb{P}[B] > 0$.

Now let X and Z be random variables taking values $x_1, \ldots, x_m$ and $z_1, \ldots, z_n$, respectively. The conditional expectation of X given $Z = z_j$ is defined as

$$y_j = \mathbb{E}[X \mid Z = z_j] = \sum_i x_i \mathbb{P}[X = x_i \mid Z = z_j],$$

where we assume $\mathbb{P}[Z = z_j] > 0$ for all j. As motivation for the general definition, we make the following observations.

- We can think of the conditional expectation as a *random variable* $Y = \mathbb{E}[X \mid Z]$ defined as follows:

$$Y(\omega) = y_j \text{ on } G_j = \{\omega \colon Z(\omega) = z_j\}.$$

- Then Y is $\mathcal{G}$-measurable where $\mathcal{G} = \sigma(Z)$.
- On sets in $\mathcal{G}$, the expectation of Y agrees with the expectation of X. Indeed, note first that

$$
\begin{aligned}
\mathbb{E}[Y; G_j] &= y_j \mathbb{P}[G_j] \\
&= \sum_i x_i \mathbb{P}[X = x_i \mid Z = z_j] \mathbb{P}[Z = z_j] \\
&= \sum_i x_i \mathbb{P}[X = x_i, Z = z_j] \\
&= \mathbb{E}[X; G_j].
\end{aligned}
$$

This is also true for all $G \in \mathcal{G}$ by summation.

We are ready to state the general definition of the conditional expectation. Its existence and uniqueness follow from the next theorem.

Theorem B.6.1 (Conditional expectation). *Let $X \in L^1(\Omega, \mathcal{F}, \mathbb{P})$ and $\mathcal{G} \subseteq \mathcal{F}$ a sub-σ-algebra. Then:*

(i) (Existence) *There exists a random variable $Y \in L^1(\Omega, \mathcal{G}, \mathbb{P})$ such that*

$$\mathbb{E}[Y; G] = \mathbb{E}[X; G] \; \forall G \in \mathcal{G}. \tag{B.6.1}$$

Such a Y is called a version of the conditional expectation *of X given $\mathcal{G}$ and is denoted by $\mathbb{E}[X \mid \mathcal{G}]$.*

(ii) (Uniqueness) *It is unique in the sense that if Y and Y' are two versions of the conditional expectation, then $Y = Y'$ almost surely.*

When $\mathcal{G} = \sigma(Z)$, we sometimes use the notation $\mathbb{E}[X \mid Z] := \mathbb{E}[X \mid \mathcal{G}]$. A similar convention applies to collections of random variables, for example, $\mathbb{E}[X \mid Z_1, Z_2] := \mathbb{E}[X \mid \sigma(Z_1, Z_2)]$, and so on.

We first prove uniqueness. Existence is proved below after some more concepts are introduced.

Proof of Theorem B.6.1 (ii) By way of contradiction, let Y, Y' be two versions of $\mathbb{E}[X \mid G]$ such that without loss of generality $\mathbb{P}[Y > Y'] > 0$. By monotonicity, there is $n \geq 1$ with $G = \{Y > Y' + n^{-1}\} \in \mathcal{G}$ such that $\mathbb{P}[G] > 0$. Then, by definition,

$$0 = \mathbb{E}[Y - Y'; G] > n^{-1} \mathbb{P}[G] > 0,$$

which gives a contradiction. $\blacksquare$

To prove existence, we use the L^2 method. In $L^2(\Omega, \mathcal{F}, \mathbb{P})$, the conditional expectation reduces to an orthogonal projection.

Theorem B.6.2 (Conditional expectation: L^2 case). *Let $X \in L^2(\Omega, \mathcal{F}, \mathbb{P})$ and $\mathcal{G} \subseteq \mathcal{F}$ a sub-σ-algebra. Then there exists an (almost surely) unique $Y \in L^2(\Omega, \mathcal{G}, \mathbb{P})$ such that*

$$\|X - Y\|_2 = \Delta := \inf\{\|X - W\|_2 \colon W \in L^2(\Omega, \mathcal{G}, \mathbb{P})\},$$

and, moreover, $\langle Z, X - Y \rangle = 0 \; \forall Z \in L^2(\Omega, \mathcal{G}, \mathbb{P})$. *In particular, it satisfies* (B.6.1). *Such a Y is called* the orthogonal projection of X on $L^2(\Omega, \mathcal{G}, \mathbb{P})$.

Proof   Take (Y_n) such that $\|X - Y_n\|_2 \to \Delta$. We use the fact that $L^2(\Omega, \mathcal{G}, \mathbb{P})$ is complete (Theorem B.4.10) and first seek to prove that (Y_n) is Cauchy. Using the parallelogram law (Theorem B.4.20), note that

$$\|X - Y_r\|_2^2 + \|X - Y_s\|_2^2 = 2 \left\| X - \frac{1}{2}(Y_r + Y_s) \right\|_2^2 + 2 \left\| \frac{1}{2}(Y_r - Y_s) \right\|_2^2.$$

The first term on the right-hand side is at least $2\Delta^2$ by definition of Δ, so taking limits $r, s \to +\infty$ we have what we need, that is, that (Y_n) is indeed Cauchy.

Let Y be the limit of (Y_n) in $L^2(\Omega, \mathcal{G}, \mathbb{P})$. Note that by the triangle inequality,

$$\Delta \le \|X - Y\|_2 \le \|X - Y_n\|_2 + \|Y_n - Y\|_2 \to \Delta$$

as $n \to +\infty$. As a result, for any $Z \in L^2(\Omega, \mathcal{G}, \mathbb{P})$ and $t \in \mathbb{R}$,

$$\|X - Y - tZ\|_2^2 \ge \Delta^2 = \|X - Y\|_2^2,$$

so that, expanding and rearranging, we have

$$-2t\langle Z, X - Y \rangle + t^2 \|Z\|_2^2 \ge 0,$$

which is only possible *for every* $t \in \mathbb{R}$ if the first term is 0.

Uniqueness follows from the parallelogram law and the definition of Δ. ∎

We return to the proof of existence of the conditional expectation. We use the standard machinery.

Proof of Theorem B.6.1 (i)   The previous theorem implies that conditional expectations exist for indicators and simple functions. Now take $X \in L^1(\Omega, \mathcal{F}, \mathbb{P})$ and write $X = X^+ - X^-$, so we can assume X is in fact non-negative without loss of generality. Using the staircase function,

$$X^{(r)} = \begin{cases} 0 & \text{if } X = 0, \\ (i-1)2^{-r} & \text{if } (i-1)2^{-r} < X \le i2^{-r} \le r, \\ r & \text{if } X > r, \end{cases}$$

we have $0 \le X^{(r)} \uparrow X$. Let $Y^{(r)} = \mathbb{E}[X^{(r)} \mid \mathcal{G}]$. Using an argument similar to the proof of uniqueness, it follows that $U \ge 0$ implies $\mathbb{E}[U \mid \mathcal{G}] \ge 0$ for a simple function U. Using linearity (which is immediate from the definition), we then have $Y^{(r)} \uparrow Y := \limsup Y^{(r)}$, which is measurable in $\mathcal{G}$. By (MON),

$$\mathbb{E}[Y; G] = \mathbb{E}[X; G], \; \forall G \in \mathcal{G}.$$

That concludes the proof. ∎

Before deriving some properties, we give a few examples.

Example B.6.3   If $X \in L^1(\Omega, \mathcal{G}, \mathbb{P})$, then $\mathbb{E}[X \mid \mathcal{G}] = X$ almost surely trivially. ◀

Example B.6.4   If $\mathcal{G} = \{\emptyset, \Omega\}$, then $\mathbb{E}[X \mid \mathcal{G}] = \mathbb{E}[X]$. ◀

Example B.6.5 Let $A, B \in \mathcal{F}$ with $0 < \mathbb{P}[B] < 1$. If $\mathcal{G} = \{\emptyset, B, B^c, \Omega\}$ and $X = \mathbf{1}_A$, then

$$\mathbb{P}[A \mid \mathcal{G}] = \begin{cases} \frac{\mathbb{P}[A \cap B]}{\mathbb{P}[B]} & \text{on } \omega \in B, \\ \frac{\mathbb{P}[A \cap B^c]}{\mathbb{P}[B^c]} & \text{on } \omega \in B^c. \end{cases}$$

◄

Intuition about the conditional expectation sometimes breaks down.

Example B.6.6 On $(\Omega, \mathcal{F}, \mathbb{P}) = ((0, 1], \mathcal{B}(0, 1], \lambda)$, let $\mathcal{G}$ be the σ-algebra of all counta-ble and co-countable (i.e., whose complement in $(0, 1]$ is countable) subsets of $(0, 1]$. Then $\mathbb{P}[G] \in \{0, 1\}$ for all $G \in \mathcal{G}$ and

$$\mathbb{E}[X; G] = \mathbb{E}[\mathbb{E}[X]; G] = \mathbb{E}[X]\mathbb{P}[G],$$

so that $\mathbb{E}[X \mid \mathcal{G}] = \mathbb{E}[X]$. Yet, $\mathcal{G}$ contains all singletons and we seemingly have "full infor-mation," which would lead to the wrong guess $\mathbb{E}[X \mid \mathcal{G}] = X$.   ◄

We show that the conditional expectation behaves similarly to the ordinary expectation. In what follows, all X and X_is are in $L^1(\Omega, \mathcal{F}, \mathbb{P})$ and $\mathcal{G}$ is a sub σ-algebra of $\mathcal{F}$.

Lemma B.6.7 (cLIN). *If* $a_1, a_2 \in \mathbb{R}$, *then* $\mathbb{E}[a_1X_1 + a_2X_2 \mid \mathcal{G}] = a_1\mathbb{E}[X_1 \mid \mathcal{G}] + a_2\mathbb{E}[X_2 \mid \mathcal{G}]$ *a.s.*

Proof Use the linearity of expectation and the fact that a linear combination of random variables in $\mathcal{G}$ is also in $\mathcal{G}$.   ∎

Lemma B.6.8 (cPOS). *If* $X \geq 0$, *then* $\mathbb{E}[X \mid \mathcal{G}] \geq 0$ *a.s.*

Proof Let $Y = \mathbb{E}[X \mid \mathcal{G}]$ and assume for contradiction that $\mathbb{P}[Y < 0] > 0$. There is $n \geq 1$ such that $\mathbb{P}[Y < -n^{-1}] > 0$. But that implies, for $G = \{Y < -n^{-1}\}$,

$$\mathbb{E}[X; G] = \mathbb{E}[Y; G] < -n^{-1}\mathbb{P}[G] < 0,$$

a contradiction.   ∎

Lemma B.6.9 (cMON). *If* $0 \leq X_n \uparrow X$, *then* $\mathbb{E}[X_n \mid \mathcal{G}] \uparrow \mathbb{E}[X \mid \mathcal{G}]$ *a.s.*

Proof Let $Y_n = \mathbb{E}[X_n \mid \mathcal{G}]$. By (cLIN) and (cPOS), $0 \leq Y_n \uparrow$. Then letting $Y = \limsup Y_n$, by (MON),

$$\mathbb{E}[X; G] = \mathbb{E}[Y; G],$$

for all $G \in \mathcal{G}$.   ∎

Lemma B.6.10 (cFATOU). *If* $X_n \geq 0$, *then* $\mathbb{E}[\liminf X_n \mid \mathcal{G}] \leq \liminf \mathbb{E}[X_n \mid \mathcal{G}]$ *a.s.*

Proof Note that, for $n \geq m$,

$$X_n \geq Z_m := \inf_{k \geq m} X_k \uparrow \in \mathcal{G},$$

so that $\inf_{n \geq m} \mathbb{E}[X_n \mid \mathcal{G}] \geq \mathbb{E}[Z_m \mid \mathcal{G}]$. Applying (cMON),

$$\mathbb{E}[\lim Z_m \mid \mathcal{G}] = \lim \mathbb{E}[Z_m \mid \mathcal{G}] \leq \lim \inf_{n \geq m} \mathbb{E}[X_n \mid \mathcal{G}].$$

∎

Lemma B.6.11 (cDOM). *If $X_n \leq V \in L^1(\Omega, \mathcal{F}, \mathbb{P})$ and $X_n \to X$ a.s., then*

$$\mathbb{E}[X_n \mid \mathcal{G}] \to \mathbb{E}[X \mid \mathcal{G}] \text{ a.s.}$$

Proof Applying (cFATOU) to $W_n := 2V - |X_n - X| \geq 0$,

$$\begin{aligned}
\mathbb{E}[2V \mid \mathcal{G}] &= \mathbb{E}[\liminf_n W_n \mid \mathcal{G}] \\
&\leq \liminf_n \mathbb{E}[W_n \mid \mathcal{G}] \\
&= \mathbb{E}[2V \mid \mathcal{G}] - \liminf_n \mathbb{E}[|X_n - X| \mid \mathcal{G}],
\end{aligned}$$

so we must have

$$\liminf_n \mathbb{E}[|X_n - X| \mid \mathcal{G}] = 0.$$

Now use that $|\mathbb{E}[X_n - X \mid \mathcal{G}]| \leq \mathbb{E}[|X_n - X| \mid \mathcal{G}]$ (which follows from (cPOS)). ∎

Lemma B.6.12 (cJENSEN). *If f is convex and $\mathbb{E}[|f(X)|] < +\infty$, then*

$$f(\mathbb{E}[X \mid \mathcal{G}]) \leq \mathbb{E}[f(X) \mid \mathcal{G}].$$

In addition, we highlight (without proof) the following important properties of the conditional expectation.

Lemma B.6.13 (Taking out what is known). *If $X \in L^1(\Omega, \mathcal{F}, \mathbb{P})$ and $Z \in m\mathcal{G}$ is bounded or if X is bounded and $Z \in L^1(\Omega, \mathcal{G}, \mathbb{P})$, then $\mathbb{E}[ZX \mid \mathcal{G}] = Z\,\mathbb{E}[X \mid \mathcal{G}]$. This is also true if $X, Z \geq 0$, $\mathbb{E}[X] < +\infty$ and $\mathbb{E}[ZX] < +\infty$, or $X \in L^2(\Omega, \mathcal{F}, \mathbb{P})$ and $Z \in L^2(\Omega, \mathcal{G}, \mathbb{P})$.*

Lemma B.6.14 (Role of independence). *If $X \in L^1(\Omega, \mathcal{F}, \mathbb{P})$ is independent of $\mathcal{H}$ then $\mathbb{E}[X \mid \mathcal{H}] = \mathbb{E}[X]$. In fact, if $\mathcal{H}$ is independent of $\sigma(\sigma(X), \mathcal{G})$, then $\mathbb{E}[X \mid \sigma(\mathcal{G}, \mathcal{H})] = \mathbb{E}[X \mid \mathcal{G}]$.*

Lemma B.6.15 (Conditioning on an independent random variable). *Suppose X, Y are independent. Let ϕ be a function with $\mathbb{E}|\phi(X, Y)| < +\infty$ and let $g(x) = \mathbb{E}(\phi(x, Y))$. Then,*

$$\mathbb{E}(\phi(X, Y) \mid X) = g(X).$$

Lemma B.6.16 (Tower property). *If $\mathcal{H} \subseteq \mathcal{G}$ is a σ-algebra and $X \in L^1(\Omega, \mathcal{F}, \mathbb{P})$,*

$$\mathbb{E}[\mathbb{E}[X \mid \mathcal{G}] \mid \mathcal{H}] = \mathbb{E}[\mathbb{E}[X \mid \mathcal{H}] \mid \mathcal{G}] = \mathbb{E}[X \mid \mathcal{H}].$$

TOWER
PROPERTY

That is, the "smallest σ-algebra wins."

An important special case of the latter, also known as the law of total probability or the law of total expectation, is $\mathbb{E}[\mathbb{E}[X \mid \mathcal{G}]] = \mathbb{E}[X]$.

One last useful property:

Lemma B.6.17 *Let $(\Omega, \mathcal{F}, \mathbb{P})$ be a probability space. If $Y_1 = Y_2$ a.s. on $B \in \mathcal{F}$, then $\mathbb{E}[Y_1 \mid \mathcal{F}] = \mathbb{E}[Y_2 \mid \mathcal{F}]$ a.s. on B.*

B.7 Filtered Spaces

Finally we define stochastic processes. Let E be a set and let $\mathcal{E}$ be a σ-algebra defined over E.

PROCESS **Definition B.7.1** *A* stochastic process *(or* process*) is a collection* $\{X_t\}_{t \in \mathcal{T}}$ *of* $(E, \mathcal{E})$-*valued random variables on a probability space* $(\Omega, \mathcal{F}, \mathbb{P})$, *where* $\mathcal{T}$ *is an arbitrary* index set.

Here is a typical example.

Example B.7.2 When $\mathcal{T} = \mathbb{Z}_+$ (or $\mathcal{T} = \mathbb{N}$ or $\mathcal{T} = \mathbb{Z}$) we have a *discrete-time process*, in which case we often write the process as a sequence $(X_t)_{t \geq 0}$. For instance:

- $X_0, X_1, X_2, \ldots$ i.i.d. random variables;
- $(S_t)_{t \geq 0}$, where $S_t = \sum_{i \leq t} X_i$ with X_i as above.

We let

$$\mathcal{F}_t = \sigma(X_0, X_1, \ldots, X_t),$$

which can be thought of as "the information known up to time t." For a fixed $\omega \in \Omega$, $(X_t(\omega) : t \in \mathcal{T})$ is called a *sample path*. ◀

SAMPLE PATH

Definition B.7.3 *A* random walk on $\mathbb{R}^d$ *is a process of the form:*

$$S_t = S_0 + \sum_{i=1}^{t} X_i, \quad t \geq 1,$$

where the X_is are i.i.d. in $\mathbb{R}^d$, independent of S_0. The case X_i uniform in $\{-1, +1\}$ is called simple random walk on $\mathbb{Z}$.

Filtered spaces provide a formal framework for time-indexed processes. We restrict ourselves to discrete time. (We will not discuss continuous-time processes in this book.)

Definition B.7.4 *A* filtered space *is a tuple* $(\Omega, \mathcal{F}, (\mathcal{F}_t)_{t \in \mathbb{Z}_+}, \mathbb{P})$ *where:*

- $(\Omega, \mathcal{F}, \mathbb{P})$ *is a probability space;*
FILTRATION
- $(\mathcal{F}_t)_{t \in \mathbb{Z}_+}$ *is a* filtration, *that is,*

$$\mathcal{F}_0 \subseteq \mathcal{F}_1 \subseteq \cdots \subseteq \mathcal{F}_\infty := \sigma(\cup_t \mathcal{F}_t) \subseteq \mathcal{F},$$

where each $\mathcal{F}_i$ is a σ-algebra.

ADAPTED **Definition B.7.5** *Fix* $(\Omega, \mathcal{F}, (\mathcal{F}_t)_{t \in \mathbb{Z}_+}, \mathbb{P})$. *A process* $(W_t)_{t \geq 0}$ *is* adapted *if* $W_t \in \mathcal{F}_t$ *for all t.*

Intuitively, in the previous definition, the value of W_t is "known at time t."

PREDICTABLE **Definition B.7.6** *A process* $(C_t)_{t \geq 1}$ *is* predictable *if* $C_t \in \mathcal{F}_{t-1}$ *for all $t \geq 1$.*

Example B.7.7 Continuing Example B.7.2. The collection $(\mathcal{F}_t)_{t \geq 0}$ forms a filtration. The process $(S_t)_{t \geq 0}$ is adapted. On the other hand, the process $C_t = \mathbf{1}\{S_{t-1} \leq k\}$ is predictable. ◀

Bibliography

[Abb18]   E. Abbe. Community detection and stochastic block models. *Found. Trends Commun. Inf. Theory*, 14(1–2):1–162, June 2018. Publisher: Now Publishers, Inc.

[ABH16]   E. Abbe, A. S. Bandeira, and G. Hall. Exact recovery in the stochastic block model. *IEEE Transactions on Information Theory*, 62(1):471–487, January 2016.

[Ach03]   D. Achlioptas. Database-friendly random projections: Johnson–Lindenstrauss with binary coins. *J. Comput. Syst. Sci.*, 66(4):671–687, 2003.

[AD78]   R. Ahlswede and D. E. Daykin. An inequality for the weights of two families of sets, their unions and intersections. *Z. Wahrsch. Verw. Gebiete*, 43(3):183–185, 1978.

[AF]   D. Aldous and J. A. Fill. Reversible Markov chains and random walks on graphs. www.stat.berkeley.edu/~aldous/RWG/book.html.

[AGG89]   R. Arratia, L. Goldstein, and L. Gordon. Two moments suffice for Poisson approximations: The Chen–Stein Method. *Ann. Probab.*, 17(1):9–25, January 1989.

[AJKS22] A. Agarwal, N. Jiang, S. M. Kakade, and W. Sun. Reinforcement learning: Theory and algorithms. https://rltheorybook.github.io/rltheorybook_AJKS.pdf, 2022.

[AK97]   N. Alon and M. Krivelevich. The concentration of the chromatic number of random graphs. *Combinatorica*, 17(3):303–313, 1997.

[Ald83]   D. Aldous. Random walks on finite groups and rapidly mixing Markov chains. In *Seminar on Probability*, XVII, volume 986 of Lecture Notes in Math., pages 243–297. Springer, Berlin, 1983.

[Ald90]   D. J. Aldous. The random walk construction of uniform spanning trees and uniform labelled trees. *SIAM J. Discrete Math.*, 3(4):450–465, 1990.

[Ald97]   D. Aldous. Brownian excursions, critical random graphs and the multiplicative coalescent. *Ann. Probab.*, 25(2):812–854, 1997.

[Alo03]   N. Alon. Problems and results in extremal combinatorics. I. *Discrete Math.*, 273(1–3):31–53, 2003.

[AMS09]   J.-Y. Audibert, R. Munos, and C. Szepesvári. Exploration–exploitation tradeoff using variance estimates in multi-armed bandits. *Theor. Comput. Sci.*, 410(19):1876–1902, 2009.

[AN04]   K. B. Athreya and P. E. Ney. *Branching Processes*. Dover, Mineola, 2004. Reprint of the 1972 original [Springer, New York, MR0373040].

[ANP05]   Dimitris Achlioptas, Assaf Naor, and Yuval Peres. Rigorous location of phase transitions in hard optimization problems. *Nature*, 435:759–764, 2005.

[AS11]   N. Alon and J. H. Spencer. *The Probabilistic Method*. Wiley Series in Discrete Mathematics and Optimization. Wiley, Hoboken, NJ, 2011.

[AS15]   E. Abbe and C. Sandon. Community detection in general stochastic block models: Fundamental limits and efficient algorithms for recovery. In Venkatesan Guruswami, ed., *IEEE 56th Annual Symposium on Foundations of Computer Science, FOCS 2015, Berkeley, 17–20 October, 2015*, pages 670–688. IEEE Computer Society, 2015.

[Axl15]   S. Axler. *Linear Algebra Done Right*. Undergraduate Texts in Mathematics. Springer, Cham, 3rd ed., 2015.

[AZ18]   M. Aigner and G. M. Ziegler. *Proofs from The Book*. Springer, Berlin, 6th ed., 2018.

[Azu67]   K. Azuma. Weighted sums of certain dependent random variables. *Tōhoku Math. J. (2)*, 19(2):357–367, 1967.

[BA99]   A.-L. Barabási and R. Albert. Emergence of scaling in random networks. *Science*, 286(5439):509–512, 1999.

[BBFdlV00] D. Barraez, S. Boucheron, and W. Fernandez de la Vega. On the fluctuations of the giant component. *Combin. Probab. Comput.*, 9(4):287–304, 2000.

[BC03]   B. Brinkman and M. Charikar. On the impossibility of dimension reduction in L1. In *Proceedings of the 44th Annual IEEE Symposium on Foundations of Computer Science*, IEEE Computer Society, page 514, 2003.

[BCB12]   S. Bubeck and N. Cesa-Bianchi. *Regret Analysis of Stochastic and Nonstochastic Multi-armed Bandit Problems*. Now, 2012. Google-Books-ID: Rl2skwEACAAJ.

[BCMR06]   C. Borgs, J. T. Chayes, E. Mossel, and S. Roch. The Kesten–Stigum reconstruction bound is tight for roughly symmetric binary channels. In *FOCS*, pages 518–530, 2006. doi:10.1109/FOCS.2006.76

[BD97]   R. Bubley and M. E. Dyer. Path coupling: A technique for proving rapid mixing in Markov chains. In *38th Annual Symposium on Foundations of Computer Science, FOCS '97, Miami Beach, Florida, October 19–22, 1997*, pages 223–231. IEEE Computer Society, 1997. doi: 10.1109/SFCS.

[BDDW08]   R. Baraniuk, M. Davenport, R. DeVore, and M. Wakin. A simple proof of the restricted isometry property for random matrices. *Constr. Approx.*, 28(3):253–263, 2008.

[BDJ99]   J. Baik, P. Deift, and K. Johansson. On the distribution of the length of the longest increasing subsequence of random permutations. *J. Amer. Math. Soc.*, 12(4):1119–1178, 1999.

[Ber14]   N. Berestycki. Lectures on mixing times: A crossroad between probability, analysis and geometry. https://homepage.univie.ac.at/nathanael.berestycki/wp-content/uploads/2022/05/mixing3.pdf, 2014.

[Ber46]   S. N. Bernstein. *Probability Theory (in Russian)*. M.-L. Gostechizdat, 1946.

[BH57]   S. R. Broadbent and J. M. Hammersley. Percolation processes. I. Crystals and mazes. *Proc. Cambridge Philos. Soc.*, 53:629–641, 1957.

[BH16]   A. S. Bandeira and R. Handel. Sharp nonasymptotic bounds on the norm of random matrices with independent entries. *Ann. Probab.*, 44(4):2479–2506, 2016.

[Bil12]   P. Billingsley. *Probability and Measure*. Wiley Series in Probability and Statistics. Wiley, Hoboken, NJ, 2012.

[BKW14]   I. Benjamini, G. Kozma, and N. Wormald. The mixing time of the giant component of a random graph. *Random Structures Algorithms*, 45(3):383–407, 2014.

[BLM13]   S. Boucheron, G. Lugosi, and P. Massart. *Concentration Inequalities: A Nonasymptotic Theory of Independence*. Oxford University Press, Oxford, 2013.

[Bol81]   B. Bollobás. Random graphs. In *Combinatorics (Swansea, 1981)*, volume 52 of London Math. Soc. Lecture Note Ser., pages 80–102. Cambridge University Press, Cambridge, 1981.

[Bol84]   B. Bollobás. The evolution of random graphs. *Trans. Amer. Math. Soc.*, 286(1):257–274, 1984.

[Bol98]   B. Bollobás. *Modern Graph Theory*, volume 184 of Graduate Texts in Mathematics. Springer-Verlag, New York, 1998.

[Bol01]   B. Bollobás. *Random Graphs*, volume 73 of Cambridge Studies in Advanced Mathematics. Cambridge University Press, Cambridge, 2nd ed., 2001.

[BR06a]   B. Bollobás and O. Riordan. *Percolation*. Cambridge University Press, New York, 2006.

[BR06b]   B. Bollobás and O. Riordan. A short proof of the Harris–Kesten theorem. *Bull. London Math. Soc.*, 38(3):470–484, 2006.

[Bre17]   P. Bremaud. *Discrete Probability Models and Methods*, volume 78 of Probability Theory and Stochastic Modelling. Springer, Cham, 2017. Probability on graphs and trees, Markov chains and random fields, entropy and coding.

[Bre20]   P. Bremaud. *Markov Chains—Gibbs Fields, Monte Carlo Simulation and Queues*, volume 31 of Texts in Applied Mathematics. Springer, Cham, 2020. 2nd ed.

[Bro89]   A. Z. Broder. Generating random spanning trees. In *FOCS*, pages 442–447. IEEE Computer Society, 1989. doi: 10.1109/SFCS.1989.63516.

[BRST01]   B. Bollobás, O. Riordan, J. Spencer, and G. Tusnády. The degree sequence of a scale-free random graph process. *Random Struct. Algorithms*, 18(3):279–290, 2001.

[BS89]   R. Boppona and J. Spencer. A useful elementary correlation inequality. *J. Combin. Theory Ser. A*, 50(2):305–307, 1989.

[Bub10]   S. Bubeck. *Bandits Games and Clustering Foundations*. Ph.D. thesis, Université des Sciences et Technologie de Lille – Lille I, June 2010.

[BV04]   S. P. Boyd and L. Vandenberghe. *Convex Optimization*. Berichte über verteilte messysteme. Cambridge University Press, Cambridge, 2004.

[Car85]   T. K. Carne. A transmutation formula for Markov chains. *Bull. Sci. Math. (2)*, 109(4):399–405, 1985.

[CDL+12]   P. Cuff, J. Ding, O. Louidor et al. Glauber dynamics for the mean-field Potts model. *J. Stat. Phys.*, 149(3):432–477, 2012.

[CF07]   C. Cooper and A. Frieze. The cover time of sparse random graphs. *Random Struct. Algorithms*, 30(1–2):1–16, 2007.

[Che52]   H. Chernoff. A measure of asymptotic efficiency for tests of a hypothesis based on the sum of observations. *Ann. Math. Statistics*, 23:493–507, 1952.

[Che70]   J. Cheeger. A lower bound for the smallest eigenvalue of the Laplacian. In *Problems in Analysis (Sympos. in Honor of Salomon Bochner, Princeton Unviersity, Princeton, 1969)*, pages 195–199. Princeton University Press, Princeton, 1970.

[Che75]   L. H. Y. Chen. Poisson approximation for dependent trials. *Ann. Probab.*, 3(3):534–545, 1975.

[Chu97]   F. R. K. Chung. *Spectral Graph Theory*, volume 92 of CBMS Regional Conference Series in Mathematics. Published for the Conference Board of the Mathematical Sciences, Washington, DC; by the American Mathematical Society, Providence, 1997.

[CL06]   F. Chung and L. Lu. *Complex Graphs and Networks*, volume 107 of CBMS Regional Conference Series in Mathematics. Published for the Conference Board of the Mathematical Sciences, Washington, DC; by the American Mathematical Society, Providence, 2006.

[CR92]   V. Chvatal and B. Reed. Mick gets some (the odds are on his side) [satisfiability]. In *Foundations of Computer Science, 1992. Proceedings., 33rd Annual Symposium on*, pages 620–627, 1992. doi: 10.1109/SFCS.1992.267789.

[Cra38]   H. Cramér. Sur un nouveau théorème-limite de la théorie des probabilités. *Actualités Scientifiques et Industrielles*, 736:5–23, 1938.

[CRR+89]   A. K. Chandra, P. Raghavan, W. L. Ruzzo, R. Smolensky, and P. Tiwari. The electrical resistance of a graph captures its commute and cover times (detailed abstract). In David S. Johnson, ed., *STOC*, pages 574–586. Association for Computing Machinery, New York, 1989.

[CRT06a]   E. J. Candès, J. Romberg, and T. Tao. Robust uncertainty principles: Exact signal reconstruction from highly incomplete frequency information. *IEEE Trans. Inform. Theory*, 52(2):489–509, 2006.

[CRT06b]   E. J. Candès, J. K. Romberg, and T. Tao. Stable signal recovery from incomplete and inaccurate measurements. *Comm. Pure Appl. Math.*, 59(8):1207–1223, 2006.

[CT05]   E. J. Candès and T. Tao. Decoding by linear programming. *IEEE Trans. Inform. Theory*, 51(12):4203–4215, 2005.

[CW08]   E. J. Candès and M. B. Wakin. An introduction to compressive sampling. *Signal Process. Mag., IEEE*, 25(2):21–30, 2008.

[Dev98]   L. Devroye. Branching processes and their applications in the analysis of tree structures and tree algorithms. In Michel Habib, C. McDiarmid, J. Ramirez–Alfonsin, and B. Reed, eds., *Probabilistic Methods for Algorithmic Discrete Mathematics*, volume 16 of Algorithms and Combinatorics, pages 249–314. Springer Berlin Heidelberg, 1998.

[Dey]   P. Dey. Lecture notes on "Stein–Chen method for Poisson approximation." https://faculty.math.illinois.edu/~psdey/414CourseNotes.pdf

[DGG$^+$00]   M. Dyer, L. A. Goldberg, C. Greenhill, M. Jerrum, and M. Mitzenmacher. An extension of path coupling and its application to the Glauber dynamics for graph colourings (extended abstract). In *Proceedings of the Eleventh Annual ACM-SIAM Symposium on Discrete Algorithms (San Francisco, 2000)*, pages 616–624. Association for Computing Machinery, New York, 2000.

[dH]   F. den Hollander. Probability theory: The coupling method, 2012. http://websites.math.leid enuniv.nl/probability/lecturenotes/CouplingLectures.pdf.

[Dia88]   P. Diaconis. *Group Representations in Probability and Statistics*, volume 11 of Institute of Mathematical Statistics Lecture Notes—Monograph Series. Institute of Mathematical Statistics, Hayward, 1988.

[Dia09]   P. Diaconis. The Markov chain Monte Carlo revolution. *Bull. Amer. Math. Soc. (N.S.)*, 46(2):179–205, 2009.

[Die10]   R. Diestel. *Graph Theory*, volume 173 of Graduate Texts in Mathematics. Springer, Heidelberg, 4th ed. 2010.

[DKLP11]   J. Ding, J. H. Kim, E. Lubetzky, and Y. Peres. Anatomy of a young giant component in the random graph. *Random Struct. Algorithms*, 39(2):139–178, 2011.

[DMR11]   C. Daskalakis, E. Mossel, and S. Roch. Evolutionary trees and the Ising model on the Bethe lattice: A proof of steel's conjecture. *Probab. Theory Related Fields*, 149:149–189, 2011. https://doi.org/10.1007/s00440-009-0246-2.

[DMS00]   S. N. Dorogovtsev, J. F. F. Mendes, and A. N. Samukhin. Structure of growing networks with preferential linking. *Phys. Rev. Lett.*, 85:4633–4636, 2000.

[Doe38]   W. Doeblin. Exposé de la théorie des chaînes simples constantes de markoff à un nombre fini d'états. *Rev. Math. Union Interbalkan*, 2:77–105, 1938.

[Don06]   D. L. Donoho. Compressed Sensing. *IEEE Trans. Inform. Theory*, 52(4):1289–1306, 2006.

[Doo01]   J. L. Doob. *Classical Potential Theory and Its Probabilistic Counterpart*. Classics in Mathematics. Springer, Berlin, 2001.

[DP11]   J. Ding and Y. Peres. Mixing time for the Ising model: A uniform lower bound for all graphs. *Ann. Inst. Henri Poincaré Probab. Stat.*, 47(4):1020–1028, 2011.

[DS84]   P. G. Doyle and J. L. Snell. *Random Walks and Electric Networks*. Carus Mathematical Monographs. Mathematical Association of America, Washington, DC, 1984.

[DS91]   P. Diaconis and D. Stroock. Geometric bounds for eigenvalues of Markov chains. *Ann. Appl. Probab.*, 1(1):36–61, 1991.

[DSC93a]   P. Diaconis and L. Saloff-Coste. Comparison techniques for random walk on finite groups. *Ann. Probab.*, 21(4):2131–2156, 1993.

[DSC93b]   P. Diaconis and L. Saloff-Coste. Comparison theorems for reversible Markov chains. *Ann. Appl. Probab.*, 3(3):696–730, 1993.

[DSS22]   J. Ding, A. Sly, and N. Sun. Proof of the satisfiability conjecture for large k. *Ann. of Math. (2)*, 196(1):1–388, 2022.

[Dur85]   R. Durrett. Some general results concerning the critical exponents of percolation processes. *Z. Wahrsch. Verw. Gebiete*, 69(3):421–437, 1985.

[Dur06]   R. Durrett. *Random Graph Dynamics*. Cambridge Series in Statistical and Probabilistic Mathematics. Cambridge University Press, 2006.

[Dur10]   R. Durrett. *Probability: Theory and Examples*. Cambridge Series in Statistical and Probabilistic Mathematics. Cambridge University Press, Cambridge, 2010.

[Dur12]   R. Durrett. *Essentials of Stochastic Processes*. Springer Texts in Statistics. Springer, New York, 2nd ed. 2012.

[DZ10]   A. Dembo and O. Zeitouni. *Large Deviations Techniques and Applications*, volume 38 of Stochastic Modelling and Applied Probability. Springer-Verlag, Berlin, 2010. Corrected reprint of the 2nd ed. (1998).

[Ebe]   A. Eberle. Markov Processes. 2021. https://uni-bonn.sciebo.de/s/kzTUFff5FrWGAay.

[EKPS00]   W. S. Evans, C. Kenyon, Y. Peres, and L. J. Schulman. Broadcasting on trees and the Ising model. *Ann. Appl. Probab.*, 10(2):410–433, 2000.

[ER59]   P. Erdős and A. Rényi. On random graphs. I. *Publ. Math. Debrecen*, 6:290–297, 1959.

[ER60]   P. Erdős and A. Rényi. On the evolution of random graphs. *Magyar Tud. Akad. Mat. Kutató Int. Közl.*, 5:17–61, 1960.

[Fel71]   W. Feller. *An introduction to Probability Theory and Its Applications. Vol. II*. 2nd ed. John Wiley, New York-London-Sydney, 1971.

[FK16]   A. Frieze and M. Karoński. *Introduction to Random Graphs*. Cambridge University Press, Cambridge, 2016.

[FKG71]   C. M. Fortuin, P. W. Kasteleyn, and J. Ginibre. Correlation inequalities on some partially ordered sets. *Comm. Math. Phys.*, 22:89–103, 1971.

[FM88]   P. Frankl and H. Maehara. The Johnson–Lindenstrauss lemma and the sphericity of some graphs. *J. Combin. Theory Ser. B*, 44(3):355–362, 1988.

[Fos53]   F. G. Foster. On the stochastic matrices associated with certain queuing processes. *Ann. Math. Statistics*, 24:355–360, 1953.

[FR98]   A. M. Frieze and B. Reed. Probabilistic analysis of algorithms. In M. Habib, C. McDiarmid, J. Ramirez–Alfonsin, and B. Reed, eds., *Probabilistic Methods for Algorithmic Discrete Mathematics*, volume 16 of Algorithms and Combinatorics, pages 36–92. Springer, Berlin, 1998.

[FR08]   N. Fountoulakis and B. A. Reed. The evolution of the mixing rate of a simple random walk on the giant component of a random graph. *Random Struct. Algorithms*, 33(1):68–86, 2008.

[FR13]   S. Foucart and H. Rauhut. *A Mathematical Introduction to Compressive Sensing*. Applied and Numerical Harmonic Analysis. Birkhäuser, Basel, 2013.

[FV18]   S. Friedli and Y. Velenik. *Statistical Mechanics of Lattice Systems*. Cambridge University Press, Cambridge, 2018. A Concrete Mathematical Introduction.

[GC11]   A. Garivier and O. Cappé. The KL-UCB algorithm for bounded stochastic bandits and beyond. In *Proceedings of the 24th Annual Conference on Learning Theory*, pages 359–376. JMLR Workshop and Conference Proceedings, December 2011. ISSN: 1938-7228.

[GCS+14]   A. Gelman, J. B. Carlin, H. S. Stern et al. *Bayesian Data Analysis*. Texts in Statistical Science Series. CRC Press, Boca Raton, 3rd ed., 2014.

[Gil59]   E. N. Gilbert. Random graphs. *Ann. Math. Statist.*, 30:1141–1144, 1959.

[GL06]   D. Gamerman and H. F. Lopes. *Markov Chain Monte Carlo*. Texts in Statistical Science Series. Chapman & Hall/CRC, Boca Raton, 2nd ed. 2006. Stochastic simulation for Bayesian inference.

[Gri97]   G. Grimmett. Percolation and disordered systems. In *Lectures on Probability Theory and Statistics (Saint–Flour, 1996)*, volume 1665 of Lecture Notes in Math., pages 153–300. Springer, Berlin, 1997.

[Gri10a]   G. Grimmett. *Probability on Graphs*, volume 1 of Institute of Mathematical Statistics Textbooks. Cambridge University Press, Cambridge, 2010.

[Gri10b]   G. R. Grimmett. *Percolation*. Grundlehren der mathematischen Wissenschaften. Springer, Berlin, 1999.

[Gri75]   D. Griffeath. A maximal coupling for Markov chains. *Z. Wahrscheinlichkeitstheorie und Verw. Gebiete*, 31:95–106, 1974/75.

[GS20]   G. R. Grimmett and D. R. Stirzaker. *Probability and Random Processes*. Oxford University Press, Oxford, 2020. 4th ed.

[Ham57]   J. M. Hammersley. Percolation processes. II. The connective constant. *Proc. Cambridge Philos. Soc.*, 53:642–645, 1957.

[Har]   N. Harvey. Lecture notes for CPSC 536N: Randomized Algorithms. www.cs.ubc.ca/~nickhar/W12/.

[Har60]   T. E. Harris. A lower bound for the critical probability in a certain percolation process. *Proc. Cambridge Philos. Soc.*, 56:13–20, 1960.

[Har63]   T. E. Harris. *The Theory of Branching Processes*. Die Grundlehren der mathematischen Wissenschaften, Band 119. Springer-Verlag, Berlin; Prentice Hall, Englewood Cliffs, 1963.

[Haz16]   E. Hazan. Introduction to online convex optimization. *Found. Trends Opt.*, 2(3–4):157–325, 2016.

[HJ13]   R. A. Horn and C. R. Johnson. *Matrix Analysis*. Cambridge University Press, Cambridge, 2nd ed., 2013.

[HLW06]   S. Hoory, Nathan Linial, and Avi Wigderson. Expander graphs and their applications. *Bull. Amer. Math. Soc. (N.S.)*, 43(4):439–561 (electronic), 2006.

[HMRAR98]   M. Habib, C. McDiarmid, J. Ramirez-Alfonsin, and B. Reed, eds. *Probabilistic Methods for Algorithmic Discrete Mathematics*, volume 16 of Algorithms and Combinatorics. Springer-Verlag, Berlin, 1998.

[Hoe63]   W. Hoeffding. Probability inequalities for sums of bounded random variables. *J. Amer. Statist. Assoc.*, 58:13–30, 1963.

[HS07]   T. P. Hayes and Alistair Sinclair. A general lower bound for mixing of single-site dynamics on graphs. *Ann. Appl. Probab.*, 17(3):931–952, 2007.

[IM98]   P. Indyk and R. Motwani. Approximate nearest neighbors: Towards removing the curse of dimensionality. In Jeffrey Scott Vitter, ed., *STOC*, pages 604–613. Association for Computing Machinery, New York 1998.

[Jan90]   S. Janson. Poisson approximation for large deviations. *Random Struct. Algorithms*, 1(2):221–229, 1990.

[JH01]   G. L. Jones and J. P. Hobert. Honest exploration of intractable probability distributions via Markov chain Monte Carlo. *Statist. Sci.*, 16(4):312–334, 2001.

[JL84]   W. B. Johnson and J. Lindenstrauss. Extensions of Lip–schitz mappings into a Hilbert space. In *Conference in Modern Analysis and Probability (New Haven, Conn., 1982)*, volume 26 of Contemp. Math., pages 189–206. Amer. Math. Soc., Providence, 1984.

[JLR11]   S. Janson, T. Luczak, and A. Rucinski. *Random Graphs*. Wiley Series in Discrete Mathematics and Optimization. Wiley, Hoboken, NJ, 2011.

[JS89]   M. Jerrum and A. Sinclair. Approximating the permanent. *SIAM J. Comput.*, 18(6):1149–1178, 1989.

[Kan86]   M. Kanai. Rough isometries and the parabolicity of Riemannian manifolds. *J. Math. Soc. Japan*, 38(2):227–238, 1986.

[Kar90]   R. M. Karp. The transitive closure of a random digraph. *Random Struct. Algorithms*, 1(1):73–93, 1990.

[Kes80]   H. Kesten. The critical probability of bond percolation on the square lattice equals $\frac{1}{2}$. *Comm. Math. Phys.*, 74(1):41–59, 1980.

[Kes82]   H. Kesten. *Percolation Theory for Mathematicians*, volume 2 of Progress in Probability and Statistics. Birkhäuser, Boston, 1982.

[KP]   J. Komjáthy and Y. Peres. Lecture notes for Markov chains: Mixing times, hitting times, and cover times, 2012. Saint-Petersburg Summer School. www.win.tue.nl/~jkomjath/SPBlecturenotes.pdf.

[KRS]   M. J. Kozdron, L. M. Richards, and D. W. Stroock. Determinants, their applications to Markov processes, and a random walk proof of Kirchhoff's matrix tree theorem, 2013. http://arxiv.org/abs/1306.2059.

[KS66a]   H. Kesten and B. P. Stigum. Additional limit theorems for indecomposable multidimensional Galton–Watson processes. *Ann. Math. Statist.*, 37:1463–1481, 1966.

[KS66b]   H. Kesten and B. P. Stigum. A limit theorem for multi-dimensional Galton–Watson processes. *Ann. Math. Statist.*, 37:1211–1223, 1966.

[KS67]   H. Kesten and B. P. Stigum. Limit theorems for decomposable multi-dimensional Galton–Watson processes. *J. Math. Anal. Appl.*, 17:309–338, 1967.

[KS05]   G. Kalai and S. Safra. Threshold phenomena and influence: Perspectives from Mathematics, Computer Science, and Economics. In *Computational Complexity and Statistical Physics*. Oxford University Press, Oxford, December 2005. _eprint: https://academic.oup.com/book/0/chapter/354512033/chapter-pdf/43716844/isbn-9780195177374-book-part-8.pdf.

[KSK76]   J. G. Kemeny, J. L. Snell, and A. W. Knapp. *Denumerable Markov Chains*. Springer-Verlag, New York-Heidelberg-Berlin, 2nd ed., 1976. With a chapter on Markov random fields, by David Griffeath, Graduate Texts in Mathematics, No. 40.

[Law05]   G. F. Lawler. *Conformally Invariant Processes in the Plane*, volume 114 of Mathematical Surveys and Monographs. American Mathematical Society, Providence, 2005.

[Law06]   G. F. Lawler. *Introduction to Stochastic Processes*. Chapman & Hall/CRC, Boca Raton, 2nd ed. 2006.

[Led01]   M. Ledoux. *The Concentration of Measure Phenomenon*. Mathematical Surveys and Monographs. American Mathematical Society, Providence, 2001.

[Lin02]   T. Lindvall. *Lectures on the Coupling Method*. Dover, Mineola, 2002. Corrected reprint of the 1992 original.

[LL10]   G. F. Lawler and V. Limic. *Random Walk: A Modern Introduction*. Cambridge Studies in Advanced Mathematics. Cambridge University Press, Cambridge, 2010.

[Lov83]   L. Lovász. Submodular functions and convexity. In *Mathematical Programming: The State of the Art (Bonn, 1982)*, pages 235–257. Springer, Berlin, 1983.

[Lov12]   L. Lovász. *Large Networks and Graph Limits*, volume 60 of American Mathematical Society Colloquium Publications. American Mathematical Society, Providence, 2012.

[LP16]   R. Lyons and Y. Peres. *Probability on Trees and Networks*, volume 42 of Cambridge Series in Statistical and Probabilistic Mathematics. Cambridge University Press, New York, 2016.

[LP17]   D. A. Levin and Y. Peres. *Markov Chains and Mixing Times*. American Mathematical Society, Providence, 2017. 2nd ed., with Contributions by Elizabeth L. wilmer, with a chapter on "Coupling from the past" by James G. Propp and David B. Wilson.

[LPW06]   D. A. Levin, Y. Peres, and E. L. Wilmer. *Markov Chains and Mixing Times*. American Mathematical Society, Providence, 2006.

[LR85]   T. L Lai and H. Robbins. Asymptotically efficient adaptive allocation rules. *Adv. Appl. Math.*, 6(1):4–22, 1985.

[LS88]   G. F. Lawler and A. D. Sokal. Bounds on the L^2 Spectrum for Markov Chains and Markov Processes: A Generalization of Cheeger's Inequality. *Trans. Amer. Math. Soc.*, 309(2):557–580, 1988.

[LS12]   E. Lubetzky and A. Sly. Critical Ising on the square lattice mixes in polynomial time. *Comm. Math. Phys.*, 313(3):815–836, 2012.

[LS20]   T. Lattimore and C. Szepesvári. *Bandit Algorithms*. Cambridge University Press, Cambridge, 2020.

[Lu90]   T. Łuczak. Component behavior near the critical point of the random graph process. *Random Struct. Algorithms*, 1(3):287–310, 1990.

[Lug]   G. Lugosi. Concentration-of-measure inequalities, 2004. www.econ.upf.edu/~lugosi/anu .pdf.

[LuPW94]   T. Łuczak, B. Pittel, and J. C. Wierman. The structure of a random graph at the point of the phase transition. *Trans. Amer. Math. Soc.*, 341(2):721–748, 1994.

[Lyo83]   T. Lyons. A simple criterion for transience of a reversible Markov chain. *Ann. Probab.*, 11(2):393–402, 1983.

[Lyo90]   R. Lyons. Random walks and percolation on trees. *Ann. Probab.*, 18(3):931–958, 1990.

[Mat88]   P. Matthews. Covering problems for Markov chains. *Ann. Probab.*, 16(3):1215–1228, 1988.

[Mau79]   B. Maurey. Construction de suites symétriques. *C. R. Acad. Sci. Paris Sér. A-B*, 288(14):A679–A681, 1979.

[McD89]   C. McDiarmid. On the method of bounded differences. In *Surveys in Combinatorics, 1989 (Norwich, 1989)*, volume 141 of London Math. Soc. Lecture Note Ser., pages 148–188. Cambridge University Press, Cambridge, 1989.

[ML86]   A. Martin–Löf. Symmetric sampling procedures, general epidemic processes and their threshold limit theorems. *J. Appl. Probab.*, 23(2):265–282, 1986.

[ML98]   A. Martin–Löf. The final size of a nearly critical epidemic, and the first passage time of a Wiener process to a parabolic barrier. *J. Appl. Probab.*, 35(3):671–682, 1998.

[MNS15a]   E. Mossel, J. Neeman, and A. Sly. Consistency thresholds for the planted bisection model. In Rocco A. Servedio and Ronitt Rubinfeld, eds., *Proceedings of the Forty-Seventh Annual ACM on Symposium on Theory of Computing, STOC 2015, Portland, June 14–17, 2015*, pages 69–75. Association for Computing Machinery, New York, 2015.

[MNS15b]   E. Mossel, J. Neeman, and Allan Sly. Reconstruction and estimation in the planted partition model. *Probab. Theory Related Fields*, 162(3-4):431–461, 2015.

[Mor05]   F. Morgan. *Real Analysis*. American Mathematical Society, Providence, 2005.

[Mos01]   E. Mossel. Reconstruction on trees: beating the second eigenvalue. *Ann. Appl. Probab.*, 11(1):285–300, 2001.

[Mos03]   E. Mossel. On the impossibility of reconstructing ancestral data and phylogenies. *J. Comput. Biol.*, 10(5):669–678, 2003.

[Mos04]   E. Mossel. Phase transitions in phylogeny. *Trans. Amer. Math. Soc.*, 356(6):2379–2404, 2004.

[MP03]   E. Mossel and Y. Peres. Information flow on trees. *Ann. Appl. Probab.*, 13(3):817–844, 2003.

[MR95]   R. Motwani and P. Raghavan. *Randomized Algorithms*. Cambridge University Press, Cambridge, 1995.

[MS86]   V. D. Milman and G. Schechtman. *Asymptotic Theory of Finite-Dimensional Normed Spaces*, volume 1200 of Lecture Notes in Mathematics. Springer-Verlag, Berlin, 1986. With an appendix by M. Gromov.

[MT06]   R. Montenegro and P. Tetali. Mathematical aspects of mixing times in Markov chains. *Found. Trends Theor. Comput. Sci.*, 1(3): 237–354, 2006.

[MT09]   S. Meyn and R. L. Tweedie. *Markov Chains and Stochastic Stability*. Cambridge University Press, Cambridge, 2nd ed., 2009. With a prologue by Peter W. Glynn.

[MU05]   M. Mitzenmacher and E. Upfal. *Probability and Computing: Randomized Algorithms and Probabilistic Analysis*. Cambridge University Press, New York, 2005.

[Nic18]   B. Nica. *A Brief Introduction to Spectral Graph Theory*. EMS Textbooks in Mathematics. European Mathematical Society (EMS), Zürich, 2018. https://doi.org/10.4171/188.

[Nor98]   J. R. Norris. *Markov Chains*, volume 2 of Cambridge Series in Statistical and Probabilistic Mathematics. Cambridge University Press, Cambridge, 1998. Reprint of 1997 original.

[NP10]   A. Nachmias and Y. Peres. The critical random graph, with martingales. *Israel J. Math.*, 176:29–41, 2010.

[NW59]   C. St. J. A. Nash-Williams. Random walk and electric currents in networks. *Proc. Cambridge Philos. Soc.*, 55:181–194, 1959.

[Ott49]   R. Otter. The multiplicative process. *Ann. Math. Statistics*, 20:206–224, 1949.

[Pem91]   R. Pemantle. Choosing a spanning tree for the integer lattice uniformly. *Ann. Probab.*, 19(4):1559–1574, 1991.

[Pem00]   R. Pemantle. Towards a theory of negative dependence. *J. Math. Phys.*, 41(3):1371–1390, 2000. Probabilistic techniques in equilibrium and nonequilibrium statistical physics.

[Per]   Y. Peres. Course notes on Probability on trees and networks, 2004. http://stat-www.berkeley.edu/~peres/notes1.pdf.

[Per09]   Y. Peres. The unreasonable effectiveness of martingales. In *Proceedings of the Twentieth Annual ACM-SIAM Symposium on Discrete Algorithms*, SODA '09, pages 997–1000, Society for Industrial and Applied Mathematics, Philadelphia, 2009.

[Pet]   G. Pete. Probability and geometry on groups. Lecture notes for a graduate course. www.math.bme.hu/~gabor/PGG.html.

[Pey08]   R. Peyre. A probabilistic approach to Carne's bound. *Potential Anal.*, 29(1):17–36, 2008.

[Pit76]   J. W. Pitman. On coupling of Markov chains. *Z. Wahrscheinlichkeitstheorie und Verw. Gebiete*, 35(4):315–322, 1976.

[Pit90]   B. Pittel. On tree census and the giant component in sparse random graphs. *Random Struct. Algorithms*, 1(3):311–342, 1990.

[RAS15]   F. Rassoul-Agha and T. Seppäläinen. *A course on Large Deviations with an Introduction to Gibbs Measures*, volume 162 of Graduate Studies in Mathematics. American Mathematical Society, Providence, 2015.

[RC04]   C. P. Robert and G. Casella. *Monte Carlo Statistical Methods*. Springer Texts in Statistics. Springer-Verlag, New York, 2nd ed. 2004.

[Res92]   S. Resnick. *Adventures in Stochastic Processes*. Birkhäuser Boston, Boston, 1992.

[Roc10]   S. Roch. Toward extracting all phylogenetic information from matrices of evolutionary distances. *Science*, 327(5971):1376–1379, 2010.

[Rom15]   D. Romik. *The Surprising Mathematics of Longest Increasing Subsequences*, volume 4 of Institute of Mathematical Statistics Textbooks. Cambridge University Press, New York, 2015.

[RS17]   S. Roch and A. Sly. Phase transition in the sample complexity of likelihood-based phylogeny inference. *Probability Theory and Related Fields*, 169(1):3–62, 2017.

[RT87]   W. T. Rhee and M. Talagrand. Martingale inequalities and NP-complete problems. *Math. Oper. Res.*, 12(1):177–181, 1987.

[Rud73]   W. Rudin. *Functional Analysis*. McGraw-Hill Series in Higher Mathematics. McGraw-Hill Book, New York-Düsseldorf-Johannesburg, 1973.

[Rus78]   L. Russo. A note on percolation. *Z. Wahrscheinlichkeitstheorie und Verw. Gebiete*, 43(1):39–48, 1978.

[SE64]   M. F. Sykes and J. W. Essam. Exact critical percolation probabilities for site and bond problems in two dimensions. *J. Mathematical Phys.*, 5:1117–1127, 1964.

[SJ89]   A. Sinclair and M. Jerrum. Approximate counting, uniform generation and rapidly mixing Markov chains. *Inform. and Comput.*, 82(1):93–133, 1989.

[Spi56]   F. Spitzer. A combinatorial lemma and its application to probability theory. *Trans. Amer. Math. Soc.*, 82:323–339, 1956.

[Spi12]   D. A. Spielman. Lecture notes on spectral graph theory. www.cs.yale.edu/homes/spielman/561/2012/index.html, 2012.

[SS87]   E. Shamir and J. Spencer. Sharp concentration of the chromatic number on random graphs $G_{n,p}$. *Combinatorica*, 7(1):121–129, 1987.

[SS03]   E. M. Stein and R. Shakarchi. *Fourier Analysis*, volume 1 of Princeton Lectures in Analysis. Princeton University Press, Princeton, 2003.

[SS05]   E. M. Stein and R. Shakarchi. *Real Analysis*, volume 3 of Princeton Lectures in Analysis. Princeton University Press, Princeton, 2005.

[SSBD14]   S. Shalev-Shwartz and S. Ben-David. *Understanding Machine Learning: From Theory to Algorithms*. Cambridge University Press, Cambridge, 2014.

[Ste]   J. E. Steif. A mini course on percolation theory, 2009. www.math.chalmers.se/~steif/perc.pdf.

[Ste72]   C. Stein. A bound for the error in the normal approximation to the distribution of a sum of dependent random variables. In *Proceedings of the Sixth Berkeley Symposium on Mathematical Statistics and Probability (University California, Berkeley, 1970/1971), Vol. II: Probability theory*, pages 583–602, University of California Press, Berkeley, 1972.

[Ste97]   J. Michael Steele. *Probability Theory and Combinatorial Optimization*, volume 69 of CBMS-NSF Regional Conference Series in Applied Mathematics. Society for Industrial and Applied Mathematics (SIAM), Philadelphia, 1997.

[Ste98]   G. W. Stewart. *Matrix Algorithms. Vol. I*. Society for Industrial and Applied Mathematics, Philadelphia, 1998.

[Str65]   V. Strassen. The existence of probability measures with given marginals. *Ann. Math. Statist.*, 36:423–439, 1965.

[Str14]   Daniel W. Stroock. Doeblin's Theory for Markov Chains. In *An Introduction to Markov Processes*, pages 25–47. Springer Berlin Heidelberg, Berlin, Heidelberg, 2014.

[SW78]   P. D. Seymour and D. J. A. Welsh. Percolation probabilities on the square lattice. *Ann. Discrete Math.*, 3:227–245, 1978. Advances in graph theory (Cambridge Combinatorial Conf., Trinity College, Cambridge, 1977).

[Tao]   T. Tao. Open question: deterministic UUP matrices. https://terrytao.wordpress.com/2007/07/02/open-question-deterministic-uup-matrices/.

[Var85]   N. Th. Varopoulos. Long range estimates for Markov chains. *Bull. Sci. Math. (2)*, 109(3):225–252, 1985.

[vdH10]   R. van der Hofstad. Percolation and random graphs. In *New perspectives in Stochastic Geometry*, pages 173–247. Oxford University Press, Oxford, 2010.

[vdH17]   R. van der Hofstad. *Random Graphs and Complex Networks. Vol. 1*. Cambridge Series in Statistical and Probabilistic Mathematics. Cambridge University Press, Cambridge, 2017.

[vdHK08]   R. van der Hofstad and Michael Keane. An elementary proof of the hitting time theorem. *Amer. Math. Monthly*, 115(8):753–756, 2008.

[Vem04]   S. S. Vempala. *The Random Projection Method*. DIMACS Series in Discrete Mathematics and Theoretical Computer Science, 65. American Mathematical Society, Providence, 2004. With a foreword by Christos H. Papadimitriou.

[Ver18]   R. Vershynin. *High-Dimensional Probability: An Introduction with Applications in Data Science*. Cambridge University Press, 2018. Google-Books-ID: TahxDwAAQBAJ.

[vH16]   R. van Handel. Probability in high dimension. www.princeton.edu/~rvan/APC550.pdf, 2016.

[Vil09]   Cédric Villani. *Optimal Transport*, volume 338 of Grundlehren der Mathematischen Wissenschaften [Fundamental Principles of Mathematical Sciences]. Springer-Verlag, Berlin, 2009.

[Wen75]   J. G. Wendel. Left-continuous random walk and the Lagrange expansion. *Amer. Math. Monthly*, 82:494–499, 1975.

[Whi32]   H. Whitney. Non-separable and planar graphs. *Trans. Amer. Math. Soc.*, 34(2):339–362, 1932.

[Wil91]   D. Williams. *Probability with Martingales*. Cambridge Mathematical Textbooks. Cambridge University Press, Cambridge, 1991.

[Wil96]   D. B. Wilson. Generating random spanning trees more quickly than the cover time. In Gary L. Miller, ed., *STOC*, pages 296–303. Association for Computing Machinery, New York, 1996.

[YP14]   S.-Y. Yun and A. Proutière. Community detection via random and adaptive sampling. In Maria-Florina Balcan, Vitaly Feldman, and Csaba Szepesvári, eds., *Proceedings of The 27th Conference on Learning Theory, COLT 2014, Barcelona, Spain, June 13–15, 2014*, volume 35 of JMLR Workshop and Conference Proceedings, pages 138–175. JMLR.org, PMLR, 2014.

[Yur76]   V. V. Yurinskiĭ. Exponential inequalities for sums of random vectors. *J. Multivariate Anal.*, 6(4):473–499, 1976.

Index

Printed in the United States
by Baker & Taylor Publisher Services